The ruminant immune system in health and disease

T0296926

The ruminant immune system

in health and disease

EDITED BY W. IVAN MORRISON

International Laboratory for Research on Animal Diseases
Nairobi, Kenya

The right of the
University of Cambridge
to print and sell
all manner of books
was granted by
Henry VIII in 1534.
The University has printed
and published continuously
since 1584.

CAMBRIDGE UNIVERSITY PRESS

Cambridge

New York   New Rochelle

Melbourne   Sydney

CAMBRIDGE UNIVERSITY PRESS
Cambridge, New York, Melbourne, Madrid, Cape Town, Singapore, São Paulo, Delhi

Cambridge University Press
The Edinburgh Building, Cambridge CB2 8RU, UK

Published in the United States of America by Cambridge University Press, New York

www.cambridge.org
Information on this title: www.cambridge.org/9780521115476

First published 1986
Reprinted 1988
This digitally printed version 2009

A catalogue record for this publication is available from the British Library

Library of Congress Cataloguing in Publication data

Main entry under title:

The Ruminant immune system in health and disease

1. Veterinary immunology.   2. Ruminants – Immunology.
I. Morrison, W. Ivan.
SF757.2R86 1986   636.2′089′6079   85–31350

ISBN 978-0-521-32443-4 hardback
ISBN 978-0-521-11547-6 paperback

Contents

**Role of immune responses in protection
against infectious diseases**

List of authors

Adams, T.E.
Cold Spring Harbor Laboratory, Cold Spring Harbor, New York, USA.

Beh, K.J.
Commonwealth Scientific and Industrial Research Organization, Division of Animal Health, McMaster Laboratory, Parramatta Road, Private Bag No. 1, P.O. Glebe, NSW 2037, Australia.

Beverley, P.C.L.
Imperial Cancer Research Fund, Human Tumour Immunology Group, School of Medicine, University College London, University Street, London WC1E 6JJ, England.

Black, S.J.
International Laboratory for Research on Animal Diseases, P.O. Box 30709, Nairobi, Kenya.

Blanden, R.V.
Department of Microbiology, John Curtin School of Medical Research, P.O. Box 334, Canberra, ACT 2601, Australia.

Brandon, M.R.
Department of Veterinary Preclinical Sciences, School of Veterinary Science, University of Melbourne, Parkville, Victoria 3052, Australia.

Christensen, A.K.
Department of Anatomy and Cell Biology, The University of Michigan Medical School, Ann Arbor, Michigan 48109, USA.

Christensen, N.
Department of Pathology, University of Auckland School of Medicine, Private Bag, Auckland, New Zealand.

Davies, A.J.S.
Chester Beatty Research Institute, Institute of Cancer Research, Royal Cancer Hospital, Fulham Road, London SW3 6JB, England.

Davis, W.C.
Department of Veterinary Microbiology and Pathology, Washington State University, Pullman, Washington 99164-7040, USA.

De Baetselier, P.
Vrije Universiteit Brussel, Instituut voor Molekulaire Biologie, Paardenstraat G5, 1640 St-Genesius-Rode, Brussels, Belgium.

Emery, D.L.
Commonwealth Scientific and Industrial Research Organization, Division of Animal Health, Private Bag No. 1, Parkville, Victoria 3052, Australia.

Fossum, C.
Department of Virology, The Royal Veterinary College, Biomedical Centre, Box 585, S-751 23 Uppsala, Sweden.

Gallery, F.
Department of Comparative and Experimental Pathology, College of Veterinary Medicine, University of Florida, Box J-145, Gainesville, Florida 32601, USA.

Gauntlett, D.
Department of Comparative and Experimental Pathology, College of Veterinary Medicine, University of Florida, Box J-145, Gainesville, Florida 32601, USA.

Germain, R.N.
Laboratory of Immunology, Department of Health and Human Services, National Institutes of Health, Bethesda, Maryland 20205, USA.

Goddeeris, B.M.
International Laboratory for Research on Animal Diseases, P.O. Box 30709, Nairobi, Kenya.

Groocock, C.M.
Veterinary Research Department, Kenya Agricultural Research Institute, Kikuyu, Kenya.

Hall, J.G.
Chester Beatty Research Institute, Institute of Cancer Research, Block X, Clifton Avenue, Belmont, Sutton, Surrey SM2 5PX, England.

Hamers, R.
Vrije Universiteit Brussel, Instituut voor Molekulaire Biologie, Paardenstraat G5, 1640 St-Genesius-Rode, Brussels, Belgium.

Hoang-Xuan, M.
Laboratoire Immunologie et Virologie Des Tumeurs, Groupe INSERM V 152, Hopital Cochin, 27 Rue du Fauborg Saint-Jacques, 75674 Paris Cedex 14, France.

Knisley, K.A.
Department of Microbiology and Environmental Health, College of Veterinary Medicine and Biomedical Sciences, Colorado State University, Fort Collins, Colorado 80523, USA.

Lalor, P.A.
International Laboratory for Research on Animal Diseases, P.O. Box 30709, Nairobi, Kenya. Present address: Department of Genetics, Stanford University School of Medicine, Stanford, California 94305, USA.

Lascelles, A.K.
Commonwealth Scientific and Industrial Research Organization, Division of Animal Health, McMaster Laboratory, Parramatta Road, Private Bag No. 1, P.O. Glebe, NSW 2037, Australia.

Lawman, M.P.
Department of Comparative and Experimental Pathology, College of Veterinary Medicine, University of Florida, Box J-145, Gainesville, Florida 32601, USA.

Levy, D.
Laboratoire Immunologie et Virologie Des Tumeurs, Groupe INSERM V 152, Hopital Cochin, 27 Rue du Fauborg Saint-Jacques, 75674 Paris Cedex 14, France.

Mahoney, D.F.
Division of Tropical Animal Sciences, Commonwealth Scientific and Industrial Research Organization, Long Pocket Laboratories, Private Bag No. 3, P.O. Indooroopilly, Queensland 4068, Australia.

Marbrook, J.
Department of Pathology, University of Auckland School of Medicine, Private Bag, Auckland, New Zealand.

McGuire, T.C.
Department of Veterinary Microbiology and Pathology, Washington State University, Pullman, Washington 99164-7040, USA.

Miller, H.R.P.
Animal Diseases Research Association, Department of Pathology, Moredun Institute, 408 Gilmerton Road, Edinburgh EH17 7JH, Scotland.

Morein, B.
Department of Virology, Biomedical Centre, The National Veterinary Institute, Biomedicum, Box 585, S-751 23 Uppsala, Sweden.

Morris, B.
Department of Immunology, John Curtin School of Medical Research, Australian National University, Canberra, ACT 2601, Australia.

Morrison, W.I.
International Laboratory for Research on Animal Diseases, P.O. Box 30709, Nairobi, Kenya.

Mukkur, T.K.
Commonwealth Scientific and Industrial Research Organization, Division of Animal Health, McMaster Laboratory, Parramatta Road, Private Bag No. 1, P.O. Glebe, NSW 2037, Australia.

Musoke, A.J.
International Laboratory for Research on Animal Diseases, P.O. Box 30709, Nairobi, Kenya.

Nantulya, V.M.
International Laboratory for Research on Animal Diseases, P.O. Box 30709, Nairobi, Kenya.

Osburn, B.I.
School of Veterinary Medicine, University of California, Davis, California 95616, USA.

Pearson, L.D.
Department of Microbiology and Environmental Health, College of Veterinary Medicine and Biomedical Sciences, Colorado State University, Fort Collins, Colorado 80523, USA.

Pedersen, N.C.
Department of Immunology, John Curtin School of Medical Research, Australian National University, Canberra, ACT 2601, Australia. Present address: School of Veterinary Medicine, University of California, Davis, California 95616, USA.

Perryman, L.E.
Department of Veterinary Microbiology and Pathology, Washington State University, Pullman, Washington 99164-7040, USA.

Rurangirwa, F.R.
Small Ruminant Collaborative Research Program, Veterinary Research Laboratories, P.O. Kabete, Kenya.

Sachs, J.A.
Department of Immunology, The London Hospital Medical College, Turner Street, London El 2AD, England.

Scollay, R.G.
The Walter and Eliza Hall Institute, P.O. Royal Melbourne Hospital, Melbourne, Victoria 3050, Australia.

Skinner, M.A.
Department of Pathology, University of Auckland School of Medicine, Private Bag, Auckland, New Zealand.

Spooner, R.L.
Animal Breeding Research Organization, West Mains Road, Edinburgh EH9 3JQ, Scotland.

Stagg, D.A.
Veterinary Research Department, Kenya Agricultural Research Institute, Kikuyu, Kenya.

Teale, A.J.
International Laboratory for Research on Animal Diseases, P.O. Box 30709, Nairobi, Kenya.

Trevella, W.
Department of Immunology, John Curtin School of Medical Research, Australian National University, Canberra, ACT 2601, Australia.

Van Der Loo, W.
Vrije Universiteit Brussel, Instituut voor Molekulaire Biologie, Paardenstraat G5, 1640 St-Genesius-Rode, Brussels, Belgium.

Watson, D.L.
Commonwealth Scientific and Industrial Research Organization, Division of Animal Health, Pastoral Research Laboratory, Armidale, NSW, Australia.

Webster, P.
International Laboratory for Research on Animal Diseases, P.O. Box 30709, Nairobi, Kenya.

Zilber, M.T.
Laboratoire Immunologie et Virologie Des Tumeurs, Groupe INSERM V 152, Hopital Cochin, 27 Rue du Faubourg Saint-Jacques, 75674 Paris Cedex 14, France.

Preface

Ruminant animals have been kept under domestication for several thousand years. During this time, they have been subjected to intensive selection for meat and milk production, so that they now contribute approximately 25% of world protein consumption. Domestic ruminants are kept under a variety of environmental conditions and husbandry systems in various parts of the world and they are exposed to many stresses and disease-causing organisms against which they have had little opportunity to acquire resistance by natural selection. It is not surprising, therefore, that immunological research in ruminants has concentrated on diseases caused by infectious or parasitic organisms. Despite a relatively superficial understanding of the immune system of ruminants, this research has led to the development of successful immuno-prophylactic measures against a number of infectious diseases. However, many other important diseases have remained unchecked.

In the past decade, technological advances in immunology and molecular biology have led to rapid progress in our knowledge of the structure and function of the immune system. Thus, it is now possible, in outbred species, to dissect in some detail the cellular and molecular basis of various types of immune response and to use similar techniques to identify important antigenic components of pathogenic organisms. Moreover, there is now greater awareness of the fact that individuals within an outbred species exhibit significant heterogeneity in their capacity to mount specific immune responses and to resist disease. There is good reason to believe that studies of immunology in ruminants will help to elucidate the mechanisms of innate and acquired resistance to disease and thus aid in the development of control measures for diseases which hitherto have not been amenable to prophylaxis.

This book presents the proceedings of an international conference on the Application of Ruminant Immunology to the Control of Bovine Diseases, held in Nairobi, Kenya, in September 1983. The principal aims of the conference were to provide a review of current information on the immune system of ruminants and to discuss how studies of ruminant immunology can be applied to the development of improved disease control measures. Papers were presented on ruminant immunology and on a number of topics in other species which were considered to be of comparative interest. The papers presented were in no way intended to provide a comprehensive coverage of ruminant immunology, but were selected to emphasize areas in which there have been significant advances in the past few years.

The conference was organized by the International Laboratory for Research on Animal Diseases (ILRAD), with assistance from the Australian Centre for International Agricultural Research, the Commonwealth Foundation, the Ford Foundation, the Government of the Netherlands and May and Baker Ltd. ILRAD is one of 13 centres in a worldwide agricultural research network sponsored by the Consultative Group on International Agricultural Research. Financial support is provided by the World Bank, the United Nations Development Program and the governments of Australia, Belgium, Canada, Denmark, the Federal Republic of Germany, France, Italy, Japan, the Netherlands, Norway, Saudi Arabia, Sweden, Switzerland, the United Kingdom and the United States of America.

Many people contributed to the preparation of this book. In particular, I wish to thank Ms Sidney Westley and Ms Valerie Howe who spent many hours patiently reading, correcting and editing the manuscripts. I am also most grateful to my colleagues Drs Cynthia Baldwin, Sam Black, Max Murray, Antony Musoke, Onesmo ole MoiYoi and Alan Teale for editorial assistance.

> Ivan Morrison
> ILRAD
> Nairobi, Kenya.

1

Evolution and functions of the immune system

A.J.S. DAVIES

The conventional view of the immune response is that it represents a defence mechanism concerned with resistance to invasion by pathogenic organisms. This notion will be considered critically and an alternative hypothesis will be advanced that immunity in vertebrates has evolved as an accommodation device. The suggestion will be made that the ability to achieve symbiotic associations is of considerable evolutionary significance.

Introduction

Amongst immunologists there is the belief that the immune response is a defence mechanism against attack by foreign organisms. This common sense view provides a warm glow of job satisfaction to its protagonists and, in addition, it accords with the Darwinian notion that nature is hostile and only the fit survive. The emphasis is placed on individual survival in which the capability of defending oneself against infection looks to have an obvious selective advantage.

It is almost cavalier to query such a neat and tidy view of the world, but progress in science is not made by unquestioning acceptance of what seems to be fact. Indeed Karl Popper would have us believe that it is only by the search for, and discovery of, exceptions that we can advance. What is intended in this mildly heretical essay is to propose that, in addition or alternatively to being a defence mechanism against attack, the immune response can be an accommodation device to facilitate the development of relatively peaceful associations which, in some instances, could be or could become symbiotic. The argument will be advanced that the pathological consequences of acceptance of foreign organisms can be of little evolutionary significance and may be outweighed by the long-term advantages of being able to acquire genetic material and/or its products from other organisms. This alternative view is not put forward on the grounds of perversity but because the defence hypothesis in its strict form could lead to some wrong strategies of intervention.

General definition of immunity

The word 'immune' according to the dictionary (*Chambers 20th century* 1977 edition) means 'exempt: not liable to danger, especially of infection'. Immunologists have arrogated the word to imply something special with exquisite specificity in which a highly sophisticated mechanism is brought to bear upon material introduced to the body from outside. Much of the thinking of immunologists is, however, based on the study of a variety of laboratory artefacts that provide adequate exercize in analytic method but from which it is sometimes difficult to recognize the basic phenomenon under investigation.

In its general sense, immunity is concerned with organismal interactions in which, as the dictionary says, there is the possibility of danger. A further implication in the definition is that the danger in question is that of invasion of a large, multicellular organism by a smaller one and that a state of parasitization is involved. (In the present context, the epithet 'parasite' will be applied to viruses, bacteria, fungi, protozoans and metazoans alike when the organisms concerned invade and potentially or actually harm their hosts.)

Symbiosis in plants

In nature there are many examples of organisms living together in which no danger is recognized. Sometimes such associations require a considerable degree of adaptation on the part of the organisms concerned which sometimes have little or no independent capacity for survival. Lichens, which consist of a fungal and a blue-green algal component, exemplify just such obligate symbiosis. The fungus provides the thallus and the alga, having chlorophyll, the capacity to fix energy from sunlight. The resultant ubiquitous chimaeras are capable effectively of colonizing otherwise bare surfaces in a wide variety of climates. Whether lichens evolved as a consequence of an initial close physical association or whether they derive from a state in which the fungus was parasitic on and dangerous to the alga, or *vice versa*, is not known, but it would be facile to think of either immunity or resistance as a significant component of an apparently mutually beneficial association.

Partial parasitism in plants

In contrast, and again using an example from the plant kingdom, *Viscum album*, the mistletoe, grows on and extensively invades the trees on which it is an obligate parasite. Mistletoe has many adaptations to its hemiparasitic mode of life including an apparent diminution in its chlorophyll content, the capacity to produce an haustorial root system that penetrates and absorbs nutrients from the host tree, and sticky, nasty-tasting fruits, which greedy birds often wipe from their bills onto neighbouring tree branches. These adaptations are doubtless the outcome of a prolonged evolutionary process.

Mistletoe grows slowly, and multiple infections, if that is the right word, can eventually debilitate and kill the host tree. There are several points to be made here. Firstly, mistletoe is relatively host-tree species specific. Secondly, although it is in the long term deleterious to its host, it is doubtful if it brings about a degree of impairment of reproductive capacity which is of evolutionary significance. This illustrates what is seen repeatedly throughout the plant and animal kingdom; parasites often neither kill nor seriously harm their hosts. Thirdly, neither immunity nor resistance to mistletoe seems to exist; the extent to which mistletoe fails to infect all species of trees is more safely attributed to host specificity and it should be noted at this stage that this is a factor which must be taken into consideration when considering the interactions of mammals and the various lower organisms with which they come into contact. There is one final point which is highly relevant to ruminant immunology, mistletoe commonly and seriously parasitizes cultivated apple trees and in old orchards it can appear to be their principal enemy. This is not to say that naturally growing trees, which incidentally are uncommon in the British Isles, are never parasitized, but simply that the severity of parasitization is often higher in circumstances of cultivation in which natural selection has been prevented or has only occurred for what is by evolutionary standards, a short time.

Catastrophic parasitization in plants

Lastly, before leaving the plant kingdom, it might be useful to consider the example of *Phytophthora infestans*. During the mid-19th century, as is well known, the potato crop in Ireland failed due to the depredations of this fungus. Many thousands of people died as an indirect consequence of an unusual example of parasitization. It was unusual in its high lethality to the host organism which, it should be noted, was cultivated. In time, plant breeders produced resistant varieties and on many grounds, at least in Ireland, such a catastrophe is unlikely to recur. Contemporary varieties of potato are relatively resistant because the fungal hyphae are unable to gain access to the host plant during the early stages of its growth. In common parlance, the resistant potato varieties are said to be immune to *P. infestans* but the immunity is a passive process and involves no active adaptation on the part of the host such as can occur in vertebrates subjected to parasitic invasion. If the infection of the potato had taken place in natural circumstances the extermination of the host species could have been an evolutionary dead end for the parasite.

In summary, these examples of organismal interaction from the plant kingdom have been used to illustrate the phenomena of obligate symbiosis, host specificity in parasitization, the high degree of adaptation that can be seen in some parasites, the unusual susceptibility of cultivars to parasitization, the vested interests of parasites in the survival of their hosts and the distinction between passive and active immunity. The reason for giving these examples is to emphasize certain principles in the interaction between organisms in which there is no elaborate immunological apparatus. These principles still hold in interactions between organisms that have more sophisticated immunological devices, and in trying to derive a biological *raison d'être* for immune reactions, it is as well to remember the existence of the more basic reactional systems.

A special meaning of immunity

The specialized *Dictionary of Immunology* (Herbert & Wilkinson 1977) compiled by and for immunologists defines immunity as follows: 1. 'nonsusceptibility to the invasive or pathogenic effects of foreign organisms or to the toxic effects of antigenic substances'— this is rather similar to the general dictionary definition previously quoted — and 2. 'a stage of heightened responsiveness to antigen such that antigen is bound or eliminated more rapidly than in the nonimmune state; thus inclusive of all types of humoral and cell-mediated immunity'. The specificity of immunity in the second definition is implicit in the use of the word antigen which is definable only in relation to the capacity to elicit complementary binding molecules, i.e. antibodies.

Neither of these two definitions imposes any taxonomic limits on the responding organisms but, it is widely believed, the second one relates primarily to vertebrates and the first concerns various mechanisms which do not engender specifically complementary cells and/or molecules. The nonspecific systems include the phagocytic cells which circulate in the blood systems of all triploblastic animals. Nonspecific is in some sense a misnomer in that the cells concerned can distinguish self from nonself and they are to a degree inducible in both numbers and activity. However, without the involvement of cells with specific receptors or their products, phagocytic cells are unable to react exclusively to specific stimuli. Seen thus, specific immunity depends on the restriction of particular classes of cells, usually lymphocytes as will be seen, to interact with just one kind of foreignness rather than foreignness *per se*. Nonspecific immunity is regarded with a somewhat jaundiced eye by proper immunologists but, as it can be argued that it is the basic defence mechanism of all triploblastic organisms, it is worthwhile to consider it in some detail, starting with its origins.

Self-recognition

No self-respecting *Amoeba* will attempt to phagocytose its sibs unless perhaps they happen to be dead. If an encountered organism is not self, then it may be phagocytosed and eaten. If it is self and the circumstances otherwise are propitious, mating may occur. The point to be made here is that self-recognition by primitive eukaryotic cells relates to feeding, or perhaps to sex, but probably not to defence from attack. Parenthetically, precise recognition of genetically controlled allotypic differences has also been adopted as a means of ensuring outbreeding in some sexually reproducing unicellular organisms.

In truly metazoan, as opposed to colonial, multicellular organisms feeding became restricted to specialized cells of the body. In the adult form of diploblastic coelenterates, such as *Hydra*, most cells are in fixed positions and the capacity to ingest food material is associated with the amoebocytes that are one of the cell populations lining the enteron. Digestive enzymes are secreted into the enteron from the other, glandular, cells. Presumably recognition of self takes place, but whether this is accomplished at the level of the whole organism and/or each of its constituent cells is not clear. The main issue is that during the course of evolution, as feeding became associated with particular parts of the multicellular body, the capacity for self-recognition, however achieved, probably no longer remained a formal necessity as part of the feeding process. Self-recognition probably remained an integral part of the process of embryogenesis. Indeed, in the myxomycetes, which are multicellular organisms at some stage of their life cycle, cellular recognition mechanisms are known to be a necessary part of the final assembly of the mature body. Similarly, it has been known for many years that if different species of sponge are dissociated into their component cells and mixed, reaggregation of the parental bodies will occur in time (Wilson & Penny 1930). Self-recognition at the cellular level is the key to such a capacity for reassociation.

In the rather larger coelenterates such as sea anemones there are phagocytic cells which can migrate within the mesoglea. Whether these cells are concerned similarly with feeding or with picking up and engulfing invading organisms, or both, is not known. The nematocysts possessed by many coelenterates might look like a defence mechanism but are probably primarily for the immobilization and killing of small potential food organisms.

Thus it appears that little can be said about defence mechanisms in either protozoans or diploblasts except that, as the organisms in question not only exist but are very abundant, they have had adequate resistance to those things that have endangered them. The resistance could have been based on the passive possession of devices for making them unpalatable to higher organisms. If this is the case, such apparatus must have developed secondarily as, presumably, in the early stages of evolution there were no threatening larger animals. Alternatively, and in an even more passive sense, the capacity for reproduction might have been adjusted to keep pace with whatever reductions in population occurred as a result of predation. The general principles of this process were enunciated by Malthus many years ago.

Ecological balance

Coexistence of organisms of different kinds in a steady state indicates the achievement of a balance which is perhaps best thought of as ecological and not necessarily either unstable or involving the preservation of all the individuals of a particular kind. Many of the larger coelenterates, for example, carry with them huge and varied colonies of associated organisms some of which die within or are killed by their host. The carried organisms for their part live, as well as they may, on the food which drops from the master's table. The lesson from this is that symbiotic associations need not necessarily preserve all the component individuals for mutual benefit to the various species to be deducible.

Life strategies

Students of evolution discuss these things, rather teleologically, as indicating various life strategies in which individual survival can be subordinate to the survival of the species. The well known episodic suicide of the lemmings, presumably related to reduction of a population density, unacceptable to the survival of a significant population of the species, exemplifies one such stratagem. Wynne-Edwards (1967), in his account of *Animal dispersion in relation to social behaviour*, gives many more illustrations of a similar kind.

There are difficulties in understanding such phenomena in that individuals of a species which die without exercize of full reproductive potential carry genetic material which is lost from the gene pool of the species and is thus selected against. A way of getting round this theoretical problem, as has been pointed out previously (Davies et al. 1980), is to suppose the ubiquitous possession of genes determining a selectively advantageous course of action in possible environmental conditions. The adoption of the relevant activity, with death of individual members of the species, can then be of advantage to the species as a whole without total loss of the genetic material responsible. As will be seen, it can be argued that, in some instances, death of individuals of particular species of mammals due to infection has advantage to the species concerned. In the human species, it is the avowed role of the medical practitioner to preserve the lives of individuals when possible but, in doing so in circumstances of infection, it could be argued that there might be disadvantage to the species as a whole. This is not an argument which is ever likely to influence the pattern of activity of the medical profession, and neither should it, but the biologist should be able to state it without fear of being accused of eugenic thinking.

The defence mechanisms of triploblasts

Triploblastic organisms have a vascular system and fluid-filled body spaces within which are peripatetic phagocytic cells. There are in addition various components of the body fluid or secretions which have the capacity to kill or repel invading organisms. It is not known whether the phagocytic cells are metabolically dependent on the materials they pick up, nor is it certain to what extent they are scavengers of body waste as opposed to catchers and killers of invaders. Common sense seems to dictate that phagocytes, and the various humoral cytocidal agents, are concerned at least to some extent with limiting or preventing systematization of foreign organisms which happen to gain access to the body. Perhaps the possession of an alimentary canal which has, in addition to food, many inhabiting bacteria, viruses and protozoa makes it necessary to have some additional safeguards against breakdown of the barrier of the gut wall.

The 'immune' response of invertebrates

Far too little is known about disease in invertebrates to make anything other than superficial comments about their mechanisms for resistance to infection. It has been shown that there exist molecules in the haemolymph which have some degree of specific binding capacity following introduction of antigens (Cooper 1974). Similarly, transplant rejection has been demonstrated in a variety of worms (Cooper 1968).

In neither instance is it known if the mechanisms evolved have any relationship to resistance to disease. What seems clear is that the apparatus concerned with immune responses in invertebrates is far less elaborate than in vertebrates. It would probably be too sweeping a generalization to argue that invertebrates lack specific immune mechanisms and yet survive in a putatively hostile world, as has been noted previously for plants. Neither would it be legitimate, in the present state of knowledge, however tempting, to draw the corollary that if invertebrates can survive without an immune response, using largely nonspecific mechanisms to defend themselves, then it is unnecessary to postulate that the immune response of vertebrates is primarily a defence mechanism. Despite these reservations, Occam's razor is a sharp instrument and its use in relation to the defensive theory of immunity seems at least partly justified by the fact that many organisms survive without well-recognized immunological responses.

The immunological apparatus in vertebrates

In vertebrates, there are recognized various organs of the body which are concerned with the generation and housing of the populations of lymphocytes. It may be that invertebrates possess some comparable structures but, if so, they pass at present almost entirely unrecognized. The populations of lymphocytes in the vertebrate body are regarded as the cells largely responsible for the exercize of the property of specific immunity and it is on them that the minds of contemporary immunologists centre. The various lymphoid organs are seen as, either the sites in which the generation of immune responses occur, and/or locations for the generation of lymphocytes. Whilst there can be little quarrel with such statements of fact, there are differences of emphasis of interpretation among the interested parties and some uncertainties of imputation of function.

Lymph and lymphatics

For example, it is generally argued by lymphologists that the system of lymphatics which is to be found in vertebrates is designed in the first instance for the return of body proteins lost from the high-pressure vascular system. Lymphologists for their part tend to ignore the cellular component of lymph, which they see as irrelevant to most of their thinking. In their turn, the lymphocytologists will, if they can, ignore the fluid component.

The spleen

The spleen is found in nearly all vertebrates but its precise function is not understood. Congenitally asplenic mice have certain immunological deficiencies, some of which seem to be associated with the existence of the splenic framework rather than any cells which move through the spleen (Lozzio & Wargon 1974; Oster, Koontz & Wyler 1980). Surgical removal of the spleen in adult small rodents seems to have relatively little effect on the maintenance of the capacity to respond to antigens, though in man it can lead to potentially lethal pneumococcal infections. It is clear that we have still much to learn about the spleen.

The thymus

The thymus is certainly a provider of cells to the peripheral lymphocyte pools in vertebrates and is perhaps in addition a source of hormones that influence peripheral lymphocytes, although this has been questioned (Davies 1975). The T cells that derive from the thymus supposedly have two main functions: firstly, to help (or suppress) antibody production by B cells. Secondly, T cells can be demonstrated *in vitro* to have cytotoxic potential and by inference this property is advantageous *in vivo* in effecting the cell-mediated immunity which might be important in a variety of parasitic diseases. That the thymus and T cells are a reinforcement mechanism of the humoral component of the immunological apparatus is apparent, but why such reinforcement is necessary is not so clear.

Bonemarrow and lymph nodes

Neither bonemarrow nor lymph nodes are ubiquitous in the vertebrates. However, in the lower forms aggregates of haemopoietic tissue exist in sites other than bone shafts, and lymphocyte clusters, sometimes quite large, are found in, for example, amphibians.

Evolution of the immunological apparatus

It is not clear why the apparatus housing the lymphocyte populations has changed during the evolution of the vertebrates, although it is possible to guess. The system of immunoglobulin isotypes, for example, shows a progressive elaboration along what we suppose to be an evolutionary sequence. The early cartilaginous fishes have a limited array of 19S immunoglobulin molecules, whereas the birds and the mammals have a fair number of isotypically distinct immunoglobulins. The adoption of a predominantly terrestrial mode of life and homoiothermy may have been significant in creating a requirement for more sophisticated defence mechanisms, in which the emphasis has been placed more on membranes in constant contact with dust and germ-laden air. Yet it would be dangerous to make too many generalizations in the present state of our knowledge. Though we can see that changes have occurred, their significance is not always obvious.

Specific reinforcement of nonspecific immunity

In vertebrates there is a highly elaborate complement system which plays a part in a wide variety of inflammatory conditions, not all of which require an immunological trigger. It is also known that certain types of antibody can facilitate phagocytosis and thus lend a degree of acquired specificity to some of the phagocytic cells. It could sensibly be argued that this reinforcement of a primarily nonspecific and destructive defence system is the *raison d'être* of the populations of lymphocytes. The argument is not compelling, however, taken in conjunction with the clear evidence that many very foreign and obviously antigenic moieties live in and on vertebrate organisms. Neither can we be certain at present that comparable reinforcement will not be found in invertebrates which lack lymphocytes.

Dover (1982) has argued that there are molecular drives in evolution which push change along particular paths that have little *a priori* selective advantage. Alternatively, or in addition, it could be argued that more sophisticated defence mechanisms evolved *pari passu* with increase in size and elaboration of the vertebrate body. In the present context, the proposition is that the selective advantages of coexistence led to the elaboration of mechanisms concerned with its evolution.

Self-recognition and lymphocytes

Before leaving this general consideration of the defence systems of triploblasts and turning to the complexities of the conventional immune mechanisms in mammals, it is worthwhile giving a brief mention again to self-recognition. Peripatetic phagocytic cells in the triploblast body must be equipped to distinguish self from nonself, otherwise they would not be able to function as they do. Likewise, as has already been indicated, many interactions involving cell-to-cell recognition during embryogenesis are known. Thus recognition of either self or nonself is by no means a unique property of the population of lymphocytes. This conclusion does not negate an interest in seeking to ascertain how lymphocytes recognize other lymphocytes and nonlymphocytic agencies. It does perhaps again point to the fact that the recognitional systems of individual lymphocytes, or their clones, are restricted so that they can respond specifically. The receptors which mediate these restricted responses are, of course, the main subject of study by contemporary immunologists. Immunologists argue that the system of lymphocytes as a whole has an infinite capacity to respond, based on the existence of large numbers of clones of specifically reactive cells. Whilst the response of any clone is specific, the collective response capability makes it possible to respond to any kind of foreignness.

The vertebrate immune response

The immune response, as usually studied by immunologists, involves a phased and complex reaction to one or more nonreplicating antigenic stimuli delivered to an intact laboratory rodent or a cell population growing in tissue culture. These so-called model studies are difficult to interpret. Firstly, there are relatively few antigenic stimuli *in vivo* in the real world comparable with those deriving from the hypodermic syringe of the immunologists — insect bites and food antigens perhaps qualify. Further, great care must be taken with the results of *in vitro* experiments because the circumstances of the reaction may well lack many of the homeostatic constraints that pertain in living animals.

Persistent antigenic stimuli

It may prove that living organisms, which can and do replicate on gaining entry to the vertebrate body, may evoke a response from the host which is qualitatively different from that elicited by, for example, a single injection of sheep red cells, but there is no *a priori* evidence for supposing that this will be so. In a limited set of experiments in mice, it was shown by the simple criterion of determination of the duration of T-cell mitotic response that repeated injections of T cells led to a slight prolongation of response in comparison with a single injection (Davies 1969). By the same token the response to *Plasmodium yoelii* was long relative to that following a single injection of sheep red cells but stopped well before the level of parasitaemia diminished (Jayawardena et al. 1975). Such studies illustrate that there may well be regulatory devices built into the cells that respond to antigenic stimuli which prevent continuation of the response if the stimulus persists, but they do not predicate the view that responses to acute and to chronic antigenic stimuli differ other than in duration. Thus, whilst it behoves the immunologists to remember the artificiality of the systems they study, there is little reason for them to suppose, on evidence such as that so far presented, that their conclusions carry no weight in the world of parasitic disease.

Adaptational immunity and immunological memory

The essence of specific immunity is that there is an adaptational response involved which is related to the quality of the stimulus concerned. The responding organism is often left in a permanent state of heightened capacity to respond to a repetition of the same stimulus. Most contemporary theoreticians would argue that the capacity to respond, initially, exists in the form of small numbers of specifically reactive cells and that the response involves an increase in their number and organization. Thus any immunologically active condition is seen as augmentation of an existing state rather than the consequences of some instructional mechanism.

The response to antigen can sometimes continue long after the first stimulus. Such a phenomenon can be associated with existence of a nonbiodegradable antigenic stimulus or, in the instance of parasitic infection, the antigenic stimulus often persists because the parasite is alive and well and growing despite the immune response which, by conventional wisdom, is designed to remove it. It is in this respect that the responses to living and nonliving antigens differ. Introduced 'dead' antigen may be degraded but it has no capacity for active change. In contrast the parasite can change its numbers or its physical condition either as part of its life cycle or in response to the immune activity of the host organism. This is particularly true of both protozoan and metazoan eukaryotic parasites. Host-parasite relationships involving hosts with immunological responses are in fact highly complex systems of mutual adaptation.

The triploblastic invertebrates probably have a relatively limited capacity to undergo adaptation to parasitic invasion via any specific immune responses in comparison with the vertebrates, and this is a major difference between the two groups. It is a corollary of this argument that invertebrates are unlikely to be able to influence their parasites in such an elaborate manner as do vertebrate hosts.

Rejection of parasites in vertebrates

In vertebrates the outcome of invasion by a foreign organism can be outright rejection of the interloper. In these circumstances, apparently, the vertebrate has successfully operated a defence mechanism and the parasite has failed. If the parasite has been housed in the host organism long enough to enact its particular life style, then sterile immunity as an outcome need not be an evolutionary dead end for the parasite. It could benefit the host in that second contact with the same parasite need not lead to even transient and possible detrimental invasion. It is difficult to determine whether sterile immunity following parasite rejection is a common phenomenon. Small numbers of small parasites could be extremely hard to find without efficient tests for infectivity, and these often do not exist or have not been undertaken. Dineen and Szenberg (1961) showed many years ago in the artificial circumstances of skin grafting in mammals that the persistence of immunity was associated with the retention of a small but viable residue of donor cells. Such residues could well be common in parasitic infections in which persistent immunity is evident.

Qualified acceptance of parasites

Despite these uncertainties, it is clear that in many instances an invading parasite is not rejected absolutely, although its numbers may be much reduced. From the point of view of the parasite, this could be regarded as a successful outcome to the interaction if the number of residual parasites is appropriate to its life style. The host, in contrast might be thought to have failed if rejection of the parasite was the aim of the immune response. On the other hand, it is quite possible that the resistance to further infection by the same or even a different invader could be augmented by the existence of the retained parasites. Such a notion underlies the use of some of the immunological adjuvants.

The prevailing theory of immune responses dictates the pre-existence of specifically responding cells, and immunologists argue that an increase in the number of these cells occurs during the course of the response. In addition, the pattern of products of a stimulated cell population differs from the starting pattern of, for example, the isotypes of immunoglobulin molecules on the virgin B-cell array. Thus memory responses may differ not only in their speed but in their pattern from the primary response. All this written, it could be that amplification of an ongoing, albeit low-level, response could be a quicker and more effective means of stepping up the defence system than evocation of 'sterile' memory.

It is not yet certain how ongoing responses to chronic antigenic stimulation differ from memory responses when the original stimulus has been totally removed from the reacting system and the initial response has subsided. It is possible that this situation rarely occurs because in a live animal in conventional circumstances there will always be nonspecific stimuli which dust, as it were, the memory cells in the immunological library. The result could be a limited display of the response to previous experiences, without further contact with the stimulating antigen. On theoretical grounds it seems difficult to predict whether ongoing responses would be more effective than 'sterile' memory; either could be useful.

Lethal parasite infections

A third outcome of parasitization which should be considered is that the host dies. From the view point of the individual host, this failure to survive could be due to a genetic incapacity to respond in a manner adequate either to repel the parasite or to retain it in a nonlethal form. To the host species the genetic loss could mean the removal of an unfit individual, i.e. negative selection, which need not necessarily be disastrous because according to Malthusian principles the breeding rate could maintain a stable population size. Removal of the susceptible genotype could be advantageous to the species in some circumstances. As far as the human species is concerned, loss of the individual is nearly always regarded as a tragedy but it is as well to remember that from an evolutionary standpoint there are several interpretations of the significance of death following encounter with a parasite.

From the point of view of the parasite, death of the host can be a stage of the life cycle. Alternatively, if the parasite perishes with the host, it could indicate a lack of adaptation to the host in question. The failure could, as has been argued for the host, be associated with the lack of appropriate genes to respond to the physiological milieu and/or the immune response of the host, and thus constitute an example of negative selection. Whatever the case in a particular situation, the point to remember is that death of the host can indicate simply the incapacity to respond, on the part of the host and/or the parasite, to a set of circumstances in which mutual adaptation, with survival of both, is probably in most instances the desirable and, in time, the likely outcome.

Some specific examples of parasitization

The argument so far, with the exception of the botanical examples, has been almost entirely theoretical and it will be as well to consider various instances of parasitization to see how the phenomenon of mutual adaptation, involving the immune response or reactive changes in the parasite, holds up in the real world. The selection of examples will derive to some extent from the author's experience working in conjunction with a number of parasitologists. In some of the examples to be used the consequences of contact with a parasite will be considered with or without the host having an intact immune response. The emphasis in such examples will usually concern T-cell deprivation.

Viral infections
Childhood viruses

In Western civilizations there are a number of childhood viral illnesses such as chicken pox, measles, mumps and German measles. Contact with the causative agents usually occurs within the first 10 years of life and, although there are occasional complications, the usual outcome is an acute period of disease, during which time vast amounts of virus may be produced in the body of the host. There often follows full recovery and subsequent resistance to further contact with the same infection. It is believed that the viruses concerned often or always persist in the body of the host following the acute phase of infection, but maintain a low profile. Persistent chicken pox virus, many years later, can give rise by recrudescence to the painful condition of shingles. Similarly it has been tentatively suggested that the persistence of measles virus can rarely lead to subacute sclerosing panencephalitis or even multiple sclerosis. Measles can kill undernourished children and Burnet in particular has argued that this is due to the failure of T cells to control the virus, which indeed it may be. Also measles epidemics in children of remote people who have never previously been in contact with the disease can initially lead to the death of many individuals. It is not certain in populations of children that do not die as a consequence of contact with childhood viruses whether the rare complications that occur are caused by the immune response, i.e. immunopathological, or are a consequence of failure to respond in a particular way.

In relation to poliomyelitis it has been reported that in children that live in poor and crowded conditions paralytic polio is a rare disease, ostensibly because the virus is transmitted to the children repeatedly and early in their lives. Under such circumstances the virus is encountered and probably retained asymptomatically. In higher socioeconomic classes, contact with the virus is less likely and tends to take place later in life with more frequent paralytic consequences. It is not clear why late and infrequent contact should lead to pathology whereas early and heavy infection can be asymptomatic. Such phenomena are not easily predictable from standard immunological theory.

The general lesson to be learned from this glance at some viral diseases of childhood is that the majority of the individuals in the host population encounter the viruses in question and most recover without significant ill effects. If the immune response is intact and if the genetic capacity to respond in an appropriate manner has been selected for, the outcome of the interaction is that the viruses are accommodated. The physiological effects the retained viruses may have on the body of their hosts are not known, neither is the prevalence nor condition of the residual invader clearly documented. There are rare outcomes of the common condition of carrying these childhood viruses but the exact way in which this involves the immune response, if at all, is not known. It is not believed that viruses such as those at present under discussion infect germline cells such that their possession becomes a genetic characteristic, nor is there evidence of matrilineal transfer in the milk of previously infected mothers. It is worth noting that there are many arthropod-borne viruses, known mainly in less well developed parts of the world, that are associated with quite specific human diseases. In most instances the diseases concerned are nonlethal and further infections with the same organisms are rare. Thus the childhood virus diseases of Western civilizations are by no means unique in their pattern of infection.

Herpes viruses

The herpes group, aside from the varicella-zoster virus, includes cytomegalovirus (CMV), the Epstein Barr virus (EBV) and herpes I and II. Most adult members of human society in the Western world possess one or other or all of these agents. It is not known how, or even if, immunological control is exercized over these viruses (Roizman 1969). In various experimental studies of CMV in mice, it has so far not proved easy to demonstrate the degree of involvement of either T or B cells in controlling the infections which can, in any instance, often be low key and relatively nonpathogenic. CMV can cause problems in humans, but largely in physiologically debilitated adult or extremely young patients. Debilitation or youth may well involve the reduction or failure of development of the immune response against the virus, but it is not certain that this is the significant factor in any virally related pathology that develops. Most telling of all perhaps, in relation to herpes I, is that it is perfectly possible to have a healing skin lesion on the face, for example, with an acute and developing lesion in the genital area. In such circumstances it seems that the systemic immune response is ineffective, even though it may be possible to demonstrate high levels of *in vitro* neutralizing antibody. There is considerable epidemiological evidence linking herpes virus infections with carcinomas both of cervix and larynx. Despite this suspicion, it is not known why some individuals should have problems nor even is it certain that the apparent involvement of herpes is a consequence rather than a cause of malignancy or its antecedent hyperplasias.

EBV is known to infect B cells and it is likely that the infected B cells excite a response from T cells which can be part of the disease known as glandular fever or infectious mononucleosis. There are many interesting features of the infection not least of which are the pathognomonic heterophile antibodies. These, perhaps, show what happens if the memory repertoire of lymphocytes is disturbed. There are other possible outcomes of infection with EBV, such as various neoplasias within the lymphoid system, but these are rare and there is little indication as to why tumours arise in some individuals of a population of people in which the virus is almost ubiquitous.

Speculation about host-virus interactions

Perhaps at this point a note of speculation would be in order. The childhood viruses and those others of the herpes group occur almost universally but rarely cause significant trouble. In relation to the childhood viruses, there is probably some involvement of the immune response in the host-virus interaction, but there is less evidence that this is so with some of the herpes viruses. None of these viruses is transmitted from infected individuals to their offspring via the gametes, but the whole array of viruses is usually acquired relatively early in life. It could be supposed that there are considerable biological advantages in the possession of these viruses, not always to the individual but possibly to the species. The childhood diseases might represent a beneficial exercize for the lymphoid system at an early stage of development with a residue of useful reactivity when the viruses persist. Persistent viruses could also be responsible for altering the physiological milieu in those parts of the body that they infect, not necessarily in a deleterious manner. In either instance, irrespective of whether advantage to the host accrues from the possession of the virus, the normal immune response may well be conducive to living in relative peace with the virus and thus an accommodation, not a rejectional device.

Rabies and myxomatosis viruses

The rabies virus is one of the most feared of all by *Homo sapiens* because of the nasty symptoms of infection and the inexorably lethal outcome. There are, however, a number of species of animal, such as foxes and bats, which are commonly infected by the virus and nevertheless survive. In vampire bats, it has been noted that flare up of the disease occurs when the host population reaches high density suggesting that the species concerned is apparently using the virus as a population control device (Anon. 1970). By the argument used earlier, advantage to the species could be said to accrue from the presence of the virus. In turn the immunological mechanism required to come to a relatively stable living arrangement with the virus can only be effective if it accommodates rather than rejects.

In the last 30 years or so, myxoma virus has wreaked havoc among the populations of wild rabbits in the British Isles and Australia. In both environments, the epidemics concerned largely affected a genetically unselected population of animals. Rabbits breed quickly and, at least in the British Isles, they have recently tended to rely less on burrows where transmission of the virus, by fleas, is more likely. Whatever the mechanism, it seems that within a short time the rabbit populations as a whole have come to terms with a highly lethal virus and are now not only relatively resistant to its pathogenic effects, but are thought to carry the virus usually without disease signs (Fenner 1968). The virus itself has changed and it is likely that selection for the ability of the host to resist has occurred by exercize of the postulated immunological accommodation device. There may or may not be any advantage to the host species due to the possession of the virus, but its acquisition has been accomplished with only a minor check in evolutionary time.

Oncogenes

One last point in relation to viruses. It is becoming increasingly apparent that specific genetic sequences are associated with malignant phenotypes in mammalian cells (Klein 1983). Contemporary molecular biologists are unravelling the tangle of evidence and it is clear that in time they will be able to describe accurately the kinds and modes of action of the genes involved. It is also believed that some of the sequences can be thought of as viral, in so far as they either are or have been infectious. In the instance of the retroviruses, germline transfer of proviruses can occur (Weiss et al. 1981), and matrilineal passage of oncogenetic virus particles, during the process of lactation, has also been recorded (Bittner 1936).

In a few instances in domestic animals, limitation and reduction of the oncogenetic consequences of the possession of a virus can occur apparently using immunological mechanisms. Marek's disease virus falls into this category, in that it can cause tumours and a degree of protection can be gained against it by the artificial stratagem of vaccination (P.M. Biggs personal communication).

Some, if not all, of the viruses associated with oncogenesis seem to have the capacity, directly or indirectly to augment cell replication rates and it is not beyond the bounds of possibility that this faculty might be beneficial to the host organism (Todaro 1978). The idealized concept of the vertebrates as consisting of organisms capable of rejecting foreign viruses by the immune response not only could be wrong in fact, but is arguably wrong in its somewhat xenophobic assumption that foreignness is a bad thing and should be avoided if possible.

Bacterial infections
Escherichia coli

All triploblastic organisms (except those in the plastic isolators of experimentalists) and the embryos of mammals and birds carry bacteria both on their external and some of their internal surfaces. The relationships of these ubiquitous organisms to their hosts have been all too little studied, but it is apparent that they exist in the gut of vertebrates as a complex of species which is relatively stable and which can be of considerable benefit to its possessors. Just how the immune response modulates the build up of some or all components of this bacterial gemisch is not known, but one example from the bovine perhaps illustrates the principles.

Colostrum-deprived calves get no anti-*Escherichia coli* antibodies from their dams and can easily succumb to overwhelming enteritis due to *E. coli* in the immediate postpartum period (Penhale et al. 1973). On the other hand *E. coli* plays a necessary role in the breakdown of plant materials in the ruminant gut. The moral of this story, which is well known to the veterinarians, is that antibody provides a means by which *E. coli* can be acquired by an organism without problems; accommodation of the foreign organism and not its rejection is beneficial.

Salmonella species

In experimental infections with *Salmonella typhimurium*, T-cell deprivation has relatively little effect on the ID_{50} following parenteral introduction to a 'resistant' strain of mice such as CBA (O'Brien & Metcalf 1982). Knowing this, the argument that sensitive strains lack the T cells required to respond to the bacterium is weakened. It should be stressed that T cells could be involved in generating whatever resistance to secondary or sublethal infection exists in mice but they seem to play only a small role in primary, potentially lethal infections. Such evidence as there is suggests that susceptibility and resistance are properties of the host macrophage populations.

One other bacterial infection exemplifies a different facet of host-parasite interactions. *S. livingstone* is found infrequently in humans and it can cause mild enteric symptoms. It was also found in a breeding colony of mice where it was associated with some preweaning deaths (Simmons & Simpson in preparation). The source of the infection was never discovered but it could have been one of the animal-handling staff. *S. livingstone* was isolated from the infected mice and attempts were made to infect clean mice in isolation by gastric intubation. It was possible to produce a transient gut infection in animals between 3 and 12 weeks old but not outside these times. Susceptibility to infection, such as was discovered, was in any instance asymptomatic and apparently independent of the presence of T cells. If mice were treated with streptomycin before the introduction of *S. livingstone*, asymptomatic permanent infection resulted. It seemed in this instance that ease of infection depended more on the availability of an ecological niche than any reactive capacity on the part of the host. If the niche was not available, resistance to infection was likely to have been a property of the bacteria already present in the gut. Such examples as this, for which there are precedents (Bohnhoff & Miller 1962; Bohnhoff, Miller & Martin 1964), strengthen the argument that in parasite infection the immune response can be irrelevant or accommodating, but need not be defensive if this has to mean rejectional.

General considerations on bacterial infections

It is tempting but premature to advance the argument that the specific immune response is largely irrelevant to the accommodation of intracellular bacterial parasites, even though it might have a role in controlling the tempo of disease or resistance to superinfection. Whatever the answer, it is at least worthwhile considering that the interaction concerned is not primarily of an immunological nature.

In all seriously debilitated animals, human or otherwise, bacterial infection can cause death but in many of these instances it could be argued that what has heretofore been an ecologically balanced situation, in which host and bacterium have coexisted possibly with mutual benefit, is disturbed. The disturbance could involve relaxation of an immunological defence mechanism, but could also in many instances be due to less specific causes, such as failure or malfunction of the phagocytic system associated with, for example, high levels of circulating steroids. Certainly, clinical bacteriologists caring for seriously immuno-compromised patients tend to be more concerned with the maintenance and restoration of neutrophil counts than with the state of the lymphoid system.

Protozoan infections

The protozoa include many parasites, some of which kill large numbers of both humans and domestic animals each year. In nearly all instances, the individuals that die constitute a small proportion of those infected. Some of the exceptions are of particular interest to immunological parasitologists who study the bovine. *Theileria parva*, the causal agent of East Coast fever in cattle, can cause large-scale mortality in undipped herds. The infection has many interesting features, not the least of which is that *Theileria* inhabits and transforms the host lymphocytes (Purnell 1977) which should constitute, according to the belief of immunologists, the defence mechanism. Some of the game animals in the geographical regions in which cattle herds can be decimated are infected with *Theileria* spp. without obvious pathological consequences. Cattle that recover from *Theileria* or those artificially immunized are also healthy carriers.

Various species of trypanosome can also cause serious disease in cattle. Again the causal agents are known in what appears to be a carrier state in game animals. As has been widely discussed, cattle, whether *Bos indicus* or *B. taurus*, are not indigenous species in Africa. Certain types of *B. taurus*, which are believed to have been on the continent some 2,000 years longer than *B. indicus*, are ostensibly more resistant to trypanosomal infection. Their resistance is only relative and is referred to as trypanotolerance, by which is implied a capacity to carry trypanosomes and to be able to sustain moderate field challenges without disastrous effects (Murray, Morrison & Whitelaw 1982).

From a practical point of view, in relation to theilerial and trypanosomal infections, cattle herders are faced with the potential depredations of parasites upon animals which have not been adequately selected for the capacity to live in harmony with the parasites. The question seems to be whether in such circumstances an immune response which at best is ineffective can be turned into one which is effectively accommodational or even rejectional. It might be worth trying to determine whether the antibody array in infected but asymptomatic game animals has any particular features which might be mimicked in domestic cattle. This search, whilst yielding fascinating results, could, nevertheless, leave us wondering how to take the next step, as our present capacity to engineer immune responses in a particular way is limited. In time it may prove possible to extract and use genetic information relating to resistance to trypanosomes, but at the moment this is probably not feasible.

In cattle, it has been demonstrated on a number of occasions that skilled and intensive husbandry, involving the use of regular prophylactic and therapeutic dosage with appropriate chemicals, is compatible with rearing healthy animals of the susceptible varieties in areas where both *T. parva* and potentially pathogenic trypanosomes occur (Bourn & Scott 1978). Such a demonstration illustrates what can be done, but the various expedients involved are unlikely to be widely adopted in Africa in the near future, not because they are scientifically implausible but because they are difficult in practical terms. Nevertheless, it is as well not to forget what can be done within the existing framework of knowledge.

Trypanosomes in laboratory rodents

Some of the species of trypanosome which affect and harm cattle and members of the species *Homo sapiens* can be grown in laboratory rodents, where they are in most instances quickly lethal. Much effort has been put into working with these experimental models and a great deal has been discovered about trypanosomal infections in mice, but it should not be forgotten that most of the infections that have been studied are artefactual both in respect of the inappropriateness of the host and the high lethality. The rodent-trypanosome combinations can show little of the mutual adaptation and relatively peaceful coexistence that, as has already been stressed, can characterize parasitic infections.

T. musculi, which is believed by some to be a natural parasite of mice, is exceptional in that, following introduction of as few as one parasite into a normal CBA mouse, a parasitaemia arises that lasts for a week or so and then disappears. Resistance to repeated infection with the same parasite appears to be absolute and high levels of circulating antibody are found in the recovered hosts. Animals with a major quantitative deficiency of T cells similarly infected develop a much more severe parasitaemia and eventually die with fulminant disease (Viens 1972). This seems a straightforward story of T-cell control of a defensive response that can lead to sterile immunity. Closer examination, however, revealed that recovered normal animals have a residue of infection in their kidneys (Viens & Targett 1971). There, in special capillary loops, viable parasites were always found bathing in a fluid which should have been antipathetic to their existence. It may well be that many parasites were being lost to attack by antibody but, as far as has been determined, the host/parasite association in its chronic condition is stable, except perhaps during pregnancy in females, and seems to cause little harm to the host. On the face of it, as transmission occurs by blood-feeding ectoparasites, it would seem that localization in the kidney is of little benefit to the parasite. The possibility that benefit accrues from the presence of the parasite and that association could be symbiotic has not yet been explored.

Antigenic variation in trypanosomes

One facet of certain trypanosome infections in both man and the bovine is the fantastic capacity of the parasites to vary their antigenicity. This phenomenon seems so obviously to exemplify a response by the parasite to the defence mechanisms of the host that the alternative explanation of maladaptation seems not to have been considered. It can be argued that if, in evolutionary terms, relatively peaceful mutual coexistence is the aim, then antigenic variation is an attempt on the part of the parasite to manipulate the immune response of the host to produce an appropriate accommodatory pattern of antibodies. That this can occur is evident from *T. musculi* in which no antigenic variation is known at present but antibody production is evident. It may either be that the ancestors of *T. musculi* never knew the trick of antigenic variation or, alternatively, the capacity for antigenic variation has been subject to genetic erosion beyond the time of its usefulness. Explanations of this kind may seem bizarre to those who have spent much of their lives engaged in attempting to understand the genetic basis or range of antigenic variation in trypanosomes but to suppose that the trypanosomes and their hosts are *a priori* at war, rather than engaged in trying to achieve peace, is to polarize the approach in a way that is arguably unproductive. On this score, more information is needed on the extent of antigenic variation in trypanosomes existing in wild, as compared with domestic, ungulates.

Plasmodium chabaudi infections in mice

The protozoan parasite *Plasmodium chabaudi* has been subject to considerable laboratory scrutiny. Briefly, when introduced to C57Bl mice there follow three peaks of parasitaemia, the first higher than the second and the second higher than the third. It is supposed that these peaks represent the kind of cyclicity of antigenic change often seen in trypanosomal infections. The prediction on conventional wisdom is that relaxation of the immune response in such circumstances would lead to fulminant infection by parasites having the starting antigenicity, from which there is no immunological pressure to change. What happens in fact, when *P. chabaudi* is introduced into T cell-deprived mice, is that the parasitaemic cyclicity becomes protracted over many months (Leke, Viens & Davies 1981). Most of the host animals eventually die. It can be argued in the circumstances of these experiments that the parasite was attempting to elicit a reaction from a host which, by virtue of its experimentally imposed immunoparesis, was unable to make the appropriate accommodatory response. Reconstitutional experiments with the appropriate array of antibodies could perhaps resolve this problem.

Leishmania

Mouse strains vary widely in respect of their susceptibility to a number of protozoa. One of the best researched examples is that of *Leishmania tropica*. This intracellular parasite can produce either a resolving skin lesion when introduced intradermally into mice or, alternatively, a lethal visceral disease. The same genus in man gives rise to similar pathologies. In CBA mice, T-cell deprivation alters the pattern of response to intradermal introduction of leishmanial promastigotes so that the duration and size of the resulting skin lesion is increased and some metastatic ulcers occur (Preston et al. 1972). T-cell deprivation in CBA mice did not lead to the visceral disease, which occurs in immunologically intact but genetically susceptible animals. In normal CBA mice with resolved lesions there is resistance to further development of ulcers following introduction of promastigotes by the intradermal route. It is not certain but it is suspected that persistence of the parasite is associated with this immunity.

There is a further interesting aspect of leishmanial disease in man in that Greenblatt and colleagues have shown that *Leishmania* organisms can develop antigens on their surfaces which are identical to the ABO blood group antigens. The paradox is that epidemiological studies show that the antigens, which can be found on surveying leishmanial isolates from particular geographical localities, are not those which correspond to the predominant blood groups in the human populations in the same regions (Greenblatt et al. 1981). If antigenicity predisposes to rejection, the parasite seems to have adopted a suicidal mode of life. Alternatively, if the parasite requires the isoantibodies of the host in order to modulate a benign course for its infection without great impairment of the vigour of the host, then the quality of its acquired antigenicity makes some sense.

Whatever is subsequently discovered about leishmanial infections, and there is much still unknown, on present evidence it almost seems that the serious pathological consequences of contact with the parasite derive from a lack of the immune responses and/or the quality of macrophages required to react with or retain the organism relatively asymptomatically. Again it can be argued that the optimal immune response to *Leishmania* is accommodatory, not rejectional. The issue of whether the immune status of those humans who have had a dermal lesion is associated with persistence of the parasite is again not resolvable on present evidence.

Fungal infections

There are some fungal infections known in *Homo sapiens*. *Candida albicans*, for example, is an almost ubiquitous component of the investing flora of the external surface of human beings and is usually also found on some of the mucous internal surfaces. It can cause a variety of distressing but nonlethal disorders and, much more rarely, is associated with potentially lethal systemic disease. Human beings certainly recognize candidal antigens immunologically but it is not clear whether the relevant responses are a significant part of the host-parasite interface. In a limited experimental study involving the intradermal introduction of *Candida* in normal and T-cell deprived mice, there was little evidence of a major difference in the pattern of response (J. Domer, D.C. Dumonde & A.J.S. Davies unpublished). In both instances an ulcer developed which, if anything, was larger in the normal mice than in the deprived animals. There was no pathological systematization in either group.

Cryptococcus neoformans is frequently found in the faeces of feral pigeons in cities such as London and New York. There is no evidence that the pigeons suffer any untoward effects from this infection. Humans in areas in which pigeon faeces abound inhale crytococcal spores which can occasionally cross the lung epithelium and into the bloodstream. The means and frequency of penetration are unknown. With the exception of one tramp who is said to have fallen asleep in a pigeon loft with his nostrils close to a large pile of pigeon faeces (D.W.R. MacKenzie personal communication), there are few records of normal individuals suffering pathological consequences of contact with *C. neoformans* spores. However, opinions on this problem are still in a state of flux. It is known that patients with Hodgkin's disease, or sarcoidosis, can develop crytococcal meningitis which is often lethal. AIDS patients are also known to have succumbed to the same disease (Waterson 1983). Such experimental work as has been done in mice suggests that neither T (Cauley & Murphy 1979) nor B (Monga et al. 1979) cells make a critical contribution in preventing death after systematic introduction of cryptococcal spores. In the instance of *Cryptococcus* and, possibly, *Candida* and *Salmonella typhimurium*, the specific immune response seems to be of little significance in relation to primary host resistance. Such a possibility is intriguing and somewhat at variance with the view of immunologists that the immune response is an allpurpose reactional mechanism designed to respond specifically to any foreign entity. The truth seems to be that from the point of view of parasitic infection, an appropriate, accommodating or rejectional, response is not made in all circumstances, and that when it is there is a genetic precondition arrived at by natural selection.

Metazoan parasites

A number of metazoan organisms, largely invertebrates, have adopted a parasitic mode of existence. Vertebrate parasites are rare and tend to be either ectoparasites such as various cyclostomatous fish or social parasites such as the cuckoo.

Endoparasitic worm infestations are fairly common and they cause a number of serious diseases in man, his domestic animals and probably in many wild species of animal. *Schistosoma mansoni*, which can cause serious debilitating disorders in humans in parts of Africa, has been widely studied both in its natural state and in experimental animals. In human populations in endemic areas, repeated infection can occur from an early age. There is a slow rise in faecal egg count up to the age of 20 to 30 and then a slow decline. It may be that a degree of acquired resistance builds up in humans over time, but this is a controversial issue. There are various possible lethal complications of schistosome infection, but the majority of the infected individuals do not die as a simple consequence. In mice, which are not the natural hosts, the same parasite produces an infection which is often lethal. The role of the immune response in such experimental circumstances is equivocal, but it has been suggested that immune mechanisms can actually facilitate egg excretion in normal mice but less so in T-cell deprived animals (Dunne et al. 1983). It has also been suggested in these experimental infections in mice that the development of the parasite is abnormal in the absence of an immune response (Harrison & Doenhoff 1983) and that, in any instance, the early stages of infection elicit little if any specific and effective reaction from the lymphoid system (Harrison, Bickle & Doenhoff 1982). Warren (1973) argued cogently that, in relation to schistosome infections, there are factors other than immunity, such as lack of room in the host for more parasites and keratin thickening on the soles of the feet, which restrict larval access and should be given more credibility in relation to anti-parasite resistance than such immunological mechanisms as cell-mediated and/or humoral immunity. How far arguments such as these are valid in the general context of host-parasite interactions is a matter for speculation.

Trichinella spiralis, an intestinal roundworm in both rodents and man, can cause a persistent infection based on muscle-encysted larvae. In the absence of T cells in rodents, the infection builds up progressively and is usually eventually fatal (Walls 1971). There are various interesting facets of the disease, including an eosinophilia, which seem to be regulated by T-lymphocyte activation. During the period of infection, there is resistance to superinfection which may well be mediated immunologically, but the principal point to be made is that in normal, immunologically quite competent individuals of many species of animal the severity of infection is largely proportional to the level of contact or challenge by the parasites rather than apparently any other factors.

Most of the worm diseases can be avoided by human beings and most of those that cannot can be treated satisfactorily provided suitable medical services are available. It may prove possible to vaccinate against infection using irradiated larvae, as it is with the bovine lung worm. In this instance, what develops is a kind of immunologically mediated resistance which leads to rejection of further infection during the time of field challenge. Repeated vaccination is now adopted as the best means of maintaining resistance but it may prove in time that what is required to give safe and long-lasting resistance is some kind of 'trickle' infection, with live parasites present in small numbers throughout the life of the host animal.

Conclusions

The variety of host-parasite combinations is bewildering and it is difficult to arrive at any satisfactory generalizations about them without it being possible to cite paradoxical exceptions. Nevertheless, it must be remembered in trying to achieve an academic understanding of the evolution of immune processes that, in domestic animals and many tribes of the human species, the process of natural selection has quite deliberately or accidently been subjected to erosion of its naturalness. Many of the diseases that humans regard as undesirable, either in themselves or their domestic animals, represent conditions which in the wild state would probably lead either to extinction of the host or mutual adaptation. Our present understanding of and approach to the diseases in question represent a kind of impatience with the slow tempo of evolution. It may indeed prove possible to intervene, either by artificial selection or by vaccinational protocols that hasten or circumvent the processes of mutual adaptation. Yet the unnaturalness of what is being attempted should not be allowed unduly to affect our interpretation of what has been argued here to be the real quality of host-parasite relationships.

Our epidemiological knowledge of parasitic disease is extensive. Our understanding of the mechanisms by which control of disease is effected in nature is fragmentary. In Benenson's (1970) lists of communicable diseases in humans, the overwhelming majority of disorders are not associated with innate resistance to infection. Susceptibility seems to be almost universal in uninfected individuals. In many of the diseases listed, the causal organisms can be carried asymptomatically with high mortality due to infection in only few instances, and then in special circumstances. It is well within the bounds of possibility that susceptibility to infection carries with it sufficient long-term advantages to vertebrates and particularly to mammals to outweigh the adverse consequences of occasional death due to infection. Darwin was quite right in his interpretation that natural selection leads to the evolution of fit species. He was arguably wrong in attributing to nature any kind of hostility.

Human beings have always tended to view the natural world with an anthropocentric bias in which individual survival has been at a premium. Such a view has tended to obscure a larger truth in nature: that in constant environmental circumstances a state of ecological balance will usually arise between the living inhabitants of a particular region. This stable state, which will usually be dynamic and may well involve a balanced loss of many individual organisms, by predation for example can be seen again and again. Its disturbance can have unpredictable consequences and restoration of a new and sometimes totally different equilibrium may well take a long time to evolve.

In host-parasite relationships, the immune responses may play a mediatory role in some instances, achieving a state of balanced equilibrium between the host and the parasite communities based on mutual coexistence which, for a given set of environmental circumstances, could be long lasting. Many, but not all, of the diseases which crop up involve alteration of the environment which can destroy these fine balances. Famine and war promote disease, presumably not because of their primary effects on T or B cells but because of the profound physiological disturbances to the populations of humans concerned which, in their turn, both increase access to disease organisms and alter existing ecological balances. The problems that arise are sometimes ameliorable by immunological engineering. For example, anti-typhoid vaccination is of great assistance in preventing outbreaks of typhoid fever in urbanized populations of humans whose water supply has become contaminated. Yet at the end of the day many of the solutions which are derived will be sociological, rather than immunological.

Acknowledgements
I am most grateful to my many friends and colleagues who have listened to my questions, often provided answers and been patient with my idiosyncracies. Without them I could not have prepared this paper, but its infelicities accrue to me alone. My particular thanks are due to Dr Bob Ashman of the John Curtin School of Medical Research, Canberra, Mr Nick Bradley and Dr Thea Connell of the Institute of Cancer Research, Prof. Edwin Cooper of the University of California, Mrs Pat Davies of Croydon Technical College, Dr Mike Doenhoff of the London School of Hygiene and Tropical Medicine, Dr Beryl Jamieson of the Royal Marsden Hospital, Prof. Bill Jarrett of the University of Glasgow, Dr Eckhardt Kolsch of the University of Munster, Prof. Donald MacKenzie of the London School of Hygiene and Tropical Medicine, Dr Spedding Micklem of the University of Edinburgh, Dr Max Murray of the International Laboratory for Research on Animal Diseases, Nairobi, Mr Derek Simmons of the Institute of Cancer Research, Dr Geoff Targett of the London School of Hygiene and Tropical Medicine, Prof. Derek Wakelin of the University of Nottingham and Dr Robin Weiss of the Institute of Cancer Research. My secretary, Mrs Marjorie Kipling, has, as always, been a lighthouse of strength in the sea of confusion that derives from my trying to write about fancy rather than fact.

References

Anon. (1970). *Report of the committee of enquiry on rabies.* Command Paper 4696. London: Her Majesty's Stationery Office.

Benenson, A.S. ed. (1970). *Control of communicable diseases in man.* Washington: American Public Health Association.

Bittner, J.J. (1936). Some possible effects of nursing on the mammary gland tumour incidence in mice. *Science,* **84,** 162.

Bohnhoff, M. & Miller C.P. (1962). Enhanced susceptibility to *Salmonella* infection in streptomycin-treated mice. *Journal of Infectious Diseases,* **111,** 117-27.

Bohnhoff, M., Miller, C.P. & Martin, W.R. (1964). Resistance of the mouse's intestinal tract to experimental *Salmonella* infection. 1. Factors which interfere with the initiation of infection by oral inoculation. *Journal of Experimental Medicine,* **120,** 805-16.

Bourn, D. & Scott, M. (1978). The successful use of work oxen in agricultural development of tsetse-infected land in Ethiopia. *Tropical Animal Health & Production,* **10,** 191-203.

Cauley, L.K. & Murphy, J.W. (1979). Response of congenitally athymic (nude) and phenotypically normal mice to *Cryptococcus neoformans* infection. *Infection & Immunity,* **23,** 644-51.

Cooper, E.L. (1968). Transplantation immunity in annelids. *Transplantation,* **6,** 322-37.

Cooper, E.L. ed. (1974). *Invertebrate immunology.* Contemporary topics in immunobiology 4. New York: Plenum Press.

Davies, A.J.S. (1969). The thymus and the cellular basis of immunity. *Transplantation Reviews,* **1,** 43-91.

Davies, A.J.S. (1975). Thymus hormones? *Annals of the New York Academy of Sciences,* **249,** 61-67.

Davies, A.J.S., Hall, J.G., Targett, G.A.T. & Murray, M. (1980). The biological significance of the immune response with special reference to parasites and cancer. *Journal of Parasitology,* **66,** 705-21.

Dineen, J.K. & Szenberg, A. (1961). Regeneration of elements of donor origin in orthotopically grafted skin following the homograft response. *Nature,* **191,** 153-55.

Dover, G. (1982). Molecular drive: a cohesive mode of species evolution. *Nature,* **299,** 111-16.

Dunne, D.W., Hassounah, O., Musallam, R., Lucas, R., Pepys, M.B., Baltz, M. & Doenhoff, M. (1983). Mechanisms of *S. mansoni* egg excretion: parasitological observations in immunosuppressed mice reconstituted with immune serum. *Parasite Immunology,* **5,** 47-60.

Fenner, F. (1968). *The biology of animal viruses,* vols. 1 & 2. New York: Academic Press.

Greenblatt, C.L., Kark, J.D., Schnur, L.F. & Slutzky, G.M. (1981). Do leishmania serotypes mimic human blood group antigens? *Lancet,* **1,** 505-6.

Harrison, R.A. & Doenhoff, M.J. (1983). Retarded death of *S. mansoni* in immunosuppressed mice. *Parasitology,* **86,** 429-38.

Harrison, R.A., Bickle, Q. & Doenhoff, M.J. (1982). Factors affecting the acquisition of resistance against *S. mansoni* in mice: evidence that the mechanisms which mediate resistance during early patent infections may lack immunological specificity. *Parasitology,* **84,** 93-110.

Herbert, W.J. & Wilkinson, P.C. eds. (1977). *A dictionary of immunology.* 2nd ed. Oxford: Blackwell Scientific Publications.

Jayawardena, A.N., Targett, G.A.T., Leuchars, E., Carter, R.L., Doenhoff, M.J.

& Davies, A.J.S. (1975). T-cell activation in murine malaria. *Nature*, **258**, 149-51.

Klein, G. (1983). Specific chromosomal translocations and the genesis of B cell-derived tumours in mice and men. *Cell*, **32**, 311-15.

Leke, R., Viens, P. & Davies, A.J.S. (1981). Interaction between *Plasmodium chabaudi* and C57BL mice with particular reference to the immune response. *Clinical & Experimental Immunology*, **45**, 627-32.

Lozzio, B.B. & Wargon, L.B. (1974). Immune competence of hereditarily asplenic mice. *Immunology*, **27**, 167-78.

Monga, D.P., Kumar, R., Mohapatra, L.N. & Malaviya, A.N. (1979). Experimental cryptococcosis in normal and B cell-deficient mice. *Infection & Immunity*, **26**, 1-3.

Murray, M., Morrison, W.I. & Whitelaw, D.D. (1982). Host susceptibility to African trypanosomiasis: trypanotolerance. *Advances in Parasitology*, **21**, 2-57.

O'Brien, A.D. & Metcalf, E.S. (1982). Control of early *Salmonella typhimurium* growth in innately *Salmonella*-resistant mice does not require functional T lymphocytes. *Journal of Immunology*, **129**, 1349-51.

Oster, C.N., Koontz, L.C. & Wyler, D.J. (1980). Malaria in asplenic mice: effects of splenectomy, congenital asplenia and splenic reconstitution on the course of infection. *American Journal of Tropical Medicine & Hygiene*, **29**, 1138-42.

Penhale, W.J., Logan, E.F., Selman, I.E., Fisher, E.W. & McEwan, A.D. (1973). Observations on the absorption of colostral immunoglobulins by the neonatal calf and their significance in colibacillosis. *Annales de Recherches Vétérinaires*, **4**, 223-33.

Preston, P.M., Carter, R.L., Leuchars, E., Davies, A.J.S. & Dumonde, D.C. (1972). Experimental cutaneous leishmaniasis. 3. Effects of thymectomy on the course of infection of CBA mice with *Leishmania tropica*. *Clinical & Experimental Immunology*, **10**, 337-57.

Purnell, R.E. (1977). East Coast fever: some recent research in East Africa. *Advances in Parasitology*, **15**, 83-132.

Roizman, B. (1969). *The herpes virus: a biochemical definition of the group.* Current Topics in Immunobiology 49, pp. 1-79. Berlin: Springer Verlag.

Todaro, G.J. (1978). RNA tumour-virus genes and transforming genes: patterns of transmission. *British Journal of Cancer*, **37**, 139-58.

Viens, P. (1972). Immunological responses of mice to *Trypanosoma musculi* infection. Ph.D. thesis. University of London.

Viens, P. & Targett, G.A.T. (1971). *Trypanosoma musculi* infection in intact and thymectomized CBA mice. *Transactions of the Royal Society of Tropical Medicine & Hygiene*, **65**, 424-29.

Walls, R.S. (1971). Some immunological aspects of eosinophilia. Ph.D. thesis. Oxford University.

Warren, K.S. (1973). Regulation of the prevalence and intensity of schistosomiasis in man: immunology or ecology. *Journal of Infectious Diseases*, **127**, 595-609.

Waterson, A.P. (1983). Acquired immune deficiency syndrome. *British Medical Journal*, **286**, 743-46.

Weiss, R.A., Teich, N.M., Varmus, H.E. & Coffin, J.M. eds. (1981). *Molecular biology of tumour viruses.* RNA tumour viruses. Part 3. 2nd ed. Cold Spring Harbor, New York: Cold Spring Harbor Laboratory.

Wilson, H.V. & Penny, J.T. (1930). The regeneration of sponges (Microciona) from dissociated cells. *Journal of Experimental Zoology*, **56**, 73-147.

Wynne-Edwards, V.C. (1967). *Animal dispersion in relation to social behaviour.* Edinburgh: Oliver & Boyd.

Cellular components of the immune system

2

Lymphocyte subpopulations in the mouse

R. SCOLLAY

A large number of markers have been used to define lymphocyte subpopulations in the mouse. The majority of the most useful are antibodies against membrane antigens, but lectin binding, cell size and density, and morphology have also proved useful. Probably over 50 different antibodies recognize lymphocyte determinants, but only some are well defined and perhaps only a dozen or so are very useful at present. The best known are surface immunoglobulin which defines all mature B cells, and Thy 1 which defines all or most T cells. These two markers define almost all cells with lymphocyte morphology in thymus, lymph nodes and spleen, exceptions being the small population of null cells in spleen (about 5%) and lymph nodes (less than 1%) and a very small population of Thy 1-negative cells in the thymus (less than 0.5%). The nature of the Ig-negative Thy 1-negative cells is still in doubt. A number of stages in B-cell maturation can now be well defined, including several which precede the appearance of surface immunoglobulin. There is also clearly heterogeneity among mature B cells in terms of surface phenotype and triggering requirements, but whether these represent independent lineages or different stages of maturation of a single lineage remains controversial. Among mature T cells there are two well defined, independent lineages: the suppressor/cytotoxic lineage (MHC class I restricted) and the helper/inducer lineage (MHC class II restricted). Difficulties begin when one moves back to the thymus where at least five distinct Thy 1-positive subpopulations are easily defined. The interrelationships of these and the stage at which the mature subpopulations separate from a common precursor (if at all) are still unknown.

The field of immunology can be divided broadly into three categories: the immunology of animals of economic or veterinary importance, human immunology and the immunology of laboratory animals. All three areas can add to our basic understanding of the immune system but each is to some extent distinct. The study of animals of economic importance is generally aimed at manipulation or treatment at the population level. Individuals are usually of little importance, genetic selection is a possible and often necessary approach, and diagnosis or treatment must be possible 'in bulk'. For human immunologists, the individual is of key importance, and a great deal of time, money and expertise may be spent (rightly or wrongly) on a single individual. Knowledge is therefore considered of value regardless of its general applicability if it may eventually help an individual. In stark contrast, the immunology of laboratory animals (mostly rodents) is of limited value in its own right. The treatment of disease in field mice or sewer rats is a task with fairly low priority. Rodent immunology is primarily of value as a model system in which experimental manipulation can be performed which would be difficult or impossible in humans or large animals.

An adequate understanding of the immune responses involved in any disease process requires detailed knowledge of the cells involved in that response: what controls them, where they come from, how they are interrelated. It may be useful then to review our current understanding of the murine lymphocyte populations and see to what extent this information can be of use in delving into the complexities of lymphocytes in outbred animals such as cattle. I shall briefly review our understanding of the murine lymphocyte subpopulations in this context. A comparison of the detailed information available for mice, rats and humans, along with the less complete data for other species, such as sheep and cattle, allows a reasonable guess at which mouse-derived information is likely to be broadly applicable and which is not.

The marker explosion

In the last few years, the number of surface antigens which can be identified on lymphocytes has risen dramatically. Most have been detected because they are polymorphic in the species and hence one strain will make an antibody against another which expresses a slightly different molecule (see McKenzie & Potter 1979; Hogarth & McKenzie 1983; Sutton, Hogarth & McKenzie in press; Ahmed & Smith 1982a; 1982b). Many have now been defined by monoclonal antibodies which give a precision that was often lacking in earlier studies with allo- and heteroantisera. Among these are the better known antigenic determinants such as those on immunoglobulins (isotypes, allotypes, idiotypes), molecules such as Thy 1 and Lyt 1, 2 and 3, and the major histocompatibility (MHC) antigens. Well over 30 antibodies have been described in the Ly series. Altogether, more than 60 determinants have been described and, although there is some overlap, with two names given to the same molecule, at least a large proportion of this number represents separate gene products. For many of these, little is known apart from

Table 1. Cell populations in murine lymphoid tissues. Values are approximate and vary with age, strain, degree of immunization and other factors.

	T cells	B cells	Null cells[a]	Other cells[b]
Bonemarrow	1-2%	15%	15%	70%
Thymus	>99.5%	<0.1%	<0.5%	<0.1%
Lymph node	75%	25%	0-5%	1%
Spleen	25%	55%	3-13%	5-15%

[a]Thy 1-negative, Ig-negative cells with lymphoid morphology.
[b]Granulocytes, macrophages, haemopoietic cells, etc.

Figure 1. Two-colour staining of murine lymph node cells, presented as correlated two-parameter contour plots. The darkest areas indicate where the major subpopulations occur. Cells were analysed by flow cytometry on a FACS II using a single laser. Box A is control unstained cells. Box B is cells stained with rhodamine-anti-Ig. B cells show as red positive. Box C is cells stained with rhodamine-anti-Ig and fluorescein-anti-Thy 1. T cells now show as green positive; there are no cells which are Thy 1-positive and Ig-positive but some double negative null cells can be seen. Box D is cells stained with rhodamine-anti-Ig, rhodamine-anti-Lyt 2 and fluorescein-anti-Thy 1. Lyt 2-positive and Lyt 2-negative cells can now be seen among the Thy 1-positive cells. Among the Thy 1-positive cells the distinction between Lyt 2-positive and Lyt 2-negative is clear, while both Thy 1 and Ig show a continuous distribution from positive to negative, making an absolute distinction between positive and negative cells impossible. Details of the staining protocols can be found in Scollay and Shortman (1983).

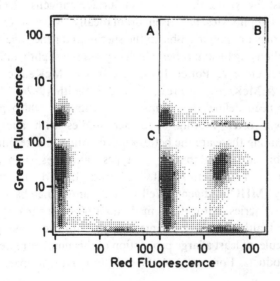

strain distribution (important in defining separate determinants) and limited information on tissue distribution.

Among these antibodies one can find a huge variety of tissue distributions. Some are on many kinds of haemopoietic cells, some are limited to very small subpopulations (e.g. supressor T cell-specific), others are on some T cells, some B cells and some nonlymphoid cells. Some could represent important molecules in immune regulation (e.g. receptors for differentiation signals or cell interaction molecules), but many may represent more general functions such as membrane transport molecules or structural components. Therefore, deciding which of them will be useful to an immunologist interested in understanding and manipulating the immune system will be a lengthy task.

The large number of these polymorphic (or at least dimorphic) surface molecules complicates the situation in outbred animals. The overall set of molecules which show variation within a species may well vary from species to species, so the mouse experience may be of limited value in some cases. In addition, heterologous (interspecies) immunization may select for a different set of antigens from those seen in intraspecies immunization. In this context, the human situation is closer to that of sheep and cattle since most of the monoclonal antibodies against human cell subsets have been made in mice. Nonetheless, some of the important and useful markers, such as Ig isotypes and markers of functional T-cell subsets, have proven to be extremely consistent in various species. We will concentrate on these in the following pages, with particular emphasis on T cells, partly because T cells are important in regulating immune responses and partly because this is where my interest and experience lie.

Major subpopulations in lymphoid tissue

Table 1 shows the approximate breakdown of murine lymphoid tissues. The data are based on a wide reading of the literature and personal experience. All the numbers vary somewhat with age, strain and degree of antigen stimulation. T cells are those expressing detectable Thy 1, while B cells are those expressing surface immunoglobulin. However, it must be remembered that in mixed tissues, such as spleen and lymph nodes, there is a continuum between positive and negative cells even with the best reagents and the most sophisticated analysis (see Figure 1). Thus the null-cell population certainly includes a few T cells, with Thy 1 below the detectable level (limit: a few thousand molecules per cell), and similarly a few B cells. This can be demonstrated in some cases. For example, on T cells there is an approximately inverse relationship between expression of Thy 1 and Lyt 1 with the lowest Thy 1-positive cells being strongly

Lyt 1-positive. Some of the Thy 1-negative cells are strongly Lyt 1-positive (and Ig-negative), and part of a continuous group of cells expressing Lyt 1 and Thy 1. Thus, defining T cells as Ig-negative cells expressing either Lyt 1 or Thy 1 allows the inclusion of some null cells in the T-cell class (Ledbetter et al. 1980; R. Scollay unpublished).

On the other hand, there clearly are some Thy 1-negative, Lyt 1-negative, Ig-negative cells in mouse lymphoid tissues, particularly spleen (usually about 10%). The exact nature of these remains controversial but they have been variously identified as natural killer (NK) cells, pre-T cells, pre-B cells and haemopoietic stem cells. It is possible that the null-cell compartment includes all of these (see for example Twomey & Kouttab 1981).

Nonetheless, it is clear that most cells with lymphocyte morphology in lymphoid tissues can be clearly defined as T (Thy 1-positive) or B (Ig-positive) (see Figure 1). In general there is little overlap between these two classes but very careful analysis reveals some mixed cells. Thus for example, in spleen and lymph nodes there is clearly a small population of cells positive for both Ig and the 'T-cell marker', Lyt 1 (Lanier et al. 1981; Manohar et al. 1982; R. Scollay & K. Shortman unpublished).

In terms of the mouse as a model, Thy 1 is a bad example since the homologous molecule in man and rats has a different distribution and in these species is not particularly useful as a T-cell marker. However, Lyt 1 has a close equivalent in the human, Leu 1 or T 1 (Ledbetter et al. 1981), and of course Ig expression is a universal marker of mature B cells.

The B cell

The common B cell expresses surface immunoglobulin of both IgM and IgD isotypes and is the most common type in lymph nodes and spleen. The more differentiated cells expressing surface IgG, IgA or IgE are relatively rare in resting lymphoid tissues. The early stages of B-cell differentiation, before surface Ig is detectable, are less well defined but progress is being made. Cells with cytoplasmic, but not surface, Ig are clearly the immediate precursors of B cells and several monoclonal antibodies have now been described which define even earlier stages in the B-cell lineage (see articles in *Immunological Reviews*, **69**, 1983). The first surface Ig-positive cells appear in foetal liver at about day 16 to 17 of gestation, and they mostly express IgM. They later express Ia and then IgD so B cells in normal adult tissues are mostly IgM-positive, Ia-positive and IgD-positive. The first immunocompetent B cells appear at around the time of birth, although rare reactive cells can be detected several days earlier.

Table 2. Immunoglobulin expression and gene rearrangement in the B-cell lineage. From Coffman 1983.

	Ig synthesis	Surface Ig	Heavy chain genes[a]	Light chain genes[a]
Pluripotent stem cell	0	no	G/G	G/G
Large pre-B	μ	no	R/R	G/G
Small pre-B	μ	no	R/R	G/G or G/R or R/R
B cell	μL, δL	yes	R/R	G/R or R/R
Plasma cell	μL, γL, αL, etc.	yes	R/R	G/R or R/R

[a] G = Germline configuration. The area of DNA which codes for the immunoglobulin chain has not yet been altered. This is the usual situation in all cells except B cells. R = rearranged genes. The DNA coding for the immunoglobulin chain has been shuffled so that the various sections of DNA coding for different parts of the protein have been brought close together. This shuffle varies slightly from cell to cell, the variation being the source of the multitude of binding sites on the expressed immunoglobulin. Only rearranged genes are transcribed.

Knowledge of some of the molecular events involved in Ig production and the generation of diversity in B cells is far in advance of our understanding of the related cell biology. The expression of Ig requires a complex series of genetic rearrangements, first of the heavy chain locus and later of the light chain locus (see Table 2). Following these two events IgM and IgD are expressed but after antigenic stimulation the B cell must further rearrange its heavy chain genes to allow expression of the more mature isotypes — IgG, IgE and IgA. The details of these genetic events are beyond the scope of this review but can be found in Adams and colleagues (1981) and Tonegawa (1983).

Perhaps more controversial is the question of B-cell heterogeneity. Although it is clear that B cells are heterogeneous in physical characteristics, marker expression and requirements for activation, it is unclear whether these differences indicate degrees of maturation of cells in a single lineage or whether a number of different lineages can be identified. The subsets defined by the presence or absence of the antigen Lyb 5 are the best, but not the only, example at this stage. For instance, the interaction of Lyb 5-positive B cells with T-helper cells is not MHC restricted and can be replaced by soluble factors, while Lyb 5-negative cells must interact directly with the helper cells and the interaction is MHC restricted (Hodes, Hathcock & Singer 1983).

It is clear that the mouse model has yielded considerable information concerning the molecular events involved in the production of immunoglobulins. This information seems to have wide applicability and its extension to other species does not seem to present any difficulties.

38   R. Scollay

Table 3. Functional subpopulation markers.

Lineage	MHC restriction	mouse[a]	Marker rat[b]	man[c]
Cytotoxic/suppressor	class I	Lyt 2	MRC Ox-8	T8, Leu 2
Helper/inducer	class II	L3T4	W3/25	T4, Leu 3

[a]Boyse et al. 1968; Kisielow et al. 1975; Dialynas et al. 1983.
[b]Mason et al. 1983.
[c]Reinherz & Schlossman 1980.

The T cell

As we have seen in the mouse, Thy 1 is a relatively good marker for T cells: 99.5% of thymocytes and 98% of peripheral T cells express this marker. Although many nonlymphoid cells, such as brain cells, fibroblasts and many normal bonemarrow cells, express Thy 1 at very low levels (Basch & Berman 1982), it is generally an adequate marker for thymus-derived cells in lymphoid tissues. The homologous molecule in rats and humans is much less useful as a T cell marker since it is not 'pan-T' and is widely expressed on haemopoietic cells. Lyt 1 is also a pan-T marker (Mathieson et al. 1979; Scollay & Shortman 1983), but is expressed on a small subset of B cells (Manohar et al. 1982; Lanier et al. 1981). It has a homologue in the human, Leu 1 or T 1 (Ledbetter et al. 1981), and so probably in other species. Pan-T reagents have proven difficult to find and so far have only been discovered in mouse and man, but a combination of markers which define functional subsets of T cells (apparently similar in many species — mouse, man, rat) may be just as useful if not more so.

These functional T-lineage markers seem to be well conserved across species with a similar distribution in mouse, rat (Mason et al. 1983) and human (Reinherz & Schlossman 1980). The definition of functional subpopulations and the biochemistry, where known, are similar in these three species, making it likely that similar molecules with similar characteristics will be found in other animals. In the mouse, the two important antigens are Lyt 2, which was first defined in the 1960s (Boyse et al. 1968) although its importance in defining functional subsets was not realized for some years (Kisielow et al. 1975), and L3T4, described only recently (Dialynas et al. 1983). Table 3 shows the equivalent markers in human and rat. Under normal circumstances, all lymph node or spleen T cells express one or other of these antigens, but not both. Lyt 2-positive, L3T4-negative cells are usually restricted to recognition of class I MHC antigens, and under normal circumstances the functions involved are cytoxicity and suppression. Lyt 2-negative, L3T4-positive cells are

Table 4. Markers which distinguish cortical and medullary thymocytes.

Marker*	Cortex (85%)	Medulla (15%)	Peripheral T
Thy 1	high	low	low
PNA	high	low	low
LL	low	high	high
TL	mostly positive	negative	negative
H2-K	low or negative	high	high
H2-D	low	high	high

*PNA = peanut agglutinin.
LL = lobster lectin.
TL = thymus leukaemia antigen.

restricted to recognition of class II MHC antigens and the functions involved are usually induction, help and production of soluble mediators. However, it is clear that restriction of recognition to class I or class II MHC antigens is a better definition of the lineages than function (reviewed in Swain 1983), since Lyt 2-positive, L3T4-negative cells can function as helpers in a class I-restricted manner, while Lyt 2-negative, L3T4-positive cells can be cytotoxic if appropriate class II-positive targets are used. These two lineages have been regarded as separate and non-interchangeable for many years (e.g. Huber et al. 1976), and I think it is reasonable to regard them as true lineages. However, under certain circumstances, mature cells may be able to change lineages (see Thomas & Calderon 1982) or may appear to be restricted to the wrong class of MHC antigen (Pawelec, Schneider & Wernet 1983). These are unusual events, seen mainly with cell lines maintained *in vitro*, and they may not reflect the physiological situation. For the purposes of the following discussion, I will assume that T cells can in fact be divided into two distinct lineages.

The separation into functional lineages is not the only subdivision. There are now a number of markers which further subdivide these lineages. For example, Lyt 22 (Chan et al. 1983) can apparently distinguish suppressor cells from cytotoxic cells within the Lyt 2-positive lineage. Similarly, it has been claimed that the antigen Qa 1 can be used to separate Lyt 1-positive helper cells which are activated in the presence of B cells (Qa 1-positive) from those which require only macrophages (Qa 1-negative) (McDougal et al. 1982), and that sensitivity to depletion with anti-Ia antibodies distinguishes late-acting, nonantigen-specific helper cells from early-acting, antigen-specific helper cells (Keller et al. 1980). It seems likely that further antibodies will be developed which can delineate even smaller populations with precisely defined functions, since any unique function must involve a unique set of receptors or surface interaction molecules.

The source of peripheral T cells

The thymus is a major source of peripheral T cells. Neonatal thymectomy in mice results in severe, life-long T-cell deficiency. The nature of the cells produced by the thymus remains somewhat controversial, although in my view there is now little doubt that cells which leave the thymus are phenotypically mature (Scollay 1982), already separated into functional subpopulations (Scollay et al. 1978) and able to function, at least *in vitro* (Scollay, Chen & Shortman, 1984). They are like medullary thymocytes and quite unlike cortical thymocytes, and thus appear to come from a medullary-type pool in the thymus.

The exact relationships of the thymic cortex and medulla to each other, and their relative roles in the production of Lyt 2-positive, L3T4-negative and Lyt 2-negative, L3T4-positive T cells and in determining restriction, remain a puzzle (for detailed discussion see Scollay 1983). On the other hand, the distribution of markers in the thymus is relatively clear (see Scollay & Shortman 1983). Quite a large number of markers distinguish cortical and medullary cells. These include Thy 1, peanut agglutinin (PNA), H-2K/D, lobster lectin (LL) and thymus leukaemia antigen (TL). Table 4 shows the distribution of these in the thymus. Figure 2 shows an example of a correlation between Thy 1 and PNA. Clearly only two major populations are defined by these markers and frozen section data show that they correlate with the cortex and medulla.

The distribution of Lyt 2 and L3T4 is more complex (Table 5). Cells in the medulla are mostly positive for one or other marker, while a few are negative for both. In the cortex most cells are positive for both markers, while a few are negative for both. The cortex is the only place where double positives are found in any number in normal animals (Dialynas et al. 1983; Ceredig et al. 1983; R. Scollay & K. Shortman unpublished). In general, the distribution of the equivalent markers in man and rat are similar and it is possible that existing differences are technical rather than real. Although the differentiation sequence from double negative cortical, to double positive cortical, to single positive medullary seems a logical one at first glance and is the one most often quoted in the literature, a closer analysis reveals that the situation is much more complex. It is now clear that most cortical thymocytes die *in situ* and never give rise to medullary or peripheral T cells. The cortical population turns over about every 3 h, which means almost 30% of the thymocytes disappear each day, but the total number of exported cells only constitutes about 1% of thymocytes per day. In fact, this 1% figure is about equivalent to the estimated cell loss from the medulla each day, as 15% of the thymus is medullary and the turnover time is about 2

Table 5. Distribution of lineage-specific markers.

Lymphocyte population	Percentage cells of each phenotype[a]			
	Lyt 2−/L3T4−	Lyt 2+/L3T4+	Lyt 2−/L3T4+	Lyt 2+/L3T4−
Thymic cortex	4%	80%	abs	abs
Thymic medulla	1%	abs	10%	5%
Peripheral T cells	abs	abs	60%	40%

[a]Percentages of thymic cells refer to total thymocytes.
abs = absent or present at very low frequency.

weeks, so cell loss is approximately 1% of thymocytes per day. There is no evidence at all that typical Lyt 2-positive, L3T4-positive cortical cells ever give rise to medullary cells or peripheral cells. Whether the cortex contains a small number of cells which can migrate elsewhere remains a controversial question. The relationship between cortex and medulla and the reasons for the huge daily loss of cells from the cortex (e.g. selection against self-reacting cells or for appropriate restriction elements) remain among the fascinating puzzles in the field of immunology.

Figure 2. A two-colour contour plot showing the correlation between expression of Thy 1 and the peanut agglutinin (PNA) receptor on mouse thymocytes. Thymocytes, cortisone-resistant thymocytes (CRT) and lymph node cells (LN) were stained with red anti-Thy 1 and green PNA. In thymus, two major populations can be seen, one expressing low levels of both markers and one expressing high levels of both. The former is the same as CRT (a representative sample of medullary cells) and very similar to peripheral T cells. In the box marked LN, the population of Thy 1-negative cells consists mainly of B cells. Thus medullary thymocytes are similar to peripheral T cells, while cortical thymocytes are quite different. Staining details and further discussion can be seen in Scollay and Shortman (1983).

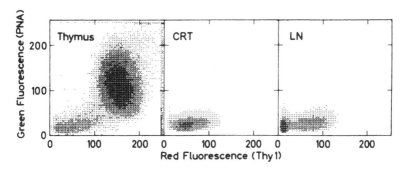

42   *R. Scollay*

A note on Lyt 1

Historically, Lyt 1 has been widely used in the mouse and the Lyt 1-negative, 2-positive and Lyt 1-positive, 2-negative subsets have generally correlated with functional division. However, it is now clear that most thymocytes and T cells express Lyt 1, and the so-called Lyt 1-negative, 2-positive population is really only low in Lyt 1 expression (Ledbetter et al. 1980; Mathieson et al. 1979). Thus the difference in Lyt 1 expression between functional lineages is relative, not absolute. The antigen L3T4 should replace Lyt 1 as the marker of choice in functional studies, since it allows absolute definition of the functional subsets, in combination with Lyt 2. Lyt 1 is more useful as a pan-T reagent than as a lineage marker.

Conclusion

Antibodies can be raised against many antigens expressed on mouse lymphoid cells. At least some of these antigens have proven extremely valuable in defining subpopulations of lymphocytes and in dissecting the complexities of immune responses. Some of these antigens are well conserved in different species and many of the human antigens can be identified with mouse antibodies. The consistency of these antigens between species and the clear demonstration that heteroantibodies can be raised against them should encourage the search for similar markers in sheep and cattle. The use of these markers is essential for understanding and probably useful for manipulating the immune response in livestock diseases.

References

Adams, J.M., Kemp, D.J., Bernard, O., Gough, N., Webb, E., Tyler, B., Gerondakis, S. & Cory, S. (1981). Organization and expression of murine immunoglobulin genes. *Immunological Reviews*, **59**, 5-32.

Ahmed, A. & Smith, A.H. (1982a).Surface markers, antigens and receptors on murine T and B cells. Part 1. *Critical Reviews in Immunology*, **3**, 331.

Ahmed, A. & Smith, A.H. (1982b). Surface markers, antigens and receptors on murine T and B cells. Part 2. *Critical Reviews in Immunology*, **4**, 19.

Basch, R.S. & Berman, J.W. (1982). Thy 1 determinants are present on many murine haemopoietic cells other than T cells. *European Journal of Immunology*, **12**, 359-64.

Boyse, E.A., Miyazawa, M., Aoki, T. & Old, L.J. (1968). Lyt A and Lyt B: two systems of lymphocyte isoantigens in the mouse. *Proceedings of the Royal Society of Biology*, **170**, 174.

Ceredig, R., Dialynas, D., Fitch, F.W. & MacDonald, H.R. (1983). Precursors of T-cell growth factor producing cells in the thymus. *Journal of Experimental Medicine*, **158**, 1654-71.

Chan, M., Tada, T., Kimura, S., Hoffman, M., Miller, R., Stutman, O. & Hammerling, U. (1983). Characterization of T-lymphocyte subsets with

monoclonal antibodies: discovery of a distinct marker Ly M22 of
suppressor T cells. *Journal of Immunology,* **130,** 2075-78.

Coffman, R.L. (1983). Surface antigen expression and immunoglobulin gene
rearrangement during mouse pre-B-cell development. *Immunological
Reviews,* **69,** 5-23.

Dialynas, D., Wilde, D.B., Marrack, P., Pierres, A., Wall, K.A., Havran, W.,
Otten, G., Loken, M.R., Pierres, M., Kappler, J. & Fitch, F.W. (1983).
Characterization of the murine antigenic determinant, designated L3T4a,
recognized by monoclonal antibody GK1-5. *Immunological Reviews,* **74,** 29-
56.

Hodes, R.J., Hathcock, K.S. & Singer, A. (1983). Major histocompatibility
complex restricted self-recognition by B cells and T cells in responses to
TNP-Ficoll. *Immunological Reviews,* **69,** 25-50.

Hogarth, M. & McKenzie, I.F.C. (1983). Lymphocyte antigens. In
Fundamental immunology (ed. W. Paul). New York: Raven Press.

Huber, B., Cantor, H., Shen, F. & Boyse, E. (1976). Independent
differentiative pathways of Lyt 1 and Lyt 23 subclasses of T cells. *Journal
of Experimental Medicine,* **144,** 1128-33.

Keller, D.M., Swierkosz, J.E., Marrack, P. & Kappler, J. (1980). Two types
of functionally distinct, synergizing helper T cells. *Journal of Immunology,*
124, 1350-70.

Kisielow, P., Hirst, J., Shiku, H., Beverley, P.C.L., Hoffman, M., Boyse, E. &
Oettgen, H. (1975). Lyt antigens as markers for functionally distinct
subpopulations of thymus-derived lymphocytes of the mouse. *Nature,* **253,**
219.

Lanier, L., Warner, N.L., Ledbetter, J.A. & Herzenberg, L.A. (1981).
Expression of Lyt 1 antigen on certain murine B-cell lymphomas. *Journal
of Experimental Medicine,* **153,** 998-1003.

Ledbetter, J.A., Rouse, R.V., Micklem, H.S. & Herzenberg, L.A. (1980). T-
cell subsets defined by expression of Lyt 1, 2, 3 and Thy 1 antigens. *Journal
of Experimental Medicine,* **152,** 280-95.

Ledbetter, J.A., Evans, R.L., Lipinski, M., Cunningham-Rundles, C., Good,
R.A. & Herzenberg, L.A. (1981). Evolutionary conservation of surface
molecules that distinguish T-lymphocyte helper/inducer and cytotoxic/
suppressor subpopulations in mouse and man. *Journal of Experimental
Medicine,* **153,** 310-23.

Manohar, V., Brown, E., Leiserson, W.M. & Chused, T.M. (1982).
Expression of Lyt 1 by a subset of B lymphocytes. *Journal of Immunology,*
129, 532-38.

Mason, D.W., Arthur, R.P., Dallman, M.J., Green, J.R., Spickett, G.P. &
Thomas, M.L. (1983). Functions of rat T-lymphocyte subsets isolated by
means of monoclonal antibodies. *Immunological Reviews,* **74,** 57-82.

Mathieson, B.J., Sharrow, S.O., Campbell, P.S. & Asofsky, R. (1979). An
Lyt-differentiated thymocyte subpopulation detected by flow
microfluorometry. *Nature,* **277,** 478-80.

McDougal, J., Shen, F., Cort, S. & Bard, J. (1982). Two Lyt 1 helper-cell
subsets distinguished by Qa-1 phenotype. *Journal of Experimental Medicine,*
155, 831-38.

McKenzie, I.F.C. & Potter, T. (1979). Murine lymphocyte surface antigens.
Advances in Immunology, **27,** 179-338.

Pawelec, G., Schneider, E.M. & Wernet, P. (1983). Human T-cell clones with
multiple and changing functions: indications of unexpected flexibility in
immune response networks. *Immunology Today,* **4,** 275-78.

Reinherz, E.L. & Schlossman, S.F. (1980). The differentiation and function of human T lymphocytes. *Cell*, **19**, 821-27.

Scollay, R. (1982). Thymus-cell migration: cells migrating from thymus to peripheral lymphoid organs have a mature phenotype. *Journal of Immunology*, **128**, 1566-70.

Scollay, R. (1983). Intrathymic events in the differentiation of T lymphocytes. *Immunology Today*, **4**, 282-86.

Scollay, R. & Shortman, K. (1983). Thymocyte subpopulations: an experimental review, including flow cytometric cross-correlations between the major murine thymocyte markers. *Thymus*, **5**, 245-95.

Scollay, R., Chen, W.-F. & Shortman, K. (1984). The functional capabilities of cells leaving the thymus. *Journal of Immunology*, **132**, 25-30.

Scollay, R., Kochen, M., Butcher, E. & Weissman, I. (1978). Lyt markers on thymus-cell migrants. *Nature*, **276**, 79-80.

Sutton, V.R., Hogarth, M. & McKenzie, I.F.C. (in press). Surface antigens of lymphocytes. In *The lymphocyte: structure and function* (ed. J.J. Marchalonis). New York: Marcel Dekker.

Swain, S.L. (1983). T-cell subsets and the recognition of MHC class. *Immunological Reviews*, **74**, 129-42.

Thomas, D. & Calderon, R. (1982). T-helper cells change their Lyt 1, 2 phenotype during an immune response. *European Journal of Immunology*, **12**, 16.

Tonegawa, S. (1983). Somatic generation of antibody diversity. *Nature*, **302**, 575-81.

Twomey, J.J. & Kouttab, N.M. (1981). The null-cell compartment of the mouse spleen. *Cellular Immunology*, **63**, 106-17.

3

Definition of lymphocyte subpopulations in man

P.C.L. BEVERLEY

The development in recent years of *in vitro* techniques to assay lymphocyte function and methods to identify and separate human leucocyte subsets has advanced understanding of the human immune system. Both conventional heteroantisera and autoantibodies have been used to identify human leucocytes, although conventional antisera can be made specific for leucocyte subsets only by extensive absorption, and are therefore invariably of low titre. Variation among batches is an additional problem. Autoantibodies, while not requiring absorption, are of limited availability and variable specificity. A number of rosette methods detect specific receptors on human leucocytes and these have been and continue to be useful. In addition, a large number of monoclonal antibodies to human cell-surface antigens are now available, including antibodies which identify thymus-derived (T) lymphocytes, bonemarrow-derived (B) lymphocytes, myeloid cells and null cells, and many antibodies which separate subsets of these major lineages. Among T cells, two major subsets — the T-helper/inducer and T-cytotoxic/suppressor cells — have been identified. These have very different functions *in vitro* and disturbances of each correlate with different diseases *in vivo*. Similar heterogeneity is being revealed among the other lymphocyte populations. Monoclonal antibodies can be used to identify and isolate molecules, as well as cells, and in future subpopulations of cells will be categorized in terms of the known functions of their surface molecules. Several of the present generation of anti-T-cell monoclonal antibodies interfere in T-cell assays and the function of the molecules identified by these antibodies is thus becoming clear. In the longer term, it may be possible to manipulate immune responses *in vivo* using antibodies directed at specific surface antigens or soluble products (lymphokines) of human lymphocyte subsets.

Introduction

Understanding of human immune responses has advanced rapidly in recent years because of the development of *in vitro* techniques for assaying lymphocyte function and methods for identifying and separating human leucocyte subsets. In addition, those studying human immunology have had the benefit of access to the results of earlier studies in the mouse. This combination of new methods and clues from the mouse has led to such rapid progress that many recent fundamental advances in immunology have been derived from experiments on the human immune system.

The discovery that lymphocytes are phenotypically and functionally heterogeneous was of fundamental importance. The demonstration that lymphocytes producing antibody (Glick, Chang & Jaap 1956) and mediating cellular immunity (Miller 1961) belong to separate bursa-derived (B) and thymus-derived (T) lineages prompted a search for means of identifying and separating these cell types. In the mouse, alloantisera against Thy 1 (Reif & Allen 1965; Kisielow et al. 1975) Lyt 1 and Lyt 2 (Kisielow et al. 1975) proved extremely useful for identification of T cells and T-cell subsets, and heteroantisera to mouse immunoglobulin could be used to identify B cells. In man it is difficult to raise alloantisera to differentiation antigens which do not also contain antibodies against major histocompatibility (MHC) antigens, so other methods have been used to identify human lymphocyte subsets.

The ability of human T cells to form rosettes with sheep red blood cells (E rosettes) has provided a standard method for enumerating T cells, while production of heteroantisera by immunization with purified human immunoglobulin has allowed the identification of B cells bearing surface membrane immunoglobulin. In order to study heterogeneity within these major subsets, particularly among T cells, other methods were required. Three main approaches were used before the development of monoclonal antibodies.

Many authors showed that human mononuclear cells carry receptors for IgM and IgG, but the demonstration that these receptors are present on two non-overlapping subpopulations of T cells opened the way to functional studies of human T-cell subpopulations. It was rapidly demonstrated that T cells with receptors for IgM (T_M) can help antibody response to pokeweed mitogen (PWM), while T cells with receptors for IgG (T_G) cannot (Moretta et al. 1977a). Indeed T_G cells suppress this antibody response. This methodology was rapidly applied to disease states to show disturbances in the T_M and T_G subsets in a variety of conditions (Moretta et al. 1977b).

At about the same time, others were using two types of antisera to separate different human cell types. It was shown that heteroantisera produced by immunization with human thymocytes can be made specific for T cells by absorption with appropriate nonT cells (e.g. red cells, B cells from patients with chronic lymphocytic leukaemia or liver homogenates) (Janossy 1981). By using mature T-cell leukaemias for further selective absorption, such heteroantisera can be made specific for T-cell subsets. Once again two major subsets were defined: Th1-positive cells which mediate help for antibody production and Th2-positive cells which are suppressive (Evans et al. 1977). Antisera obtained from patients with the autoimmune disease, juvenile rheumatoid arthritis, were also shown to identify a subpopulation of T lymphocytes which is suppressive (Strelkauskas et al. 1978).

Yet all these methods suffer from a number of problems. IgM and IgG receptors are present on a variety of cells other than T lymphocytes and are not stable markers, particularly after *in vitro* culture. Heteroantisera and autoantisera differ in specificity from batch to batch and are often of low titre because of the extensive absorption required to render them specific. Nevertheless, the data obtained with these reagents have been important in several respects. They firmly established the existence of human T-cell subsets with distinct surface properties and led to the concept of two major subsets of helper/inducer and suppressor/cytotoxic cells. Finally, methods of immunization and absorption using tumour cells and cell lines suggested strategies for immunization and screening which have led directly to the production of most of the present monoclonal antibodies against human leucocytes.

Figure 1. Twenty-eight hybridoma culture supernatants were assayed by indirect radiolabeled antiglobulin binding to peripheral blood T cells or a B lymphoblastoid line. Results are expressed as a percentage of the binding of control anti-common leucocyte antigen supernatant. Horizontal lines show background binding of the labeled antiglobulin, and x identifies two T-specific supernatants (Beverly 1980).

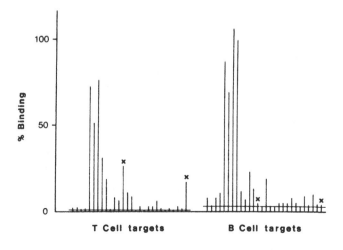

Production of monoclonal antibodies to human leucocyte subsets

Most researchers have adopted a 'shotgun' strategy for the production of monoclonal antibodies to human leucocyte differentiation antigens. In most cases researchers do not know the nature of the antigens to which they wish to raise antibodies, so they commonly use whole cells as the immunogen. Cells must be chosen of an appropriate differentiated type. In raising antibodies against human T lymphocytes, for example, it would be appropriate to use thymocytes or peripheral T lymphocytes isolated by E-rosette formation. Clearly the immunizing cells bear many antigens shared by all human cells. It has been estimated that 75% of the genes expressed by lymphocytes are also expressed by fibroblasts and 98% of the genes expressed by B lymphocytes are also expressed by T lymphocytes. However, 2% of the genome is approximately 200 genes, although it is not known what proportion of these are cell-surface molecules (Davis et al. 1982).

Most of the hybridomas produced following fusion of cells from an animal immunized with T cells will produce antibodies which react with all human cells. Thus, an appropriate screening strategy should make it possible to identify rapidly the few hybridomas producing antibody of the desired specificity among many irrelevant ones. Two main approaches have been used.

With the first approach, the hybridoma supernatants are screened for antibody against the immunizing cell type, thymocytes or T cells in this example, and all supernatants which react are then tested on a different cell type. This might be purified B lymphocytes if anti-T-cell monoclonal antibodies are desired or fibroblasts if antibodies of broader haemopoietic specificity are required. An example of data from such a T-cell *versus* B-cell screen is given in Figure 1. This example illustrates several points. First, since large numbers of supernatants need to be screened, the screening assay must be simple and rapid. We detect antibody bound to viable cells with a radiolabeled anti-mouse immunoglobulin antibody (Beverley 1980). Second, different cell types are required in quantity for screening. These may not always be readily available from normal tissues so it is often more convenient to use either tumour cell lines grown *in vitro* or fresh or cryopreserved leukaemia cells. In Figure 1, the source of B cells is a B lymphoblastoid cell line transformed by Epstein-Barr virus, but cells from the peripheral blood of a patient with chronic lymphocytic leukaemia might equally well be used. Antibodies giving distinct reactions in paired tests of this type usually detect useful differentiation antigens, so we would normally reclone the hybridoma without further testing.

An alternative strategy is to test each hybridoma supernatant on a mixed population of cells. This can be done either on cells in suspension using indirect immunofluorescence or on frozen tissue sections using the immunoperoxidase method (Mason et al. 1982). If cells are tested in suspension, only antibodies which stain a subset of the mixed cell suspension are selected. Figure 2 illustrates the immunoperoxidase method. On this section of a tonsil, it is clear that only a proportion of the cells are stained, in this case the T cells. This methodology, though perhaps initially more laborious than cell-binding assays, provides in one assay information on the cellular and tissue distribution of an antigen defined by a monoclonal antibody. Both strategies can, however, readily detect antibodies to lymphocyte subsets which may represent 10% or less of the immunizing population, and both have been used successfully to generate useful anti-human monoclonal antibodies.

The major subsets of the human immune system
T cells

Characterization of human lymphocyte subsets depends on the means available for identifying and separating cells and for testing their function. In man, functional assays must in the main be carried out *in*

Figure 2. Frozen section of part of a tonsil stained with UCHT1 (anti-T3) monoclonal antibody by the indirect immunoperoxidase method. T lymphocytes in the paracortical zone show membrane staining while most cells in the central germinal centre are unstained. Magnification x 25.

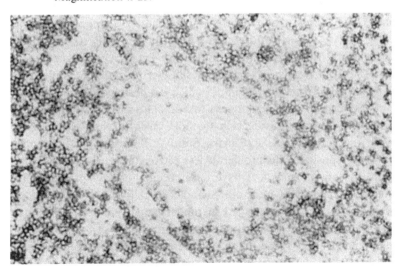

Table 1. Function of T4-positive and T8-positive cells. The symbols indicate the strength of response: + is a vigorous response, − is no response, ± is a weak response.

Function/assay	T4-positive	T8-positive
Proliferative responses to:		
Soluble antigen	+	±
MHC class I	−	+
MHC class II	+	−
Mitogens: PHA, Con A	+	+
Helper activity for:		
PWM response	+	−
Specific antibody response	+	−
Suppressor activity for:		
PWM response	+	+
Specific antibody response	?	+
Cytotoxicity against:		
MHC class I	−	+
MHC class II	+	−

vitro, so the *in vivo* function of cell subsets can only be inferred. Much of the data on lymphocyte function has also been obtained using mitogens to activate the lymphocytes, and it is clear that such polyclonal responses differ in some respects from responses to specific antigens. Thus, some caution should be exercized in using such terms as 'helper/inducer' or 'suppressor/cytotoxic' cell.

Nevertheless there is now overwhelming evidence that in man, as in rodents, there are two major subsets of peripheral T lymphocytes which differ greatly in phenotype and function as well as in their microenvironment (Janossy et al. 1980). Table 1 summarizes data on the properties of the two non-overlapping subsets of T cells identified by anti-T4 and anti-T8 monoclonal antibodies. T4-positive cells proliferate well in response to soluble antigens and alloantigens (Van Wauwe & Goossens 1981a) and provide help for both PWM-driven (Reinherz et al. 1979) and antigen-specific (Callard, Smith & Beverley 1982) antibody responses. T8-positive cells proliferate less actively in response to soluble antigens and alloantigens and have never been shown to provide help for antibody responses. On the other hand, T8-positive cells can suppress the PWM response and in earlier experiments appeared to contain most cytotoxic effector cells. This led to the categorization of T4-positive cells as helper/inducer and T8-positive as suppressor/cytotoxic cells. However, several lines of evidence suggest that this is an oversimplification.

Perhaps the most telling results are those which show that cytotoxic T cells may be either T4-positive or T8-positive. It is clear, however, that T4-positive and T8-positive killer cells differ in specificity. The T4-positive killer cells identified so far are invariably directed at MHC class II antigens, while the T8-positive killer cells are directed at class I antigens (Biddison et al. 1982; Reinherz, Meuer & Schlossman 1983). This finding, plus evidence that anti-T4 only blocks cytotoxicity by T4-positive killer cells while anti-T8 blocks T8-positive killer cells, has prompted the suggestion that T4 and T8 antigens may function as recognition units for MHC class II and class I molecules respectively (Biddison et al. 1982; Reinherz, Meuer & Schlossman 1983). Thus, the two major subsets of T4-positive and T8-positive cells and their equivalents in the mouse (Swain et al. 1981) are not committed to particular functions, but rather to recognizing antigen in association with different MHC molecules.

Much of the foregoing discussion is based on the analysis of the specificity and phenotype of T cells activated in mixed leucocyte cultures (MLC) and directed at alloantigens. In recognition of a foreign antigen — for example a virus — in association with self-antigen, it appears that the predominant population of T-killer cells is T8-positive and recognizes virus plus self-MHC class I antigen (McMichael et al. 1977). Nevertheless, it is important to recognize that T4-positive is not synonymous with help and T8-positive with suppression or cytotoxicity. Furthermore, the finding that MHC class II alloantigen can stimulate both helper and killer cells while MHC class I only stimulates killer cells may explain why matching for MHC class II is much more important than class I in transplantation.

Another approach which has cast doubt on the simple association of T4- and T8-positive cells with helper or suppressor/cytotoxic function is the use of other monoclonal antibodies to reveal additional heterogeneity among T cells and new functional subsets. A number of monoclonal antibodies (Table 2) have been produced which react with a proportion of T cells (Lum et al. 1982; Reinherz et al. 1982; Gatenby et al. 1982; Corte et al. 1982). All of these appear to subdivide both the T4-positive and T8-positive subsets. While the functional analysis of subsets carrying these new monoclonal antibody-defined antigens is far from complete, a number of points have emerged. The first is that the T4-positive subset is extremely heterogeneous. While this subset is responsible for much of the proliferative response to soluble antigen and alloantigen (MHC class II) as well as providing help for PWM responses and specific antibody responses, the first three functions are mediated by a small 5/9-positive subset of T4-positive cells. Whether some of the remaining T4-positive

Table 2. Distribution and functional properties of T-cell subsets defined by monoclonal antibodies. The symbols indicate which subset responds: + indicates antigen-positive cells respond, − antigen-negative cells respond, ± both subsets respond, + > − antigen-positive subset responds more vigorously than antigen-negative subset.

	Antibody			
	9.3	TQ1	Leu 8	5/9
% of T cells positive	70%	60%	70%	15%
Antigen present on nonT cells	no	yes	?	yes
Help for PWM response	+	−	−	+
Suppression of PWM response	−			
Response in autologous MLC	−	+>−	+>−	
Suppression of autologous MLC			±	
Response in allogeneic MLC	+>−	±		+
Response to soluble antigen		±		+
Cytotoxicity	+			−

cells represent the subset of T4-positive suppressor cells identified with the OKT17 antibody (Thomas et al. 1982) remains unclear. Data derived from studies of human T4-positive interleukin-2 (IL-2)-dependent T-cell clones also indicate considerable heterogeneity. At least two major classes of clone have been identified, one which mediates antigen-specific help and does not produce IL-2 and a second which does not mediate antigen-specific help but does produce IL-2 (Lamb et al. 1983).

Table 2 illustrates two other points. The first is that in many cases more than one subset of cells performs a particular function. For instance, two subsets synergize in generating suppression of an autologous MLC. This, of course, agrees with much older data in mice demonstrating multiple interactions of different subsets in the generation of suppressors and helpers (Feldmann et al. 1977). Finally, the 9.3 antibody promises to be useful in separating suppressor and cytotoxic cells on the basis of their phenotype, a manoeuvre which has so far proved difficult by other means although heterogeneity among T8-positive cells has been demonstrated using anti-T1 monoclonal antibodies (Callard et al. 1981).

So far I have discussed the question of heterogeneity among T cells but have avoided the question of defining T cells. Conventionally the formation of E rosettes has been considered a hallmark of human T lymphocytes. In vitro functional assays and data from immunodeficient children have supported the view that E-positive cells are derived from thymus, while E-negative cells include surface immunoglobulin-positive (B) lymphocytes and accessory cells (monocytes). Some confusion was thrown into this area, however, when it was demonstrated that both E-

Table 3. B-cell differentiation and markers. A + symbol indicates presence of the marker at a given stage of differentiation. ALL is acute lymphoblastic leukaemia, CLL is chronic lymphocytic leukaemia and PLC is prolymphocytic leukaemia.

	Different stages of B-cell maturation					
	Stem cell	B-progenitor	Pre-B cell	Immature B cell	Mature B cell	Plasma cell
Immunoglobulin	–	–	cIg	sIg	sIg	cIg
Leukaemic equivalent	null ALL	common ALL	pre-B ALL	CLL	PLL	myeloma
Reactivity with monoclonal antibodies:						
CALLA		+	+			
BA2 (p24)			+	+		
FMC1				+	+	
FMC7					+	
B1			+	+	+	
B2			+	+		
BA1		+	+	+	+	
class II MHC		+	+	+	+	+
activation antigens						+

positive and E-negative cells include cells with natural killer activity (Herberman 1981). Studies using the pan-T monoclonal anti-T3 antibody have shown that E-positive T3-positive cells function in several assays as would be expected of T cells. In contrast, 5 to 10% of E-positive cells which are T3-negative are inert in T-cell assays but have high levels of natural killer activity (Beverley & Callard 1981). Thus, T3 appears to be more firmly associated with mature T-cell function than does the E receptor, a finding in accord with data showing that T3 is closely associated with T cell-specific receptors for antigens (Reinherz, Meuer & Schlossman 1983; Zanders et al. 1983).

In summary, it is apparent from studies of human T-cell heterogeneity that a single antigen is seldom, if ever, associated with a particular function assayed *in vitro*. Conclusions regarding *in vivo* immune function based on counts of T4-positive and T8-positive or other cells should therefore be regarded with caution. This is largely because at present the function of most of the surface molecules identified by monoclonal antibodies is unknown. When a better understanding of molecular function is attained, surface phenotyping will be more meaningful.

Non-T cells

While phenotypic markers have been used a great deal to separate subsets of T cells for functional analysis, it has been emphasized that few markers are firmly linked to particular functions. This is also true for the majority of antigens defined by monoclonal antibodies on non-T cells. Table 3 presents some information on B-cell antigens (McKenzie & Zola 1983). The only marker listed in the table with a known function is immunoglobulin itself. Nevertheless, the existence of markers which divide peripheral B cells into categories should make it possible to determine whether all B cells belong to a single lineage with different functional properties expressed at different stages of maturation, or whether there are functionally distinct subsets of B cells, as is the case with T cells. Data obtained using separation based on the isotype of surface immunoglobulin expressed demonstrate that not all peripheral blood B cells can respond to PWM, even in the presence of T cells (Kuritani & Cooper 1982).

Many of the B-cell antigens exemplify another feature of monoclonal antibody defined-antigens — they are not completely specific for B cells. Although some antibodies separate B cells from other mononuclear cells, class II MHC antigens are expressed on monocytes and activated T cells as well as on some endothelial and epithelial cells, CALLA is weakly expressed on granulocytes, and several other B-cell antigens are present on renal tubular cells (Bernard et al. 1984). The significance of these shared antigens is unclear but presumably implies shared functions.

Accessory cells play an important role in immune responses both in presenting antigen and as effector cells. Three major categories are recognized (Poulter 1983). Phagocytic cells of the monocyte/macrophage series are widely distributed in blood and tissues: they have lysosomal enzyme activity and a proportion express MHC class II antigens. Follicular dendritic cells are restricted to germinal centres and appear to present antigen to B cells. The remaining category of accessory cell appears to function in presenting antigen to T-helper cells (Janossy et al. 1980; Poulter 1983). This class of cells includes Langerhans cells of the skin, interdigitating cells and dendritic cells, all of which express high levels of class II MHC antigens. There are phenotypic differences between these types of cells but their relationship to each other is unclear. Furthermore the functions of different categories of accessory cells are not well understood, although data obtained *in vivo* in experimental animals suggest that antigen presented on different accessory cell types can induce very different immune responses (Poulter 1983). Futhermore, different accessory cells accumulate in different pathological conditions.

Thus, the data suggest, as in the case of T and B cells, that accessory cells are heterogeneous in phenotype and function and that further progress in understanding their heterogeneity will come from better *in vitro* assays and methods for determining the function of accessory cell-specific markers detected by monoclonal antibodies.

The molecules of human lymphocytes

A major problem in the interpretation of the heterogeneity of lymphocytes has been a lack of understanding of the function of the molecules detected by monoclonal antibodies. Recent experiments using IL-2-dependent T-cell clones and monoclonal anti-T-cell antibodies have now begun to throw light on the function of at least some lymphocyte membrane antigens. I have already alluded to the fact that anti-T4 and anti-T8 antibodies block the cytotoxicity of T4-positive and T8-positive cytotoxic clones. This only occurs if anti-T4 or -T8 antibody is present during the cytotoxic assay, and evidence from target cell-binding studies suggests that inhibition occurs at the stage of recognition of the target by the cytotoxic cell (Landegren et al. 1982). Anti-T3 antibodies can also block cytotoxic activity but the mechanism in this case appears to be very different from that of anti-T4 or anti-T8 antibodies. In the first place, anti-T3 antibodies block when they are present during the cytotoxic assay and when they are preincubated with the effector T cells and washed before the assay. Target-binding studies show that anti-T3 does not appear to prevent recognition of target cells but acts at a later stage, at least if the effector cells have been exposed to the antibody only briefly (Landegren et al. 1982).

The interpretation of these data becomes clearer if some additional findings are taken into account. It has been shown that T cells exposed to anti-T3 antibody rapidly lose (modulate) surface T3. Once modulation has occurred, the cells are no longer cytotoxic and do not respond to alloantigen by proliferation (Reinherz, Meuer & Schlossman 1983). Furthermore, in the same series of experiments, it was shown that modulation of T3 also led to modulation of a clonotypic molecule defined by a monoclonal antibody. The clonotypic molecule almost certainly represents the T-cell receptor for antigen, so the data suggest that T3 and the receptor for antigen are functionally associated on the cell surface, a conclusion supported by the observation that T-cell clones exposed to a high dose of specific antigen modulate T3 and then are unable to respond to a subsequent stimulatory dose of antigen on antigen-presenting cells (Zanders et al. 1983). Anti-T3 antibodies are also mitogenic for fresh human T cells (Van Wauwe & Goossens 1981b), so it seems likely that

T3 acts as a transducer of the signal generated when antigen is bound to the T-cell receptor. Direct perturbation of T3, as by binding of antibody, bypasses the receptor for antigen and can stimulate a resting T cell or block the function of a more differentiated effector cell.

While the function and structure of the T3 molecule are only partially delineated (Borst et al. 1983), the antigen recognized by the anti-Tac monoclonal antibody seems very clearly to be the T-cell receptor for IL-2 (Leonard et al. 1982). Thus, at first sight, the function of the Tac antigen is straightforward. Following antigen or mitogen stimulation, T cells express Tac and proliferation of the cells (expansion of clones) is then driven by the binding of IL-2 to its receptor, although the molecular events which occur following IL-2 binding have yet to be described. Unfortunately, there appears to be an alternate pathway for maintenance of T-cell proliferation. We have shown that there are at least two major types of T4-positive T-cell clones and these respond differently when stimulated with influenza virus in the presence of accessory cells. Both types proliferate in response to the stimulus but one produces IL-2 while the second does not (Lamb et al. 1983). Confirmation that the second type is able to proliferate without the involvement of IL-2 has been obtained by showing that anti-Tac does not block the response, while clones which produce IL-2 are readily inhibited (Lamb et al. unpublished). These results demonstrate the complexity of lymphocyte response mechanisms even at the clonal level and emphasize the importance of identifying the signals and receptors involved.

Conclusion

In this review, I have discussed briefly the heterogeneity of human mononuclear cells. I have set out to show that while phenotype and function are associated, individual markers of phenotype do not correlate well with a given function assayed *in vitro*. Thus, terms such as 'T4-positive helper/inducer cells' are misleading since this population includes helpers, suppressors and cytotoxic cells. The lack of correlation between phenotype and function is mainly due to the paucity of information regarding the function of the molecules identified by monoclonal antibodies. Phenotyping cells will only become meaningful as a means of identifying function when this deficit is remedied, most likely by a combination of the techniques of cellular immunology and molecular biology.

Acknowledgements

I am grateful to Susan Chandler for typing this manuscript and to Deborah Rowe for providing Figure 2.

References

Bernard, A., Boumsell, L., Dausset, J., Milstein, C. & Schlossman, S.F. eds. (1984). *Leucocyte typing: human leucocyte differentiation antigens detected by monoclonal antibodies. Proceedings of first international reference workshop on human leucocyte differentiation antigens*, 838pp. Berlin: Springer Verlag.

Beverley, P.C.L. (1980). Production and use of monoclonal antibodies in transplantation immunology. In *Transplantation and clinical immunology* (eds. J.L. Touraine, J. Traeger, H. Betuel, J. Brochier, J.M. Dubernard, J.P. Revillard & R. Triau), vol. 11, pp. 87-94. Amsterdam: Excerpta Medica.

Beverley, P.C.L. & Callard, R.E. (1981). Distinctive functional characteristics of human 'T' lymphocytes defined by E rosetting or a monoclonal anti-T-cell antibody. *European Journal of Immunology*, **11**, 329-34.

Biddison, W.E., Rao, P.E., Talle, M.A., Goldstein, G. & Shaw, S. (1982). Possible involvement of the OKT4 molecule in T-cell recognition of class II HLA antigens. *Journal of Experimental Medicine*, **156**, 1065-76.

Borst, J., Alexander, S., Elder, J. & Terhorst, C. (1983). The T3 complex on human T lymphocytes involves four structurally distinct glycoproteins. *Journal of Biological Chemistry*, **258**, 5135-41.

Callard, R.E., Smith, C.M. & Beverley, P.C.L. (1982). Phenotype of human T-helper and suppressor cells in an *in vitro* specific antibody response. *European Journal of Immunology*, **12**, 232-36.

Callard, R.E., Smith, C., Worman, C., Linch, D., Cawley, J. & Beverley, P.C.L. (1981). Unusual phenotype and function of an expanded subpopulation of T cells in patients with haemopoietic disorders. *Clinical & Experimental Immunology*, **43**, 497-505.

Corte, G., Mingari, M.C., Moretta, A., Damiani, G., Moretta, L. & Bargellesi, A. (1982). Human T-cell subpopulations defined by a monoclonal antibody. 1. A small subset is responsible for proliferation to allogeneic cells or to soluble antigens and for helper activity for B-cell differentiation. *Journal of Immunology*, **128**, 16-19.

Davis, M.M., Cohen, D.I., Neilsen, E.A., de Franco, A.L. & Paul, W.E. (1982). The isolation of B and T cell specific genes. In *UCLA symposia on molecular and cellular biology* (ed. E. Vitetta), vol. 20, pp. 261-66. New York.

Evans, R.L., Breard, J.M., Lazarus, H., Schlossman, S.F. & Chess, L. (1977). Detection, isolation and functional characterization of two human T-cell subclasses bearing unique differentiation antigens. *Journal of Experimental Medicine*, **145**, 221-33.

Feldmann, M., Beverley, P.C.L., Woody, J. & McKenzie, I. (1977). T-T interactions in the induction of suppressor and helper T cells: analysis of membrane phenotype of precursor and amplifier cells. *Journal of Experimental Medicine*, **145**, 793-801.

Gatenby, P.A., Kansas, G.S., Xian, C.Y., Evans, R.L. & Engleman, E.G. (1982). Dissection of immunoregulatory subpopulations of T lymphocytes within the helper and suppressor sublineages in man. *Journal of Immunology*, **129**, 1997-2000.

Glick, B., Chang, T.S. & Jaap, R.G. (1956). The bursa of Fabricius and antibody production. *Poultry Science*, **35**, 224.

Herberman, R.B. (1981). Significance of natural killer (NK) cells in cancer research. *Human Lymphocyte Differentiation*, **1**, 63-75.

Janossy, G. (1981). Membrane markers in leukaemia. In *The leukaemic cell* (ed. D. Catovsky), pp 129. Edinburgh: Churchill Livingstone.

58   *P. C. L. Beverley*

Janossy, G., Tidman, N., Selby, W.S., Thomas, J.A., Grainger, S., Kung, P.C. & Goldstein, G. (1980). Human T lymphocytes of inducer and suppressor type occupy different microenvironments. *Nature*, **288**, 81-84.

Kisielow, P., Hurst, J.A., Shiku, H., Beverley, P.C.L., Hoffman, E.A. & Oettgen, H.F. (1975). Lyt antigens as markers for functionally distinct subpopulations of thymus-derived lymphocytes of the mouse. *Nature*, **253**, 219-20.

Kuritani, T. & Cooper, M.D. (1982). Human B-cell differentiation. 2. Pokeweed mitogen-responsive B cells belong to a surface immunoglobulin D-negative subpopulation. *Journal of Experimental Medicine*, **155**, 1561-66.

Lamb, J.R., Zanders, E.D., Feldmann, M., Eckels, D.D., Woody, J.N., Lake, P. & Beverley, P.C.L. (1983). The dissociation of interleukin-2 production and antigen-specific helper activity by clonal analysis. *Immunology*, **50**, 397-405.

Landegren, U., Ramstedt, U., Axberg, I., Ullberg, M., Jondal, M. & Wigzell, H. (1982). Selective inhibition of human T-cell cytotoxicity at levels of target recognition or initiation of lysis by monoclonal OKT3 and Leu-2a antibodies. *Journal of Experimental Medicine*, **155**, 1579-84.

Leonard, W.J., Depper, J.M., Uchiyama, T., Smith, K.A., Waldmann, T.A. & Greene, W.C. (1982). A monoclonal antibody that appears to recognize the receptor for human T-cell growth factor; partial characterization of the receptor. *Nature*, **300**, 267-69.

Lum, L.G., Orcutt-Thordarson, N., Seigneuret, M.C. & Hansen, J.A. (1982). *In vitro* regulation of immunoglobulin synthesis by T-cell subpopulations defined by a new human T-cell antigen (9.3). *Cellular Immunology*, **72**, 122-29.

Mason, D.Y., Maiem, M., Abdulaziz, Z., Nash, J.R.G., Gatter, K.C. & Stein, H. (1982). Immunohistological applications of monoclonal antibodies. In *Monoclonal antibodies in clinical medicine* (eds. A.J. McMichael & J.W. Fabre), pp. 585-635. London: Academic Press.

McKenzie, I.F.C. & Zola, H. (1983). Monoclonal antibodies to B cells. *Immunology Today*, **4**, 10-15.

McMichael, A.J., Ting, A., Zweerink, H.J. & Askonas, B.A. (1977). HLA restriction of cell-mediated lysis of influenza virus-infected human cells. *Nature*, **270**, 524-26.

Miller, J.F.A.P. (1961). Immunological function of the thymus. *Lancet*, **2**, 748-49.

Moretta, L., Webb, S.R., Grossi, C.E., Lydyard, P.M. & Cooper, M.D. (1977a). Functional analysis of two human T-cell subpopulations: help and suppression of B-cell responses by T cells bearing receptors for IgM or IgG. *Journal of Experimental Medicine*, **146**, 184-200.

Moretta, L., Mingari, M.C., Webb, S.R., Pearl, E.R., Lydyard, P.M., Grossi, C.E., Lawton, A.R. & Cooper, M.D. (1977b). Imbalances in T-cell subpopulations associated with immunodeficiency and autoimmune syndromes. *European Journal of Immunology*, **7**, 696-700.

Poulter, L.W. (1983). Antigen presenting cells *in situ*: their identification and involvement in immunopathology. *Clinical & Experimental Immunology*, **53**, 513-20.

Reif, A.E. & Allen, J.M.V. (1965). The AKR thymic antigen and its distribution in leukaemias and nervous tissue. *Journal of Experimental Medicine*, **120**, 413-33.

Reinherz, E.L., Meuer, S.C. & Schlossman, S.F. (1983). The delineation of antigen receptors on human T lymphocytes. *Immunology Today*, **4**, 5-8.

Reinherz, E.L., Kung, P.C., Goldstein, G. & Schlossman, S.F. (1979). Further characterization of the human inducer T-cell subset defined by monoclonal antibody. *Journal of Immunology,* **123**, 2894-96.

Reinherz, E.L., Morimoto, C., Fitzgerald, K.A., Hussey, R.E., Daley, J.F. & Schlossman, S.F. (1982). Heterogeneity of T4-positive inducer T cells defined by a monoclonal antibody that delineates two functional subpopulations. *Journal of Immunology,* **128**, 463-68.

Strelkauskas, A.J., Schauf, V., Wilson, B.S., Chess, L. & Schlossman, S.F. (1978). Isolation and characterization of naturally occurring subclasses of human peripheral blood T cells with regulatory functions. *Journal of Immunology,* **120**, 1278-82.

Swain, S.L., Dennert, G., Wormsley, S. & Dutton, R.W. (1981). The Lyt phenotype of long-term allospecific T-cell line: both helper and killer activities to Ia are mediated by Lyt 1 cells. *European Journal of Immunology,* **11**, 175-80.

Thomas, Y., Rogozinski, L., Irigoyen, O.H., Shen, H.H., Talle, M.A., Goldstein, G. & Chess, L. (1982). Functional analysis of human T-cell subsets defined by monoclonal antibodies. 5. Suppressor cells within the activated OKT4-positive population belong to a distinct subset. *Journal of Immunology,* **50**, 397-405.

Van Wauwe, J. & Goossens, J. (1981a). Monoclonal anti-human lymphocyte antibodies: enumeration and characterization of T-cell subsets. *Immunology,* **42**, 157-64.

Van Wauwe, J. & Goossens, J. (1981b). Mitogenic actions of orthoclone OKT3 on human peripheral blood lymphocytes: effects of monocytes and serum components. *International Journal of Immunology,* **3**, 203-8.

Zanders, E.D., Lamb, J.R., Feldmann, M., Green, N. & Beverley, P.C.L. (1983). Tolerance of T-cell clones is associated with membrane antigen changes. *Nature,* **303**, 625-27.

4

Definition of lymphocyte populations in cattle using lectins

C. FOSSUM

Lectins are proteins or glycoproteins that bind to carbohydrate moieties and are able to agglutinate cells or precipitate glycoconjugates. Most lectins are extracted from the seeds of plants or from snails or fish. The capacity of lectins to bind to specific sugars makes them useful in research on membranes. The binding of lectins to glycoproteins or glycolipids in the cell membrane can be inhibited by the specific sugar. Lectins have been used for blood group typing and for identification of lymphocyte subpopulations. In man and mouse, the most widely used lectins are peanut agglutinin (PNA), which binds to immature T cells and germinal centre B cells in both species, *Helix pomatia* A haemagglutinin (HPA), which binds to mouse T cells, and soya bean agglutinin (SBA), which binds to mouse and human B cells and human T-helper cells. Binding of lectins to bovine lymphocytes has been investigated in the search for a reliable T-cell marker. Of a number of lectins tested, several have been found to bind to bovine lymphocytes. The findings with PNA and HPA have been of particular interest, since double-labeling experiments indicate that these lectins bind to nonB cells. Both PNA and HPA bind to approximately 60% of bovine peripheral blood lymphocytes. HPA, covalently coupled to sepharose, has been used to separate bovine lymphocytes into populations enriched or depleted with T cells. Bovine T cells can be further divided into subpopulations according to the presence of Fc receptors for IgG and IgM in their cell membranes.

Introduction

Lectins have been defined as 'sugar-binding proteins or glycoproteins that agglutinate cells or precipitate glycoconjugates' (Goldstein et al. 1980). The ability of lectins to agglutinate erythrocytes has been known for a long time. As early as 1888, Stillmark made an extract from the beans of *Ricinus communis* (castor bean) which was found to agglutinate erythrocytes from different species, and in 1891 Hellin extracted a lectin from *Abrus precatorius*. Both these lectins — ricin and abrin — were highly toxic. Also in 1891, Paul Ehrlich used ricin and abrin in a study which led to the discovery of the specificity of antibodies. Ehrlich (1957) found that after repeated inoculation with low doses of ricin, mice could tolerate an otherwise lethal dose. The same reaction pattern was found with abrin. The most important observation was that a mouse which had acquired tolerance to ricin could not tolerate abrin, and *vice versa*. Furthermore, serum from a tolerant mouse could transfer the tolerance to a mouse not previously exposed to the lectin. Ehrlich also demonstrated that immunity to the toxins is transferred from the mother to the offspring during pregnancy. From these findings, he described the specificity of antibodies.

More recently, other lectins have been extracted from different plants, and it was observed in 1948 that different lectins have different specificities in their ability to agglutinate erythrocytes. Some lectins were found to be blood-group specific. An association between blood-group specificity and sugars was established in 1952 by Watkins and Morgan, who found that agglutination could be inhibited by specific sugars. This work provided the first evidence for the presence of sugars on cell membranes.

Each molecule of a lectin has two or more binding sites which are necessary for agglutination. A lectin-binding site can be described as a cleft or a groove in which a sugar residue can fit. The attachment between lectin and sugar can be compared to that between antigen and antibody or between enzyme and substrate. However, in the case of lectins and sugars the binding is weak, formed by hydrophobic interactions and hydrogen bonds, but without the formation of covalent bonds. The binding is also reversible and can be competitively inhibited by the specific sugar in solution.

A model of the structure of cell membranes, put forward in 1972 by Singer and Nicholson, is still valid today. It is called the fluid mosaic model and describes the cell membrane as a phospholipid bilayer, with the hydrophilic polar head groups of the phospholipids exposed outwards while the hydrophobic fatty acids are hidden inside. Proteins are inserted in the membrane, some of which are glycoproteins with their sugar residues exposed outwards. These exposed sugars act as targets for the lectins. The cell membrane is sufficiently fluid to allow lateral movement of membrane lipids and proteins.

Methodology

The most commonly used methods for detecting lectin-binding sites on the surface of cells are fluorescence microscopy, using fluorochrome-labeled lectins, and electron microscopy using lectins labeled with ferritin or colloidal gold. The binding of fluorochrome-labeled lectins is usually carried out with live cells. Under these circumstances, when the membrane receptors become cross-linked by a multivalent lectin (or an antibody), they start to redistribute in the plane of the membrane and form clusters, patches and finally caps at one pole of the cell. The movement in the cell membrane is temperature dependent and requires metabolic activity of the cell. The movement is, therefore, inhibited by lowering the temperature to 4°C or by adding the metabolic inhibitor sodium azide.

The phenomenon of cap formation can be used to distinguish two different structures on the cell surface in double-labeling experiments. The first marker, for example a lectin, is allowed to interact with the cell under capping conditions, while the second marker is added to the system under noncapping conditions. From such experiments it can be ascertained whether the two markers bind to independent receptors. In the case of co-capping, the markers might bind to the same receptor or to two different, but linked receptors, since independent receptors may co-cap (Bourguignon et al. 1978).

A method has been developed to identify and quantify specific lectin-binding molecules in cell membranes using human erythrocyte membranes (Robinson et al. 1975). Membranes are solubilized in detergent and the proteins are separated by polyacrylamide gel electrophoresis. Excess aldehyde groups are blocked with sodium borohydride and the gel is incubated with radiolabeled lectin. The lectin-binding bands are visualized by autoradiography. Another method for screening lectin receptors on membranes has been applied to chicken erythrocytes by Brogren and Bisati (1981). The solubilized membranes are melted into a gel at the bottom of a plate. Gel containing antiserum against the whole membranes is placed on the upper part of the plate. In an intermediate gel, wells are punched out and filled with 1 to 10 μl of the lectins to be tested. During electrophoresis, the membrane fragments have to pass the row of lectins before entering the polyvalent antibody gel. If the lymphocyte membrane has receptors for the lectin tested, it will be absorbed, which results in a dip in the precipitation line in front of the well containing the specific lectin.

Lectins can also be used to separate different cell populations. As a result of their agglutinating capacity, the easiest separation is achieved by adding lectins to a cell suspension. Agglutinated cells are separated from nonagglutinated cells by sedimentation at unit gravity in a viscous medium, such as foetal bovine serum. Dissociation of the agglutinated cells is achieved by adding the competitive sugar. Selective agglutination has been used for murine thymocytes with peanut agglutinin (PNA) (Reisner, Linker-Israeli & Sharon 1976) and for murine splenocytes with soya bean agglutinin (SBA) (Reisner, Ravid & Sharon 1976).

Subpopulations of leucocytes with receptors for a lectin that also binds to erythrocytes can be isolated by a rosetting technique. Mixing erythrocytes, lectin and the lymphocyte suspension leads to rosette formation, with the lectin-binding cells. The rosettes can be separated from nonrosette-forming cells on a cushion of Hypaque-Ficoll. Dissociation of the rosettes is achieved by lysing the red blood cells by osmotic shock or by adding the competitive sugar. For example, Reisner and colleagues (1978) have separated PNA-positive murine splenocytes by rosette formation with rabbit erythrocytes.

Affinity chromatography of cell suspensions can be performed using lectins covalently bound to a solid phase, preferably a gel matrix packed in a column. The cell suspension is added to the column and nonbound cells are removed by washing with buffer. Bound cells are eluted with increasing amounts of the specific sugar. By coupling HPA to sepharose, it is possible to separate B from T lymphocytes of mice, rats, humans and cattle (Hammarstrom et al. 1978; Hellstrom et al. 1976; Morein et al. 1979).

The use of lectins for identification of lymphocyte subpopulations
Mouse

The thymus contains two distinct subpopulations of lymphocytes — the immature form in the cortical region and the mature form in the medullary region. The cortical region contains 85 to 90% of total thymocytes. These cells have no cell-mediated immune (CMI) reactivity: they are not responsive to mitogens, they do not react in mixed lymphocyte reactions (MLR), and they do not induce graft-*versus*-host reactions (GVH). The remaining 10 to 15% of thymocytes are found in the medullary region. These cells are mature and possess CMI reactivity (Droege & Zucker 1975; Greaves, Owen & Raff 1974). Reisner, Linker-Israeli and Sharon (1976) showed that the major thymocyte subpopulation can readily be separated from the minor one by agglutination with PNA. The agglutinated cells can be recovered as single viable cells after dissociation with D-galactose but show no CMI reactivity, thus being similar to immature cortical thymocytes. The unagglutinated cells comprise 5 to 10% of the total thymocytes and possess CMI reactivity. However, PNA-negative thymocytes become positive after treatment with neuraminidase, indicating that their PNA receptors are masked by sialic acid residues. Similarly, 76% of the peripheral blood lymphocytes become PNA-positive after neuraminidase treatment (Newman & Boss 1980). Double-labeling with anti-Ig reveals that only Ig-negative blood lymphocytes expose PNA receptors after removal of the sialic acid, even though PNA-positive B cells have been demonstrated in germinal centres (Rose et al. 1980). From these findings, it has been suggested that PNA receptors probably are present on a large number of immature murine cells, but are masked by sialic acid during the maturation of T cells and are gradually lost during the maturation of B cells.

Using SBA in an affinity chromatography system, mouse spleen cells can be separated into two populations. The SBA-negative cells are mostly T cells, while the SBA-positive cells are mostly B cells with some cross-reaction (Reisner, Ravid & Sharon 1976; Reisner et al. 1980a). SBA agglutination gives better separation than conventional methods based on adherence to nylon wool or lysis with anti-T-cell sera and complement.

HPA binds to 85 to 95% of neuraminidase-treated murine T lymphocytes, while most B cells are negative (Mattes & Holden 1981). Binding studies with different concentrations of HPA suggest that at least two kinds of HPA receptors exist which differ in their affinity for the lectin. Due to this difference in affinity, neuraminidase-treated mouse splenocytes can be fractionated by application to a column loaded with HPA bound to sepharose and elution with different concentrations of the competitive sugar. Elution with 0.1 mg/ml N-acetyl-D-galactose-amine gives a cell population highly enriched for natural killer cell activity (Haller et al. 1978).

Human

PNA and SBA have also been used as markers for human lymphoid cells and 60 to 80% of human thymocytes have been shown to bind PNA. After pretreating thymocytes with neuraminidase, all the cells become positive for PNA. As in the mouse system, the PNA-positive cells are immature and are found in the cortical region of the thymus, while the PNA-negative cells are found in the medullary region. Only 1% of peripheral blood lymphocytes bind PNA (Reisner et al. 1979), while SBA agglutinates all B lymphocytes and the majority of T cells in human peripheral blood. The SBA-positive T cells have helper activity, while the SBA-negative T cells have suppressor activity (Reisner et al. 1980b).

Cattle

The successful use of lectins for identification of human and murine lymphocyte populations has stimulated similar work in cattle. There is an urgent need for a reliable T-cell marker in cattle since the conventional method based on rosette formation with sheep red blood cells (SRBC) requires modification of the SRBC. Variable numbers of rosette-forming cells, ranging from 4 to 70%, have been reported among bovine peripheral blood mononuclear cells (PBM), depending on how the SRBC were pretreated (Grewal, Rouse & Dabiuk 1976; Fruchtmann, Uhlenbruck & Schmid 1977; Higgins & Stack 1977; Wardley 1977; Reeves & Renshaw

Table 1. Binding of lectins to bovine PBM. The results are expressed as the arithmetic mean, with the standard deviation given in brackets. For WGA the two individual values are given. From Pearson et al. 1979.

Number of animals	Number of experiments	Proportion of lymphocytes of each type (%)			
8	12	PNA+/Ig− 45.5 (7.9)	PNA+/Ig+ 1.4(1.2)	PNA−/Ig+ 34.4 (12.8)	PNA−/Ig− 19.0 (6.6)
7	10	SBA+/Ig− 17.2 (11.7)	SBA+/Ig+ 0.4 (0.7)	SBA−/Ig+ 42.4 (13.8)	SBA−/Ig− 40.1 (6.7)
2	2	WGA+/Ig− 20, 4	WGA+/Ig+ 13, 4	WGA−/Ig+ 27, 6	WGA−/Ig− 40, 5
5	5	Con A+ >95			
5	5	UEA+ 0			

1978; Paul, Brown & Miller 1979). In 1979, Pearson and colleagues tested the mitogenic and binding capacities of several lectins with bovine PBM. Concanavalin A (Con A), phytohaemagglutinin (PHA) and pokeweed mitogen (PWM) were found to stimulate bovine lymphocytes to mitosis, while SBA, wheatgerm agglutinin (WGA), *Ulex europeus* (UEA) and HPA did not stimulate cell division. The binding of fluorochrome-labeled lectins to bovine PBM was studied in double-labeling experiments with anti-bovine-Ig. A summary of the results is presented in Table 1. These studies showed that PNA binds selectively to T lymphocytes.

We have tested several lectins for binding to bovine PBM isolated from four animals. *Crotalaria juncea* bound to 28%, *Lens culinaris* to 80%, *Vicia ervilia* to 100% and HPA to 40%. In further work, we focused on HPA, since this lectin has been shown to be a T-cell marker in the human, mouse and rat systems. Following neuraminidase treatment, 40% of bovine PBM was positive for HPA. None of the HPA-positive cells carried membrane-bound immunoglobulins. Since the proportion of T lymphocytes in most species is larger than 40%, we suspected that a subpopulation of T cells was not stained by the direct method, possibly due to a small number of receptors or to low-affinity receptors for HPA. To increase the sensitivity of the fluorescence test for detection of the HPA receptor, an indirect method was developed with F(ab′)$_2$ fragments of antibodies against HPA in a second layer. With the more sensitive indirect method, 60% of the cells were positively labeled with HPA. None

66   C. Fossum

Table 2. Proportion of lymphocytes in PBM positive for the marker in the vertical column labeled with FITC and also positive for the marker in the horizontal column, labeled with TRITC: percentage mean values from five experiments.

FITC	TRITC			
	HPA	anti-HPA[a]	PNA	anti-Ig
HPA	-	n.d.	100	<1
Anti-HPA[a]	n.d.	-	100	<1
PNA	60	90	-	2
Anti-Ig	0	0	5	-

[a]An indirect test was employed using unlabeled HPA followed by labeled F(AB)₂ anti-HPA antibody.

of the HPA-positive cells were positive for immunoglobulins, indicating that the HPA-positive cells were T lymphocytes.

To confirm further the absence of Ig-bearing cells among the HPA-positive cells, the total cell population was deprived of B lymphocytes. Wigzell's (1976) method for separating mouse lymphocytes was used, modified for bovine lymphocytes. Briefly, glass beads with a mean size of 0·2 mm were activated to bind protein by treatment with acid (HCl, H_2SO_4). After washing with PBS, the beads were incubated with bovine immunoglobulins overnight. The nonbound immunoglobulins were removed by washing with PBS and the beads were incubated with anti-bovine rabbit sera. After washing again with PBS, the cells were applied to a column and allowed to pass with a flow rate of approximately 2 ml/min. Immunoglobulin-bearing cells were retained in the column as confirmed by fluorescence microscopy (<1% Ig-positive cells, with $n = 6$, in the eluted fraction). The eluted fraction was enriched for HP positive cells (65% ± 6·9 ($n = 6$), as detected by the direct method (Morein et al. 1979). Bovine PBM depleted of monocytes by treatment with carbonyl iron were also separated on a sepharose column with covalently bound HPA. Approximately 80 to 85% of the cells applied to the column were recovered. The neuraminidase-treated cells were incubated for 15 min at room temperature in the column before nonbound cells were eluted with buffer. The unbound fraction was enriched in Ig-bearing cells (38% Ig+) and strongly depleted of HPA-positive cells (4%). The cells that had bound to the lectin column were eluted sequentially with 0·1 mg/ml and 1 mg/ml of the competitive sugar D-Ga l NAc. The fraction eluted with the highest concentration of sugar was enriched for HPA-positive cells (86%) and depleted of Ig-bearing cells (6%) (Morein et al. 1979).

Since HPA and PNA seemed to identify the same cell population among

Table 3. Percentage of bovine PBM with specific markets. The proportion of PBM positive for PNA, ATS and Ig was determined in 14 animals. Double-labeling was performed in three animals. When determining positive cells, latex ingesting monocytes were excluded. From Usinger & Splitter 1981.

PNA±	Ig+	ATS+	PNA+/ATS−	PNA−/ATS+	PNA+/ATS+
61 ± 1.4	27 ± 1.9	62 ± 1.2	1	1	98

bovine PBM, double-labeling experiments were performed with the two lectins and with rabbit antibodies against bovine Ig. As seen in Table 2, all HPA-positive cells were also positive for PNA, but not for anti-Ig. Of the PNA-positive cells, 60% were also positive for HPA using the direct method and 90% were positive for HPA using the indirect method. Small numbers of cells positive for both PNA and Ig were detected (Johansson & Morein 1983).

PNA has also been compared with an anti-thymocyte serum (ATS) for identification of bovine T lymphocytes (Usinger & Splitter 1981). The ATS was raised in rabbits against foetal bovine thymocytes and adsorbed with bovine red blood cells and a suspension of tumour cells from bovine mesenteric lymphoma tissue. Double-labeling experiments performed under capping and noncapping conditions revealed that the anti-thymocyte serum and PNA bound to the same cells but to different receptors. Less than 1% of the cells were positive for only ATS or PNA, while 98% of the positive cells carried both markers. Double-labeling experiments with ATS, PNA and anti-Ig excluded the possibility that B lymphocytes were labeled by ATS or PNA (Table 3). Both ATS and PNA were shown to bind to 98% of thymocytes in suspension.

Other methods for identifying lymphocyte subpopulations

In most of this work double labeling has been performed to exclude the possibility that the T-cell markers are also present on B lymphocytes. Bovine B lymphocytes are identified by their binding of antibody to bovine Ig. However, some of the T lymphocytes have F_c receptors for immunoglobulins in their cell membrane. This makes it advisable to use $F(ab')_2$ fragments of the anti-Ig serum to avoid binding via the F_c receptors to nonB cells. From our studies (Johansson & Morein 1983), no difference was observed in the proportion of B lymphocytes detected using either direct or indirect immunofluorescence, indicating that the number of immunoglobulins in the cell membrane of B lymphocytes is sufficient to be detected by the direct method. The most striking influence on the proportion of Ig-bearing cells observed was the presence of cytophilic antibodies. These antibodies were lost if the cells were incubated for 1 h in serum-free medium (Watson 1976) before staining. The proportion of Ig-positive cells decreased from 25% without preincubation to 18% after preincubation (Johansson & Morein 1983).

Yang (1981) also tested ATS as a T-lymphocyte marker, raised in goat against calf thymocytes and absorbed with packed bovine erythrocytes, bovine serum and bovine bonemarrow cells. The ATS reacted with 95% of the thymocytes in suspension, with 5% of bonemarrow cells and with 73% ($\pm$ 5·7) of bovine PBM. This ATS was then tested by Canning, Kramer and Kaberle (1983) in combination with binding by various bacteria (*Escherichia coli, Brucella melitensis, Corynebacterium pyogenes*) for identification of lymphocyte subsets in bovine PBM. They detected 70% ($\pm$ 2·2) of the PBM positive for ATS, and the population could be divided into five subpopulations based upon ability to bind the different bacteria. This may be useful in the identification of subpopulations of T cells with different regulatory functions.

Surovas and colleagues (1982) have detected bovine T cells with receptors for the Fc part of IgG and IgM, Tγ and Tμ cells respectively. The proportion of Tγ and Tμ cells was determined in healthy cattle and in cattle with chronic lymphocytic leukaemia (CLL). The normal proportion of Tγ cells in PBM in healthy cattle was 9% $\pm$ 2 ($n = 11$), with values ranging from 1 to 18%. In CLL cattle, the proportion increased to 58·5% $\pm$ 3·8 ($n = 13$). The proportion of Tμ cells in PBM from normal cattle was 25% $\pm$ 4 ($n = 11$), ranging from 6 to 37%, while in PBL from CLL cattle it was only 4% $\pm$ 1 ($n = 13$).

We have found a higher proportion of cells with Fc receptors for IgG (Fcγ-positive cells) among PBM isolated from healthy cattle than reported by Surovas and colleagues (1982). Among PBM from 46 animals, 16% $\pm$7.1 of the cells were identified as Fcγ-positive by their ability to form rosettes with IgG-coated erythrocytes. Before examination by fluorescence microscopy, the nucleated cells were counter-stained with acridine orange to facilitate identification of lymphocytes hidden in erythrocyte complexes.

The proportion of monocytes among the PBM isolated from 38 of the animals was found to be 6% $\pm$ 2·4, using flow cytometric evaluation of the proportion of cells ingesting fluorescent latex beads. This would indicate that 60 to 70% of the Fcγ-positive cells are lymphocytes. Double-labeling with HPA showed that 65% of the Fcγ-positive cells were HPA-positive. Evidence has been obtained that these Fcγ-positive cells have an important immunoregulatory function. Their removal from the total cell population increases the response of the depleted population to both mitogenic and antigenic stimulation. Addition of increasing proportions of Fcγ-positive cells to the depleted cell population decreases the response to the initial level observed with intact PBM. These results indicate that the Fcγ-positive cells have a suppressive function (Fossum, Bergman & Morein in press).

Concluding remarks

A great deal more work is required in order to define the various lymphocyte subpopulations which make up the bovine immune system. From results reported here, it seems possible to identify bovine T lymphocyte with lectins or ATS. It is now of great importance to find markers for the subpopulations of cells, both for identification and for separation from the total cell population into functional subsets. Lectins will be useful tools for such work, although monoclonal antibodies will also play an increasingly important role.

References

Bourguignon, L.Y.W., Hyman, R., Trowbridge. I. & Singer, S.J. (1978). Participation of histocompatibility antigens in capping of molecularly independent cell-surface components by their specific antibodies. *Proceedings of the National Academy of Sciences of the USA*, **75**, 2406-10.

Brogen, C.H. & Bisati, S. (1981). In *Lectins biology, biochemistry and clinical biochemistry* (ed. T.C. Bog-Hansen), pp. 375-85. Berlin: Walter de Gruyter.

Canning, P.C., Kramer, T.T. & Kaberle, M.L. (1983). Identification of bovine lymphocyte subpopulations by combined bacterial adherence and fluorescent antibody technique. *American Journal of Veterinary Research*, **44**, 297-300.

Droege, W. & Zucker, R. (1975). Lymphocyte subpopulations in the thymus. *Transplantation Reviews*, **25**, 3-25.

Ehrlich, P. (1957). *The collected papers of Paul Ehrlich*. London: Pergamon Press.

Fruchtmann, R., Uhlenbruck, G. & Schmid, D.O. (1977). Zur Charakterisierung von T- und B-lymphozyten beim Rind. *Zentralblatt für Veterinärmedizin*, Reihe B, **241**, 486-96.

Fossum, C., Bergman, R. & Morein, B. (1985). Suppressor activity of Fcγ-positive cells during a persistent infection with *Mycobacterium avium*. *Research in Veterinary Science*, **38**, 270-8.

Goldstein, I.J., Hughes, R.C., Monsigny, M., Osawa, T. & Sharon, N. (1980). What should be called a lectin? *Nature*, **285**, 66.

Greaves, M.F., Owen, J.J.T. & Raff, M.C. (1974). *T and B lymphocytes: origins, properties and roles in immune responses*, 316pp. Amsterdam: Excerpta Medica.

Grewal, A.S., Rouse, B.T. & Dabiuk, L.A. (1976). Erythrocyte rosettes: a market for bovine T cells. *Canadian Journal of Comparative Medicine*, **40**, 298-305.

Haller, O., Gidlund, M., Hellstrom, U., Hammarstrom, S. & Wigzell, H. (1978). A new surface market on mouse natural killer cells: receptors for *Helix pomatia* A haemagglutinin. *European Journal of Immunology*, **8**, 765-71.

Hammarstrom, S., Hellstrom, U., Dillner, M.L., Perlman, P., Axelsson, B. & Robertsson, E.S. (1978). In *Affinity chromatography*, p. 273. Oxford: Pergamon Press.

Hellin, H. (1891). Der giftige Eiweisskörper Abrin und dessen Wirkung auf das Blut. Inauguration Dissertation.

Hellstrom, U., Hammarstrom, S., Dillner, M.-L., Perlmann, H. & Perlmann, P. (1976). Fractionation of human blood lymphocytes on Helix pomatia A haemagglutinin coupled to sepharose beads. *Scandinavian Journal of Immunology*, **5**, 5-45.

Higgins, D.A. & Stack, M.F. (1977). Bovine lymphocytes: recognition of cells forming spontaneous (E) rosettes. *Clinical & Experimental Immunology*, **27**, 348-54.

Johansson, C. & Morein, B. (1983). Evaluation of labeling methods for bovine T and B lymphocytes. *Veterinary Immunology & Immunopathology*, **4**, 345-59.

Mattes, M.J. & Holden, M.T. (1981). The distribution of Helix pomatia lectin receptors on mouse lymphoid cells and other tissues. *European Journal of Immunology*, **11**, 358-65.

Morein, B., Hellstrom, U., Axelsson, L.-G., Johansson, C. & Hammarstrom, S. (1979). Helix pomatia A haemagglutinin, a surface marker for bovine T lymphocytes. *Veterinary Immunology & Immunopathology*, **1**, 27.

Newman, R.A. & Boss, M.A. (1980). Expression of binding sites for peanut agglutinin during murine B-lymphocyte differentiation. *Immunology*, **40**, 193-200.

Paul, P.S., Brown, T.T. & Miller, J.M. (1979). Enhancement of rosette formation between sheep erythrocytes and bovine T lymphocytes by 2-aminoethyl isothiouronium-bromide and dextran. *Immunology Letters*, **1**, 93-96.

Pearson, T.W., Roelants, G.E., Lundin, L.B. & Mayor-Withey, K.S. (1979). The bovine lymphoid system: binding and stimulation of peripheral blood lymphocytes by lectins. *Journal of Immunological Methods*, **26**, 271-82.

Reeves, J.H. & Renshaw, H.W. (1978). Surface membrane markers on bovine peripheral blood lymphocytes. *American Journal of Veterinary Research*, **39**, 917-23.

Reisner, Y., Linker-Israeli, M. & Sharon, N. (1976). Separation of mouse thymocytes into two subpopulations by the use of peanut agglutinin. *Cellular Immunology*, **25**, 129-34.

Reisiner, Y. Ravid, A. & Sharon, N. (1976). Use of soya bean agglutinin for the separation of mouse B and T lymphocytes. *Biochemical & Biophysical Research Communications*, **72**, 1582-91.

Reisner, Y., Itzicovitch, L., Meshorer, A. & Sharon, N. (1978). Haemopoietic stem cell transplantation using mouse bonemarrow and spleen cells fractionated by lectins. *Proceedings of the National Academy of Sciences of the USA*, **75**, 2933-36.

Reisner, Y., Biniaminove, M., Rosenthal, E., Sharon, N. & Remot, B. (1979). Interaction of peanut agglutinin with normal human leukaemia cells. *Proceedings of the National Academy of Sciences of the USA*, **76**, 447.

Reisiner, Y., Ikehara, S., Hodes, M.Z. & Good, R.A. (1980a). Allogenic haemopoietic stem cell transplantation using mouse spleen cells fractionated by lectins: in vitro study of cell fractions. *Proceedings of the National Academy of Sciences of the USA*, **77**, 1164-68.

Reisiner, Y., Pahwa, S., Chiao, J.W., Sharon, N., Evans, R.L. & Good, R.A. (1980b). Separation of antibody helper and antibody suppressor human T cells by using soya bean agglutinin. *Proceedings of the National Academy of Sciences of the USA*, **77**, 6778-82.

Robinson, P.J., Bull, F.G., Anderton, B.H. & Roitt, I.M. (1975). Direct autoradiographic visualization in SDS-gels of lectin-binding components of the human erythrocyte membrane. *FEBS Letters*, **58**, 330-33.

Rose, M.L., Birbeck, M.S.C., Wallis, W.J., Forrester, J.A. & Davies, A.J.S. (1980). Peanut lectin binding properties of germinal centres of mouse lymphoid tissue. *Nature*, **284**, 364-66.

Singer, S.J. & Nicholson, G.L. (1972). The fluid mosaic model of the structure of cell membranes. *Science*, **175**, 720-27.

Stillmark, K.H. (1888). Uber Ricin, ein giftiges Ferment aus den Samen von *Ricinus comm.* L. und einigen anderen Euphorbiaceen. Inauguration Dissertation.

Surovas, B., Pieskus, J., Tamosiunas, B. & Sadauskas, P. (1982). Tμ and Tγ cell subsets in normal and leukaemic cows. *Thymus*, **4**, 31-43.

Usinger, W.R. & Splitter, G.A. (1981). Two molecularly independent surface receptors identify bovine T lymphocytes. *Journal of Immunological Methods*, **45**, 209-19.

Wardley, T. (1977). An improved E-rosette technique for cattle. *British Veterinary Journal*, **133**, 432-34.

Watkins, W.M. & Morgan, W.T.J. (1952). Neutralization of the anti-H agglutinin in eel serum by simple sugars. *Nature*, **169**, 825-26.

Watson, D.L. (1976). The effect of cytophilic IgG_2 on phagocytosis by ovine polymorphonuclear leucocytes. *Immunology*, **31**, 159-65.

Wigzell, H. (1976). Specific affinity fraction of lymphocytes using glass or plastic bead columns. *Scandinavian Journal of Immunology*, **5**, 5.

Yang, T. (1981). Identification of bovine T- and B-lymphocyte subpopulations by immunofluorescence surface-marker analysis. *American Journal of Veterinary Research*, **42**, 755-57.

5

Monoclonal antibodies to bovine leucocytes define heterogenicity of target cells for *in vitro* parasitosis by *Theileria parva*

P.A. LALOR, W.I. MORRISON and

S.J. BLACK

Monoclonal antibodies have been produced which identify determinants on putative class II major histocompatibility antigens and determinants restricted to monocytes or T lymphocytes plus monocytes, within bovine peripheral blood mononuclear cells (PBM). In addition, several antibodies, which recognize determinants on activated leucocyte populations, have been derived. These antibodies, together with a monoclonal antibody to bovine IgM, have been used to purify specific populations of cells from PBM in order to test their susceptibility to infection with sporozoites of *Theileria parva* and to examine the phenotypes of cloned infected cell lines derived from such populations. *Theileria*-infected cell lines were readily generated from each of the cell populations tested, except the purified monocytes, indicating that *T. parva* is capable of infecting and transforming different populations of lymphocytes, including both T and B lymphocytes. Comparison of the phenotypes of cell lines derived from different precursor cells revealed some differences. However, cell lines derived from B cells, in most instances, did not express surface immunoglobulin; furthermore, other determinants which were restricted in expression on normal PBM were expressed on all cell lines. Thus, with the antibodies available, it is difficult to ascertain the precursor cell type from the phenotype of the parasitized cell progeny.

Introduction

East Coast fever (ECF), a tick-borne disease of cattle caused by the protozoan parasite *Theileria parva*, is endemic in East Africa (Purnell 1977; Morrison et al. 1986). Infection of cattle is initiated by deposition of sporozoites from the salivary glands of infected ticks. The sporozoites rapidly enter cells of the lymphoid system where they differentiate within 3 to 4 days to the macroschizont stage. The macroschizont can be seen by light microscope as a membrane-enclosed body usually containing 3 to 15 parasite nuclei (Stagg et al. 1981). The development of macroschizonts is associated with induction of proliferation of the infected host cells, resulting in exponential multiplication of the parasitized cell population. The macroschizont-infected cells are of central importance in the pathogenicity of infections with *T. parva* and there is evidence that they are also targets for cell-mediated immune mechanisms mediating recovery from, or immunity to infection (Morrison et al. Chapter 29, this volume).

Exposure of susceptible cattle to infection with *T. parva* results in high morbidity and mortality rates (Brocklesby, Barnett & Scott 1961). However, there is evidence that some populations of indigenous *Bos indicus* cattle, born in ECF-endemic areas and exposed to natural parasite challenge from birth, develop mild, clinically inapparent infections, with mortality rates of less than 1% (Moll, Lohding & Young 1981; Young 1981). The mechanisms mediating this resistance to the parasite are not known.

One aspect of infections with *T. parva* which is poorly understood, and which may be relevant to susceptibility to infection, is the nature of the cell which the parasite infects. Analyses of infected cell populations have been hampered by the lack of markers for bovine leucocyte populations. The studies described in this chapter were undertaken to derive monoclonal antibodies which could be used to define different populations of bovine leucocytes and to purify the cell populations in order to test their susceptibility to infection with *T. parva*. Although in no way complete in their scope, these studies have provided an experimental procedure with which to explore the potential heterogeneity of target cells for theilerial parasitosis and the pathogenic effects elicited *in vivo* by different parasitized cell types.

Background

A proportion of macroschizont-infected cells, removed from an infected host, will undergo continued exponential growth if cultured *in vitro*. These cell lines exhibit growth properties generally associated with transformed cells, growing in defined media supplemented only with low levels of foetal bovine or donor horse serum (Morrison et al. 1986) Similar cell lines can also be established by *in vitro* infection of normal bovine lymphoid cells with theilerial sporozoites isolated from infected ticks (Brown et al. 1973). The parasitized cells appear to be capable of unlimited growth potential *in vitro*. However, removal of the schizonts from the cells by either theileriocidal drugs (Pinder et al. 1981), or agents which cause unequal parasite distribution into daughter cells (J. Naessens personal communication), results in the loss of proliferative capacity of the nonparasitized cells.

Phenotypic analyses of macroschizont-infected cells which arise in diseased cattle suggest that the cells are negative for expression of both surface and cytoplasmic immunoglobulin (Ig)(Emery 1981). Parasitized cell lines, cultured *in vitro*, are also negative for surface Ig, whether derived from *in vivo* or *in vitro* infections (Duffus, Wagner & Preston 1978; Pinder, Withey & Roelants 1981). The surface phenotypes of these parasitized cell lines, defined by lectin stains and monoclonal antibody BT3/812, are similar to that of a subset of T lymphocytes, representing 7% of bovine peripheral blood mononuclear cells (PBM). Hence, it was suggested that infection of bovine cells by *T. parva* was restricted to a distinct subset of T cells (Pinder, Withey & Roelants 1981). However, because the cell lines used in this study for phenotypic analyses were produced by infection of large numbers of cells and subsequently subcultured numerous times *in vitro*, it is possible that they represented highly selected populations of cells. Furthermore, the data infer that the phenotypes of the cells preparasitosis and postparasitosis will be identical and cell type-specific. It is now known that this assumption is not valid.

Characterization of monoclonal antibodies which identify restricted populations of cells within the bovine lymphoid system

The first step towards defining target cells for *T. parva* parasitosis was the production and characterization of monoclonal antibodies which reacted with distinct cell types in the bovine lymphoid system. The BALB/ c mice, from which these monoclonal antibodies were produced, were immunized with cells from various sources, including PBM, thymocytes, and an *in vitro T. parva*-parasitized cell line (Lalor et al. in press). The monoclonal antibodies were selected for reactivity with surface determinants expressed on resting (unstimulated) or activated (stimulated with mitogens or alloantigens, or by *T. parva* parasitosis) PBM. The specificities of these antibodies have been characterized partially by definition of the cellular distribution of the determinants within the bovine lymphoid system and definition of the functional cell types in PBM, identified by the antibodies in *in vitro* proliferative response assays (Lalor et al. in press; Morrison et al. Chapter 13 this volume).

Table 1. Proportions of cells reacting with monoclonal antibodies in different populations of leucocytes derived from bovine PBM. Adherent PBM and fibroblasts were stained on glass coverslips and evaluated using a fluorescence microscope. The other cell populations were stained in suspension and analysed by the FACS. Adherent cells were isolated by culture of PBM on glass coverslips for 2 days at 37°C. Between 75% and 95% of adherent cells in these cultures were highly phagocytic for 1.1 μm latex beads. Activated blast cells were obtained by culturing PBM cells with Con A (5 μg/ml) for 3 days or in MLR cultures for 7 days. Only the fluorescent signals from the activated cells (as identified by increased forward light scatter) were analysed by the FACS. Fibroblast monolayer cultures were established *in vitro* by long-term culture of adherent cells isolated from thymus and lymph nodes, and subcultured by trypsinization. Cell cultures identical in growth properties and surface phenotype were also obtained by outgrowth of a small proportion of adherent cells from PBM. Numbers in parentheses indicate weak levels of staining.

Cell type	% cells stained with monoclonal antibody						
	P5	R1	P8	B5/4	P4	P10	P13
PBM	75	30	10	22	0	(10)	0
Adherent PBM	95	10	96	1	ND	98	0
Con A blasts	90	70	0	3	(20)	95	17
MLR blasts	95	60	0	0	(10)	95	17
Fibroblasts	100	0	0	0	100	100	0

Monoclonal antibody P8 identifies monocytes within PBM

The monoclonal antibody (MAb) P8 reacted with between 5 and 18% of bovine PBM, isolated on Hypaque-Ficoll (Pharmacia). The proportion of P8-positive cells correlated with the proportion of monocytes, defined by alpha naphthyl acetate esterase (ANAE) staining and nuclear morphology, in these PBM preparations, and in PBM preparations depleted of monocytes by defibrination and removal of plastic-adherent cells, or enriched for monocytes by centrifugation on percoll gradients (Lalor et al. in press; Goddeeris, Lalor & Morrison this volume). Analysis with the fluorescence-activated cell sorter (FACS) indicated that the P8-positive cells in PBM were considerably larger (indicated by degree of forward light scatter) than the P8-negative cells. The P8-positive and P8-negative cells in PBM comprised 97% and less than 3% respectively, of cells positive for cytoplasmic ANAE staining.

The P8 determinant was also expressed on phagocytic macrophages isolated by short-term culture of adherent cells from PBM (Table 1). The properties of large size (compared to lymphocytes), *in vitro* adherence to plastic, phagocytosis and staining for ANAE exhibited by P8-positive cells in bovine PBM are all characteristic of monocytes and macrophages in other species (Nichols & Bainton 1975; Jones 1975; Rosenberg et al. 1981). We therefore propose that the MAb P8 reacts specifically with monocytes in bovine PBM.

Monoclonal antibodies R1 and P2 react with determinants associated with class II MHC molecules

Within bovine PBM, the determinants identified by MAbs R1 and P2 were expressed on most, if not all, surfaces (s) IgM-positive lymphocytes, and on a proportion of sIgM-negative cells, which generally comprised between 5 and 15% of PBM. The R1-positive P2-positive sIgM-negative cells included a population of monocytes, based on large size and adherent properties. These R1-positive monocytes are the predominant cell type which elicit proliferative responses in allogeneic T lymphocytes in mixed leucocyte reactions (MLR) (Lalor et al. in press). The R1 and P2 determinants were not expressed on resting peripheral blood T lymphocytes, but were expressed on a proportion of T-cell blasts activated by mitogen or alloantigen (Lalor et al. in press).

The R1 determinant was expressed in all cattle tested, whereas the P2 determinant was expressed in approximately 30% of cattle tested. In lymphoid organs isolated from cattle which expressed the P2 determinant, both monoclonal antibodies showed identical reactivities for restricted cell types. The antibodies reacted with lymphocytes in the B-dependent areas of spleen and lymph nodes, as well as large, interdigitating cells in the T-dependent areas. Within the thymus, both R1 and P2 reacted with the epithelial cells in both the cortex and medulla but did not readily detect either medullary or cortical lymphocytes (Lalor et al. in press Morrison et al. this volume). The restricted cellular reactivities of the molecule(s) identified by these two antibodies, as described above, is characteristic of class II MHC molecules in other species (Hammerling 1976; Charron et al. 1980). We therefore propose that the MAb R1 recognizes a monomorphic determinant, and MAb P2 recognizes a host-restricted determinant associated with bovine class II MHC molecules.

Monoclonal antibody P5 identifies T lymphocytes and monocytes within PBM

In 20 cattle tested, the MAb P5 reacted with between 50% and 85% of the isolated PBM. The P5 determinant was expressed on the majority of sIgM-negative cells, but not on sIgM-positive lymphocytes, within PBM. Analysis with the FACS indicated that P5-positive PBM were composed of a population of lymphocytes and a population of large cells which probably represented monocytes. Expression of the P5 determinant on monocytes was confirmed, first by selective depletion of the P5-positive large cells by removal of plastic-adherent cells from PBM, and secondly by detection of high levels of the P5 determinant on phagocytic macrophages prepared by short-term culture of adherent PBM (Table 1).

The P5-positive (small) lymphocytes in PBM were represented by two distinct populations which differed in the level of expression of the P5 determinant. The populations are referred to as P5-low and P5-high cells (see Figure 1). The relative proportions of the P5-negative, P5-low and P5-high cells varied considerably between different cattle. In some animals, the P5-high lymphocytes did not appear to be present in PBM (Figure 1). The MAb P5 was lytic in the presence of complement, resulting in complete depletion of the P5-high lymphocytes and P5-positive monocytes, but leaving intact a proportion of P5-low cells (representing up to 30% of the residual cells in several experiments). These cells, which could not be killed by increasing concentrations of either MAb P5 or complement, were possibly resistant to lysis because of the low level of the P5 determinant expressed. Both the P5-low and P5-high cells were negative for expression of the R1 determinant (i.e. class II MHC molecules), in that complement-mediated MAb R1 lysis of PBM killed most if not all R1-positive cells, but left both P5-high and P5-low populations intact (Figure 1).

Treatment of PBM populations from several cattle with MAb P5 plus complement, severely impaired the ability of the cells to mount proliferative responses to alloantigens in MLC. In contrast, depletion of R1-positive cells, or of monocytes by plastic adherence, either had no effect on, or enhanced, proliferative responses to alloantigens (Lalor et al. in press; J . Newson & P.A. Lalor unpublished). Therefore, the ability to proliferate in response to alloantigens is a property of P5-positive R1-

negative lymphocytes, represented by the P5-high and P5-low cells within bovine PBM. Similarly, the ability to proliferate in response to the mitogens concanavalin A (Con A) and phytohaemagglutinin (PHA) are properties of the P5-high and P5-low lymphocytes (Lalor et al. in press; P.A. Lalor unnpublished). By analogy with the defined roles of murine and human T lymphocytes in proliferative responses to mitogens and alloantigens (Unanue 1981; Palacios 1982), the P5-high R1-negative and P5-low R1-negative cells in bovine PBM are likely to represent T lymphocytes. The difference in levels of P5 determinant on the T cells may relate to their maturational stage, in that the P5 determinant is expressed on the majority (approximately 90%) of thymocytes, but at levels considerably lower than on the P5-low T cells in blood (Lalor et al. in press).

Monoclonal antibodies P10, P13 and P4 identify different determinants expressed on activated cells derived from PBM
The P10 determinant was expressed at only low levels on resting PBM, but at greatly elevated levels on the majority of blast cells arising after 3 days of culture with the mitogens Con A, PHA or pokeweed mitogen (PWM) or 5 to 6 days of culture with alloantigens in MLR (Table 1). Increases in the level of P10 expression could be detected as early as 1 day after culture of PBM with mitogens. The small percentage of blast cells generated in mitogen cultures which expressed sIgM also expressed the P10 determinant. Expression of the P10 determinant was also observed on activated, phagocytic macrophages, isolated by culture of monocytes from PBM for 2 to 5 days, and on fibroblastic cell monolayers, isolated by culture of adherent cells from bovine blood, spleen, skin or thymus, and serially passaged by trypsinization (Table 1). The MAb P10, therefore, defines a determinant of general leucocyte activation, being expressed on activated T and B cells, macrophages and fibroblast cultures. In its phenotypic expression and cellular distribution, the P10 determinant shows some similarity to the transferrin receptor in humans (Haynes 1981; Haynes et al. 1981; Ward, Kushner & Kaplan 1982).
The determinant identified by MAb P13 was not detected on resting PBM, but was expressed on 5% of blast cells arising after 4 days of culture in MLR. By 7 days of culture, the proportion of P13-positive blast cells had risen to approximately 70%. The determinant was also detected on a proportion of blast cells induced by 4 to 5 days of culture of PBM with Con A or PHA. The restricted expression of the P13 determinant, together with the kinetics of appearance on alloantigen-

activated T-cell blasts, is similar to that defined for glycoproteins of large molecular weight associated with activated murine and human cytotoxic T lymphocytes (Andersson et al. 1978; Kimura & Wigzell 1978; Bach et al. 1981).

The determinant defined by MAb P4 was not detected on resting PBM, and was detected at relatively low levels on a proportion (about 20%) of mitogen-activated PBM blast cells. High levels of the determinant were expressed on the fibroblast monolayer cultures described above. The nature of both the molecule identified by MAb P4, and the cells which express it, await further definition.

Definition of target cells within bovine PBM for *in vitro* parasitosis by *Theileria parva*

The monoclonal antibodies described above, together with those that react with bovine IgM (MAb B5/4, Pinder et al. 1980) or IgG (A2, S. Kar & P. Wells unpublished), can be used to distinguish between T lymphocytes, B lymphocytes and monocytes within bovine PBM, as well

Figure 1. Representation of P5-high and P5-low T lymphocytes in bovine PBM. Each profile represents data from 40,000 cells, presented as a histogram relating the fluorescence intensity of the stained cells on the abscissa to the relative number of cells on the ordinate axis. PBM from animals 648 (a) and 427 (b) stained by immunofluorescence with MAb P5. The P5-high and P5-low cell subsets are marked H and L respectively. PBM from animal 427 (c) stained by immunofluorescence with MAb R1. PBM from animal 427 (d) treated with MAb R1 plus complement before staining with (A) MAb R1 and (B) MAb P5. Note that both the P5-high and P5-low populations of T cells are left intact.

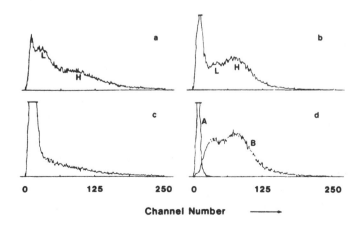

Channel Number ⟶

Table 2. *Putative specificities of monoclonal antibodies to bovine leucocyte membrane antigens.*

Monoclonal antibody	Putative specificities	Cell subsets identified within peripheral blood
R1, P2	class II MHC molecules	B cells, a proportion of monocytes, T-cell blasts
P5	leucocyte-differentiation antigen	T cells, monocytes, fibroblasts
B5/4	IgM (Pinder et al. 1980)	B cells
P8	antigen specific for monocytes and granulocytes	monocytes, activated macrophages, granulocytes
P10	leucocyte-activation antigen	activated T and B cells and macrophages, fibroblasts
P13	T-cell activation antigen	a proportion of T-cell blasts
P4	not defined	a proportion of lymphoblasts, fibroblast cultures

as to detect heterogeneity within activated cell populations derived from PBM (Table 2). These antibodies were used in the studies described below, to begin characterization of the cell types which become parasitized during infections with *T. parva.*

Different cell subsets within bovine PBM were purified using fluoresceinated monoclonal antibodies and the FACS. The isolated cell subsets were incubated for 1 h with *T. parva* (Muguga) sporozoites prepared from infected tick salivary glands, and distributed at a range of cell densities into 96-well culture plates, ranging from 10 to 1000 cells per well, 24 replicate wells per cell density. After 3 to 4 weeks of culture the cells were evaluated for establishment of macroschizont-infected cell lines. The susceptibility of each defined cell subset in PBM to theilerial infection and transformation was assessed by comparison of its relative frequency of cell-line establishment to that observed in PBM from which the defined cells had been depleted by FACS and in intact (untreated) PBM.

In a series of experiments, the cell subsets within bovine PBM defined by the membrane phenotypes P5-high R1-negative and P5-low R1-negative (T lymphocytes), R1-positive P5-negative (B lymphocytes and other as yet unidentified cell types, but excluding T cells), B5/4-positive P5-negative (B lymphocytes) and P8-positive (monocytes) were sorted, infected with *T. parva* and distributed into limiting dilution assays.

Theileria-infected cell lines were readily generated from each of the cell subpopulations tested except the P8-positive monocytes. In two experiments, the frequency of parasitized cell lines establishing from monocytes was 20- to 40-fold lower than that of cell lines derived from the sorted P8-negative (monocyte-depleted) PBM, and in fact was consistent with the fraction of P8-negative cells (approximately 3 to 5%) contaminating the sorted P8-positive cells.

The low infectivity of many of the sporozoite preparations used meant that these limiting dilution analyses could only be used with certainty to compare the relative frequencies (but not to measure the absolute frequencies) of theilerial transformation in different cell subpopulations. However, in occasional experiments in which the infectivity of the sporozoite preparation was high, more than 1 in 30 cells from either intact PBM or sorted T or B cells established parasitized cell lines.

Characterization of the membrane phenotypes of *Theileria*-infected bovine T and B lymphocytes

Twelve *T. parva*-infected cell lines derived from each of the sorted P5-high, P5-low, R1-positive, P5-negative and B5/4-positive cell subsets in PBM were expanded from wells in the limiting dilution assays. These cell lines were selected from wells in which the initial input cell density resulted in cell-line establishment in less than 63% of the wells. Thus, each selected cell line represented a clone, or the progeny of only a few initially infected cells (Fazekas de St Groth 1982).

The cell lines were examined by FACS for expression of the antigenic determinants identified by the monoclonal antibodies; a summary of their phenotypes is shown in Table 3. All parasitized cell lines expressed the P5 determinant, although the level of expression on the T-cell lines was 3 to 4 times higher than that observed on nonT-cell lines. T-cell lines derived from both P5-low and P5-high T-cell precursors expressed similar high levels of the P5 determinant. All T and nonT-cell lines expressed class II MHC molecules (defined by MAb R1). These molecules were expressed on the cell lines in a cyclical manner, being present on 60 to 80% of the cells within a clone at any time. Cell lines depleted of R1-positive cells by treatment with MAb R1 plus complement re-expressed high levels of the R1 determinant following further culture. Cycling of class II MHC molecules has also been observed on human and murine lymphomas (Sarkar et al. 1980; Lanier & Warner 1981).

All cell lines expressed the P10 determinant, although the level varied considerably between clones and was generally lowest on cell lines derived from R1-positive P5-negative (nonT cell) and B5/4-positive (B cell)

Table 3. Cell-surface phenotype of *Theileria*-infected cell lines derived by infection with sporozoites *in vitro* of different preselected populations of bovine PBM. The levels of expression were defined by immunofluorescence using the FACS. The results shown are representative of five or more clones of each type tested. Levels of expression: $+++$ high on all cells; $++$ intermediate; $+$ low; $-$ not detected. The / symbol (as in $-/++$) indicates that clones were identified which expressed either one or the other of the listed phenotypes. Only two of the cell lines examined with the B5/4 antibody exhibited the $++$ phenotype, and in one of these expression was not stable (see text).

Precursor cells for infection		Level of expression on parasitized cell lines of determinant identified by MAb						
Membrane phenotype	Lineage	P5	R1	B5/4	P8	P10	P13	P4
P5 low	T cell	$+++$	$++$	$-$	$-$	$++/+++$	$+/+++$	$-$
P5 high	T cell	$+++$	$++$	$-$	$-$	$++/+++$	$++/+++$	$-$
R1$^+$ P5$^-$	nonT cell	$+$	$++$	$-/++$	$-$	$+/++$	$-$	$-/++$
B5/4$^+$	B cell	$+$	$++$	$-/++$	$-$	$+/++$	$-$	$-/++$

precursors. All T-cell clones expressed the P13 determinant although levels of expression again were variable. The P13 determinant was not, however, expressed on any nonT-cell clone (i.e derived from R1-positive P5-negative or B5/4-positive precursor cells in PBM). Conversely, the P4 determinant was expressed at variable levels on nonT-cell, and defined B-cell clones, but not on T-cell clones. None of the clones examined expressed the monocyte-specific determinant identified by MAb P8.

Surface IgM detected by both MAb B5/4 and FITC-labeled goat anti-bovine IgM was expressed on one of the cell lines derived from each of the sorted R1-positive P5-negative and B5/4-positive cell populations. IgM was not stable on the cell line derived from the B5/4-positive population, being lost after continued subculture. IgG was not expressed on any of the cell lines. There was no detectable secretion of IgM or IgG, as measured in radioimmunoassays, into the supernatants from 48 B-cell lines (derived from B5/4-positive precursor cells) grown to high cell densities (P.A. Lalor unpublished). We are currently investigating the possibility that secretion or membrane expression of Ig by the transformed B-cell lines is inducible by B-cell maturation factors secreted by mitogen-activated T cells. By comparison of the surface phenotypes, it seems likely that many (if not all) the cell lines derived from sorted R1-positive P5-negative precursor cells are in fact B cell in origin.

The loss of sIgM expression which follows *in vitro* parasitosis of bovine B lymphocytes makes this marker of limited value in defining the precursor cell origin of pre-established parasitized leucocytes. In fact, considerable changes occur in the expression of several defined membrane antigens following infection of both T and B cells with *T. parva*. These changes, which include acquisition of blast cell antigens associated with the proliferative state of the infected cells and gain or loss of the expression of markers by which the resting cell populations within PBM were defined, highlight the difficulties in defining an infected cell type by correlation of phenotypes before and after infection. However, parasitized cells do appear to express surface phenotypes which may be characteristic of, and restricted to, the cell lineage from which they were derived.

Significance of these studies to *Theileria parva* infections in cattle

These studies provide strong evidence that the target cells for *in vitro* parasitosis by *T. parva* are heterogeneous, including at least T and B lymphocytes within peripheral blood. However, it should be emphasized that the results are relevant only in the context of the *in vitro* infection system used. The finding that monocytes do not establish parasitized cell lines does not mean that the cells within the monocytic lineage do not become parasitized during infections with *T. parva* in cattle. It is possible that parasitosis of monocytes requires obligatory activation or growth factors, provided in the animal, but not in the *in vitro* culture system, or it is possible that cells of the myelomonocytic lineage at different maturational stages may become infected.

Therefore, to define further heterogeneity of target cells for *T. parva* parasitosis, it may be necessary to modify culture conditions (e.g. by addition of soluble growth factors) and to extend the range of target cells used to include defined cell subsets isolated from bonemarrow, spleen or thymus. Definition of potential heterogeneity of target cells for infection with *T. parva*, either within the T- and B-lymphocyte lineages or within other as yet undefined cells within the lymphoid system, will also require the production of further monoclonal antibodies which can detect distinct functional subsets within these cell types. A number of monoclonal antibodies have since been produced which do detect heterogeneity within bovine T and B lymphocytes and other as yet undefined cells (Davis et al., Ch. 6; P.A. Lalor & W.I. Morrison unpublished). Indeed, the *Theileria*-transformed clones of defined precursor cell origin have provided useful tools in the definition of the specificities of new monoclonal antibodies.

The findings from these *in vitro* studies are now being extended to *in vivo* infections in order to define the possible significance of the infection of different cell types in determining the outcome of infections with *T*.

parva. As a first approach, attempts have been made to phenotype the parasitized cells which arise in cattle during the early and late phases of lethal infections or during self-limiting chronic infections. However, difficulties have been encountered; as yet, there exist no monoclonal antibodies which can be used with certainty to identify specific parasitized cell types and hence define the precursor cell origins. Despite the apparent relationship in surface phenotypes of precursor cells and *in vitro* parasitized cells, phenotypic analyses of other parasitized cell clones derived following infection of undefined precursor cells *in vitro* or from cells of infected animals indicate that there is no apparent conformity to the patterns of determinant expressions defined. For example, parasitized clones have been isolated which express the surface phenotypes defined by $P5^{+++}$ $P13^-$, $P5^+P13^{+++}$, $P5^{+++}$ $R1^-$ (J. Newson, J. Naessens & P.A. Lalor unpublished). Thus, it seems unreasonable to assume that parasitized T and B cells are uniquely defined by the phenotypes $P5^{+++}$ $P13^{+/+++}$ $R1^{++}$ and $P5^+$ $P13^-$ $R1^{++}$, respectively. It is also difficult to relate the level of determinant expression on *in vitro*—derived clones to that observed on the heterogeneous population of infected and uninfected cells present in *in vivo* infections. Attempts to culture and clone parasitized cell lines from an infected animal for phenotypic analyses have been hampered by the low frequency of macroschizont—infected cells (less than 1%) that give rise to cell lines *in vitro* (J. Newson, J. Naessens & W. Barry personal communication).

The second experimental approach has been to determine the pathogenic effects on the naive autologous hosts, of defined cell types infected *in vitro*. In a preliminary experiment, the sIgM-positive B lymphocytes from two cattle were purified by the FACS, and intact PBM from another two cattle were passed through the FACS but not separated. The cells from each animal were exposed *in vitro* to sporozoites of *T. parva* (Muguga) and washed extensively. After incubation at 37°C overnight, 2×10^5 cells were inoculated into each autologous animal as described previously (Morrison & Buscher 1983). Cells from each animal were also cultured at limiting dilution *in vitro* and were shown to give rise to *Theileria*-infected cell lines at similar frequencies. Of the two animals inoculated with 2×10^5 autologous sIgM-positive PBM, neither developed detectable infection, although one animal was found to be immune when subsequently challenged with a lethal dose of sporozoites. Both of the cattle which received 2×10^5 autologous intact PBM developed patent infections and one died. While insufficient animals were infected to draw a statistical conclusion from this experiment, the results suggest that infected B cells generate milder infections than other cell types, That is, severity of disease may be related to the nature of the cells which initially become infected with *T. parva*. These experiments will be repeated, using *in vitro*-parasitized cell clones derived from purified B cells, T cells and bulk cell lines derived from intact whole PBM. These studies should lead to improved understanding of the roles of infected cell subsets in the pathogenesis of *T. parva* infections in cattle.

Acknowledgements
We wish to thank J.G. Magondu, K.S. Logan, R.M. Jack, B. Goddeeris, B. Otim and D.A. Stagg for their help in these studies.

References

Andersson, L.C., Gahmberg, C.G., Kimura, A.K. & Wigzell, H. (1978). Activated human T lymphocytes display new surface glycoproteins. *Proceedings of National Academy of Sciences of the USA*, **75**, 3455-58.

Bach, F.H., Alter, B.J., Widmer, M.B., Segall, M. & Dunlap, B. (1981). Cloned cytotoxic and noncytotoxic lymphocytes in man and mouse: their reactivities and a large cell surface membrane protein (LMP) differentiation marker system. *Immunological Reviews*, **54**, 5-26.

Brocklesby, D.W., Barnett, S.F. & Scott, G.R. (1961). Morbidity and mortality rates in East Coast fever (*Theileria parva* infection) and their application to drug screening procedures. *British Veterinary Journal*, **117**, 529-31.

Brown, C.G.D., Stagg, D.A., Purnell, R.E., Kanhai, G.K. & Payne, R.C. (1973). Infection and transformation of bovine lymphoid cells *in vitro* by infective particles of *Theileria parva*. *Nature*, **245**, 101-3.

Charron, D.T., Engleman, E.G., Benike, C.J. & McDevitt, H.O. (1980). Ia antigens on alloreactive T cells in man detected by monoclonal antibodies. Evidence for synthesis of HLA-D/DR molecules of the responder type. *Journal of Experimental Medicine*, **152**, 127s-36s.

Duffus, W.P.H., Wagner, G.G. & Preston, J.M. (1978). Initial studies on the properties of a bovine lymphoid cell culture line infected with *Theileria parva*. *Clinical & Experimental Immunology*, **37**, 347-53.

Emery, D.L. (1981). Kinetics of infection with *Theileria parva* (East Coast fever) in the central lymph of cattle. *Veterinary Parasitology*, **9**, 1-16.

Fazekas de St. Groth, S.F. (1982). The evaluation of limiting dilution assays. *Journal of Immunological Methods*, **49**, R11-R23.

Hammerling, G.J. (1976). Tissue distribution of Ia antigens and their expression on lymphocyte subpopulations. *Transplantation Reviews*, **30**, 64-82.

Haynes, B.F. (1981). Human T-lymphocyte antigens as defined by monoclonal antibodies. *Immunological Reviews*, **57**, 127-61.

Haynes, B.F., Hemler, M., Cotner, T., Mann, D.L., Eisenbarth, G.S, Strominger, J.L. & Fanci, A. (1981). Characterization of a monoclonal antibody (5E9) that defines a human cell-surface antigen of cell activation. *Journal of Immunology*, **127**, 347-51.

Jones, T. C. (1975). Attachment and ingestion phases of phagocytosis. In *Mononuclear phagocytes in immunity, infection and pathology* (ed. R. van Furth), pp. 269-82. London: Blackwell Scientific Publications.

Kimura, A.K. & Wigzell, H. (1978). Cell-surface glycoproteins of murine cytotoxic T cells. 1. T145, a new cell-surface glycoprotein selectively expressed on Lyt 1-negative 2-positive cytotoxic T lymphocytes. *Journal of Experimental Medicine,* **147**, 1418-34.

Lalor, P.A., Morrison, W.I., Goddeeris, B.M., Jack, R.M. & Black, S.J. Monoclonal antibodies identify phenotypically and functionally distinct cell types in the bovine lymphoid system. *Veterinary Immunology and Immunopathology,* in press.

Lanier, L.L. & Warner, N.L. (1981). Cell cycle related heterogeneity of Ia antigen expression on murine B lymphoma cell line: analysis by flow cytometry. *Journal of Immunology,* **126**, 626-31.

Moll, G., Lohding, A. & Young, A.S. (1981). The epidemiology of theileriosis in the Trans-Mara division, Kenya. In *Advances in the control of theileriosis* (eds. A.D. Irvin, M.P. Cunningham & A.S. Young), Current Topics in Veterinary Medicine & Animal Science 14, pp. 56-59. The Hague: Martinus Nijhoff.

Morrison, W.I. & Buscher, G. (1983). The early events of infection with *Theileria parva* in cattle: infectivity for cattle of leucocytes incubated *in vitro* with sporozoites. *Veterinary Parasitology,* **12**, 145-53.

Morrison, W.I., Lalor, P.A., Goddeeris, B.M. & Teale, A.J. (1986). Theileriosis: Antigens and host–parasite interactions. In *Parasite Antigens: Toward New Strategies for Vaccines* (ed. T.W. Pearson) pp. 167-212. New York: Marcel Dekker.

Nichols, B.A. & Bainton, D.F. (1975). Ultrastructure and cytochemistry of mononuclear phagocytes. In *Mononuclear phagocytes in immunity, infection and pathology* (ed. R. van Furth), pp. 17-55. London: Blackwell Scientific Publications.

Palacios, R. (1982). Mechanism of T-cell activation: role and functional relationship of HLA-DR antigens and interleukins. *Immunological Reviews,* **63**, 73-110.

Pinder M., Withey, K.S. & Roelants, G.E. (1981). *Theileria parva* parasites transform a subpopulation of T lymphocytes. *Journal of Immunology,* **127**, 389-90.

Pinder, M., Musoke, A.J., Morrison, W.I. & Roelants, G.E. (1980). The bovine lymphoid system. 3. A monoclonal antibody specific for bovine cell-surface and serum IgM. *Immunology,* **40**, 339-65.

Pinder, M., Kar, S., Withey, K.S., Lundin, L.B. & Roelents, G. (1981). Proliferation and lymphocyte stimulatory capacity of *Theileria*-infected lymphoblastoid cells before and after the elimination of intracellular parasites. *Immunology,* **44**, 51-60.

Purnell, R.E. (1977). East Coast fever: some recent research in East Africa. *Advances in Parasitology,* **15**, 83-132.

Rosenberg, S.A., Ligler, F.S., Ugolini, V. & Lipsky, P.E. (1981). A monoclonal antibody that identifies human peripheral blood monocytes recognizes the accessory cells required for mitogen-induced T-lymphocyte proliferation. *Journal of Immunology,* **120**, 1473-77.

Sarkar, S., Glassy, M.C., Ferrone, S. & Jones, O.W. (1980). Cell cycle and the differential expression of HLA-A, B and HLA-DR antigens on human B-lymphoid cells. *Proceedings of the National Academy of Sciences of the USA,* **77**, 7297-301.

Stagg, D.A., Dolan, T.T., Leitch, B.L. & Young, A.S. (1981). The initial stages of infection of cattle cells with *Theileria parva* sporozoites *in vitro*. *Parasitology,* **83** , 191-97.

Unanue, E.R. (1981). The regulatory role of macrophages in antigenic stimulation. 2. Symbiotic relationship between lymphocytes and macrophages. *Advances in Immunology,* **31**, 1-136.

Ward, J.H., Kushner, J.P. & Kaplan, J. (1982). Transferrin receptors of human fibroblasts: analysis of receptor properties and regulation. *Biochemical Journal,* **208**, 19-26.

Young, A.S. (1981). The epidemiology of theileriosis in East Africa. In *Advances in the control of theileriosis* (eds. A.D. Irvin, M.P. Cunningham & A.S. Young), Current Topics in Veterinary Medicine & Animal Science 14, pp. 38-55. The Hague: Martinus Nijhoff.

6

Construction of a library of monoclonal antibodies for the analysis of the major histocompatibility gene complex and the immune system in ruminants

W.C. DAVIS, L.E. PERRYMAN and

T.C. McGUIRE

A library of cross-reactive monoclonal antibodies that can be used to study the phylogenetic and functional relations of products of the major histocompatibility complex (MHC) and products of genes coding for differentiation molecules expressed on monocytes and lymphocytes was developed by immunizing groups of mice by injection of: (1) Hypaque-Ficoll purified peripheral blood leucocytes from one or more species (cattle, goats, sheep, horses, rabbits, dogs, cats, rats), or (2) thymocytes from one or more species. Three days after a booster injection of cells, spleen cells were taken and fused with myeloma cells (NS-1, SP/2Ag14 or X63.653). Antibody-producing hybrids were detected by using complement-mediated cytotoxicity or enzyme-linked immunosorbent assays. By using cells from a species not used in the immunization protocol, as well as cells from species used in the protocol, monoclonal antibodies (MoAb) were readily detected which were cross-reactive with cell membrane antigens present in two or more species. Analysis of 120 cloned cell lines using a Becton Dickinson FACS Research Analyser has thus far demonstrated the presence of MoAb to conserved MHC class I and class II antigens present on cells in mice, humans, horses, cattle, goats, sheep, rabbits, mink, dogs, cats and chickens, as well as a series of MoAb that react with differentiation molecules expressed predominantly on monocytes, T or B cells, or react with molecules present on two or more cell types. Studies are in progress to demonstrate whether the cross-reactive MoAb detect antigenic molecules already defined by commercially available MoAb with restricted species reactivity, or detect new molecules. The development of a set of MoAb that cross-react with conserved antigenic epitopes on class I and class II MHC molecules permits the direct comparison of the role of MHC restriction in mouse, man and other species, while the development of cross-reactive antibodies to differentiation molecules permits the analysis of the mechanisms of immune regulation. The latter set of MoAb will, in addition, permit the delineation of unique species differences in the function of lymphoid cell subpopulations.

Introduction

The discovery that cells producing monoclonal antibodies (MoAb) of desired specificity can be immortalized through cell fusion has launched a new era for the use of antibodies in biological analysis (Kohler & Milstein 1975; Oi & Herzenberg 1980; Kennett, McKearn & Bechtol 1981; Hammerling, Hammerling & Kearney 1981; Davis, McGuire & Perryman 1983). This technology has been exploited in the analysis of cells involved in the immune response. Previous studies had revealed that three major groups of cells interact in a highly regulated manner, i.e. T and B lymphocytes and antigen-presenting cells (monocytes/macrophages and dendritic cells). Further analysis has provided evidence for the existence of functionally distinct subpopulations that can be defined by the expression of different constellations of surface membrane molecules, referred to as differentiation molecules (Lampson & Lenz 1980; Reinherz & Schlossman 1981; Bhan et al. 1981; Kung et al. 1981; Engleman et al. 1981a; 1981b; Ledbetter et al. 1981; Howard et al. 1981; Abo & Balch 1981; Goldstein 1982; Gatenby et al. 1982; Hanjan, Kearney & Cooper 1982). Monoclonal antibodies now provide an opportunity to detail the antigenic display of each cell type and trace its course through differentiation to functional maturity. The progress made in analysing the immune responses in humans and inbred mice has emphasized the utility of the technology.

The challenge is to extend MoAb technology to elucidate the immune response in domestic animals and to use the information obtained to control disease. This challenge involves two central issues: (1) the need to define interspecies differences in the immune system that influence host response to disease, and (2) the need to elucidate the genetic factors that affect the way an animal responds to an infectious agent. Because the primary emphasis has been placed on the immune system in mouse and humans, few well-characterized reagents exist to meet the needs of investigators interested in food animal research (Pinder, Pearson & Roelants 1980). An extensive set of MoAb is required that will permit the analysis of the major histocompatibility complex (MHC) and elucidation of its role in regulating expression of immunity.

Two approaches are available for the production of MoAb. The first is the development of MoAb to relevant antigens on a species by species basis. Although this has proven effective in the study of the human and murine lymphoid system, the problem remains to establish that the antigens identified and the cells expressing these antigens in each species are indeed homologous counterparts, and that unique differences in functional activity are attributable to species-specific modification of the immune system and not misinterpretations of comparative data. In studies with humans, a number of MoAb have been developed that appear to recognize different determinants on the same molecules and other MoAb that appear to detect different molecules with similar patterns of expression on leucocytes (Reinherz & Schlossman 1981; Engleman et al. 1981a; 1981b; Hanjan, Kearney & Cooper 1982). Without rigorous proof of identity, one is left with the question of whether investigators using different sets of MoAb are studying the same cell populations and/or differentiation antigens. The finding that a number of antigens are differentially expressed on subpopulations of one or more cell types (Pinder, Pearson & Roelants 1980; Engleman et al. 1981a; 1981b; Ledbetter et al. 1981; Howard et al. 1981; Haynes 1981; Gatenby et al. 1982; Hanjan, Kearney & Cooper 1982) emphasizes that this is not a trivial problem and that it may complicate correct interpretation of data, especially in interspecies comparisons of immune function.

The second approach is to produce cross-reactive antibodies useful in two or more species. When an animal is immunized with cells from another species two types of antibodies are formed, antibodies that recognize antigenic epitopes on molecules that are essentially species specific, and antibodies that recognize antigenic epitopes expressed on molecules in more than one species. Most of these antibodies recognize molecules with a common phylogenetic origin. Advantage has been taken of such antibodies, using both xenoantisera and MoAb, to detail the relation of proteins, most notably the immunoglobulins (Neoh et al. 1973; Vaerman, Kobayashi & Heremans 1974; Balch, Dagg & Cooper 1976) and products of the MHC (Iha et al. 1973; Kvist, Klareskog & Peterson 1978; Kvist, Ostberg & Peterson 1978; Stallcup, Springer & Mescher 1981; Figueroa, Davis & Klein 1981; Ivanyi 1981; Soloski et al. 1981; Teillaud et al. 1982; Brodsky & Parham 1982; Hoang-Xuan et al. 1982; Pierres et al. 1982; Sachs 1983; Klein, Figueroa & Nagy 1983; Hood, Steinmetz & Malissen 1983). The success with cross-reactive antibodies has suggested that it should be possible to use monoclonal antibody technology to develop a library of MoAb which react with homologous gene products derived from two or more species. The development of such a set of antibodies would: (1) permit the direct demonstration that homologous gene products have indeed been identified and that they are expressed in an identical way in the same cell subpopulations and function in a like manner, (2) permit inter- and intraspecies comparison of molecules identified by MoAb that recognize species-restricted antigenic epitopes, (3) facilitate the elucidation of the relation of families of molecules coded for by gene sets with a common phylogenetic origin,

(4) provide a means of generating MoAb to polymorphic antigenic determinants coded for by each allelic set of genes, and (5) minimize the time and redundancy of effort required to prepare MoAb for each species of interest.

The potential of cross-reactive antibodies in the biological and biomedical sciences prompted us to explore the possibility of developing an extensive library of MoAb for use in food animal research. This chapter describes the approach and methodology we have devised for this endeavour and provides the first listing of the set of MoAb reactive with leucocyte antigens in ruminants.

Strategies for the production of cross-reactive monoclonal antibodies

The strategy for developing a library of cross-reactive antibodies has been the following: (1) apply tissue culture methodology that optimizes the outgrowth of cell hybrids following fusion and permits the primary analysis and preservation of large numbers of antibody-producing hybyridomas, (2) refine methodology for optimizing the production of hybridomas producing cross-reactive MoAb, (3) develop and/or modify assays to facilitate characterization of the patterns of reactivity of cross-reactive MoAb and determine their apparent specificity, (4) determine the biochemical and functional properties of the cell membrane molecules recognized by the MoAb, (5) determine the phylogenetic relationship of the cell membrane molecules recognized by the MoAb, and (6) define the chromosomal locus of the genes coding for the molecules and the number of allelic variants present in each species.

We have found a two-phase approach most useful for producing large numbers of hybrids. One of the major difficulties encountered in the screening of cultures for hybrids producing an antibody of interest has been the short time interval between the time of assay and the time of cell transfer to a second culture vessel. This short time interval is necessary to avoid overgrowth and cell death. A second difficulty has been the management of large numbers of cultures while determining which cell lines should be preserved. As illustrated in Figure 1, by using a two-phase procedure it is possible to work with 400 to 600 primary cultures with reasonable effort. In the first phase of production, fused cells are dispersed in 96-cell plates and cultured for 10 to 14 days. Supernatants are collected and tested for antibody activity. Cell lines producing the antibodies of interest are then transferred to 24-well plates and maintained in static culture for 2 weeks, i.e. without expansion of the cell line beyond one well during long-term culture. This is accomplished by removing excess cells every 2 to 3 days. At the end of 2 weeks, a duplicate plate is prepared and maintained without thinning for a week. Supernatants are then collected and tested for antibody activity. Selected cell lines are expanded into one, two or three wells of a six-well plate and then preserved in liquid nitrogen.

Once an aliquot of cells has been preserved, the remaining cells are allowed to overgrow and die to make a final antibody-rich supernatant for further analysis. Phase one permits the maintenance of all cell lines of interest as well as selection of the more stable, rapidly growing, antibody-producing cell lines. It also permits an opportunity to determine which cell lines should be preserved for cloning before or after primary preservation.

During phase two, each antibody is analysed systematically to select which cell lines to clone for the establishment of permanent cell lines. A hybrid cell line is plated at high cell dilution to yield 50 to 100 single colonies in three to five 96-well plates. The single clones are tested for antibody activity. Positive clones are taken and carried through the same steps as in phase one. Then four to six clones are selected and preserved to establish the cell line. The second phase permits the selection and cloning of the cell lines of immediate interest and also a determination of the stability of antibody production.

The critical factor for the consistent production of hybrids has been the use of 2-mercaptoethanol (2 ME) and lipid A (Ribi Immuno-Research Hamilton MT and List Biological Laboratories, Inc., Campbell, CA) in the culture medium and the use of thymocytes as feeder cells. Comparative studies have shown an additive effect when 2 ME is used with feeder cells. Though the mechanism has not been elucidated, the addition of lipid A (at 1 µg per well) at the time the fused cells are dispensed into 96-well plates increases the yield of hybrids by 20 to 40%. Other factors, such as the myeloma cell line used as a fusion partner, the source of plastic ware, the purity of the water used to make the medium and variation in foetal bovine serum lots, are important but secondary to the culture conditions mentioned.

When mice are immunized with xenogeneic cells, they usually produce

Figure 1. Flow diagram of a two-phase procedure for producing monoclonal antibodies. See text for explanation. The diagram was prepared in part for a workshop on Priorities in Biotechnology Research for International Development, held in July 1983 in Washington D.C. and sponsored by the Board on Science and Technology for International Development, National Academy of Sciences/National Research Council.

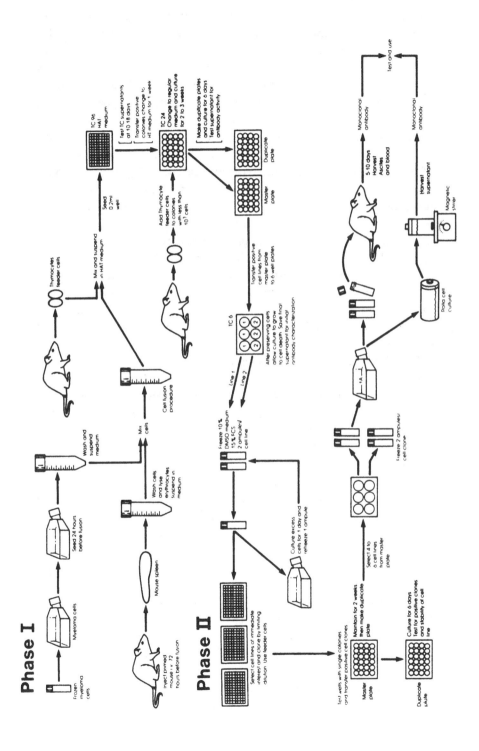

antibodies to antigenic epitopes which are either restricted to the species used for immunization or are also expressed on cells obtained from other species. If an animal has been exposed to cells from a single species, however, the predominant clonotypes present in the system produce antibody to species-restricted epitopes. We explored different strategies of immunization that might increase the number of clonotypes producing cross-reactive antibody. We found that hyperimmunization with leucocytes from multiple species (horse, cow, goat, sheep, pig, dog, rabbit, rat, human) given sequentially, or as a mixture, altered the composition of the population of B-cell clonotypes and the specificity of the antibodies being produced. The results obtained thus far indicate that the potential for producing hybrids synthesizing broadly cross-reactive antibodies increases as more species are used as cell donors.

We found that identification of cross-reactive antibodies could be facilitated by designing an immunization protocol to exclude cells from one of the species of interest. This permitted the cells from that species to be used in the primary screening for cross-reactive antibody activity. In our studies, we usually tested for antibody activity with cells from one of the species used in the immunization protocol as well as the species excluded.

Several methods of assay were used to select hybrids, the complement-mediated micro-cytotoxicity assay (CT) (Stocker & Bernoco 1979), the enzyme linked immunoassay (ELISA) (Douillard, Hoffman & Herberman 1980; Sutter, Bruggen & Sorg 1980; Daniels & Gielkins 1980) and flow cytometry (Herzenberg 1978). In the initial phases of the program, CT and ELISA were sufficient for the selection of interesting MoAb that had potential for use in genetic typing and analysis of leucocyte subsets. However, flow cytometry proved essential for final analysis of MoAb for functional studies and cross-species comparison of cross-reactivity. In the studies described here we used the FACS™ analyser (Becton Dickinson, Sunnyvale, CA). This is a new instrument similar to the laser-based fluorescence-activated cell sorters (FACS) currently in use in many laboratories. The principal differences are the mode of fluorescence excitation, volume sensing and transmission of cells through the sensing orifice. The instrument rapidly measures individual cells simultaneously for fluorescence (autofluorescence and fluorescence imparted by fluorochrome-coupled antibodies bound to cells), for single or dual colour (using a mercury-arc illuminator) and for size (as measured by electronic volume). Up to 1,000 cells can be analysed a second and the data can be displayed in two modes: a histogram profile, based on relative cell number and relative change in fluorescence intensity of unlabeled and fluorochrome-antibody tagged cells, and a dot plot profile, based on cell size and fluorescence intensity. The comparison of data displayed in these modes permits an estimation of the proportion of cells expressing a given membrane antigen, an estimate of staining intensity (dim or bright), an estimate of the size of the cells expressing the antigen and a profile of the distribution of antigen on cell subsets, as assessed by the amount of labeled antibody bound to the cells.

Our initial endeavours have focused on the selection and cloning of cell lines producing antibody to antigenically distinct leucocyte membrane molecules. The CT assay was used as the primary assay to select and establish cell lines in our initial studies. Later, both CT and ELISA were used to select cell lines for maintenance in static culture; analysis by flow cytometry was also used to distinguish cell lines producing antibody to cell subpopulations. To produce MoAb from cloned cell lines that react with bovine leucocytes, 120 cell lines have been selected from a library of 700 lines carried through static culture. These were derived from approximately 12,000 hybrids produced from 27 fusions. The following prefixes denote the origin of the hybridoma producing the MoAb listed in Table 1 and the scheme of immunization used to generate antibodies to cell membrane antigens:

B, E, G: immunized with cow, horse or goat peripheral blood leucocytes purified on Hypaque-Ficoll (HF-PBL)

HT, PT: immunized with horse or pig thymus

H: immunized with horse, rabbit, rat, dog, goat and human HF-PBL

TH: immunized with cow, horse, goat and rat thymocytes

PIg: immunized with pig and cow IgM

PORC, SPAM and CAPP: immunized with cow and pig platelets

ANA: immunized with cow erythrocytes infected with *Anaplasma marginale*

RH: immunized with rat lymph node cells and horse fibroblasts

HELP and SAG: immunized with BoLA class I antigens isolated from a membrane lysate using MOAb H58A bound to protein A from *Staphylococcus*

CA: CBA (H-2^k) mice immunized with B10.A (H-2^d) mouse lymph node cells.

Table 1. Summary of data obtained from FACS analysis of patterns of cross-reactivity of MoAb with HF-PBL derived from different species. The percentage of cells labeled is based on a statistical analysis of histogram profiles of labeled cells. The apparent specificity is indicated where sufficient data have been obtained. Where data are limited and the histograms uninformative, the designation has been left blank. Where information is limited but the histogram reveals that a distinct population of cells is labeled, a designation of subpopulation is indicated. (See next page.)

MoAb	Ig isotype	Cow	Sheep	Goat	Pig	Horse	Human	Apparent specificity
PNA	lectin	50–86	54	57	69	60–84	31	species variable
H4	IgG$_3$	15–25	15–28	10–28	15–35	15–25	15–30	MHC class II
Leu 2b	IgG$_{2a}$	–	–	–	–	15–20	28	T cytotoxic suppressor cell
Leu 8	IgG$_{2a}$	15–25	–	–	–	15–20	68	T-cell subset
H1A	IgG$_{2a}$	85	–	–	90	87	95	–
H6A	IgG$_{2a}$	85	98	50	50–90[a]	97	97	–
H11A	IgG$_{2a}$	90	87	65	62[a]	98	95	MHC class I
H17A	IgG$_{2ab}$	85[a]	–	–	90[a]	98	90[a]	MHC class I
H18A	IgM	8–10	8–11	8–11	3–14	10	15–19	monocyte
H20A	IgG$_1$	60[a]	–	–	60	74	60	–
H21A	IgM	90[a]	38–90[a]	16[a]	34–80[a]	97[a]	90[a]	MHC class I
H34A	IgG$_{2a}$	26–36	27	14–30[a]	–	34–55	20	MHC class II
H42A[b]	IgG$_{2a}$	20–28	20–29	11–20	28	32–59	19	MHC class II
H58A[b]	IgG$_{2a}$	95	94	85	83[a]	99	95	MHC class I
E4C	IgM	8–30[a]	5	3–16	10	10[a]	10[a]	T-cell subset
E18A	IgG$_{2a}$	26[a]	–	–	–	80	–	–
E23A	IgM	15	?	?	?	36	?	subpop
E40A	IgM	15–30[a]	10	12–23	11	19–37	10–20[a]	T-cell subset
E47B	IgM	10–30	9	8–20[a]	–	40–56	9–23[a]	subpop
E48C	IgM	40[a]	–	–	–	41	38[a]	subpop
E50A	IgM	40[a]	–	32	–	44	40[a]	subpop
HT53A	IgM	13–25[a]	30	–	–	32	–	subpop
B5C	IgG$_{2b}$	95[a]	94	89	–	98	90[a]	MHC class I
B16A	IgM	30	11[a]	16–21	19	20–30	9–18	monocyte + granulocytes

MoAb	Ig isotype	Cow	Sheep	Goat	Pig	Horse	Human	Apparent specificity
B18A	IgG$_3$	30[a]	—	—	12–20	20[a]	—	monocyte + granulocytes
B24A	IgM	40	—	—	—	—	—	pan-T
B26A	IgM	40–60	—	36	—	—	—	pan-T
B29A	IgG$_{2a}$	50–60[a]	—	—	30[a]	22	30	T-cell subset
G2A	IgM	30[a]	—	80	—	—	!	subpop
SAG 4A	IgG$_{2a}$	85[a]	—	60–80	—	—	—	MHC class I
SAG 7A	IgG$_{2a}$	85[a]	[a]	—	—	—	—	MHC class I
SAG 8A	IgG$_{2a}$	85[a]	35[a]	72[a]	—	—	—	MHC class I
HELP 1	IgM	85[a]	—	—	50–80[a]	—	—	MHC class I
HELP 2A	IgM, IgG$_1$	45[a]	22[a]	—	50–80[a]	—	—	MHC class I
HELP 2B	IgM	45[a]	—	—	—	—	—	MHC class I
CA 48P-A	IgM	85[a]	84[a]	50–94[a]	70–80[a]	98[a]	98[a]	MHC class I
CA-48P-A1	IgG$_{2a}$	85[a]	84[a]	50–94[a]	70–80[a]	98[a]	98[a]	MHC class I
CA 4C-A	IgM	85[a]	21[a]	—	30–80[a]	—	—	MHC class I
TH 1A	IgG$_{2a}$	20–42	—	—	—	—	—	subpop
TH 2A	IgG$_{2a}$	15–33	—	—	20–30[a]	—	—	subpop
TH 4B	IgM	20–30	12–20	20–30[a]	16–37	37	21	MHC class II
TH 11E	IgG$_{2a}$	50–80	25[a]	20[a]	30[a]	—	—	subpop
TH 12A	IgG$_{2a}$	20–45	9–25	—	—	20–41	20[a]	MHC class II
TH 14B	IgG$_{2a}$	20–30	30–33	34	—	P?	20[a]	MHC class II
TH 16A	IgG$_{2a}$	20–35[a]	24–28	34	32	—	—	MHC class II
TH 17A	IgM	86	76[a]	80	30[a]	—	90[a]	T+B lymphocytes
TH 18A	IgG$_3$	61	—	—	80[a]	—	?	
TH 21A[b]	IgG$_{2b}$	20–30	23–24	20–32	23–32	33–55	19	MHC class II

MoAb	Ig isotype	Cow	Sheep	Goat	Pig	Horse	Human	Apparent specificity
TH 22A	IgG$_1$	15–30	28–37	20	24	—	—	MHC class II
TH 31B	IgG$_1$	30–46	—	—	—	—	—	subpop
TH 57A	IgG$_{2b}$	31	—	—	19	12	—	subpop
TH 61A	IgM	20–30	10	7	57	25	22	subpop
TH 62A	IgM	10	—	—	—	—	—	subpop
TH 71A	IgM	10–20	—	28	9	27	46	subpop
TH 81A[b]	IgG$_{2a}$	30	22–30	15–40	30	—	14	MHC class II
TH 82D	IgG$_1$	12–20	23–30	10–15	—	—	—	T subpop
TH 87A	IgM	20–45	20–40	—	—	—	—	subpop
TH 90A	IgM	12–30	?	20	34	23	10	subpop
TH 92A	IgM	14–30	?	41	—	—	38	subpop
TH 97A	IgG$_{2a}$	5–10	?	6	—	—	5	subpop
PT9A	IgG$_{2a}$	10–37	15–27	15–20	35	—	32	subpop
PT25D	IgG$_3$	42	24–79	63	25–35	30–60	44	—
PT35A	IgG$_3$	40	57–70	68	25–38	21–46	40	—
PT40A	IgG$_3$	40	71	64	21–36	21–36	36	—
PT85B	IgG$_{2a}$	90	80[a]	80	85	90	90[a]	MHC class I
PIg31B	IgG$_1$	15	12–15	12–15	10–15	—	—	light chain
PIg45A	IgG$_{2b}$	15–25	12–21	10–17	10–17	12–17	—	IgM
PIg47B	IgG$_1$	15–25	10–20	10–17	10–17	12–18	—	IgM
RH 1A	IgG$_3$	21–30	23	8–18	—	20–30	32	subpop
RH 12A	IgG$_{2a}$	85[a]	—	—	—	—	—	MHC class I
RH 16A	IgG$_{2a}$	85[a]	—	—	—	—	—	MHC class I
PORC 9A	IgG$_{2a}$	17	38	40–45	19–30	—	—	subpop
SPAM 11A	IgM	8–13	5–11	5–8	7–11	—	—	monocyte?
CAPP2A	IgG$_1$	19–40	34	30–50	—	—	6	monocyte
MPGE17A	IgG$_1$	49[a]	—	—	—	—	—	subpop
ANA8A	IgG$_1$	—	—	—	—	—	—	bovine rbc
ANA13A	IgM	41	—	—	—	—	—	bovine rbc + subpop
ANA18A	IgM	31	—	15	20	—	—	bovine rbc + subpop

[a] antigen polymorphic.

Monoclonal antibodies that react with MHC gene products

To develop a useful set of antibodies to the MHC, we first identified antibodies that clearly recognized MHC gene products in humans or mice. We then used these antibodies with flow cytometry and other assays to identify additional MoAb to monomorphic and/or polymorphic MHC determinants. We screened a series of 100 MoAb prepared against products coded for by the H-2 gene complex and identified two antibodies reactive with bovine cells, CA4C and CA48P. These were cloned independently at the Abteilung Immunogenetik Max-Planck-Institut für Biologie in Tubingen, West Germany (MPI) and at Washington State University (WSU). Analysis of the sublines TI.4C and TI.48P (MPI) and CA4C-A and CA48PA (and CA48P-A1) (WSU) revealed identical patterns of reactivity for H-2D class I gene antigens (designated m84 and m89 respectively) (Figueroa, Davis & Klein 1981). Further analysis revealed that CA4C-A detects a polymorphic antigen in cattle and pigs, and CA48P-A (and CA48P-A1, a subclone producing IgG_{2a}), a polymorphic antigen in cattle, goats, sheep, pigs, horses and humans.

Subsequently, we screened 70 MoAb (from cell hybrids obtained from mice immunized with cells from one or more species) against a panel of 29 inbred strains of mice differing only at the H-2 complex (W.C. Davis, L.E. Perryman & T.C. McQuire unpublished). These endeavours yielded four antibodies that reacted with one or more strains of mice: H58A, an antibody which reacts with a class I gene product ($H-2K^{f,p,pv}$ and $H-2D^{w16}$ haplotypes) and H42A ($H-2^{w3}$ haplotype), as well as TH21A and TH81A ($H-2^{b,f,r,s,u,v,w4,w6,w7,w17,w23}$ haplotype) which react with class II gene products (Mandic et al. unpublished). As noted in Table 1, the antibodies were found to detect antigen in all species tested except for TH81A which appears to be negative for horse cells.

The specificity of these antibodies for use as standards in comparative studies was established by examining with flow cytometry using the FACS™ analyser for patterns of labeling of human HF-PBL. The labeling patterns were compared with two commercially available MoAb specific for HLA-DR, a human class II gene product: HLA-DR (donated by Becton Dickinson Monoclonal Center, Mountain View, CA) (Ledbetter et al. 1981) and H4 (donated by P. I. Terasaki and J.C. Cicciarelli, University of California at Los Angeles, Los Angeles, CA) (Lewin & Bernoco 1983). The analysis revealed distinctive patterns of labeling of cells that could be obtained by comparing histogram and dot plot displays. Regardless of whether a given antibody stained cells brightly or dimly, the patterns remained constant. As illustrated in Figure 2, HLA-

DR, H4 and H42A labeled populations of monocytes and lymphocytes, with the staining intensity more pronounced on monocytes. Further analysis revealed that HLA-DR reacts with horse HF-PBL as well as with human cells and that H4 reacts with all species tested.

When staining profiles obtained with H4 and H42A (reacted with human and bovine cells) were used as a reference to screen for additional MoAb with apparent specificity for class II antigens, dot plot profiles were found to differ between species. Intraspecies labeling profiles were found to be consistent (Figures 2 and 3). Nine antibodies were identified that react with bovine HF-PBL (H34A, H42A, TH4B, TH12A, TH14A, TH16A, TH21A, TH22A and TH81A). Except for TH12A, the MoAb were found to be broadly cross-reactive between species. Of special

Figure 2. Histogram and dot plot profiles of human HF-PBL labeled with anti-class II monoclonal antibodies. All cell preparations were stained using indirect labeling techniques as described in the Becton Dickinson *Monoclonal antibody methods manual*. HF-PBL were processed in 96-well microtitre plates, with 5 x 10⁵ cells per well in 50 μl of PBS-BSA plus azide. Cells were reacted with 50 μl of MoAb for 30 min at 4°C, washed three times, then reacted with affinity purified fluoresceinated goat anti-mouse IgM, IgG (TAGO Inc., Burlingame, CA). The cells were then washed three times, resuspended in 1.5% solution of paraformaldehyde in buffered saline and kept refrigerated until used. The fixed cells were examined within a week of preparation.

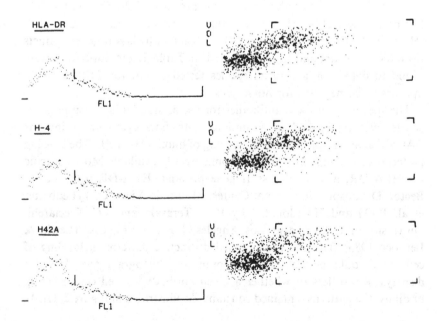

interest, H34A was found consistently to label more cells in ruminants and horses than in humans, i.e. approximately 10% more than noted with HLA-DR, H4 and the other WSU MoAb. Studies with purified T and B cells in the cow have shown that TH4B, TH12A, TH14B, TH16A, TH21A and TH22A react with B cells and monocytes (H.A. Lewin, D. Bernoco & W.C. Davis unpublished). Biochemical studies with pig cells, though preliminary, indicate that H42A, TH16A and TH21A precipitate pig class II antigens (J.K. Lunney et al. unpublished).

A number of MoAb have been identified with apparent specificity for class I antigens in cattle. These include: CA48P-A, CA48P-A1, CA4C-A, H58A, H11A, H17A, H21A, PT85B, SAG4A, SAG7A, SAG8A, HELP1A, HELP2A, RH12A, RH16A and B5C. Comparative analysis of labeling with CA48P-A and H58A revealed similar dot plot profiles, with minimal intraspecies and interspecies differences (as illustrated in Figure 3). Five of the other MoAb are of special interest since they were generated from mice immunized with antigen isolated with H58A bound

Figure 3. Histogram and dot plot profiles of bovine HF-PBL labeled with anti-MHC class I (H58A) and class II (H42A) MoAb. The profile of cells labeled with H58A is typical for labeled cells from all species. The position of the stained population of cells can vary but the pattern remains similar. The profile of cells labeled with H42A is characteristic for bovine class II MHC antigens.

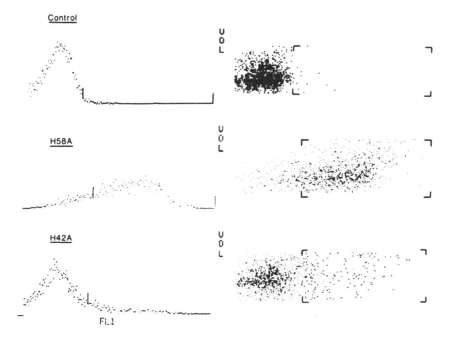

to *Staphylococcus* protein A. They differ from H58A in that they detect polymorphic determinants in cattle. Biochemical studies thus far have shown that H58A, PT85B, RH12A and H21A react with cell membrane antigens with an apparent molecular mass of 45,000 daltons.

Genetic studies with antibodies B5C, RH12A, RH16A, H17A, H21A, SAG4, SAG7A, SAG8A, HELP1A and HELP2A have yielded preliminary data consistent with the premise that they recognize bovine MHC antigens. RH12A, RH16A, SAG4U, SAG7A, SAG8A, HELP1A and HELP2A all react with the BoLA w8 class I MHC antigen (H.A. Lewin, D. Bernoco & W.C. Davis unpublished), but sufficient data are not yet available to assign haplotype specificity.

Monoclonal antibodies that recognize surface-bound IgM

Since it has been difficult to identify MoAb specific for B cells, our initial efforts to obtain a B marker have focused on identification of surface-bound IgM (sIgM). For these studies, a MoAb (DAS6) specific for bovine IgM was obtained from S. Srikumaran and R.A. Goldsby (Amherst College, CN) and a MoAb (USDA 5C9.C12.1) specific for pig IgM from P. Paul (USDA, Ames, IA). We screened 100 MoAb reactive with pig immunoglobulin by ELISA for reactivity against semipurified bovine immunoglobulins (Ig). MoAb positive by ELISA for bovine Ig were then screened for reactivity with bovine HF-PBL with the FACS analyser. Three such antibodies were identified, PIg31B, PIg45A and PIg47A. Comparative analysis with DAS6 on bovine cells and USDA 5C9,C12.1 on pig cells revealed patterns of labeling consistent with the reagents, PIg45A and PIg47A being specific for SIgM and PIg31B being specific for a light chain antigen (PIg31B reacts with all classes of pig Ig by ELISA). Further studies by FACS revealed that PIg45A and PIg47A react with goat, sheep and horse sIgM and PIg31B with goat, sheep and pig Ig light chains.

Monoclonal antibodies that recognize antigens predominantly on T cells

Several approaches have been employed to identify MoAb with apparent specificity for T cells. These include comparative analysis with the FACS analyser, cytotoxicity mediated by antibody- plus-complement and dual-fluorescence labeling of purified T and B cells and HF-PBL.

To facilitate FACS analysis of MoAb specificity, MoAb of known specificity for human T cells were obtained from Becton Dickinson (BD) (Leu 1, Leu 2b, Leu 3a, Leu 3b, Leu 4, Leu 5, Leu 7, Leu 10 and Leu

M1) and screened for patterns of reactivity with human, cow, goat, sheep, pig and horse HF-PBL. In addition, peanut agglutinin (PNA), a plant lectin found useful as a T-cell marker in mice (London, Berih & Bach 1978; London & Horton 1980) and as a monocyte marker in humans (London & Horton 1980; O'Keefe & Ashman 1982), was evaluated (Banks & Greenlee 1981; 1982). The studies revealed that two of the BD reagents (Leu 2b and Leu 8) react with horse cells and one (Leu 8) reacts with bovine cells (Table 1). The studies also revealed that PNA yields entirely different patterns of labeling of HF-PBL derived from different species. As noted in Figure 4, PNA labels predominantly monocytes in humans, T lymphocytes and some large cells (monocytes) in horses and mixed populations in ruminants and pigs (Figure 4 and Table 1).

Comparative analysis of the staining profiles of BD and WSU MoAb on human cells revealed only one similar pattern of labeling (Leu M1 and H18A). Leu M1 reacts with monocytes and granulocytes and H18A with monocytes (reactivity with granulocytes has not been established). One antibody (B29A) was found to label a small population of lymphocytes in humans, pigs and horses and a large population in cattle.

Further analysis with bovine HF-PBL using flow cytometry yielded a set of seven antibodies with apparent specificity for T cells (B1A, B24A, B26A, E4C, E40A, TH82D and MPGE17A). To date, only one of these has been firmly established as specific for a T-cell antigen, B26A. B26A detects an antigen present on the majority of T cells as assessed by antibody plus complement-mediated cytotoxicity and dual-fluorescence (H.A. Lewin, W.C. Davis & D. Bernoco, 1985). Dual-fluorescence studies and FACS analysis indicate that B24A, B26A and B29A label the majority of T cells and not B cells. FACS analysis shows that TH82D labels a small population of lymphocytes (Figure 5). TH18A, TH87A and MPG17A label populations of cells which include a large component of small lymphocytes similar to B26A (Figures 5 and 6) and some larger mononuclear cells, suggesting the antigens detected are not necessarily restricted to T cells. Preliminary studies indicate TH18A reacts with both T and B cells in cattle.

E4C and E40A usually label small populations of cells comprising both large and small cells similar to the pattern illustrated for H18A on bovine cells in Figure 5. On occasion, cells from individual animals from different species have exhibited marked increases. Evidence for E4C and E40A reacting with T cells is based on findings of immunofluorescent staining of thymus tissue. Studies in the horse show that E4C stains a small population of cells in the medulla and E40A a large population of cortical cells.

Monoclonal antibodies reactive with monocytes and monocytes/granulocytes

Seven antibodies have been identified that exhibit predominant labeling of monocytes in one or more species (Hl8A, Bl6A, Bl8A, E48C, RHlA, CAPP2A). Hl8A is the most interesting. When reacted with human HF-PBL, it labels monocytes intensely, with data displayed by dot plot corresponding closely with those obtained with PNA and Leu M1 (as illustrated in Figure 4 for PNA). With other species, however, only a small population of cells are detected (as illustrated for bovine cells in Figure 5). Comparative staining of HF-PBL and cell preparations enriched for granulocytes in the cow indicate that the antigen is represented mainly on a population of mononuclear cells.

Bl6A and Bl8A both label approximately 30% of bovine HF-PBL and exhibit similar staining profiles by histogram and dot plot analysis. However, comparison of staining profiles of bovine cell preparations enriched for neutrophils and eosinophils (80% neutrophils) shows that Bl6A labels 90 to 95% of the cells, whereas Bl8A labels only 75 to 85%.

Figure 4. Comparative analysis of horse, human and cow cells labeled directly with fluoresceinated peanut agglutinin (PNA), 1 µg PNA per 5 x 10⁵ cells. Note that the patterns differ. T lymphocytes and monocytes are labeled in the horse, monocytes in humans and multiple cell types in the cow. Dual-fluorescence staining indicates bovine B cells are not reactive with PNA.

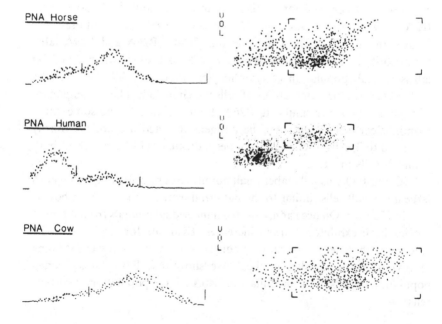

Figure 5. Comparative analysis of bovine HF-PBL labeled with
cross-reactive MoAb. Note that no distinct boundary exists between
lymphoid and monocytic cells as noted in humans. The overlap in
size is reflected in the staining profiles of cells with predominant
specificity for lymphocytes and monocytes. The staining with PIg45A
includes cells bearing sIgM and cells (monocytes) bearing IgM bound
through an Fc receptor. Cells labeled with B26A and TH82D
illustrate staining profiles obtained with MoAb with predominant
activity for T cells. Cells labeled with Hl8A illustrate one of the
staining profiles obtained with MoAb that detect cells in low
concentration in bovine HF-PBL.

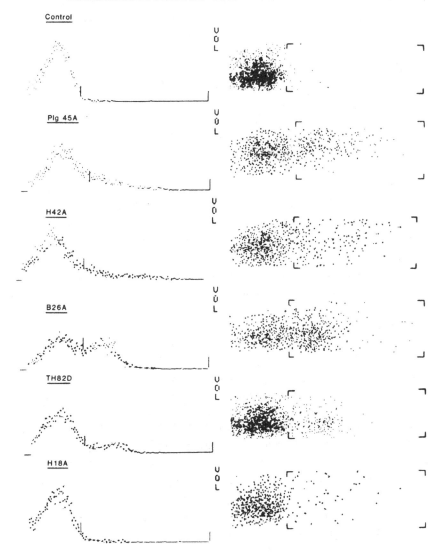

In addition, B16A detects an antigen with a broader species distribution. As with H18A, the patterns of labeling with B16A vary in different species. The antigen detected is usually present on a smaller population of cells in species other than cattle (possibly reflecting a different cell composition of HF-PBL).

E48C and E50A yield similar staining profiles when reacted with cells from cattle, horses and humans and analysed by flow cytometry. The percentage of cells labeled appears to vary proportionally with the content of monocytes present in the HF-PBL. With human cells, where monocytes and lymphocytes are readily distinguished on dot plot display, all monocytes are labeled as are a population of cells in the size range of lymphocytes. Based on intensity of staining, the expression of antigen detected with E48C and E50A varies more than the antigen detected with H18A (compare histogram profiles in Table 1). Intra- and interspecies variation in antigen expression indicate that E48C and E50A detect different antigenic epitopes.

RH1A detects an antigen that is present in multiple species. The dot plot analysis shows a complex distribution of labeled cells, suggesting variable expression of the antigen on multiple cell types, but predominantly on monocytes.

CAPP2A detects an antigen present in multiple species but, as with B16A and H18A, the antigen is present in different concentrations in HF-PBL from different species. In cattle, CAPP2A appears to detect an antigen on monocytes and some, but not all, granulocytes.

Monoclonal antibodies that recognize antigens which vary in concentration from species to species

Seven MoAb (SPAM11A, TH61A, TH62A, TH71A, TH90A, TH92A, TH97A) have been identified that recognize antigens expressed in different concentrations in different species. Dot plot analysis does not suggest that the antigens are present on a predominant cell type. The antibodies have been grouped merely for discussion. SPAM11A detects an antigen expressed on a subpopulation of HF-PBL which is usually small in ruminants and pigs. Occasionally, the number of cells detected by flow cytometry reaches 20% in cows. TH61A, TH62A, TH71A, TH90A and TH92A exhibit a wide range of reactivity in different species. Analysis thus far indicates the variation is primarily attributable to variation in the labeling of monocytes. In some species, only dim labeling occurs and

Figure 6. Comparison of histogram profiles of bovine HF-PBL labeled with cross-reactive MoAb. See text for description of apparent specificity of the antibodies.

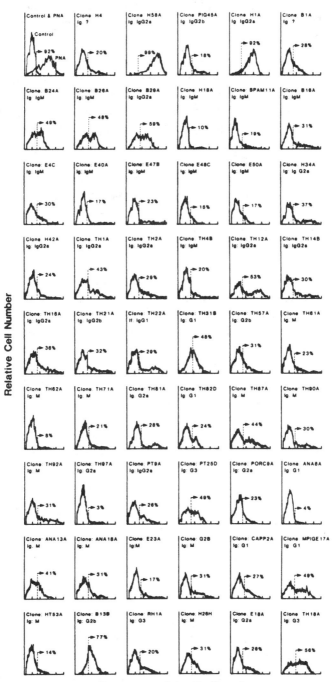

Fluorescence Intensity (log₁₀)

the shift in stained cell profiles is nominal; in others, staining is intense. Thus far, TH97A has been found to label a small population of bovine HF-PBL.

Monoclonal antibodies that react with bovine erythrocytes

Three antibodies — ANA8A, ANA13A and ANA18A — have been identified that exhibit interesting patterns of reactivity with bovine erythrocytes (rbc) and leucocytes. The hybrids producing these antibodies were selected from a group of 15 because the MoAb they produced were able to agglutinate bovine rbc. When the antibodies were tested against HF-PBL, two of the MoAb (ANA13A and ANA18A) were found to react with subpopulations of leucocytes. Only ANA8A was found to be specific for rbc. ANA13A and ANA18A react with different sized populations of HF-PBL, and the population recognized by ANA18A includes a population of granulocytes.

Monoclonal antibodies that react with mixed populations of peripheral blood leucocytes

Nineteen MoAb selected for further study yielded patterns of labeling by histogram and dot plot analysis that have not yet suggested a specificity for known cell types — H20A, H26H, B13B, E18A, E23A, E47B, HT53A G2A, TH1A, TH2A, TH11E, TH17A, TH31A, TH57A, PT9A, PT25D, PT35A, PT40A and PORC9A (Figure 6). Most of the hybrids producing these MoAb were selected because the patterns of killing noted in the CT assay appeared different. As noted in Figure 6, eight of the antibodies yielded similar histogram profiles of stained cells when reacted with bovine HF-PBL — E18A, E23A, HT53A, TH2A, TH57A, PT9A, H26H and PORC9A. However, the percentage of cells labeled by each antibody varied when reacted with HF-PBL taken from different species. B13B, TH17A and TH11E appear to stain the majority of HF-PBL taken from cattle, but in granulocyte-enriched preparations TH17A and TH11E react with different proportions of cells. TH31B reacts with both HF-PBL and granulocyte preparations with no major difference in the staining profile (Figure 6). The antigen detected appears to be in low concentration since the shift in fluorescence of labeled cells is nominal. PT25D, PT35A and PT40A yield identical histogram profiles on FACS analysis and appear to detect the same antigen. Though the profiles are similar to the one obtained with TH31B, the antibodies appear to distinguish at least two populations of cells that express different concentrations of the antigens. H20A yields a more complex pattern of labeling. The antigen detected is expressed differentially on multiple populations of monocytes, granulocytes and lymphocytes. Most lymphocytes appear to express the antigen but at a gradation from very low to very high, i.e. low and high staining intensity, with no distinct division between stained and unstained populations. Comparative studies demonstrate that TH31B, PT25D, PT35A, PT40A and H20A yield consistent patterns of labeling with HF-PBL in all species tested. E47B, TH1A, TH11E and TH17A yield contrasting patterns of labeling by FACS analysis on bovine cells (Figure 6). These MoAb also show different patterns of species cross-reactivity (Table 1). Little information is yet available to indicate their specificity.

Discussion

It is evident from the studies completed that the development of a library of cross-reactive MoAb detecting homologous gene products is possible and that the availability of such antibodies will facilitate the elucidation of the immune system in ruminants as well as other species. Effective strategies have been developed for optimizing the production of hybrids which produce cross-reactive antibodies, and the availability of fluorescence flow cytometry instrumentation has provided a rapid means of distinguishing MoAb reactive with unique leucocyte subpopulations. Moreover, the instrumentation has made it possible to analyse complex cell populations and conduct extensive comparative studies on the distribution patterns of cross-reactive MoAb on leucocytes in different species. To date, we have succeeded in identifying sets of cross-reactive MoAb specific for gene products coded for by class I and class II genes of the MHC. We have also succeeded in producing a large set of cross-reactive MoAb that react with gene products differentially expressed on leucocyte cell membranes. Though it is obvious that more work needs to be done to characterize many of the MoAb, some are immediately useful, i.e. those that detect polymorphic antigenic determinants and those that detect defined membrane molecules. The antibodies to monomorphic class I and class II antigens can be used to dissect the role of the MHC in the regulation of immunity to antigens and pathogens of interest (Stone 1981; Davis et al. 1982; Lawman 1982; Antczak 1982). The sets of antibodies are sufficiently large to permit intra- and interspecies comparisons and the demonstration of any dynamic changes in antigen expression on cells both *in vitro* and *in vivo*. The antibodies with apparent specificity for T cells and monocytes can be used to begin detailing T-cell function in relation to MHC-restriction phenomena (Klein, Figueroa & Nagy 1983; Lawman 1982; Antczak 1982).

The MoAb to monomorphic determinants (MHC and nonMHC) can also be used in two different ways to facilitate the generation of specific types of MoAb. The anti-MHC and nonMHC MoAb can be used to purify gene products for use in the generation of MoAb to antigens that define each allelic gene product present in a given species (Stallcup, Springer & Mescher 1981). In addition, MoAb specific for a given differentiation antigen expressed only in a single species can be used to purify that antigen to generate MoAb to different epitopes, on the same molecule, that are expressed on homologous molecules in other species. This strategy has already proven useful in preliminary experiments designed to make MoAb to polymorphic class I BoLA antigens. For example, SAG4A, SAG7A, SAG8A, HELP1A and HELP2A were developed with antigen isolated with MoAb H58A. These antibodies recognize polymorphic antigenic epitopes associated with the expression of BoLA w8 (D. Bernoco personal communication).

The studies completed suggest that it would be useful to screen existing MoAb reagents from other sources for cross-reactivity (Ault & Springer 1981; McMaster & Williams 1979; Lunney et al. 1983; Pierres, Rebouah & Kourilsky 1981), especially well defined MoAb that are commercially available and being used extensively in the study of the immune response in humans. Our screening of the Becton Dickinson series of MoAb has yielded three antibodies (HLA-DR, Leu 2b and Leu 8) that react with equine cells and one (Leu 8) that reacts with bovine cells. The screening of H4 obtained from P. Terasaki has yielded a broadly cross-reacting MoAb (Lewin & Bernoco 1983). Though it appears that only limited cross-reactivity will be found, each additional cross-reacting MoAb will be useful.

Our studies pose two pertinent questions. First, are the antigens detected by cross-reactivity expressed in different concentrations in different species? If so, does this reflect differences in cell function? Second, are the detected antigens closely related or composed of sets of phylogenetically related molecules (Stout et al. 1975; Ades et al. 1980; Dalchau & Fabre 1979) that share common epitopes but mediate different functions? There are no complete answers yet, but studies of the MHC provide evidence that species differences in antigen expression occur (Lunney et al. 1983) and that epitopes can be expressed on more than one phylogenetically related molecule (Klein, Figueroa & Nagy 1983). Extensive studies with cross-reactive MoAb generated against class II A and E antigens in the mouse have permitted the identification of the E homologue in the pig (Lunney et al. 1983) and humans (Pierres et al. 1982; Pierres, Rebouah & Kourilsky 1981), and possibly the A homologue. Moreover, comparative analysis of antigen distribution has shown that class II antigen is expressed at high levels on circulating T cells in the pig, in contrast to the situation in mouse and man where the homologous antigen is expressed in low concentration or is absent on resting T cells (Charron et al. 1980; Hercend et al. 1981; Mann & Sharrow 1980). Stimulation by antigens is necessary for antigen expression, suggesting a question as to whether such species variation in antigen expression reflects altered or augmented function or merely a difference unrelated to function. Analysis of the distribution of class I MHC antigens in the mouse has recently revealed the expression of a common epitope expressed on K and Qa1 molecules (Klein, Figueroa & Nagy 1983). This finding indicates that similar situations might exist in other species and suggests that cross-reactive antibodies, though specific in a given context, might detect the same and/or different molecules in cross-species comparisons (Soloski et al. 1981; Lunney et al. 1983). The fact that some of the cross-reactive MoAb found in our studies appear to detect antigens expressed in cell populations of different sizes shows that this possibility must be taken into account.

The studies with cross-reactive MoAb have revealed a different problem with important implications, i.e. the finding that cross-reactivity does not always connote close homology (Jensenius & Williams 1982; Lennox 1983). Recent studies have shown that a number of MoAb yield unusual cross-reactions, e.g. anti-Thy 1 with vimentin (Lennox 1983), anti-Thy 1 (different MoAb) with V TEPC 15 idiotype determinant (Pillemer & Weissman 1981), anti-K light chain with β_2 microglobulin (Lennox 1983), anti-tropomysin with vimentin (Blose, Matsumura & Lin 1981), anti-large T of SV40 transformed cells with normal growth related nuclear protein (Lane & Hoeffler 1980). These studies show that the detection of a useful antibody against an antigen in a given species may not be useful in another species. Also, the studies suggest that the results obtained in functional studies can include specific and nonspecific components.

In summary, we have described a set of cross-reactive MoAb that can be used for analysis of the immune response in food animals. Though further characterization is required, we are now in a position to address important questions concerning the genetic basis of disease susceptibility.

Acknowledgements

The studies reported here were supported in part by funds from the USDA-ARS-Haemoparasitic Diseases Research Unit Cooperative Agreement 58-9AHZ-2-663, USDA-ARS-Animal Disease Unit, the Department of Veterinary Microbiology and Pathology, Washington

State University, Pullman, Washington and grants USDA-SEA 82-CRSR-2-2045, USDA SEA 83-CRSR-2-2281, Formula Funding — Public Law 95-113, Agricultural Research Center Project 3073, Carnation Research Farms, National Pork Producers Council NIH-BRSG.

References

Abo, T. & Balch, C.M. (1981). A differentiation antigen of human NK and K cells identified by a monoclonal antibody (HNK-1). *Journal of Immunology*, **127**, 1024-29.

Ades, E.W., Zwerner, R.K., Acton, R.T. & Balch, C.M. (1980). Isolation and partial characterization of the human homologue of Thy 1. *Journal of Experimental Medicine*, **151**, 400-6.

Antczak, D.F. (1982). Structure and function of the major histocompatibility complex in domestic animals. *Journal of the American Veterinary Medical Association*, **181**, 1030-36.

Ault, K.A. & Springer, T.A. (1981). Cross-reaction of a rat-anti-mouse phagocyte-specific monoclonal antibody (anti-Mac-1) with human monocytes and natural killer cells. *Journal of Immunology*, **126**, 359-64.

Balch, G.M., Dagg, M.K. & Cooper, M.D. (1976). Cross-reactive T-cell antigen among mammalian species. *Journal of Immunology*, **117**, 447-49.

Banks, K.L. & Greenlee, A. (1981). Isolation and identification of equine lymphocytes and monocytes. *American Journal of Veterinary Research*, **42**, 1651-54.

Banks, K.L. & Greenlee, A. (1982). Lymphocyte subpopulations of the goat: isolation and identification. *American Journal of Veterinary Research*, **43**, 314-17.

Bhan, A.K., Nadler, L.M., Stashenko, P., McCluskey, R.T. & Schlossman, S.F. (1981). Stages of B-cell differentiation of human lymphoid tissue. *Journal of Experimental Medicine*, **154**, 737-49.

Blose, S.H., Matsumura, F. & Lin, J.J.C. (1981). Structure of vimentin 10 nm filaments probed with a monoclonal antibody that recognizes a common antigenic determinant on vimentin and tropomysin. *Cold Spring Harbor Symposia on Quantitative Biology*, **46**, 455-63.

Brodsky, F.M. & Parham, P. (1982). Evolution of HLA antigenic determinants: species cross-reactions of monoclonal antibodies. *Immunogenetics*, **15**, 151-66.

Charron, D.J., Engleman, E.G., Benike, C.J. & McDewitt, H.O. (1980). Ia antigens on alloreactive T cells in man detected by monoclonal antibodies: evidence for synthesis of HLA-D/DR molecules of the responder type. *Journal of Experimental Medicine*, **152**, 127s-37s.

Dalchau, R. & Fabre, J.W. (1979). Identification and unusual tissue distribution of canine and human homologue of Thy 1. *Journal of Experimental Medicine*, **149**, 576-91.

Daniels, D.J. & Gielkins, L.J. (1980). A simplified method for the purification of 5-aminosalicylic acid: application of the product in enzyme-linked immunosorbent assay (ELISA). *Journal of Immunological Methods*, **30**, 325-32.

Davis, W.C., McGuire, T.C. & Perryman, L.E. (1983). Biomedical and biological applications of monoclonal antibody technology in developing countries. *Periodicum Biologorum*, **85**, 259-81.

Davis, W.C., McGuire, T.C., Anderson, L.W., Banks, K.L., Seifert, S.D. & Johnson, M.I. (1982). Development of monoclonal antibodies to *Anaplasma marginale* and preliminary studies on their application. In *Proceedings of the Seventh National Anaplasmosis Conference*, pp. 285-305.

Douillard, J.Y., Hoffman, T. & Herberman, R.B. (1980). Enzyme-linked immunosorbent assay for screening monoclonal antibody production: use of intact cells as antigen. *Journal of Immunological Methods*, **39**, 309-16.

Engleman E.G., Benike, C.J., Glickman, E. & Evans, R.L. (1981a). Antibodies to membrane structures that distinguish suppressor/cytotoxic and helper T-lymphocyte subpopulations block the mixed leucocyte reaction in man. *Journal of Experimental Medicine*, **154**, 193-98.

Engleman, E.G., Warnke, R., Fox, R.I. & Levy, R. (1981b). Studies of a human T-lymphocyte antigen recognized by a monoclonal antibody. *Proceedings of the National Academy of Sciences of the USA*, **78**, 1791-95.

Figueroa, F., Davis, W.C. & Klein, J. (1981). Ten new monoclonal antibodies detecting antigen determinants on class I H-2 molecules. *Immunogenetics*, **14**, 177-80.

Gatenby, P.A., Kansas, G.S., Chen, Y.X., Evans, R.L. & Engleman, E.G. (1982). Dissection of immunoregulatory subpopulations of T lymphocytes within the helper and suppressor sublineages in man. *Journal of Immunology*, **129**, 1997-2000.

Goldstein, G. (1982). T cells and immunoregulation: a comtemporary overview. *Journal of Clinical Immunology*, **2**, 5s-7s.

Hammerling, G.J., Hammerling, U. & Kearney, J.F. eds. (1981). *Monoclonal antibodies and T-cell hybridomas: perspectives and technical advances*, 587pp. Amsterdam: Elsevier.

Hanjan, S.N.S., Kearney, J.F. & Cooper, M.D. (1982). A monoclonal antibody (MMA) that identifies a differentiation antigen on human myelomonocytic cells. *Clinical Immunology & Immunopathology*, **23**, 172-88.

Haynes, B.F. (1981). Human T-lymphocyte antigens as defined by monoclonal antibodies. *Immunological Reviews*, **57**, 127-61.

Hercend, T., Ritz, J., Schlossman, S.F. & Reinherz, E.L. (1981). Comparative expression of T9, T10 and Ia antigens on activated human T-cell subsets. *Human Immunology*, **3**, 247-59.

Herzenberg, L.A. (1978). The fluorescence-activated cell sorter (FACS): a retrospective and prospective view. In *Immunofluorescence and related staining techniques* (eds. W. Knapp et al.), pp. 99-109. Amsterdam: Elsevier.

Hoang-Xuan, M., Leveziel, H., Zilber, M.-T., Parodi, A.-L. & Levy, D. (1982). Immunochemical characterization of major histocompatibility antigens in cattle. *Immunogenetics*, **15**, 207-11.

Hood, L., Steinmetz, M. & Malissen, B. (1983). Genes of the major histocompatibility complex of the mouse. In *Annual review of immunology* (eds. W.E. Paul, C.G. Fathman & H. Metzger), vol. 1, pp. 529-68. Palo Alto, California: Annual Reviews Inc.

Howard, F.D., Ledbetter, J.A., Wong, J., Bieber, C.P., Stinson, E.B. & Herzenberg, L.A. (1981). A human T-lymphocyte differentiation marker defined by monoclonal antibodies that block E-rosette formation. *Journal of Immunology*, **126**, 2117-22.

Iha, T.H., Gerbrandt, G., Bodmer, W.F., McGary, D. & Stone, W.H. (1973). Cross-reaction of cattle lymphocytotoxic sera with HLA and other human antigens. *Tissue Antigens*, **3**, 291-302.

Ivanyi, P. (1981). Interspecies MHS relationships studies by serological and cellular cross-reactions. In *Current trends in histocompatibility: immunological profiles* (eds. R.A. Reisfeld & S. Ferrone), pp. 133-81. New York: Plenum Press.

Jensenius, J.C. & Williams, A.F. (1982). The T-lymphocyte antigen receptor — paradigm lost. *Nature*, **300**, 586-88.

Kennett, R.H., McKearn, T.J. & Bechtol, K.B. eds. (1981). *Monoclonal antibodies, hybridomas: a new dimension in biological analysis*, 423pp. New York: Plenum Press.

Klein, J., Figueroa, F. & Nagy, Z.A. (1983). Genetics of the major histocompatibility complex: the final act. In *Annual review of immunology* (eds. W.E. Paul, C.G. Fathman & H. Metzger), vol. 1, pp. 119-42. Palo Alto, California: Annual Reviews Inc.

Kohler, G. & Milstein, C. (1975). Continuous culture of fused cells secreting antibody of predefined specificity. *Nature*, **256**, 495-97.

Kung, P.C., Talle, M.-A., DeMaria, M., Ziminski, N., Look, R., Lifter, J. & Goldstein, G. (1981). Creating a useful panel of anti-T-cell monoclonal antibodies. *International Journal of Immunopharmacology*, **3**, 175-81.

Kvist, S., Klareskog, L. & Peterson, P.A. (1978). Identification of H-2 and Ia-antigen analogues in several species by immunological cross-reactions of xenoantisera. *Scandinavian Journal of Immunology*, **7**, 447-52.

Kvist, S., Ostberg, L. & Peterson, P.A. (1978). Reactions and cross-reactions of rabbit anti-H2 antigen serum. *Scandinavian Journal of Immunology*, **7**, 265-76.

Lampson, L.A. & Lenz, R. (1980). Two populations of Ia-like molecules on a human B-cell line. *Journal of Immunology*, **125**, 293-99.

Lane, D.P. & Hoeffler, W.K. (1980). SV40 large T shares an antigenic determinant with a cellular protein of molecular weight 68,000. *Nature*, **288**, 167-70.

Lawman, M.J.P. (1982). Cell-mediated immunity: induction and expression of T-cell function. *Journal of the American Veterinary Medical Association*, **181**, 1022-29.

Ledbetter, J.A., Evans, R.L., Lipinski, M., Cunningham-Rundles, C., Good, R.A. & Herzenberg, L.A. (1981). Evolutionary conservation of surface molecules that distinguish T-lymphocyte helper/inducer and T-cytotoxic/suppressor subpopulations in mouse and man. *Journal of Experimental Medicine*, **153**, 310-23.

Lennox, E.S. (1983). What can we learn about molecular homologies from cross-reactions of monoclonal antibodies? *Transplantation Proceedings*, **15**, 45-47.

Lewin, H.A. & Bernoco, D. (1983). A monomorphic class II-like determinant detected on peripheral blood B lymphocytes of cattle, horses, sheep and swine. *Federation Proceedings*, **42**, 1230.

Lewin, H.A., Davis, W.C. & Bernoco, D. (1985). Monoclonal antibodies that distinguish bovine T and B lymphocytes. *Veterinary Immunology and Immunopathology*, **9**, 87-102.

London, J. & Horton, M.A. (1980). Peanut agglutinin. 5. Thymocyte subpopulations in the mouse studied with peanut agglutinin and Lyt 6.2 antiserum. *Journal of Immunology*, **124**, 1803-7.

London, J., Berih, S. & Bach, J.F. (1978). Peanut agglutinin. 1. New tool for studying T-lymphocyte subpopulations. *Journal of Immunology*, **121**, 438-43.

Lunney, J.K., Osborne, B.A., Sharrow, S.O., Devaux, C., Pierres, M. & Sachs, D.H. (1983). Sharing of Ia antigens between species. 4. Interspecies cross-reactivity of monoclonal antibodies directed against polymorphic mouse Ia determinants. *Journal of Immunology*, **130**, 2786-93.

Mann, D.L. & Sharrow, S.O. (1980). HLA-DRw alloantigens can be detected on peripheral blood T lymphocytes. *Journal of Immunology*, **125**, 1889-96.

McMaster, W.R. & Williams, A.F. (1979). Monoclonal antibodies to Ia antigens from rat thymus: cross-reactions with mouse and human use in purification of rat Ia glycoproteins. *Immunological Reviews,* **47**, 117-37.

Neoh, S.H., Jahoda, D.M., Rowe, D.S. & Voller, A. (1973). Immunoglobulin classes in mammalian species identified by cross-reactivity with antisera to human immunoglobulin. *Immunochemistry,* **10**, 805-13.

Oi, V.T. & Herzenberg, L.A. (1980). Immunoglobulin-producing hybrid cell lines. In *Selected methods in cellular immunology* (eds. B.B. Mishell & S.M. Shiigi), pp. 351-72. San Francisco: W.H.F. Freeman and Co.

O'Keefe, D. & Ashman, L. (1982). Peanut agglutinin: a marker for normal and leukaemic cells of the monocyte lineage. *Clinical & Experimental Immunology,* **48**, 329-38.

Pierres, M., Rebouah, J.P. & Kourilsky, F.M. (1981). Cross-reactions between mouse Ia and human HLA-D/Dr antigens analysed with monoclonal antibodies. *Journal of Immunology,* **126**, 2424-29.

Pierres, M., Mercier, P., Madsen, M., Mawas, C. & Kristensen, T. (1982). Monoclonal mouse anti-I-A^k and anti-I-E^k cross-reacting with HLA-DR supertypic and subtypic determinants rather than classical DR allelic specificities. *Tissue Antigens,* **19**, 289-99.

Pillemer, E. & Weissman, I.L. (1981). A monoclonal antibody that detects a V_k-TEPC 15 idiotypic determinant cross-reactive with Thy 1 determinant. *Journal of Experimental Medicine,* **153**, 1068-79.

Pinder, M., Pearson, T.W. & Roelants, G.E. (1980). The bovine lymphoid system: derivation and partial characterization of monoclonal antibodies against bovine peripheral blood lymphocytes. *Veterinary Immunology & Immunopathology,* **1**, 303-16.

Reinherz, E.L. & Schlossman, S.F. (1981). The characterization and function of human immunoregulatory T-lymphocyte subsets. *Immunology Today,* **2**, 69-74.

Sachs, D.H. (1983). Overview of the MHC in mammalian systems. *Transplantation Proceedings,* **15**, 40-44.

Soloski, M.J., Uhr, J.W., Flaherty, L. & Vitetta, E.S. 1981. Qa2, H-2K and H-2D alloantigens evolved from a common ancestral gene. *Journal of Experimental Medicine,* **153**, 1080-93.

Stallcup, K.C., Springer, T.A. & Mescher, M.F. (1981). Characterization of anti H-2 monoclonal antibody and its use in large-scale antigen purification. *Journal of Immunology,* **127**, 923-30.

Stocker, J.W. & Bernoco, D. (1979). Technique of HLA typing by complement-dependent lympholysis. In *Immunological methods* (eds. I. Lefkovitz & B. Pernis), pp. 217-26. New York: Academic Press.

Stone, W.H. (1981). The bovine lymphocyte antigen (BoLA) system. In *The ruminant immune system* (ed. J.E. Butler), pp. 433-50. New York: Plenum.

Stout, R.D., Yutoku, M., Grossberg, A., Pressman, D. & Herzenberg, L.A. (1975). A surface membrane determinant shared by subpopulations of thymocytes and B lymphocytes. *Journal of Immunology,* **115**, 509-12.

Sutter, L., Bruggen, J. & Sorg, C. (1980). Use of an enzyme-linked immunosorbent assay (ELISA) for screening of hybridoma antibodies against cell-surface antigens. *Journal of Immunological Methods,* **39**, 407-11.

Teillaud, J.L., Crevat, D., Chardon, P., Kalil, J., Goujet-Zalc, C., Mahouy, G., Vaiman, M., Fellous, M. & Pious, D. (1982). Monoclonal antibodies as a tool for phylogenetic studies of major histocompatibility antigens and β2 microglobulin. *Immunogenetics,* **15**, 377-84.

Vaerman, J.P., Kobayashi, K. & Heremans, J. (1974). Precipitin cross-reactions between human and animal J chains. *Advances in Experimental Medicine & Biology,* **45**, 251-55.

The major histocompatibility complex

7

The human lymphocyte antigen system

J. A. SACHS

The human lymphocyte antigen (HLA) region, the major histocompatibility complex in man, consists of a number of genes of immunobiological importance on the short arm of chromosome 6. The products of these genes which are serologically detectable on cell surfaces can be classified into two groups, based on their tissue distribution and structure. The class I antigens controlled by the HLA-A, B and C genes are present on all lymphocytes and are widely distributed on other body tissues. Structurally, each class I molecule consists of a heavy chain of 45,000 daltons controlled by the polymorphic HLA-A, B or C genes, plus the monomorphic B-2 microglobulin chain of 12,000 daltons controlled by a gene on chromosome 15. The class II antigens, the HLA-DR, DQ and DP gene products, are present predominantly on B cells with a more restricted tissue distribution. The class II molecules consist of a heavy α chain of 33,000 daltons and a light β chain of 29,000 daltons. The DR and DQ antigenic products of separate genes in the D region have been extensively characterized and the two polymorphic series can be distinguished serologically. Structurally, the DR α and β chains can be distinguished from the DQ α and β chains, but it is not yet possible to differentiate the various allospecificities by immunochemical methods. Recent advances in recombinant DNA technology have made it possible to study HLA class I and II genes *per se*. Using Southern blotting techniques, class II α and β chain cDNA probes can be used to identify, by restriction fragment lengths polymorphisms (RFLP), genomic DNA from individuals with particular serologically detected class II phenotypes. Initial studies by this method, comparing RFLP of genomic DNA from diabetic patients who have HLA-associated disease-susceptibility genes and healthy individuals with the same class II phenotype, indicate differences that could be related to disease susceptibility.

Introduction

The human lymphocyte antigen (HLA) system refers to the products of a cluster of genes on the short arm of chromosome 6 in man. The HLA gene complex encompasses genes of immunobiological importance controlling, amongst other things, cell-surface determinants, transplantation antigens responsible for organ and tissue graft rejection, complement components, immune responses and a variety of different disease susceptibilities. Homologous regions have been found in all mammals investigated. Early studies in mice helped develop current ideas on the fine structure and function of HLA. Snell (1948) showed that the rejection of tumours grafted from one inbred mouse strain to another was controlled predominantly by a single chromosomal region (H-2) which was of overriding importance in tissue and organ graft rejection. This region was accordingly labeled the major histocompatibility complex (MHC). Other gene products in this region were subsequently identified by a variety of serological and cellular *in vitro* techniques. Comparable histocompatibility genes within HLA are responsible for the rejection of organ and tissue grafts in man.

Thus, the first evidence for the H-2 system was obtained from studies of the rejection of transplanted tumours between different strains of mice. By contrast, human HLA products were first detected on cell surfaces *in vitro*, and this led to the clinical association with transplant rejection. Dausset (1954) observed that antibodies in the serum of some patients agglutinated the white cells but not the red cells after multiple transfusions. Later, similar antibodies were observed reacting with white cells in some pregnant and postpartum women as a result of stimulation by foetal antigens. These alloantibodies are used as reagents to identify the highly polymorphic class I (HLA-A, B and C) and class II (HLA-DR and DC) antigenic systems. In addition to these serologically detectable HLA antigens, a group of lymphocyte-activating determinants are demonstrable on lymphocytes by the primary mixed leucocyte culture technique. These antigens are controlled by the HLA-D gene. Another group of lymphocyte-activating determinants (SB) can be identified by the secondary response of the primed lymphocyte test. The HLA genes and their products detected on cell surfaces will be reviewed in this chapter with particular emphasis on recent findings concerning their molecular biology.

Serologically defined antigens
Class I antigens — HLA-A, B and C

A series of international histocompatibility workshops was initiated in 1964, based on the distribution to participating laboratories of alloantisera reacting with white cells or lymphocytes. Analysis of the resulting data has been a major factor in the rapid advance of knowledge of the HLA system. The second and third histocompatibility workshops (van Rood 1965; Curtoni, Mattiuz & Tosi 1967) established the presence of a number of individual specificities and showed that they were controlled by several closely linked genes. After 1967, the complement-dependent micro-lymphocytotoxicity test, originally devised by Terasaki and McClelland (1964) and modified by Brand and colleagues (1970) replaced the somewhat unreliable and cumbersome agglutination reaction for the detection of these determinants.

Presentations at the fourth histocompatibility workshop (Terasaki 1970) established about 20 antigens which could be divided into two segregating groups controlled by genes, now known as HLA-A and B respectively. This grouping was based on the presence in each member of the control panel of a maximum of four specificities, two from one series (HLA-A) and two from the other (HLA-B), each parent providing one HLA-A and one HLA-B antigen. Segregation analyses of these HLA-A and B gene products in families showed that they were closely linked, i.e. individual HLA-A and B antigens in the parents segregated almost invariably together in the offspring. The combination of genes present on a single segregating chromosome is known as a haplotype (after haploid genotype). Unless a recombination event occurs between the two genes, only four combinations are possible in the offspring of two given parents. The likelihood, therefore, of two siblings being genotypically HLA identical is 1:4, of sharing one haplotype 1:2, and of being completely different 1:4. Since 1970, the number of antigens known to be controlled by the HLA-A and B loci has increased considerably.

The presence of a third gene locus with serologically detectable products on lymphocytes, HLA-C, was mooted during the 1970 histocompatibility workshop (Sandberg et al. 1970). Subsequently, a recombination event was described (Low et al. 1974) which formally separated HLA-C from HLA-B. Many individuals have either one or no detectable C-locus antigens, indicating that a number of antigens are as yet undefined. A full list of the HLA-A, B and C antigens accepted by the 1980 Histocompability Nomenclature Committee is given in Terasaki (1980).

Linkage disequilibrium

Another feature of the HLA-A, B and C antigen systems, in addition to their extreme polymorphism, is the phenomenon of linkage disequilibrium, a complex phrase which, unfortunately, clouds the rather simple concept involved. The recombination frequency between HLA-A and B is approximately 1 to 3%. Without any selective forces influencing these events, it would be expected that HLA-A and B antigens would appear together in a person according to their individual frequencies in the general population. However, some HLA-A and B antigens occur together more frequently than would be expected on the basis of such

random distribution. One in five individuals has HLA-A1, one in five
HLA-B7, and approximately one in 25 has both HLA-A1 and B7.
However, although one in four persons has HLA-A3, HLA-A3 and B7
occur together in about one in six individuals, i.e. a considerably higher
frequency than would be expected (one in 20) if these two antigens were
randomly inherited (Department of Immunology, London Hospital
Medical College unpublished). In European Caucasoid populations, the
HLA haplotyes A1-B8, A2-Bw44 and A3-B7 show highly significant
linkage disequilibrium. All the HLA-C antigens have significant linkage
disequilibrium with HLA-B antigens, whereas less that half the HLA-A
antigens do.

Distribution of HLA-A, B and C antigens in different population groups
Since HLA-A, B and C serology was confined initially to Caucasoid
panels, the 1972 histocompatibility workshop (Dausset & Colombani
1972) was primarily concerned with examining the frequencies of these
antigens in other population groups. Antisera were collected in this
workshop from participating laboratories, redistributed and used to
identify HLA-A, B and C antigens in more than 80 different ethnic
regions. Within each of the primordial groups, i.e. Caucasoid, Mongoloid
and Negroid, there was some variation in the frequencies of the HLA-
A, B and C antigens. However, between the groups there was considerably
more variation in the frequencies of the various individual antigens,
ranging from complete absence in one and a high frequency in another
to equal distribution among all three. Another interesting feature of the
workshop was the definition of new antigens, not found in Caucasoid
populations. An example of this was the identification of a new specificity,
Aw36, present only in Negroids. This specificity was identified with three
anti-HLA-A1 antisera, which in this population were found to have the
additional anti-Aw36 antibody. Surprisingly, two of the serum donors
had had Caucasoid children and the third had been tranfused with blood
from Caucasoid donors (Wolf et al. 1972). Since that workshop,
antibodies with different specificities have been found in many antisera
thought previously to be monospecific. By the same token, antigens
initially thought to be homogeneous in Caucasoids have now been
identified as two (or more) antigens, one with a relatively high frequency
in one population, and another occurring more frequently in another
population.

Linkage disequilibrium exists within all population groups, but the
specificities concerned vary considerably. The haplotypes HLA-A1-B8,
A3-B7 and Aw33-B14 are most prevalent in Caucasoids, Aw30-Bw42
and Aw26-Bw58 in Negroids and Aw33-Bw44 and Aw24-Bw52 in

Mongoloids. Comparison of the antigen frequencies in selected populations are available from the eighth histocompatibility workshop (Terasaki 1980).

Technical aspects of typing for HLA-A, B and C

The micro-lymphocytotoxicity test used for identifying HLA-A, B and C antigens is now standardized, but it is worth emphasizing some of the pitfalls associated with what is accepted by many as a simple routine procedure. The technique developed by Brand and colleagues (1970) is now generally accepted as the method of choice. Briefly, this test involves mixing microquantities of lymphocyte target cells with equal volumes of 60 appropriate HLA-A, B and C antisera for 30 min in specially designed plates with 60 wells, one for each serum. Five µl of rabbit serum is added to the mixture as a source of complement, and 1 h later the cytotoxic effect on the cells is assessed in each well by inverted light microscopy. Viable cells indicate a negative reaction and dead cells a positive reaction, i.e. the corresponding antigen is present on the dead cell. Since the antisera, usually obtained from pregnant or postpartum multiparous women, tend to be multispecific, the definition of antigens often depends on reaction patterns occurring with several sera. In our laboratory we routinely use a minimum of 180 different sera when testing for all the antigens. The identification of the majority of the antigens is relatively straightforward, but some of the less frequent ones and those forming cross-reactive groups require additional antisera which are usually only available in limited supply. It is obviously essential for proper definition that this test be reproducible but this depends on the length of incubation, the quality of the complement and the personnel involved.

Class II antigens — HLA-DR, MB (CD) and MT

Clear cut evidence for the B-cell antigen systems came from the eighth histocompatibility workshop (Bodmer et al. 1978) where several groups reported on series of locally defined antigens found on B, but not T lymphocytes. These serologically detectable antigens were the homologues in humans of the B-cell antigens controlled by the I region of the H2 complex, known as Ia, in mice.

Our knowledge of the B-cell system has developed relatively recently, although B-cell antigens are detected by essentially the same lymphocytotoxic technique used for HLA-A, B and C typing. This has been due to the small proportion of B cells (less than 20%) in the peripheral blood lymphocyte population and to the presence of additional HLA-A, B and C antibodies in the sera used for B-cell antigen testing. These two factors usually obscured B-cell reactions which may have occurred during the routine performance of the micro-lymphocytotoxic test for HLA-A, B and C antigens. The test is, therefore, performed either on a separated B-cell population or by the double immunofluorescence technique which identifies the B-cell population by fluorescent staining (Van Rood, van Leeuwen & Ploem 1976).

The seventh histocompatibility workshop in 1977 standardized the definition of antigens in different laboratories by using a common battery of antisera (Bodmer et al. 1978). Seven individual B-cell determinants were identified which were thought to be controlled by a single locus. As there was a close association of these antigens with HLA-D antigens determined by the mixed leucocyte culture technique described below, they were designated HLA-DR, i.e. related to D, and numbered according to their correlation with the pre-existing HLA-D antigens.

Two additional B-cell antigen systems have been described recently. The MB system was described by Duquesnoy, Marrari and Annen (1979) with three alleles — MB1 usually associated with DR1, 2, 6 and 8; MB2 with DR3 and 7; and MB3 with DR4, 5, 9 and 10. The strong association between the MB and DR specificities could be explained either by strong linkage disequilibrium between the DR and MB genes coding for separate molecules or by a single molecule carrying both sets of determinants. Data from two lines of investigation have unequivocally indicated a strong linkage disequilibrium, even though no recombination has been reported. Tosi and colleagues (1978) established by immunochemical means that molecules carrying DR determinants do not carry DC determinants and *vice versa*. DC is an allelic system now known to be identical to the MB system: DC1 DC3 and DC4 are equivalent to MB1, MB2 and MB3, now designated DQw1, DQw2 and DQw3 by the WHO Nomenclature Committee (Bodmer & Bodmer 1984). Kavathas, Bach and de Mars (1980) established mutant cell lines with deletions of different regions of the MHC. The mutants were induced by low-dose irradiation and different aliquots were treated with particular HLA antisera. Only those cells that had lost the particular determinant against which the alloantisera were directed continued to grow in culture. Heterozygous Epstein-Barr virus-transformed cells were used as the parent, as they express class II antigens and grow in continuous culture. The observation that some mutants lost DR but retained DC determinants indicates that the two antigens are controlled by separate genes.

The MT system (Park et al. 1980) is more complex. The MT1 specificity is recognized by the same alloantisera as DQw1. The WHO Nomenclature Committee (Bodmer & Bodmer 1984) has designated MT2 and MT3 as DRw 52 and DRw 53, with the implication that these antigens are coded for by the DR subregion and are not alleles of a separate region.

Mutant cell lines that had lost both DR and DC (MB) determinants were used to differentiate a series of monoclonal antibodies that either reacted or did not react with these cells (de Mars et al. 1983). The reacting monoclonal antibodies detected a nonDR, nonDC determinant that segregated with HLA. This suggested the presence of at least one more polymorphic locus in the HLA-D region, the products of which can be detected serologically. Immunochemical investigations indicate that this is a class II antigen (Shaw et al. 1982). This antigenic series may well be the same as the SB-lymphocyte activating determinants.

Lymphocyte-activating determinants
Mixed lymphocyte culture — HLA-D typing

When lymphocytes from two randomly selected individuals are mixed in culture, both sets proliferate after a period of a few days and give rise to lymphoblasts (Bain & Lowenstein 1964). The activating determinants of this mixed leucocyte reaction (MLR) are under the control of gene(s) in the HLA region, since cells from siblings known to be HLA-A, B and C identical usually do not stimulate each other (Amos & Bach 1968). The study of the genetic control of the MLR has been considerably helped by irradiating one set of cells, which are then capable of stimulating the untreated cells but are unable to proliferate themselves, resulting in a so-called 'one-way' MLR. Strong stimulation was found to be due to disparity at the HLA-D region of the MHC, which lies close to HLA-B but separated from it by a distance similar to that between HLA-A and B. Recent data from Sachs, Jaraquemade and Festenstein (1980) suggest that the lymphocyte-activating determinants in the HLA-D region may be controlled by more than one gene.

In a one-way culture, cells from individuals who are homozygous for HLA-D do not stimulate, or stimulate only weakly, the cells from responder homozygous or heterozygous individuals who share the HLA-D specificity with the stimulator cells (van den Tweel et al. 1973; Mempel et al. 1973; Dupont et al. 1973; Jorgensen, Lamm & Kissmeyer-Nielsen 1973) and stimulate strongly those who do not possess the antigen. Such homozygous typing cells have now been used to identify individual D-locus specificities. The most reliable source of these cells is offspring of consanguinous marriages who have inherited both haplotypes from one of their grandparents, a 1:16 possibility. Prior to work with human cells, Bradley and colleagues (1972) selected offspring in pigs which were homozygous for the MHC after testing their lymphocytes in MLR with those of their parents. Cells from individuals that failed to stimulate, but responded to parental cells in reciprocal culture could be considered homozygous for the pig equivalent of HLA-D. HLA-D typing has featured prominently in the last three histocompatibility workshops. Homozygous typing cells were collected from participating laboratories and redistributed for use against local panels of random and selected individuals. Eleven D-locus specificities (HLA-Dw1 to 11) were established at the 1975 and 1977 workshops.

A uniform MLR technique was introduced to the 1980 workshop (Dupont et al. 1980) and each experiment was repeated at least once with the same stimulator and responder cells. Although Dw1, Dw2, Dw3, Dw5, Dw8 and Dw10 were well defined, some heterogeneity was found in Dw4, Dw7 and Dw9, and Dw6 could not be established. Three new specificities were postulated. Even after three workshops, problems remain in homozygous typing for HLA-D-locus antigens. The test is long and arduous, requiring at least 6 days under culture conditions. Even after normalization of the data, it is not always possible to determine whether a given combination represents a typing response, especially if repeat cultures give disparate results (Ryder et al. 1975).

Primed lymphocyte test — SB typing

In a primed lymphocyte test, a lymphocyte population from one individual is mixed in culture for about 10 days with irradiated stimulator cells from another individual whose cells differ in respect to HLA-D products. These primed cells respond rapidly in culture when mixed with cells from the original stimulator or cells from individuals who are HLA-D/DR identical with the original stimulator.

Shaw, Johnson and Shearer (1980) identified a series of antigens that segregate independently of HLA-D/DR by sensitizing unrelated individuals with HLA-D/DR compatible stimulators. Based on the response of the various combinations to some and not others after restimulation, the polymorphic SB system was established. Recombination in informative families indicates that the locus is between HLA-D/DR/DC and GLO. The ability of mutant cell lines that have lost HLA-DR and DC to induce SB proliferation indicates that SB is different from the other HLA-D region antigens and may well be the same as an antigenic system which has been detected serologically and immunochemically. This region has recently been designated DP by the WHO Nomenclature Committee (Bodmer & Bodmer 1984).

Molecular biology of HLA genes and their products
The class I (HLA-A, B and C) antigens are expressed on almost

all nucleated cells, whereas the class II antigens have a more restricted tissue distribution, being expressed on B cells, macrophages and some endothelial cells. B cells transformed into lymphoblast cell lines with Epstein-Barr virus retain their class I and II antigens and grow readily in continuous culture, providing sufficient material to study the structure of these antigens. Nonionic detergents can be used to dissolve the plasma membranes and render the cell-surface antigens soluble. Partial purification of the HLA molecules from other cell membrane proteins can be achieved through their ability, as glycoproteins, to bind to lectins immobilized on sepharose and subsequently to be eluted with sugars.

The development of monoclonal antibodies has also contributed to elaborating the molecular biology of HLA products, notwithstanding the paucity of antibodies with allospecificity. Monomorphic antibodies directed against determinants common to all class II DR or DC molecules have greatly facilitated the purification of these molecules. Once the molecules are purified by immunoprecipitation with or adsorption to monoclonal antibodies, their molecular structure and the size of their different components can be assessed by 1-D electrophoresis in SDS polyacrylamide gels. The components can be characterized further by 2-D electrophoresis, which separates the constituent polypeptide chains according to their electrical charge as well as molecular weight. Finally, at the protein level, the aim is to compare molecules by peptide mapping and direct amino-acid sequence determinations.

At the level of the HLA genes, recombinant DNA techniques have made it possible to clone complementary DNA (cDNA) segments that encode the class I and II antigens. The first step in the production of cDNA is the isolation of specific messenger RNA (mRNA) from the total cellular poly A-RNA. By centrifugation of total mRNA on a sucrose gradient, different mRNA populations can be separated. Using *in vitro* cell-free translation or oocyte systems, the proteins synthesized by the different fractions of mRNA can be recognized by reactions with specific monoclonal antibodies. The corresponding mRNA can then be used to synthesize cDNA through the enzyme reverse transcriptase. The cDNA can be ligated to an expression vector, usually a phage or a plasmid, which is then grown in an appropriate bacterial host and cloned. The cloned cDNA can then be isolated and used to obtain pure mRNA or as a probe in hybridization studies. The capability of cDNA to hybridize with homologous genomic DNA allows the cDNA to be used as probes for identifying DNA fragments with homologous sequences in total cellular DNA extracts. The degrees of homology needed for duplex formation depends on the stringency of the washing conditions.

Class I (HLA-A, B and C)

All class I molecules have two chains — a heavy glycoprotein chain of 44,000 daltons, the product of the HLA-A, B and C genes, and a nonglycosylated protein chain of 12,000 daltons, the product of a nonMHC gene located on chromosome 15. The conformation of the heavy chain and its amino-acid sequence was deduced from biochemical analyses of glycoproteins extracted from a cell line homozygous for HLA-A2 and B7 (Strominger 1980). The heavy chain spans the plasma membrane and has three extracellular domains, each with disulphide bonds and oligosaccharides. A transmembrane domain consists of hydrophobic residues, whereas the C-terminal portion of the molecule, which comprises the cytoplasmic domain, contains hydrophilic moieties. A comparison of the amino-acid sequences of A2 and B7 molecules (Orr et al. 1979) has shown that differences are found predominantly in the 65 to 83 residues of the first domain (B2 microglobulin is homogeneous), but as yet the alloantigenic sites have not been identified.

Class II and the D region

The HLA-D region is complex. In addition to controlling the serologically detectable class II HLA-DR, MB and MT antigen systems, other genes in the region also encode lymphocyte-activating determinants of the HLA-D and SB series. These antigens originally defined by the primed lymphocyte test, are thought also to be identifiable serologically by monoclonal antibodies.

Class II antigens in both man and mouse consist of two noncovalently linked polypeptides, the heavy (α) and light (β) chains which span the plasma membrane and end with both C terminals in the cytoplasm. Their molecular masses are approximately 33,000 and 29,000 daltons. The amino-acid sequences predicted from cDNA clones of α and β chains correspond closely to that obtained directly by sequencing the protein product. Both have two external domains of approximately similar size, a transmembrane region and a C-terminal cytoplasmic region. From a complete cDNA clone of an α-DR chain, 119 residues were outside the membrane, equally distributed between the two domains, 23 spanned the membrane and 15 were inside the cell (Kaufman & Strominger 1980; Lee, Trowsdale & Bodmer 1982). One attachment site for an oligosaccharide was available for each extracellular domain. A similar overall structure, but with only one sugar side chain on the first domain, was found for the β chain.

The DR and DC α and β chains are heterogeneous within an individual. Comparing 2-D gel patterns of molecules purified by monoclonal antibodies directed at DR or DC (MB) determinants, structural differences in the β chains are apparent. Peptide mapping and partial amino-acid sequencing also indicate considerable differences within the α and β chains. Further comparisons have been drawn with α and β chains of the IA class II molecules in the H-2 system. IA and IE are two series of B cell antigens in mice, both controlled by genes in the I region. The IA α and β genes and the IE α gene are close together (in as yet undetermined order) in the IA region, whereas the IE β chain is coded for by a gene in the IE region, separated from IA by IB and IJ. Data from protein sequences and hybridization studies with appropriate α- and β-chain cDNA probes, indicate that DR is homologous to the IE product and DC to the IA product (for a review see Hurley, Giles & Capra 1983).

It is of interest to know whether the allospecificities of the class II molecules detected serologically appear as discernable differences at the molecular level. 2-D gel electrophoretic studies on several pairs of cell lines differing for DR indicate that there are differences in the β chains according to the DR specificities.

Experiments involving sequential immunoprecipitation, with two monoclonal antibodies detecting monomorphic determinants on products of the DR region, established the presence of two DR β chains (Accolla, Corte & Tosi 1983). Detergent-soluble preparations were first passaged through an immunoadsorbent column to which one monoclonal antibody was immobilized. The 'depleted' fall-through was then passed through a second immunoadsorbent column to which the second antibody was attached, and 2-D electrophoretic patterns and peptide maps showed the presence of different β chains in eluates from the different columns. When the procedure was reversed, the eluate from the column with the second antibody contained both β chains and removed all activity. These results suggest that the second monoclonal antibody recognized determinants on the two different β chains, whereas only one of the chains was recognized by the first monoclonal antibody (Accolla, Corte & Tosi 1983). This result is consistent with the presence of a second DR β chain. Indeed, a third DR β chain has been detected immunochemically which apparently does not correlate with any known, serologically detectable allospecificity.

At the genomic level, hybridization studies with cDNA probes for α and β chains of the DR, DC and SB genes indicate the presence of one α and three β chains for DR, and two α and two β chains for both DC and SB, thus accounting for the seven class II β chains derived from structural studies (Larhammar et al. 1982; Long et al. 1983). Analysis of cDNA clones encoding β chains of DR and DC confirm the differences in nucleotide sequence between these two series. Similarly, by using the cDNA probes for β chains with several different cell lines, polymorphism has been observed in both series, although at this stage no direct evidence has been obtained relating the polymorphism at the DNA level to allospecificity at the protein level (Wake, Long & Mach 1982).

The interrelationship of the serologically and structurally defined class II products and lymphocyte activating determinants controlled by the HLA-D and SB loci has been of considerable interest and controversy. A group of monoclonal antibodies that react with a mutant lacking DR and DC define a subset of class II antigens, which may identify serologically the SB product (defined by the primed lymphocyte test) encoded by a gene region centromeric to the DR/DC complex (de Mars et al. 1983). In addition, some of these monoclonal antibodies inhibit the generation of proliferation in the primed lymphocyte test. On the other hand, these 'SB' monoclonal antibodies may react with the same or a separate molecule controlled by a closely linked gene. It has not been possible to identify directly the actual molecules responsible for lymphocyte activation in the primed lymphocyte test. These considerations are particularly relevant to the recent studies on the protein structure of the D and DR components of a group of transformed cell lines homozygous for DR4. These cell lines were specifically selected on the basis of their differences for HLA-D specificities as detected by homozygous typing cells. Six D locus antigens are associated with HLA-DR4 — Dw4, Dw10, DB3, LD40, DXT and DKT2. Groner, Watson and Bach (1983), using these HLA-DR4 positive transformed cell lines, were the first to show polymorphism of the β chains associated with D-locus haplotypes by 2-D gel electrophoretic mobility patterns. At the 1983 world immunology congress held in Japan, several groups presented similar results with different DR4 cell lines. These findings suggest that the DR allodeterminants are not the same as those responsible for lymphocyte activation, although they may reside on the same molecule. In the latter event, the molecule with the DR4 determinant would have another epitope whose polymorphic structure would vary according to the HLA-D antigen involved.

Two important recent developments in the practical use of cDNA should further our knowledge of the D-region genes and their products in the immune response and HLA-associated disease susceptibility. Rabourdin-Combe and Mach (1983) transfected mouse L fibroblasts (which do not express MHC products) with genes for the DR α and β chains, DR-associated invariant chain (believed to be necessary for the development of class II molecules), and for the selection marker, the herpes simplex virus thymidine kinase. HLA-DR expression occurred as expected in the transfected cell, indicating that the four different genes were expressed inside the transformed cell, the relevant polypeptide was assembled correctly intracellularly, and the external HLA-DR products were expressed on the cell surface with the correct allospecificity. This ability to select out individual genes of the D (and other) regions should help to clarify the relationship of the different allospecificities to the actual genes and their products. The availability of these isolated HLA gene products, T-cell clones and monoclonal antibodies will afford an excellent opportunity to determine the precise functions of D-region products in the immune response.

Restriction genomic DNA maps are being examined from healthy individuals and from patients suffering from HLA-associated diseases. Owerbach and colleagues (1983) used a β-chain cDNA probe to identify differences in the hybridization patterns between DNA from healthy individuals and diabetic patients after digestion with restriction endonucleases. Significant differences were found in two fragments when patients with HLA-DR4 were compared with healthy controls: the 18 kb Pst1 fragment was present in 36 out of 37 patients but only 14 out of 22 controls, while the 3.7 kb BAMH-1 fragment was found in 10 out of 25 controls but was not found in any of the 29 patients tested. The differences may relate to disease susceptibility, and these observations also open the field for the investigation of other HLA-associated diseases by similar techniques.

Figure 1. Genetic map of HCA.

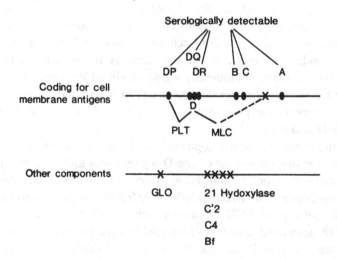

Concluding remarks

At the 1984 histocompatibility workshop, reported in Bodmer and Bodmer (1984), the class II series of antigens was designated HLA-DR, DQ (previously DC or MB) and DP (previously SB). With the addition of newly established antigens, there are now 19 HLA-A, 36 HLA-B, 8 HLA-C, 12 HLA-DR, 3 HLA-DQ and 6 HLA-DP specificities. In addition, the number of HLA-D-locus antigens defined by the mixed lymphocyte culture technique with homozygous typing cell reagents has increased to 17.

Figure 1 shows the relative positions on the short arm of chromosome 6 of the genes controlling the different series of antigens. Segregation of HLA in family studies provided a recombination frequency of 1.3% between HLA-A and HLA-C, whereas only one recombination between HLA-C and HLA-B has been reported. The frequency of recombination between HLA-B and D and HLA-D/DR and DP is similar to the frequency of recombination between HLA-A and HLA-C. There have been no definitive reports of recombination between the DR and DQ series of antigens. However, Bodmer and Bodmer (1984), on the basis of analyses of DNA clones of the D region, suggest the order and number of genes in this region as shown in Figure 2. It is apparent that the DR region consists of genes coding for one α chain and three β chains, of which one β chain is responsible for allospecificity associated with DR and two β chains are responsible for the specificities associated with DRw52 and DRw53, previously MT2 and MT3. The DQ region encodes two α and two β chains, and one of the β chains is responsible for DQ allospecificity. Similarly, the DP region has two α and two β chains, although it is not known which chain is responsible for DP specificities.

The relationship of the HLA-D series, typed by homozygous typing cells, to the serologically detectable HLA-DR series has not been established. Nor can the DP series detected by the primed lymphocyte test be considered identical to the serologically detectable class II series of antigens in the DP region. Transfection of class II genes into cells lacking these antigens can answer the question as to whether class II

Figure 2. Genetic map of the HCA-D region.

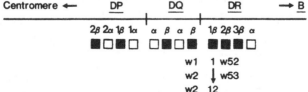

gene products detected serologically on such cells are capable of eliciting proliferative responses correlating with HLA-D or DP specificities.

Hitherto, the more frequent the recombination event between two sets of the genes in family segregation studies, the further apart they have been considered to be on the chromosome. However, Bodmer and Bodmer (1984) now suggest, based on chromosome 'walking' studies, that the apparent distance may be due to 'hot spots', i.e. sites where recombination events occur with particular frequency. The points between HLA-A and C, HLA-B and D and HLA-DR and DQ-DP may represent such regions.

References

Accolla, R.S., Corte, G. & Tosi, R. (1983). Classification of anti-class II monoclonal antibodies based on epitope and molecular assignment. *Disease Markers*, **2**, 39-46.

Amos, D.B. & Bach, F.B. (1968). Phenotypic expression of the main histocompatibility locus in man (HL-A): leucocyte antigens and mixed lymphocyte culture reactivity. *Journal of Experimental Medicine*, **128**, 623.

Bain, B. & Lowenstein, L. (1964). Genetic studies on the mixed lymphocyte reaction. *Science*, **145**, 1315-16.

Bodmer, J.G. & Bodmer, W. (1984). Histocompatibility 1984. *Immunology Today*, **5**, 251-54.

Bodmer, W.F., Batchelor, J.R., Bodmer, J.G., Festenstein, H. & Morris, P.J. eds. (1978). *Histocompatibility testing 1977*, 612pp. Copenhagen: Munksgaard.

Bradley, B.A., Edwards, J.M., Dunn, D.C. & Calne, R.Y. (1972). Quantitation of mixed lymphocyte reaction by gene dosage phenomenon. *Nature*, **240**, 54-56.

Brand, D.L., Ray, J.R., Hare, D.B., Kayhoe, D.E. & McClelland, J.D. (1970). Preliminary trials towards standardization of leucocyte typing. In *Histocompatibility testing 1970* (ed. P.I. Terasaki), pp. 357-67. Copenhagen: Munksgaard.

Curtoni, E.S., Mattiuz, P.L. & Tosi, R.M. eds. (1967). *Histocompatibility testing 1967*. Copenhagen: Munksgaard.

Dausset, J. (1954). Leucoagglutinin. 4. Leucoagglutinins and blood transfusion. *Vox Sanguinis*, **4**, 190-98.

Dausset, J. & Colombani, J. eds. (1972). *Histocompatibility testing 1972*. Copenhagen: Munksgaard.

de Mars, R., Chang, C., Marrari, M., Duquesnoy, R.J., Woreen, H., Segall, M. & Bach, F. (1983). Dissociation in expression of MB1/MT1 and DR1 alloantigens in mutants of a lymphoblastoid cell line. *Journal of Immunology*, **131**, 1318.

Dupont, B., Braun, D.W., Yunis, E.J. & Carpenter, C.B. (1980). HLA-D by cellular typing. In *Histocompatibility testing 1980* (ed. P.I. Terasaki), pp. 229-67. Los Angeles: UCLA Tissue Typing Laboratory.

Dupont, B., Jersild, C., Hansen, G., Staub-Nielsen, L., Thomsen, M. & Svejgaard, A (1973). Typing for MLC determinants by means of LD-homozygous and LD-heterologous test cells. *Transplantation Proceedings*, **5**, 1543-49.

Duquesnoy, R.J., Marrari, M. & Annen, I.C. (1979). Identification of an HLA-DR associated system of B-cell alloantigens. *Transplantation Proceedings*, **11**, 1757-60.

Groner, J.P., Watson, A.J. & Bach, F.H. (1983). Dw/LD-related molecular polymorphism of the DR4 β chain. *Journal of Experimental Medicine*, **157**, 1687-91.

Hurley, C.K., Giles, R.C. & Capra, J.D. (1983). The human MHC: evidence for multiple HLA-D region genes. *Immunology Today*, **4**, 219-26.

Jorgensen, F., Lamm, L.U. & Kissmeyer-Nielsen, F. (1973). Mixed lymphocyte cultures with inbred individuals: an approach to MLC typing. *Tissue Antigens*, **3**, 323-39.

Kaufman, J.F. & Strominger, J.L. (1980). HLA-DR light chain has a polymorphic N-terminal region and a conserved immunoglobulin-like C-terminal region. *Nature*, **297**, 694-97.

Kavathas, P., Bach, F.H. & de Mars, R. (1980). γ-ray-induced loss of expression of HLA and glyoxalase I alleles in lymphoblastoid cells. *Proceedings of the National Academy of Sciences of the USA*, **77**, 4251-55.

Larhammer, D., Schenning, L., Gustafsson. K., Wiman, K., Claesson, L., Rask, L. & Peterson, P.A. (1982). Complete amino-acid sequence of an HLA-DR antigen-like β chain as predicted from the nucleotide sequence: similarities with immunoglobulins and HLA-A, B and C antigens. *Proceedings of the National Academy of Sciences of the USA*, **79**, 3687-91.

Lee, J.S., Trowsdale, J. & Bodmer, W.F. (1982). cDNA clones coding for the heavy chain of human HLA-DR antigen. *Proceedings of the National Academy of Sciences of the USA*, **79**, 545-49.

Long, E.O., Wake, C.T., Gorski, J. & Mach, B. (1983). Complete sequence of an HLA-DR β-chain deduced from a cDNA clone and identification of multiple nonallelic DR β chain genes. *EMBO Journal*, **2**, 389-94.

Low, B., Messeter, L., Mansson, S. & Lindholm, T. (1974). Crossing-over between SD-2 (four) and SD-3 (AJ) loci of the human major histocompatibility chromosomal region. *Tissue Antigens*, **4**, 405.

Mempel, W., Grosse-Wilde, H., Baumman, P., Netzel, B., Steinbauer-Rosenthal, I., Scholz, S., Bertrams, J. & Albert, E.D. (1973). Population genetics of the MLC response: typing for MLC determinants using homozygous and heterozygous reference cells. *Transplantation Proceedings*, **5**, 401-8.

Orr, H.T., Lopez de Castro, J.A., Parham, P., Ploegh, H.L. & Strominger, J.L. (1979). Comparison of amino-acid sequences of two human histocompatibility antigens, HLA-A2 and HLA-B7: location of putative antigen sites. *Proceedings of the National Academy of Sciences of the USA*, **76**, 4395-99.

Owerbach, D., Lernmark, A., Platz, P., Ryder, L.P., Rask, L., Peterson, P.A. & Ludvigsson, J. (1983). HLA-D region β chain DNA endonuclease fragments differ between HLA-DR identical healthy and insulin-dependent diabetic individuals. *Nature*, **303**, 815-17.

Park, M.S., Terasaki, P.I., Nakata, S. & Aoki, D. (1980). Supertypic DR groups MT1, MT2 and MT3. In *Histocompatibility testing 1980* (ed. P.I. Terasaki), pp. 854-57. Los Angeles: UCLA Press.

Rabourdin-Combe, C. & Mach, B. (1983). Expression of HLA-DR antigens at the surface of mouse L cells cotransfected with cloned human genes. *Nature*, **303**, 670-74.

Ryder, L.P., Thomson, M., Platz, P. & Svejgaard A. (1975). Data reduction in LD typing. In *Histocompatibility testing 1975* (ed. F. Kissmeyer-Nielsen), pp. 557-62. Copenhagen: Munksgaard.

Sachs, J.A., Jaraquemade, A. & Festenstein, H. (1980). Intra-HLA-D-region recombinant maps HLA-Dr between HLA-B and HLA-D. *Tissue Antigens*, **17**, 43-56.

Sandberg, L., Thorsby, E., Kissmeyer-Nilsen, F. & Lindholm, A. (1970). Evidence of a third sublocus within the HL-A chromosomal region. In *Histocompatibility testing 1970* (ed. P.I. Terasaki), pp. 165-69. Copenhagen: Munksgaard.

Shaw S., Johnson A. & Shearer G.M. (1980). A new segregant series of HLA-linked B-cell alloantigens which stimulate secondary proliferative and cytotoxic responses. *Journal of Experimental Medicine,* **152**, 565-80.

Shaw, S., de Mars, R., Schlossman, S.F., Smith, P.L., Lampson, L.A. & Nadler, L.M. (1982). Serologic identification of the human secondary B-cell antigens: correlation between function, genetics and structure. *Journal of Experimental Medicine,* **156**, 731-43.

Snell, G.D. (1948). Methods for the study of histocompatibility genes. *Journal of Genetics,* **49**, 87-108.

Strominger, J.L. (1980). Structure of products of the major histocompatibility complex in man and mouse. In *Immunology 80* (eds. M. Fougereau & J. Dausset), Progress in Immunology 4, pp. 541-54. New York: Academic Press.

Terasaki, P.I. ed. (1970). *Histocompatibility testing 1970.* Copenhagen: Munksgaard.

Terasaki, P.I. (1980). In*Histocompatibility testing 1980,* (ed. P.I. Terasaki) pp. 955-93. Los Angeles: UCLA Tissue Typing Laboratory.

Terasaki, P.I. & McClelland, J.D. (1964). Microdroplet assay of human serum cytotoxine. *Nature,* **204**, 998-1000.

Tosi, R., Tanigaki, N., Centis, D., Ferrara, G.B. & Pressman, D. (1978). Immunological dissection of human Ia molecules. *Journal of Experimental Medicine,* **148**, 1592-1611.

van den Tweel, J.G., van Blusse, O., van oud Albas, A., Keuning, J.J., Goulmy, E., Termijtelen, A., Bach, M.L. & van Rood, J.J. (1973). Typing for MLC (LD). 1. Lymphocytes from cousin marriages' offspring as typing cells. *Transplantation Proceedings,* **5**, 1535-38.

van Rood, J.J. ed. (1965). *Histocompatibility testing 1965 .* Copenhagen: Munksgaard.

van Rood, J.J., van Leeuwen, A. & Ploem, J.S. (1976). Simultaneous detection of two cell populations by two-colour fluorescence and application to the recognition of B-cell determinants. *Nature,* **262**, 795-97.

Wake, C.T., Long, E.O. & Mach, B. (1982). Allelic polymorphism and complexity of genes for HLA-DR β chain: direct analysis by DNA-DNA hybridization. *Nature,* **300**, 372-73.

Wolf, E., Sachs, J.A., Oliver, R.T.D., Burke, J., Adams, E., Rondiak, G. & Festenstein, H. (1972). An HL-A variant in a Negroid population from Zambia: a new antigen cross-reactivity? In *Immunobiological standardization international symposium on HL-A reagents,* p. 242. Basel: Karger.

8

The bovine major histocompatibility complex

R.L. SPOONER

Serious studies of the bovine major histocompatibility complex (MHC) were initiated in the mid 1970s. Antibodies which lysed lymphocytes in a complement-mediated cytotoxicity assay were found in the sera of parous cows, in cattle inoculated with whole blood or lymphocytes and in cattle subject to skin grafting. An international workshop was conducted in 1978 which compared sera from nine laboratories. The sera were all to a limited number of specificities, of which 11 were agreed. These were called BoLA (bovine lymphocyte antigens) and given 'w' workshop prefixes. At the second international workshop in 1980, six more specificities were agreed. All were of the class I type, and population and family studies suggested that they were controlled by alleles at a single locus. Bovine class II antigens have been demonstrated by biochemical techniques but genetic studies of these antigens are at a very early stage. There are large differences in the frequencies of BoLA antigens between *Bos taurus* and *Bos indicus* cattle and between breeds. However, most of the workshop specificities defined in European taurine cattle are seen in both East and West Africa. BoLA antigens affect graft rejection, the immune response to human serum albumin and the hapten (T,G)-A-L and haemolytic complement levels. One blood group — M — which is closely linked to BoLA w16 is possibly related to mastitis and trypanosomiasis susceptibility. *Theileria parva-* and *T. annulata*-infected lymphoblastoid cell lines maintain BoLA w specificities. Cytolytic T lymphocytes generated in calves infected with *T. annulata* and *T. parva* show MHC restriction of their cytolytic function, and alloreactive cytolytic T lymphocytes recognize BoLA class I antigens as targets.

Introduction

The first studies of lymphocyte antigens in cattle began in laboratories whose primary interest was red-cell antigens. The presence of lymphocytotoxins in animals immunized with whole blood was reported by Borovska and Demant (1967). They suggested that the antigens detected by their sera were linked to the S system, which is one of a number of gene clusters controlling red-cell antigens. These clusters, which exhibit gene duplication and linkage disequilibrium (Grosclaude, Guerin & Houlier 1979), resemble the MHC in man and mouse. They were considered as possible candidates for the bovine MHC but this was subsequently shown not to be the case.

McGary and Stone (1970) described anti-lymphocyte antibodies in immunized and parous cattle and Iha and colleagues (1973) showed cross-reactivity between cattle anti-lymphocyte sera and some human lymphocyte antigens, suggesting a possible relationship between human and bovine lymphocyte antigens.

In 1973, Hruban and Simon presented evidence for the presence of the J blood group antigen on bovine lymphocytes. The J antigen is of similar structure to the human A antigen and is adsorbed on to red cells from the serum. Naturally occurring anti-J antibodies are found in serum. One of the J reagents used by Hruban and Simon reacted with red cells and lymphocytes from the same animal and, on the basis of this and absorption studies, they concluded that the J antigen was present on lymphocytes. Other laboratories working on bovine red-cell antigens tested sera produced by whole-blood immunization against lymphocytes and demonstrated lympholysis. Two reports showed no relationship between lymphocyte antigens and red-cell antigens (Ostrand-Rosenberg & Stormont 1974; Folger & Hines 1976).

Caldwell and colleagues (1977) presented evidence that sera from parous cattle were lymphocytotoxic. They also detected polymorphic antigenic determinants on the surface of lymphocytes. However, as their sera were multispecific and as they had only typed cows and their calves but not sires, it was not possible to determine whether the antigens they detected were controlled by single or multiple loci.

The first extensive genetic data on bovine lymphocyte antigens were described by Spooner and colleagues (1978) and Amorena and Stone (1978). We studied the inheritance of 17 antigens in 480 cattle families in France and Great Britain using 70 antisera (Spooner and colleagues 1978). The antisera were all produced by immunization between dam and calf. Of the 480 cattle families screened, only 249 were used in which the bull was positive and the dam negative for the antigen being studied. Using bulls that were heterozygous for one or other of the specificities, we found that all were inherited in a simple Mendelian ratio. The initial analysis suggested that the factors detected were the products of closely linked loci. However, subsequent studies with the same reagents showed that some of the factors were multispecific, and thus the suggestion that control was through linked loci may not have been correct (R.L. Spooner & R.A. Oliver unpublished). Spooner and colleagues (1978) also showed that there was no evidence for any relationship between lymphocyte and red-cell antigens of the B, C, FV and S systems; moreover, in only one of the sera produced by skin grafting was there any evidence of antibodies to red cells.

Concurrently with the study described above, Amorena and Stone (1978) reported the results of studies using sera from parous cattle. They defined 11 lymphocyte antigen specificities, of which 10 showed simple Mendelian segregation ratios when studied in 60 sire families. These appeared to be controlled by alleles at a single locus. Marked variation was observed between breeds in the frequency of the various specificities, although with some breeds the number of sires used was small. An important finding was that skin grafts between individuals matched for the specificities detected survived longer than skin grafts between mismatched animals. This control of graft acceptance, or histocompatibility, supports the concept that the antigens detected were in fact products of the bovine MHC. Following these initial reports, research on the bovine MHC has developed rapidly.

Class I antigens

Most research up to now has focused on class I antigens. It has been agreed to designate these antigens as 'bovine lymphocyte antigens' (BoLA). This general term will also encompass class II antigens as they are identified.

Method of testing

The test procedures used for detecting bovine MHC antigens are based on those used for detecting MHC antigens on human lymphocytes. Most laboratories have used the micro-lymphocytotoxicity test plates described by Terasaki and McClelland (1964). Lymphocytes are separated on a Hypaque-Ficoll gradient. Some laboratories use the same specific gravity (SG 1.076) as is used for separation of human lymphocytes, but we have found that a slightly lower density of 1.069 gives fewer contaminating red cells and polymorphonuclear leucocytes.

Lymphocytes at a concentration of 2×10^6 per ml are incubated in 1 µl of antiserum for 15 to 30 min. Complement, usually from rabbits, is added and the test incubated for a further 60 to 90 min. Some laboratories incubate at 37°C, others at room temperature. Some, particularly those using undiluted parous sera, use up to 5 µl of undiluted rabbit serum as a source of complement. We are using sera produced by alloimmunization, which are of higher titre than parous sera and can be used diluted. The anti-complementary effects associated with the use of undiluted serum are not a problem, and thus we normally use 1 μl of diluted serum and 1 μl of complement diluted at a ratio of 1:2. Using more diluted serum appears to allow greater swelling of the lysed cells and hence increases the contrast between live and dead cells.

Production of antisera

The majority of antisera used in the early studies were collected from parous cattle; only our studies (Spooner et al. 1978) concentrated on sera arising from skin grafting. We have compared the efficiency of producing anti-lymphocyte sera in parous cattle and in cattle following skin grafting (Spooner, Millar & Oliver 1979). Only 14% of parous cows were found to have detectable circulating antibodies (i.e. 1:2 or above), whereas after one skin graft 62%, and after two skin grafts 95%, of animals had high titres of anti-lymphocyte antibodies, with 40% having titres in excess of 1:1000. However, the cows in this study were not bled close to calving, and it has been shown in other studies (Newman & Hines 1979; Stear 1980) that transient low titre antibodies are produced close to parturition in around 30% of heifers and ewes and that the production of anti-lymphocyte antibodies increases with further pregnancies (Newman & Hines 1980). Not surprisingly, the antisera become more complex in these older animals (Hines & Newman 1981).

We used small full-thickness skin grafts at the base of the tail (Spooner et al. 1978). A simpler but equally effective technique described by Pringnitz and colleagues (1982) involves removing a small piece of skin from the base of the ear and inserting it subcutaneously on the neck of the recipient. Whatever grafting technique is used, if an antibody is produced its peak level in serum is usually attained by about 3 weeks. Thus, grafting followed by bleeding 3 weeks later provides a simple way of obtaining antisera.

The other method used to produce antisera is immunization with purified lymphocytes (Spooner et al. 1978; Amorena & Stone 1980; 1982). In general, the response is less reproducible than with skin grafting. Newman and Stear (1983) have studied the development of antibodies following both skin grafting and immunization with lymphocytes and have confirmed the greater variability following immunization with lymphocytes. As expected, they also showed that IgM antibodies are produced first and IgG antibodies later. The problems with lymphocyte immunization are mainly logistical: to purify lymphocytes from large volumes of blood and inject them into cattle takes considerably longer than skin grafting. Moreover, several lymphocyte immunizations are required, whereas a single skin graft is often adequate.

There is no evidence that the specificity of the antibodies produced by these various methods differs significantly. This is particularly supported by the international comparison tests discussed below. Where skin grafting or lymphocyte immunization is carried out between dam and offspring, the potential diversity of antibodies produced is reduced because the dam will only produce antibodies against the antigens received by her calf from the sire, and the calf will only produce antibodies to the antigens in the dam that have not been inherited. It is now possible to perform elective grafts between unrelated animals with known BoLA types aimed at producing antisera to particular specificities.

International comparison tests
    Researchers in the field of bovine lymphocyte antigens instituted a series of international comparison tests in 1978. The first international BoLA workshop took place in Edinburgh and involved the testing of sera from contributing laboratories against a panel of animals of various breeds (Spooner et al. 1979). These sera were obtained mainly from parous cattle or from cattle immunized with whole blood and a few were produced by skin grafting. Some of the sera were absorbed with lymphocytes but the majority were not. It was clear from this comparison test that antisera from several laboratories, regardless of the way in which they were produced, recognized the same set of antigenic specificities. It was agreed at the meeting that only those specificities which were detected by sera from at least two laboratories would be included as defining sera for a particular specificity. On this basis, 11 specificities were defined and designated w1, w2, w3 and so forth. The finding that no animals were positive for more than two of these specificities was consistent with the concept that they were controlled by alleles at a single locus.
    The second international comparison test was organized by laboratories in Edinburgh and Wisconsin, and the results were discussed at the second BoLA workshop, held in Wageningen in 1980 (Proceedings of the second international bovine lymphocyte antigen (BoLA) workshop 1982). In this comparison test, cells from five laboratories in the USA, one in Edinburgh and one in France were distributed to participating laboratories for testing. The criteria for the acceptance of a specificity differed slightly from those used in the first workshop: to be accepted, a specificity had to be defined by at least two sera on cells from at least two laboratories. All of the specificities defined at the first workshop were detected in the second workshop. However, w1 was not confirmed, as it was only detected by Edinburgh sera on Edinburgh cells. Six new specificities were also defined, two of which were subgroups of w6 — namely w6.1 and w6.2.

Regional workshops were held in North America and in Europe in 1982 and the results of the European workshop were published *(European workshop on bovine histocompatibility* 1983). A number of new specificities were identified and one of these, Eu 27, has since been shown in full sib families to be unlinked to the BoLA workshop specificities (R.L. Spooner & R.A. Oliver unpublished). Specificities unlinked to BoLA have been reported by D. Bernoco in the 1982 North American workshop and by Stear and Bell (1984); the latter (CA 19) was shown to be the J red-cell antigen.

Subgroups

Operationally monospecific alloantisera have been made in man and mouse, each recognizing a determinant on the products of multiple alleles of the respective histocompatibility systems. Such antisera define public specificities. Other antisera recognize so-called private specificities, i.e. those detected on the products of single alleles within the species. In the human system for example, private specificities HLA-A25 and HLA-A26 are 'splits' within the HLA-A10 public specificity (Duquesnoy & Schindler 1976).

BoLA w6 represents an example of this in cattle (Spooner & Morgan 1981). In this case, the presence of the w6 subgroup specificities is dependent on the presence of w6 itself. W6, therefore, has a higher frequency in the population than any of the subgroups and in consequence is referred to as the w6 'broad' specificity. A monoclonal antibody reacting with broad w6 has been reported (Spooner & Pinder 1983).

In the 1978 workshop, evidence was found for two subgroups of w6 (Spooner et al. 1979). We have now studied this in more detail and are able to detect at least four subgroups. One interpretation of the w6 serology is that the BoLA w6 molecules have a common antigenic site plus sites which are unique to particular subgroups (Spooner & Morgan 1981). An alternative possibility is that the w6 broad and subgroup determinants are on different molecules. According to this hypothesis, a very high degree of linkage disequilibrium would occur between w6 and the subgroup specificities.

The broad w6 specificity was detected first and the subgroups later, hence the nomenclature. More recently an operationally monospecific antiserum, ED85, has been described which reacts with all w4, w7 and w10 cells, as well as some cells negative for these three specificities. If ED85 had been found before w4, w7 and w10, these latter specificities would have been considered as subgroups of ED85.

Table 1. Frequencies (%) of workshop class I MHC specificities in six European breeds of cattle. Gene frequencies are estimated using the method of maximum likelihood and assuming Hardy-Weinberg equilibrium. From Oliver et al. 1981.

	Charolais	Simmental	Hereford	Friesian	Friesian	Ayrshire	Jersey
Number	18	18	42	129	84	22	55
Sex	M	M	M	M	F	F	F
BoLA specificity							
w1	–	–	1.2	0.8	–	–	–
w2	–	2.8	–	–	–	9.4	–
w5	–	–	9.8	–	0.6	–	1.8
w6	11.9	2.8	3.6	16.4	15.7	9.7	46.0
w7	–	5.8	12.4	0.4	1.8	27.8	0.9
w8	–	11.9	–	23.9	25.9	–	2.8
w9	–	–	22.1	0.8	–	2.4	–
w10	5.8	2.8	2.4	15.4	9.9	16.9	2.8
w11	–	8.7	1.2	10.6	14.3	2.3	–
w12	5.8	–	2.5	1.9	10.7	–	6.7
w13	–	2.8	–	0.4	–	–	1.8
w16	–	–	–	1.2	0.6	–	8.4
w20	–	–	2.4	3.2	4.7	–	–
Eu12	2.8	2.8	3.7	7.6	5.6	–	–
Null	73.7	59.5	38.8	17.4	10.2	31.7	28.7

Table 2. Frequencies of workshop class I MHC specificities in three West African breeds of cattle.

	Zebu	Baoulé	N'Dama
Number	30	70	50
BoLA specificity			
w1	-	4	-
w2	-	-	-
w3	-	-	-
w4	-	-	-
w5	-	-	-
w6	12	11	1
w7	8	11	-
w8	-	3	-
w9	4	6	6
w10	8	13	18
w11	2	4	2
w16	-	1	-
Null	55	41	70

Breed differences in the frequency of BoLA antigens

Marked differences in lymphocyte antigen frequencies between breeds were reported by Amorena and Stone (1978) and Caldwell (1979). The frequencies of internationally recognized workshop specificities in British breeds of cattle were reported by Oliver and colleagues (1981) and again major differences in frequency were seen, as shown in Table 1. When one moves from European cattle to breeds in Africa, then the differences become more marked, as might be expected. In collaboration with C. Hoste and R. Queval, we studied the frequencies of BoLA workshop specificities in an unrelated group of 70 adult Baoulé (*Bos taurus*) cattle which had been collected from many parts of Ivory Coast, with no more than four animals from any one village; a herd of 50 N'Damas (*Bos taurus*) cattle which had been kept together for some years; and 30 Zebu (*Bos indicus*) cattle of unknown relationship. The results are summarized in Table 2. The N'Dama population was notable for a high frequency of the w10 specificity. Only three other workshop specificities were detected. The Baoulé population was more representative of the breed as it occurs throughout northern Ivory Coast. In these, w1 was clearly defined and noticeably more frequent than in Europe, w2 and w3 were not detected and w4, w7 and w10 were definable, although the reactivities of the multiple sera defining them were not as similar among individual animals as in populations of European cattle. The sera defining w6 gave similar reactivities in the Baoulé as in European breeds and a similar proportion of animals gave positive reactions.

Table 3. Frequencies (%) of workshop class I MHC specificities in 351 European (*Bos taurus*) and 759 East African (*Bos indicus*) cattle. The high incidence of wz and w16 in the European group arises from the presence of many Norwegian red-and-white cattle in this sample. These specificities are at much lower frequencies in most other taurine breeds.

	European	East African
Number	351	759
BoLA specificity		
w2	18.1	2.7
w6	17.5	13.2
w6.2	9.1	0.4
w8	10.3	3.5
w9	6.8	26.4
w10	16.5	4.7
w11	8.2	0.3
w16	18.0	0.5

S. Kemp and A. Teale, working in Kenya, have studied the frequency of the workshop specificities in both *Bos taurus* and *Bos indicus* animals in East Africa. Using the sera prepared in Edinburgh and additional sera prepared in *B. indicus* animals in Kenya, they have shown variations in frequencies of some specificities between the *indicus* and European *taurus* populations: w2, w8, w10, w11 and w16 occur at significantly lower frequencies in the *indicus* cattle than in the *taurus* cattle, while w16 varies widely among different *B. taurus* breeds (Table 3). The null allele (undetected) occurred much more frequently in *indicus* than in *taurus* cattle.

Genetics

The family data reported by Spooner and colleagues (1978) and Amorena and Stone (1978; 1980), as well as further data reported by Caldwell (1979), indicated that the antigens detected were inherited in a simple Mendelian fashion and were controlled either by a single locus or a group of very closely linked loci. Not all of these specificities correspond with the specificities agreed in the first and second international workshops. However, family studies carried out more recently with sera defining all of the internationally agreed specificities have confirmed that these fit a single locus model of control (Oliver et al. 1981). Although there have been claims by some laboratories, no unequivocal demonstration of more than one class I BoLA locus has been published. Caution is required here. The experience with the horse MHC — ELA — is worth noting: publications reported more than one

Table 4. Association of BoLA w16 with blood group M in British and Norwegian cattle.

	Number of cattle tested	Number positive for BoLA w16		
		M positive	M negative	Total
Britain	774	18	1	19
Norway	576	190	15	205

ELA locus, but these were not supported by data from two international workshops (Bull 1983; Bailey et al. 1984).

Linkage with blood groups

Borovska and Demant (1967) indicated a possible relationship of lymphocyte antigens to the S blood group system. Hruban and Simon (1973) also described a relationship with the J blood group, whereas others found no relationship with red-cell antigens (Ostrand-Rosenberg & Stormont 1974; Folger & Hines 1976; Spooner et al. 1978). More recently, however, a linkage has been demonstrated between the M blood group locus and BoLA (Leveziel 1983). Our own studies of British and Norwegian cattle (Table 4) indicate that the M blood group and BoLA w16 are either the same or are very closely linked (R.L. Spooner, O. Lie & R.A. Oliver unpublished). This would also appear to be the case in Australian cattle (M.J. Stear personal communication).

Cross-reactivity with other species

Iha and colleagues (1973) have shown reactivity between bovine anti-lymphocyte sera and human lymphocytes. There was some indication of activity in some of the bovine sera correlating with HLA types. Defining sera for all of the BoLA workshop specificities have been tested on human lymphocytes, but no clearcut relationships have been shown between these and any HLA specificity (R.L. Spooner & H. Dick unpublished; R.L. Spooner & J.G. Bodmer unpublished). Brodsky, Stone and Parham (1981), Chardon and colleagues (1983) and Spooner and Ferrone (1984) have shown crosss-reactions between monoclonal antibodies to HLA class I and II products and bovine cells. Studies with alloantisera to sheep lymphocytes and with bovine sera have shown that when unabsorbed sera from cattle are tested on sheep lymphocytes, or *vice versa*, there is a considerable amount of reactivity not related to any known BoLA or OLA specificity. When the bovine sera are absorbed with bovine lymphocytes, or the ovine sera absorbed with ovine lymphocytes, the majority of the cross-species reactivity disappears, suggesting that the cross-reactive antibodies are directed against determinants common to a number of different BoLA or OLA specificities and that these are the ones that are maintained across species (D. McBride, M.J. Stear & R.L. Spooner unpublished).

Biochemistry

Using monoclonal antibodies to human class I antigens, it has been possible to identify glycoproteins similar in structure to human and mouse MHC products. An anti-class I monoclonal antibody precipitated two polypeptides of 44,000 and 12,000 daltons (Hoang-Xuan et al. 1982b). This is discusssed in detail by D. Levy et al. (this volume).

Class II antigens

Proliferation in mixed leucocyte culture (MLC) between leucocytes from different cattle was described by Usinger, Curie-Cohen and Stone (1977). They reported one-way and two-way MLC responses between related cattle and, on the basis of the hierarchy of the responsiveness, suggested that control was exerted by at least two loci. Spooner and colleagues (1978) studied MLC reactions between four sets of three full sibs and their respective dams. Within each set, at least one pair of animals did not cross-stimulate in MLC. In one set, none of the offspring showed cross-stimulation, but they were all strongly stimulated by their mother's cells. In this set, it was also possible to study the serological reactions: these showed that the three offspring had received the same haplotypes from their sire and probably from their dam also. These findings suggest that the genes coding for determinants which stimulate proliferation in MLC are closely linked to class I MHC loci. Usinger and colleagues (1981) later extended these findings to studies of full sib families produced by embryo transfer and confirmed that the major control of mixed leucocyte responsiveness is closely linked to the class I locus or loci. Monoclonal antibodies to human class II antigens have been shown to precipitate products of 27,000 and 34,000 daltons from bovine lymphocyte membranes (Hoang-Xuan et al. 1982a).

The serological definition of bovine class II antigens is at a very early stage. Preliminary studies were reported by Cwik, Cichon and Schmid (1978). Then Newman, Adams and Brandon (1982), using two-colour fluorescence micro-lymphocytotoxicity, detected polymorphic antigenic determinants which were expressed primarily on B lymphocytes and not on platelets or red cells. This is typical of the distribution of class II antigens in other species. Using a panning technique for separating T and B lymphocytes and a lymphocytotoxicity test, we have also studied B cell-specific antibodies in bovine alloantisera (R.L. Spooner, P. Millar & A.G. Morgan unpublished). Sera have been identified which will lyse B cells from individual animals, but not T cells. Monoclonal antibodies reacting with monomorphic determinants on human class II antigens react with bovine B cells, but not with bovine T cells. However, their pattern of reactivity is broad, with at least 70% of the animals tested being positive for any particular monoclonal antibody. In some cases, all animals are positive (Spooner & Ferrone, 1984). A monoclonal antibody reacting with bovine class II antigens has also been reported (Letesson et al. 1983).

Asssociations of BoLA type with immune responsiveness

The relationships between MHC antigens and the immune response of cattle to the hapten (T,G)-A--L and human serum albumin have been studied (O. Lie, H. Solbu, H.J. Larsen & R.L. Spooner unpublished). BoLA w2 was associated with lower and w16 with higher responses to HSA, whilst w8 and w20 were associated with significantly higher responses to (TG)-A--L. There is also some evidence of relationships between BoLA alleles and total haemolytic complement levels (R.L. Spooner & O. Lie unpublished).

BoLA and MHC restriction

Emery, Tenywa and Jack (1981) and Eugui and Emery (1981) have reported evidence that cytolytic cells are generated in cattle undergoing immunization against *T. parva* or in immune animals undergoing *T. parva* challenge, which kill autologous (self) cell lines infected with *T. parva*, but not infected cell lines from other animals (allogeneic). These workers presumed that the cytolytic activity was restricted by MHC antigens. Preston, Brown and Spooner (1983) have reported studies of the specificity of cytolytic cells generated during experimental *T. annulata* infection. In these experiments, cytolytic cells were detected at approximately 2 to 3 weeks following infection. They killed autologous and MHC A locus-matched, *T. annulata*-infected lymphoblastoid lines more efficiently than mismatched cell lines. The mean level of cytolytic activity increased with the increasing number of antigens shared between the cytolytic cells and the target cell line. In fact, the relationship may be closer than that shown by Preston, Brown and Spooner (1983) because some of the lines that were taken as mismatched were lines where only one BoLA allele could be identified. It is thus possible that the unidentified allele was in fact shared between effector and target. An earlier report that cell-mediated killing of bovine cells infected with infectious bovine rhinotracheitis virus or vaccinia virus was not genetically restricted (Rouse & Babiuk 1977) may be erroneous because the target cells were trypsinized and, being infected, their capacity to resynthesize MHC antigen removed by the trypsin may have been impaired.

Cell-mediated lympholysis

Teale (1983) and Teale and colleagues (1984) have demonstrated the generation of alloreactive cytolytic cells *in vitro* in the MLC. Here, effector cells sensitized by BoLA w6 subgroups w7, w8, w10 and w11 kill lymphoblastoid target cells only when they share the relevant BoLA antigen. This topic is dealt with in much greater detail by A.J. Teale and others (this volume).

BoLA expression on lymphoblastoid lines

Theileria parva and *Theileria annulata*, the causative organisms of East Coast fever and tropical theileriosis, can infect lymphocytes *in vitro* and transform them into lymphoblastoid cell lines. These cell lines can be maintained *in vitro* apparently indefinitely. It has been shown that infected cell lines retain the BoLA class I antigens of the lymphocytes from which they originated, in some cases for more than 200 subcultures (Spooner & Brown 1980). Spooner and Brown (1980) described the BoLA phenotypes of cell lines infected with *T. parva* or *T. annulata*, derived from a steer known to be chimaeric. Interesting differences were found in the lines produced. The BoLA type of the *T. parva* line differed from that of the predominant lymphocyte population in the animal, whereas that of the *T. annulata* line was the same. It was proposed that *T. parva* had established infection in the minor population and that *T. annulata* had infected both the minor and major populations. To test this, the *T. annulata* line was treated with an antiserum against a BoLA determinant present on this line, but not on the *T. parva* line, in an attempt to destroy selectively the major population. From this treated culture, a cell line was produced which had an identical BoLA type to that of the *T. parva* line. This observation indicates that BoLA antibodies can be used for separating cell populations and revealing minor cellular components (Teale et al. 1983).

Further infections of cells from this chimaeric animal were attempted with *T. parva*. This proved extremely difficult, but four cell lines were prepared; three with the BoLA type of the minor population and one with the BoLA type of the major population (C.G.D. Brown & R.L.

Spooner unpublished). As the minor population accounted for approximately 10% of the chimaerism, the infection of this population on three out of four occasions is significantly higher than what one would expect by chance. It is not yet known whether *T. parva* preferentially transforms cells of particular MHC haplotypes *in vitro* or whether some other, as yet undefined characteristic of these cells is responsible for the bias.

Role of the MHC in vaccination against theileriosis

It has been known since the early part of the century that immunizing cattle against infection with *T. parva*, using blood or tissues from infected animals, gives highly variable results (reviewed by Cunningham 1977). Using large numbers of cells from lymphoblastoid cell lines as immunogens, Brown and colleagues (1971; 1978) were able to immunize a large percentage of experimental cattle, though a few died as a result of the procedure and others were not protected. Teale (1983) and Dolan and colleagues (1984) reported experiments in which *T. parva*-infected lymphoblastoid cell lines of known BoLA type were inoculated into the same (autologous) animal, and a BoLA A locus-matched, a BoLA A locus-halfmatched and a BoLA A locus-mismatched individual. Six groups were studied: three groups were inoculated with 10^3 cells and the remaining three groups with 10^5 cells. This is well below the dose required to induce immunity using randomly chosen allogeneic cell lines, as described by Brown and colleagues (1978). One of the recipients of 10^3 autologous infected cells developed East Coast fever and died. Three weeks later, the surviving animals were challenged with *T. parva* sporozoites. All of the autologous animals, three out of the five fully matched and one out of the six halfmatched animals survived. The remaining animals, including those selected as BoLA mismatched, were fully susceptible to infection. A subsequent experiment, using a single cell line inoculated into groups of fully matched, halfmatched and mismatched recipients at 10^6 cells per animal, gave less encouraging results, although a similar trend was observed. Half of the fully matched animals survived challenge, whereas all of the halfmatched and mismatched animals underwent severe reactions and those that were not treated died (Teale 1983). These experiments support the hypothesis that a graft rejection episode following initial inoculation of lymphoblastoid cells could, if very rapid, destroy all immunogenicity. A less efficient graft response would allow the parasite to invade the host cells in sufficiently large numbers to result in the death of recipient.

Concluding remarks

In this paper I have shown that the class I MHC region is becoming increasingly well defined serologically in cattle, with at least 20 alleles controlled by one locus. Other loci have already been postulated but they have not so far received international recognition. Work on class II is at an early stage but close linkage with class I is suggested by the MLC data. Structurally, proteins identical to human class I and II have been demonstrated in the bovine lymphocyte membrane. There is also evidence that class I products are targets for alloreactive cytolytic T cells and restrict the T-cell killing of *Theileria*-infected cells.

In many respects the bovine MHC appears similar to the equivalent systems in man and mouse. In these species the MHC controls and directs the immune response. It is not, therefore, surprising that disease associations with MHC type have been found. The possibility that this may also be true in cattle is a principal reason for study of the BoLA system. Indeed there are some interesting pointers to BoLA association with diseases. The apparent relationship between w16 and high responses to HSA (O. Lie, H. Solbu, H.J. Larsen & R.L. Spooner unpublished) and susceptibility to mastitis (Solbu, Spooner & Lie 1982) is supported by the finding of a relationship between the BoLA w16-linked blood group M and mastitis susceptibility (Larsen et al. 1983). It is also interesting that, in one study carried out in Zambia, a higher frequency of blood group M was found in a low trypanosome challenge environment than in a high trypanosome environment (Carr et al. 1974). Furthermore, animals in both high and low challenge areas with blood group M were significantly more likely to have diagnosable parasitaemia than those without it. Other aspects of BoLA association with disease are dealt with in detail by Adams & Brandon (this volume).

In addition to the potential for providing markers of disease resistance, the MHC is also of fundamental importance in the response to vaccines. It could be said that many successful vaccines have been made without any knowledge of the MHC; however, with the current trend towards vaccines which rely on the recognition of limited numbers of determinants, it is possible that variation in response attributable to the MHC will assume increasing importance. Detailed study of the bovine MHC must be continued.

References

Amorena, B. & Stone, W.H. (1978). Serologically defined (SD) locus in cattle. *Science*, **201**, 159-60.

Amorena, B. & Stone, W.H. (1980). Serologic, genetic and histocompatibility studies of cattle SD antigens. *Tissue Antigens*, **16**, 212-15.

Amorena, B. & Stone, W.H. (1982). Sources of bovine lymphocyte antigen (BoLA) typing reagents. *Animal Blood Groups & Biochemical Genetics*, **13**, 81-90.

Bailey, E., Antczak, D.F., Bernoco, D., Bull, R.W., Fister, R., Guerin, G., Lazary, S., Mathews, S., McClure, J., Meyer, J., Mottironi, V.-D. & Templeton, J. (1984). Joint report of the second international workshop on lymphocyte alloantigens of the horse. *Animal Blood Groups & Biochemical Genetics*, **15**, 123-32.

Borovska, M. & Demant, P. (1967). Specificity of cytotoxic antibodies in typing sera against cattle blood group antigens. *Folia Biologica – Praha*, **13**, 473-75.

Brodsky, F.M., Stone, W.H. & Parham, P. (1981). Of cows and men: a comparative study of histocompatibility antigens. *Human Immunology*, **3**, 143-52.

Brown, C.G.D., Malmquist, W.A., Cunningham, M.P. & Burridge, M.J. (1971). Immunization against East Coast fever: inoculation of cattle with *Theileria parva* schizonts grown in cell culture. *Journal of Parasitology*, **57**, 59-60.

Brown C.G.D., Crawford, J.G., Kanhai, G.K., Njuguna, L.M. & Stagg, D.A. (1978). Immunization of cattle against East Coast fever with lymphoblastoid cell lines infected and transformed by *Theileria parva*. In *Tick-borne diseases and their vectors* (ed. J.K.H. Wilde), pp. 331-33. University of Edinburgh Press.

Bull, R.W. ed. (1983). Joint report of the first international workshop on lymphocyte alloantigens of the horse. *Animal Blood Groups & Biochemical Genetics*, **14**, 119-37.

Caldwell, J. (1979). Polymorphism of the BoLA system. *Tissue Antigens*, **13**, 319-26.

Caldwell, J., Bryan, C.F., Cumberland, P.A. & Weseli, D.F. (1977). Serologically detected lymphocyte antigens in Holstein cattle. *Animal Blood Groups & Biochemical Genetics*, **8**, 197-207.

Carr, W.R., MacLeod, J., Woolf, B. & Spooner, R.L. (1974). A survey of the relationship of genetic markers, tick-infestation level and parasitic diseases in Zebu cattle in Zambia. *Tropical Animal Health & Production*, **6**, 203-14.

Chardon, P., Kali, J., Leveziel, H., Colombani, J. & Vaiman, M. (1983). Monoclonal antibodies to HLA recognize monomorphic and polymorphic epitopes on BoLA. *Tissue Antigens*, **22**, 62-71.

Cunningham, M.P. (1977). Immunization of cattle against *Theileria parva*. In *Theileriosis: report of a workshop* (eds. J. B. Henson & M. Campbell), pp. 66-75. Ottawa: International Development Research Centre.

Cwik, A., Cichon, M.U. & Schmid, D.O. (1978). Serologically defined B-lymphocyte antigens in cattle. In *16th international conference on animal blood groups and biochemical polymorphism*, vol. 1, pp. 71-73. Leningrad.

Dolan, T.T., Teale, A.J., Stagg, D.A., Kemp, S.J., Cowan, K.M., Young, A.S., Groocock, C.M., Leitch, B.L., Spooner, R.L. & Brown, C.G.D. (1984). A histocompatibility barrier to immunization against East Coast fever using *Theileria parva*-infected lymphoblastoid cell lines. *Parasite Immunology*, **6**, 243-50.

Duquesnoy, R.J. & Schindler, T.E. (1976). Serological analysis of the HLA 10 complex. *Tissue Antigens*, **7**, 65.

Emery, D.L., Tenywa, T. & Jack, R.M. (1981). Characterization of the effector cell that mediates cytotoxicity against *Theileria parva* (East Coast fever) in immune cattle. *Infection & Immunity*, **32**, 1301-4.

Eugui, E.M. & Emery, D.L. (1981). Genetically restricted cell-mediated cytotoxicity in cattle immune to *Theileria parva. Nature,* **290**, 251-54.
European workshop on bovine histocompatibility (1983). Les colloques de l'INRA 14, pp. 1-134. Jouy-en-Josas, France: Institut National de la Recherche Agronomique.
Folger, R.L. & Hines, H.C. (1976). Bovine lymphocyte antigens: serological relationships with erythrocyte and spermatozoan antigens. *Animal Blood Groups & Biochemical Genetics,* **7**, 137-45.
Grosclaude, F., Guerin, G. & Houlier, G. (1979). The genetic map of the B system of cattle blood groups as observed in French breeds. *Animal Blood Groups & Biochemical Genetics,* **10**, 199-218.
Hines, H.C. & Newman, M.J. (1981). Production of foetally stimulated lymphocytotoxic antibodies by multiparous cows. *Animal Blood Groups & Biochemical Genetics,* **12**, 201-6.
Hoang-Xuan, M., Charron, D., Zilber, M.-T. & Levy, D. (1982a). Biochemical characterization of class II bovine major histocompatibility complex antigens using cross-species reactive antibodies. *Immunogenetics,* **15**, 621-24.
Hoang-Xuan, M., Leveziel, H., Zilber, M.-T., Parodi, A.-L. & Levy, D. (1982b). Immunological characterization of major histocompatibility antigens in cattle. *Immunogenetics,* **15**, 207-11.
Hruban, V. & Simon, M. (1973). Lymphocytotoxic antibodies against cattle J antigen. *Animal Blood Groups & Biochemical Genetics,* **4**, 183-84.
Iha, T.H., Gerbrandt, G., Bodmer, W.T., McGary, D. & Stone, W.H. (1973). Cross-reactions of cattle lymphocytotoxic sera with HL-A and other human antigens. *Tissue Antigens,* **3**, 291-302.
Larsen, B., Jensen, N.E., Madsen, P., Nielsen, S.M., Klastrup, O. & Madsen, P.S. (1983). Blood groups in relation to bovine mastitis. In *Proceedings of the 34th annual meeting of the European Association for Animal Production,* vol. 1, p. 359. Madrid.
Letesson J.J., Coppe, T., Lostrie-Trussart, N. & Depelchin, A. (1983). A bovine Ia-like antigen detected by a xenogeneic monoclonal antibody. *Animal Blood Groups & Biochemical Genetics,* **14**, 239-50.
Leveziel, H. (1983). Linkage between BoLA and the M blood group system. In *European workshop on bovine histocompatibility,* Les colloques de lINRA 14, pp. 101-3. Jouy-en-Josas, France: Institut National de la Recherche Agronomique.
McGary & Stone (1970). Immunogenetic studies of cattle lymphocytes. *Federation Proceedings,* **29**, 508.
Newman, M.J. & Hines, H.C. (1979). Production of foetally stimulated lymphocytotoxic antibodies in primiparous cows. *Animal Blood Groups & Biochemical Genetics,* **10**, 87-92.
Newman, M.J. & Hines, H.C. (1980). Stimulation of maternal anti-lymphocyte antibodies by first gestation bovine foetuses. *Journal of Reproduction & Fertility,* **60**, 237-41.
Newman, M.J. & Stear, M.J. (1983). The antibody response to bovine lymphocyte alloantigens. *Veterinary Immunology & Immunopathology,* **4**, 615-29.
Newman, M.J., Adams, T.E. & Brandon, M.R. (1982). Serological and genetic identification of a bovine B-lymphocyte alloantigen system. *Animal Blood Groups & Biochemical Genetics,* **13**, 123-29.
Oliver, R.A., McCoubrey, C.M., Miller, P., Morgan, A.L.G. & Spooner, R.L. (1981). A genetic study of bovine lymphocyte antigens (BoLA) and their

frequency in several breeds. *Immunogenetics*, **13**, 127-32.

Ostrand-Rosenberg, S. & Stormont, C. (1974). Bovine leucocyte antigens. *Animal Blood Groups & Biochemical Genetics*, **5**, 231-37.

Preston, P.M., Brown, C.G.D. & Spooner, R.L. (1983). Cell-mediated cytotoxicity in *Theileria annulata* infection of cattle with evidence for BoLA restriction. *Clinical & Experimental Immunology*, **53**, 88-100.

Pringnitz, D.J., McLaughlin, K., Benforado, I., Strozinski, I. & Stone, W.H. (1982). A simple technique using skin implants to produce histocompatibility (BoLA) typing sera. *Animal Blood Groups & Biochemical Genetics*, **13**, 91-96.

Proceedings of the second international bovine lymphocyte antigen (BoLA) workshop (1982). *Animal Blood Groups & Biochemical Genetics*, **13**, 33-53.

Rouse, B.T. & Babiuk, L.A. (1977). The direct anti-viral cytotoxicity by bovine lymphocytes is not restricted by genetic incompatibility of lymphocytes and target cells. *Journal of Immunology*, **118**, 618-24.

Solbu, H., Spooner, R.L. & Lie, O. (1982). A possible influence of the bovine major histocompatibility complex (BoLA) on mastitis. In *Proceedings of the second world congress on genetics applied to livestock production*, vol. 7, pp. 368-71. Madrid: Editorial Garsi.

Spooner, R.L. & Brown, C.G.D. (1980). Bovine lymphocyte antigens (BoLA) of bovine lymphocytes and derived lymphoblastoid lines transformed by *Theileria parva* and *Theileria annulata*. *Parasite Immunology*, **2**, 163-74.

Spooner, R.L. & Morgan, A.L.G. (1981). Analysis of BoLA w6: evidence for multiple subgroups. *Tissue Antigens*, **17**, 178-88.

Spooner, R.L. & Pinder, M. (1983). Monoclonal antibodies to bovine MHC products. *Veterinary Immunology & Immunopathology*, **4**, 453-58.

Spooner, R.L. & Ferrone, S. (1984). Cross-reaction of monoclonal antibodies to human classs I and class II products with bovine lymphocyte subpopulations. *Tissue Antigens*, **24**, 270-7.

Spooner, R.L., Millar, P. & Oliver, R.A. (1979). The production and analysis of anti-lymphocyte sera following pregnancy and skin grafting. *Animal Blood Groups & Biochemical Genetics*, **10**, 99-106.

Spooner, R.L., Leveziel, H., Grosclaude, F., Oliver, R.A. & Vaiman, M. (1978). Evidence for a possible major histocompatibility complex (BoLA) in cattle. *Journal of Immunogenetics*, **5**, 335-46.

Spooner, R.L., Oliver, R.A., Sales, D.I., McCoubrey, C.M., Millar, P., Morgan, A.G., Amorena, B., Bailey, E., Bernoco, D., Brandon, M., Bull, R.W., Caldwell, J., Cwik, S., van Dam, R.H., Dodd, J., Gahne, B., Grosclaude, F., Hall, J.G., Hines, H., Leveziel, H., Newman, M.J., Stear, M.J., Stone, W.H. & Vaiman, M. (1979). Analysis of alloantisera against bovine lymphocytes: joint report of the 1st international bovine lymphocyte antigen (BoLA) workshop. *Animal Blood Groups & Biochemical Genetics*, **10**, 63-68.

Stear, M.J. (1980). Lymphocyte antigens in sheep. Ph.D. thesis. University of Edinburgh.

Stear, M.J. & Bell, T.K. (1984). Relationships between the bovine major histocompatibility system and commonly recognized erythrocyte and serum polymorphisms. *Animal Blood Groups & Biochemical Genetics*, **15**, 231-36.

Teale, A.J. (1983). The major histocompatibility complex of cattle with particular reference to some aspects of East Coast fever. Ph.D. thesis. University of Edinburgh.

Teale, A.J., Kemp, S.J., Young, F. & Spooner, R.L. (1983). Selection, by major histocompatibility type (BoLA), of lymphoid cells derived from a bovine chimaera and transformed by *Theileria* parasites. *Parasite Immunology,* **5**, 329-35.

Terasaki, P.I. & McClelland, J.D. (1964). Microdroplet assay of human serum cytotoxins. *Nature,* **204**, 998.

Usinger, W.R., Curie-Cohen, H. & Stone, W.H. (1977). Lymphocyte-defined loci in cattle. *Science,* **196**, 1017-18.

Usinger, W.R., Curie-Cohen, M., Benforado, K., Pringnitz, D., Rowe, R., Splitter, G.A. & Stone, W.H. (1981). The bovine major histocompatibility complex (BoLA): close linkage of the genes controlling serologically defined antigens and mixed lymphocyte reactivity. *Immunogenetics,* **14**, 423-28.

9

Biochemical analysis of products of the bovine major histocompatibility complex

D.LÉVY, M.HOANG-XUAN and M.T.ZILBER

In cattle, 17 class I major histocompatibility (MHC) antigens coded for by codominant alleles at a single locus (the A locus) have been defined based on the use of alloantisera. Studies on bovine mixed leucocyte reactions indicate the existence of two or more loci coding for class II MHC antigens. It has become particularly important to delineate the biochemical characteristics of MHC products with the aim of studying structure-function relationships. The results of our studies, in which lymphocyte antigens were precipitated with bovine alloantisera, indicate that the bovine MHC encodes class I cell-surface products of conventional basic structure, consisting of a glycosylated 45,000-dalton heavy chain noncovalently associated with a 12,000-dalton β_2-microglobulin light chain. Similar experiments were not possible with the bovine class II antigens due to the lack of specific alloantisera. However, taking advantage of the well known cross-reactivities of MHC antigenic determinants in different species, we have studied the biochemical properties of bovine class II molecules. Precipitation of bovine lymphocyte extracts by antibodies reacting with murine Ia or human HLA-DR molecules yielded a noncovalently associated complex of two glycoprotein chains, of 29,000 and 34,000 daltons, associated with a 31,000-dalton invariant chain. Bovine class II antigens were found to possess determinants in common with two different class II MHC antigens (IA and IE) in the mouse.

Introduction

The genetics, serology and immunology of the major histocompatibility complex (MHC) have been the subjects of other reports (Murphy 1981). In all species of mammals and birds examined, it has been found that this major genetic region, determining transplantation rejection, codes for two groups of highly polymorphic cell-surface antigens against which most of the antibodies generated during allogeneic stimulation are directed. This chapter aims to present the current view of the structure and biochemistry of these molecules and to summarize briefly the more recent developments in this field. Most of our knowledge is derived from studies of the MHCs of mouse (H-2) and human (HLA); these studies have provided different and complementary approaches. Work with MHCs in other species has generally confirmed the principles established in these two systems. The first part of this report will be devoted to the classic transplantation antigens, also referred to as the serologically defined (SD) or class I antigens. They are expressed on a wide variety of tissues, and studies of their structure are more advanced than studies of other MHC-encoded products. The class I antigens are defined as noncovalently associated complexes of a 45,000-dalton MHC-encoded glycoprotein and β_2-microglobulin, a 12,000-dalton, nonMHC encoded protein. This definition extends to the Qa and TL antigens as well, even though these antigens demonstrate different tissue distribution.

The bovine class II antigens, the Ia antigens of mice and DR antigens of humans will also be discussed. They comprise a set of structurally distinct MHC-encoded antigens of more restricted tissue distribution; they have also been referred to as B cell-specific alloantigens and they are correlated with lymphocyte-activating determinants (LAD) or lymphocyte-defined (LD) antigens. The class II antigens consist of two noncovalently associated, membrane-bound glycoproteins of approximately 32,000 to 35,000 daltons (α chain) and 25,000 to 28,000 (β chain). Under certain conditions, a third chain (Ii) of 31,000 daltons, which is structurally invariant regardless of haplotype, can be co-isolated with I-region molecules. In regard to function, the class I gene products differ from those of class II in that the former guide cytolytic and the latter regulatory (helper and suppressor) T lymphocytes.

MHC class I products
Bovine antigens

The characterization of bovine MHC products was recently made possible by the availability of specific antisera obtained by skin grafting between dam-calf pairs in order to obtain haplo-identity (Spooner et al. 1978). Up to now, 17 different specificities have been defined and family studies have been consistent with their genetic control by alleles at a single locus (Oliver et al. 1981). In an effort to gain further insight into the nature of the bovine MHC equivalent, we have undertaken a study of the biochemical characteristics of the alloantigens detected on the surface of bovine lymphocytes.

Isolation of a biological activity requires a convenient assay with which to monitor purification procedures. Of the various immunological assay systems that can be used to detect MHC-encoded antigens, only those based on the use of antibodies fulfil this requirement. Alloantibodies have been most commonly used, although with improvement in the isolation of these antigens, it has been possible to obtain both xenoantisera and, more recently, monoclonal antibodies of appropriate specifities. The assay systems used have been inhibition of either haemagglutination, cytotoxicity or binding of antibody measured by radioimmunoassay (RIA). Because of personal experience, we used RIA to characterize and titrate the bovine alloantisera. All preliminary experiments showed a strict correlation between RIA and cytotoxicity results (Lévy et al. 1981).

Figure 1. SDS-PAGE of alloantigens from radioiodinated 910 lymphocytes after immunoprecipitation with serum 72273 and serum 701. Molecular weight markers in this and subsequent figures are bovine serum albumin (BSA: 68K), ovalbumin (OVA: 43K), carbonic anhydrase (CARB.AN: 30K), trypsin inhibitor (TRYP.INH: 21K) and lysozyme (Lys: 14K).

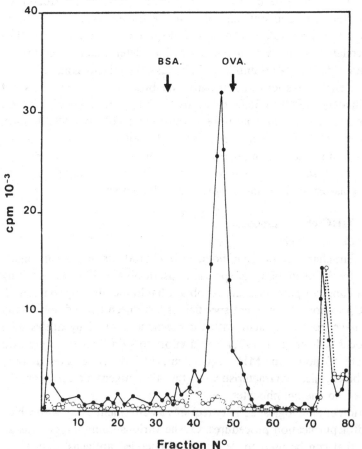

The system under study involved a serum (72273) directed against the internationally defined w16 specificity and reacting in cytotoxicity and radioimmunassay on cells from cow 910. Lymphocytes from this animal were either surface labeled with ^{125}I or metabolically labeled with ^{3}H-leucine; in some instances the cell extracts were partially purified by affinity chromatography over a column of *Lens culinaris* lentil lectin coupled to sepharose. Indirect immunoprecipitation of the extracts was performed using bovine antisera followed by fixed *Staphylococcus aureus*.

Figure 2. SDS-PAGE of alloantigens from 910 lymphocytes after immunoprecipitation with serum 72273 unabsorbed (●) or previously absorbed on 910-relevant (▲) or 615-irrelevant cells (○).

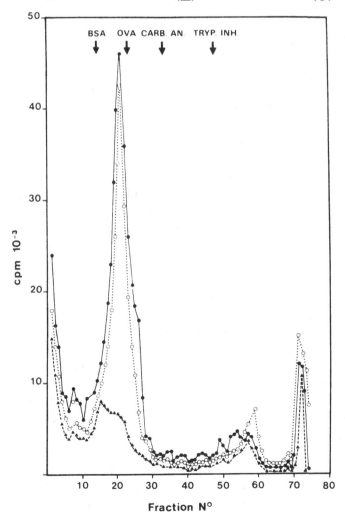

Antigen-antibody complexes eluted from the *Staphylococcus* were resolved on sodium dodecyl sulphate (SDS) 10% polyacrylamide gel electrophoresis (PAGE) under reducing conditions. Figure 1 shows gel profiles obtained by immune precipitation with a relevant (72273) and a nonrelevant serum (701): a peak running at 44,000 daltons appeared with serum 72273, but not with serum 701. Passage of this material on a column of *Lens culinaris*-sepharose suggested the glycoprotein nature of this cell-surface antigen.

The detergent-solubilized, 44,000-dalton alloantigen appeared to be associated with a smaller polypeptide of 12,000 daltons, comparable with that of the B_2-microglobulin molecule associated with human HLA or murine H-2 alloantigens. SDS-PAGE of immunoprecipitates in the absence of reducing agent also revealed the smaller polypeptide band, indicating that it was not bound to the larger polypeptide by a disulphide bridge.

To ascertain the specificity of the reaction, serum 72273 was absorbed, either on cells bearing the w16 specificity (no. 910) or on an equal number of irrelevant cells (no. 615). As assessed in a cellular radioimmunoassay (Lévy et al. 1981), complete absorption was achieved at a ratio of 16 x 10^6 cells per μl of undiluted serum. Figure 2 shows that previous absorption on w16-positive cells totally inhibits the capacity of serum 72273 to precipitate the 44,000-dalton cell-surface protein, whereas absorption on irrelevant cells has no effect. In addition, as bovine sera frequently contain high levels of antibodies specific for bovine leukaemia virus (BLV) (Lévy et al. 1977), absorption experiments were performed with the aim of eliminating the possible role of such antibodies: absorption of 50 μl serum on 2 mg purified BLV, which completely absorbs the anti-viral antibodies of a lymphosarcomatous cow (data not shown), did not remove the serum precipitating activity for bovine cell-surface antigens. When a w16 monospecific antiserum was first reacted with labeled alloantigens from 910 cells, resulting in the precipitation of a 44,000-dalton moiety, the sequential addition of serum 72273 to the remaining supernatant did not result in the precipitation of additional molecules. This suggests that sera 72273 and anti-w16 recognize the same molecules on the surface of 910 cells. When the monomorphic anti-HLA monoclonal antibody w6/32 was reacted on bovine cells, the same heavy and light chains were precipitated.

It was interesting to demonstrate that in the bovine system, as well as in the murine, human, pig, rat and hamster systems, class I MHC molecules consist of a 44,000-dalton glycoprotein noncovalently associated with a smaller component. The results of the present study imply that cattle possess an MHC which encodes class I cell-surface products of a conventional basic structure. The functional role of these individual components, however, remains to be determined.

Murine and human MHC class I products

Studies on the detailed structure of class I bovine antigens will be possible on animals whose genetic background is well defined by using monospecific antisera. However, one may already ascertain that these proteins will follow the schematic model provided by our current understanding of the mouse and human MHCs. Initial structural studies on class I antigens consisted of comparative peptide mapping analyses. These studies were followed by extensive protein sequence analyses of a number of H-2 and HLA molecules. More recently, the advent of recombinant DNA techniques has led to the molecular cloning of class I genes which has further advanced our appreciation of the structure and genetic organization of the MHC antigens.

Studies on the structure of major transplantation antigens have had to overcome two difficult problems. First, H-2 and HLA molecules are integral membrane proteins tightly associated with the cell membrane lipid bilayer and, hence, are not water soluble. Second, these molecules represent a small fraction of the total cellular components and therefore must be purified from a complex biological mixture. Multiple approaches have been devised by investigators to deal with these problems.

Sequence determination

H-2K^b and HLA-B7 were the first class I molecules whose primary structures were studied. The strategy used for determining the complete sequence of the H-2K^b molecule relied on initial isolation of fragments obtained by CNBr cleavage at methionine residues. Subsequent amino-acid sequence analyses and use of tryptic overlap peptides permitted the alignment of these fragments. The amino-acid sequences of small CNBr fragments (less than 30 amino-acid residues) were determined on the intact fragments. Larger CNBr fragments were reduced to smaller fragments by enzymatic cleavage (thrombin, trypsin, V-8 protease) in order to obtain appropriately sized peptides for sequence determination. The alignment of the CNBr peptides of H-2K^b is depicted in Figure 3. Structural analyses of similarly derived CNBr peptides from other H-2 molecules revealed strong conservation of certain features: the methionine residues are highly conserved, the size and location of the disulphide loops are apparently identical and the positions of carbohydrate groups are conserved. Also as shown in Figure 3, HLA molecules have overall structural features similar to those of H-2 molecules, thus permitting a similar strategy for sequence analyses. HLA-B7 is somewhat exceptional in that it has fewer methionine residues than other class I molecules which have been studied.

Primary structure

H-2K^b is 348 residues long, whereas HLA-B7 has 339 residues. HLA-B7 and H-2K^b have identical carbohydrate attachment sites on the asparagine residue at position 86. H-2K^b has a second glycosylation site on the asparagine residue at position 176. Also, both molecules have internal disulphide loops which are identical in size and location. The first disulphide loop encompasses residues 101 to 164 and the second spans positions 203 to 259. H-2K^b has two cysteine residues that are not involved in disulphide bond formation (position 121 and position 337). There are also two free cysteine residues in HLA-B7, both in the carboxy-terminal portion of the molecule (positions 308 and 325). In both molecules, a hydrophobic region extends approximately from residue 284 to residue 304 and is followed by a cluster of charged, basic residues (positions 308 to 312). The papain cleavage site, or more correctly, the position at which

Figure 3. Schematic aspects of some H-2 and HLA molecules. S-S indicates the presence of a disulphide bond and thus the site of CNBr cleavage. CHO locates a carbohydrate moiety.

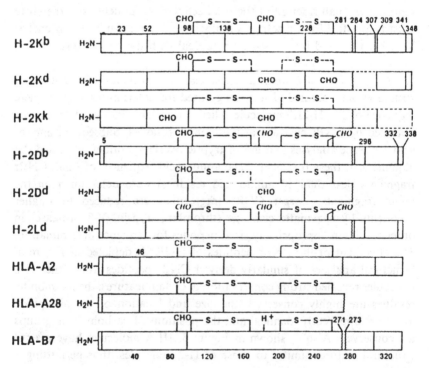

papain ceases to digest further the nondenatured molecules, has been identified as valine at position 281 in H-2K^b and arginine at position 273 in HLA-B7.

Molecular organization

The structural information obtained for class I antigens led to the derivation of the schematic model shown in Figure 4. Such molecules consist of three regions: an extracellular region (residues 1 to 283), a hydrophobic, transmembrane region (residues 284 to 307) and a hydrophilic cytoplasmic region (residue 308 to the carboxy terminal). The extracellular region can be further subdivided into three domains (N, C1 and C2 in the mouse and a1, a2 and a3 for human molecules). Postulation of the existence of such domains was prompted by evidence of primary structural homologies of class I heavy chains and β_2-microglobulin to IgG constant region domains. Other evidence which supports the existence of such domains includes: (1) the nearly identical size and linear

Figure 4. Schematic outline of a class I molecule in the cell membrane.

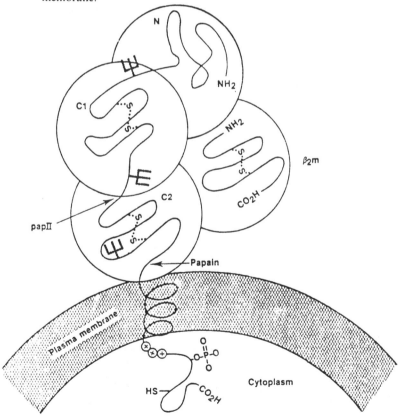

arrangement of both disulphide loop domains, C1 and C2, (2) the alignment of carbohydrate moieties in the N and C1 domains, and (3) the statistically significant homology (about 30%) between the amino-terminal domain and the first disulphide loop domain. An additional observation which lends support to the proposed globular domain structure is the fact that limited digestion with papain occurs between C2 and the transmembrane region (Tm), and between C1 and C2 (pap II).

More recently, data supporting the organization of class I heavy chains into domains have come from studies on the genetic organization of class I genes. As shown in Figure 5, genomic clones of murine class I heavy chains are organized into eight exons (Steinmetz et al. 1981) which almost coincide with the aforementioned domain structure. Exon 1 corresponds to a 21-residue precursor sequence (signal peptide) and exons 2, 3 and 4 correspond to the three extracellular domains. The membrane-binding region is encoded in exon 5 and the cytoplasmic segment of H-2 molecules is encoded by three small exons (6,7,8). Current evidence indicates that human class I molecules lack the exon analogous to exon 8 of the mouse and therefore have only seven exons (Malissen et al. 1982).

Figure 4 shows the presence of three carbohydrate moieties on class I heavy chains. The HLA molecules which have been studied have only one in the amino-terminal domain (al domain). All H-2 molecules which have been analysed have oligosaccharides attached to both the N and C1 domains, on the asparagine residues at positions 86 and 176 respectively. Some H-2 antigens, e.g. H-2K^d, are known to carry a third carbohydrate group attached to an asparagine residue at position 256 in the C2 domain.

The membrane-binding region is about 24 residues long in H-2 and HLA, approximately spanning residues 248 to 307. This length of a polypeptide, in an α-helix conformation, is capable of spanning a lipid bilayer 35 angstroms deep. No charged or polar amino acids are found in this region, which is similar to the situation with analogous regions in other integral proteins (von Heijne 1981).

The cytoplasmic region consists of the carboxy-terminal portion of the molecule and contains a large percentage of hydrophilic residues. At the boundary between the membrane and the cytoplasmic domains there is a cluster of basic amino acids which are thought to anchor the molecule in the lipid bilayer via charge interactions with the negatively charged cytoplasmic surface of the membrane. Analogous clusters of basic residues have been observed in other membrane-bound molecules.

The C-terminal region contains a prosthetic group that in HLA-A2 has been shown to be phosphoserine. H-2K^k is also phosphorylated in this part of the molecule. The exact role of cytoplasmic region phosphate groups in H-2 and HLA has not been elucidated, but it is possible that class I molecules interact with cytoskeletal elements through conformational changes occurring via such phosphate moieties.

Comparison of primary structures

A major goal of primary structural studies on histocompatibility antigens is to determine regions of diversity and similarity, by means of sequence comparisons, in the hope of determining which portions of the molecules are responsible for various biological reactivities. For such comparisons, extensive primary protein structure is now available for six H-2 antigens.

The variability plot for H-2 molecules shows that differences among class I molecules are not randomly distributed but tend to occur in clusters. One region of major diversity occurs between residues 61 and 83 (exon 2, first external domain). The third exon (second external domain) contains three clusters of notable diversity in the residue sequences 95 to 99, 114 to 116 and 152 to 157.

The third extracellular domain (C2) of murine class I antigens has two regions of sequence diversity, in the residue sequences 193 to 198 and 255 to 268. The second disulphide loop located in domain C2 is the most highly conserved region in class I molecules, probably because of a requirement to preserve an important biological function. Thus, this region has been proposed to contain a segment which binds β_2-microglobulin. HLA molecules are nearly invariant in the A$_3$ domain.

Figure 5. Exon-intron organization of gene 27.1 in the mouse aligned with corresponding protein domains (from Steinmetz et al. 1981).

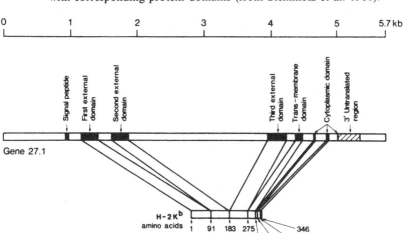

The transmembrane domain, encoded by exon 5, has strong intraspecies, but low interspecies homology. If its function is to anchor the molecule into the membrane, it probably is not necessary to retain particular amino acids, but only hydrophobic residues.

Residues in the region encoded by exon 6 are identical or highly homologous in all H-2 molecules. HLA molecules are also nearly identical to one another but differ significantly from H-2 molecules in this region. The fact that this region is so highly conserved suggests that it may be critical to the biological activity of class I molecules.

MHC class II products

Bovine class II antigens

Similar experiments to those performed with class I antigens were not possible with the putative bovine class II antigens, due to the lack of specific alloantisera. However, taking advantage of the well known cross-reactivities of MHC antigenic determinants in various species, we have studied the biochemical properties of bovine class II molecules precipitated by antibodies reacting with murine Ia or human HLA-DR molecules.

Hypaque-Ficoll-purified lymphocytes were obtained from the blood of two different cows: 910 (normal) and M-428, which had a persistent

Figure 6. Gradient slab gel electrophoresis of ^{35}S-methionine labeled M.428 lymphocytes after immunoprecipitation with: (A) A.TL serum, (B) monoclonal HLA-DR-specific antibody, (C) B10.S (8R) absorbed A.TL (anti-I-E^k), (D) B10.A. (4R) absorbed A.TH anti-A.TL (anti-I-E^k), (E) B10.HTT absorbed A.TH anti-A.TL (anti-I-A^k), (F) monoclonal I-A^k-specific antibody and (G) monoclonal anti-blood coagulation factor, anti-factor VIII as a negative control. The numbers on the left indicate the approximate molecular masses.

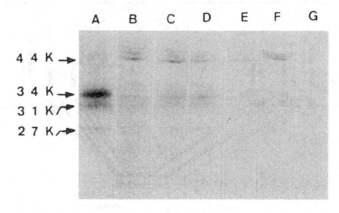

lymphocytosis due to infection with bovine enzootic leucosis virus. The latter animal had a very large number of circulating B lymphocytes. The cells were metabolically labeled by ^{35}S-methionine or ^{75}Se-methionine. After preclearing with normal nonimmune antibody, the cell extracts were immunoprecipitated and the immune complexes were analysed on 10% acrylamide gels: both conventional A.TH anti-A.TL antiserum (anti-I-A^k) and monoclonal HLA-DR-specific (Ab 1.35) antibody precipitated proteins in the region of 30,000 daltons. The same antibodies were used to characterize further the putative bovine class II molecules by analysing the immune precipitates on 5 to 15% polyacrylamide gradient slab gels. In addition to actin (44,000 daltons), three bands were obtained, corresponding respectively to 34,000, 31,000 and 27,000 daltons (Figure 6).

To evaluate whether bovine class II antigens were precipitated by antibodies specific for I-A or I-E molecules, the A.TH anti-A.TL serum was absorbed with spleen and lymph node cells of either B10.HTT (I-A^s, I-E^k), B10.A (4R) (I-A^k, I-E^b) or B10.S (8R) (I-A^k, I-E^s) mice.

Figure 7. NEPHGE-2D profiles of class II BoLA molecules immunoprecipitated from ^{35}S-methionine-labeled M.428 lymphocytes with (a) A.THJ anti-A. TH (anti-I-A^k) serum or (b) monoclonal HLA-DR-specific antibody. The basic end of the NEPHGE gel is on the left. Actin is indicated by the letter A. The arrow indicates the spot probably equivalent to the invariant chain of murine and human class II antigens.

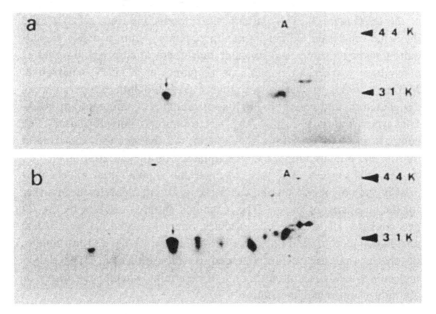

Complete absorption was assessed by retesting these absorbed antisera in a cellular radioimmunoassay (CRTA). As shown in Figure 6, C and D, the serum absorbed to remove anti-I-A^k activity precipitated three molecules in the same molecular mass range as the unabsorbed A.TH anti-A.TL serum and the Ab.1.35. With the serum absorbed to remove anti-I-E^k activity, immunoprecipitates generated a major 27,000 dalton band and two weak bands in the 31,000 to 34,000 dalton range (Figure 6E). Identical results (Figure 6F) were obtained with an ascitic fluid containing monoclonal antibody specific for I-A^k, kindly provided by Dr L.A. Herzenberg. This antibody, as well as A.TH anti-H.TL serum and Ab 1.35, strongly reacted in a CRIA on lymphocytes of cows 910 and M.428.

Such cross-reactivities with murine and human class II antigens strongly support the existence of bovine class II molecules sharing some antigenic determinants with human HLA-DR and murine I-A and I-E molecules. Furthermore, the molecules appear to be made up of three distinct polypeptide chains as previously demonstrated in the other two species (Charron & McDevitt 1980; Jones et al. 1979).

Further studies of class II molecules were performed by two-dimensional (2D)-PAGE, using for the first dimension a nonequilibrium pH gradient electrophoresis (NEPHGE) which resolves highly basic proteins. The 2D-gel pattern obtained with monoclonal HLA-DR-specific (1.35) antibodies or with an A.TH anti-A.TL-specific (anti-I-A^k) alloantiserum consisted of a set of very heterogeneous spots within the 27,000 to 34,000 dalton range, spread from the acidic to the basic end of the gel (Figure 7). This pattern was reproducible from gel to gel for the same cell lysate extract immunoprecipitate. Actin, a major 44,000-dalton acidic molecule, is a constant constituent of all gels and serves as a reference marker. In previous work (Jones et al. 1979; Charron & McDevitt 1980), it was shown that a 31,000-dalton basic molecule, designated 'Ii' and present in immunoprecipitates of different Ia or HLA-DR types, did not display any electrophoretic polymorphism. The comparison of the positions of the actin and the invariant molecule in murine and human lymphocyte precipitates with the pattern obtained on bovine cells suggests that such an invariant 31,000-dalton moiety may exist in the bovine system and can serve as an internal reference marker for the basic area of the gel. The remaining antigens resolved by the 2D gels consisted of two sets of molecules, one very acidic between 31,000 and 34,000 daltons, and one more basic of about 27,000 daltons. Studies of other animals are obviously needed to establish that the 31,000-dalton molecule is invariant in cattle and to determine whether the heavy or the light chain is polymorphic.

On the whole, these results show that: (1) class II MHC molecules can be detected in cattle, (2) they reveal biochemical characteristics very similar to those previously described in mice (Ia) and humans (HLA-DR), (3) they share antigenic determinants with both Ia and HLA-DR antigens, and (4) they possess common determinants not only with I-E but also with I-A molecules. Further work is necessary to prove the existence in cattle of different products encoded by an I-like region and to locate the parts of the molecules that have been conserved between species. Such data are important for functional studies of the MHC class II molecules.

Murine and human MHC class II products

Until recently, there has been little information regarding the primary structure of HLA-DR and H-2-Ia antigens: the structure of the DR antigen was explored using limited proteolysis of the native molecule. Current interest in the molecular biology of the MHC has resulted in studies using recombinant DNA methods: the primary structure of the HLA-DR chains has been elucidated by analysing the genes that encode them.

The nucleotide sequence clearly indicates that the HLA-DR heavy chain is split into five exons: exon 1 contains the 5′ untranslated region and a sequence encoding a leader peptide containing 25 amino acids. In contrast to class I genes, the sequences in class II genes encoding the first two amino-terminal residues are contained in exon 1. Exon 2 encodes amino-acid residues 3 to 84 and may represent the first external domain

Figure 8. Comparative schematic outline of cell-surface immunoglobulin, class I and II MHC molecules and Thy 1. Immunoglobulin-like domains are represented by circles.

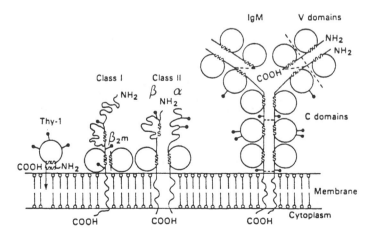

of the 34,000-dalton heavy chain (or α chain). Nucleotide sequences for residues 85 to 178 are contained in exon 3, which comprises the second external domain of the molecule. Exon 4 contains sequences coding for residues 179 to 229 and a stop codon at position 230, with a stretch of hydrophobic amino acids (residues 193 to 214) followed by 15 amino acids most of which are hydrophilic. By analogy with the structure of class I genes, this region corresponds to the transmembrane and cytoplasmic domains of the α chain, that are fused in a single exon. The 3′ untranslated region is located in exon 5. From these data, the amino-acid sequence of the mature α chain may be predicted, with a disulphide bridge between the cysteine residues at positions 107 and 163 and two carbohydrate moieties, one in each of the two extracellular domains.

Data on HLA-DRβ chain amino-acid sequence are scarce. The DRβ chain contains four cysteines which form two disulphide bridges (residues 15 to 79 and 117 to 172) and one carbohydrate attached to the asparagine residue at position 19. It comprises two external, one transmembrane and one intracytoplasmic domain.

Interestingly, the C-terminal half of the DRβ chain and of the DRα chain, the third domain of HLA-B7, and human β₂-microglobulin all have significant sequence homology (30%) to Ig-constant domains, especially at the conserved residues important in the secondary and tertiary folding of the Ig domain. All five sequences have one disulphide loop of approximately the same size. In addition, HLA-A and -B antigens and β₂-microglobulin have B-sheet secondary structure. These data support the hypothesis that these domains, as well as others such as Thy 1, arose by gene duplication and sequence divergence of a common ancestral gene (Figure 8).

References

Charron, D.J. & McDevitt, H.O. (1980). Characterization of HLA-D region antigens by two-dimensional gel electrophoresis: molecular genotyping. *Journal of Experimental Medicine*, **152**, 18s-36s.

Jones, P.P., Murphy, D.B., Hewgill, D. & McDevitt, H.O. (1979). Detection of a common polypeptide chain in I-A and I-E subregion immunoprecipitates. *Molecular Immunology*, **16**, 51-60.

Lévy, D., Hoang-Zuan, M., Colombani, M.J., Zilber, M.T., Leclerc, J.C. & Lévy, J.P. (1981). Typing of murine cell-surface antigens by cellular radioimmunoassay. *Journal of Immunological Methods*, **41**, 333-41.

Lévy, D., Deshayes, L., Parodi, A.L., Lévy, J.P., Stephenson, J.R., Devare, S.G. & Gilden, R.V. (1977). Bovine leukaemia virus-specific antibodies among French cattle. 2. Radioimmunoassay with the major structural protein (BLV p24). *International Journal of Cancer*, **20**, 543-50.

Malissen, M., Malissen, B. & Jordan, B.R. (1982). Exon-intron organization and complete nucleotide sequence of an HLA gene. *Proceedings of the National Academy of Sciences of the USA*, **79**, 893-97.

Murphy, D.B. (1981). Genetic fine structure of the H-2 gene complex. In *The role of the major histocompatibility complex in immunobiology* (ed. M.E. Dorf), pp. 1-31. New York: Garland Press.

Oliver, R.A., McCoubrey, C.M., Millar, P., Morgan, A.L.G. & Spooner, R.L. (1981). A genetic study of bovine lymphocyte antigens (BoLA) and their frequency in several breeds. *Immunogenetics*, 13, 127-32.

Spooner, R.L., Leveziel, H., Grosclaude, F., Oliver, R.A. & Vaiman, M. (1978). Evidence for a possible major histocompatibility complex (BLA) in cattle. *Journal of Immunogenetics*, 5, 335-46.

Steinmetz, M., Moore, K.W., Frelinger, J.G., Sher, B.J., Shen, F.W., Boyse, E.A. & Hood, L. (1981). A pseudogene homologous to mouse transplantation antigens: transplantation antigens are encoded by eight exons that correlate with protein domains. *Cell*, 25, 683-92.

von Heijne, G. (1981). Membrane proteins: the amino-acid composition of membrane-penetrating segments. *European Journal of Biochemistry*, 120, 275-78.

10

The role of the major histocompatibility complex in immune responses and susceptibility to disease

R.V. BLANDEN

T lymphocytes that are responsible for cytotoxicity, delayed-type hypersensitivity and helper functions in immune responses recognize cell-surface antigenic patterns which depend upon both the foreign infectious agent and self major histocompatibility complex (MHC) antigens. However, the need for self-tolerance results in the deletion or suppression of T-cell clones that respond to self-MHC antigens alone. The qualitative nature of MHC antigens and the quantity expressed on cell-surface membranes of particular tissues can influence the antigen-receptor repertoire of the T-lymphocyte pool, the efficiency of induction of T-cell responses against foreign antigens, and the efficacy of the effector activity of these responses. Thus, the MHC can influence susceptibility to infectious diseases in which mechanisms of recovery from infection are dependent upon MHC-restricted T-cell responses. Experimental models illustrating these points are discussed.

Major histocompatibility complex restriction and T-cell activation

In vertebrate species studied thus far, the major histocompatibility gene complex (MHC) contains genes that code for the polypeptide chains of glycoprotein antigens which are expressed as transmembrane proteins on cell surfaces. MHC genes and antigens are of two classes. Class I genes code for antigens of about 45,000 daltons that are expressed on virtually all cell classes. Class II genes code for two different antigens of about 28,000 and 33,000 daltons that are expressed as heterodimers, mainly on certain lymphoreticular cells such as dendritic cells, macrophages and B lymphocytes. Over the last decade or so it has been clearly established that the MHC antigens serve as self-markers for thymus-derived (T) lymphocytes, a phenomenon termed MHC restriction (Zinkernagel & Doherty 1974a; Doherty, Blanden & Zinkernagel 1976; Zinkernagel & Doherty 1979). Thus, resting mature

T cells of the recirculating pool in secondary lymphoid tissues are responsive to cell-surface membranes displaying antigenic patterns dependent upon a foreign antigen together with a self-MHC antigen. In general it seems that cytolytic T cells are restricted to class I MHC antigens and T cells that mediate help or delayed hypersensitivity are restricted to class II MHC antigens, though there are minor exceptions.

T-cell recognition of self-MHC is apparently immunological and not physiological, i.e. it is 'learned' and not intrinsically determined by the MHC genotype of the T cell. Thus T cells from (P1 x P2) F_1 hybrids responding to X foreign antigen are of two general categories: those recognizing P1 MHC + X, and those recognizing P2 MHC + X. There is no evidence of individual F_1 T cells capable of recognizing both P1 MHC + X and P2 MHC + X, as would be expected from a physiological model (Zinkernagel & Doherty 1974b). Further, T cells of P1 genotype that recognize P2 MHC + X can be found in irradiation chimaeras produced by reconstituting lethally irradiated (P1 x P2) F_1 hybrids with P1 stem cells, a clear demonstration that T cells of P1 origin can 'learn' to recognize P2 MHC as self, provided they differentiate and are stimulated to respond in a thymic environment displaying P2 MHC antigens (Doherty, Blanden & Zinkernagel 1976). More precisely stated, such experiments indicate that during the process of T-cell ontogeny in the thymus, P1 T cells can arise which express antigen-receptors capable of binding antigen patterns dependent upon P2 MHC + X, while not binding to either P2 MHC or P1 MHC, criteria required for self-tolerance. The MHC-restricted T-cell antigen-receptor repertoire is biased towards recognition of thymic MHC antigens, suggesting that recognition of thymic MHC is involved in T-cell ontogeny, but the precise mechanisms await definition (Zinkernagel & Doherty 1979).

MHC-linked immune response genes

Immune response (Ir) genes were first discovered and studied in guinea pigs by Benacerraf and coworkers, and in mice by McDevitt and coworkers in the early 1960s (for an historical review see Benacerraf & McDevitt 1972). These genes were shown to influence quantitatively both T cell-mediated and humoral immune responses (through helper T cells), and the linkage of some of them to the MHC was established by 1970.

As a consequence of the discovery of MHC restriction and the partial definition of the immunobiological processes involved, new concepts of the mechanisms of action of MHC-linked Ir genes could be formulated and to some extent tested. First, initial theories that MHC genes coded for T-cell antigen receptors (Benacerraf & McDevitt 1972) could be discarded when it became apparent that T cells of nonresponder MHC genotype could express responder phenotype if they differentiated and responded in an environment displaying responder MHC antigens (Zinkernagel & Doherty 1979; Müllbacher & Blanden 1979). Other theories about MHC-linked Ir genes have derived from the fact that effector and helper T cells must recognize MHC antigens in order to become activated. This means that MHC structural genes can be Ir genes, so there is no need to postulate additional MHC-linked genes.

There are currently two general theories on MHC Ir genes, one concerning T-cell antigen receptors and the other concerning antigen presentation. The first theory can be further subdivided into so-called preclusion models and cross-reactive tolerance models.

Preclusion models postulate two antigen receptors per T cell, one for MHC and a second for foreign antigen (X) (Langman 1978; von Boehmer, Haas & Jerne 1978), along with several other assumptions. It is assumed that the gene coding for the anti-MHC receptor of a maturing precursor T cell comes from a germline repertoire, and that further maturation only follows self-MHC recognition in the thymus. It is also assumed that the anti-X receptor of an individual T cell is coded by one out of a limited repertoire of somatically generated variants derived from certain permitted germline genes, other germline genes being precluded from participating in the somatic variation process. According to one hypothesis (von Boehmer, Haas & Jerne 1978), only a second copy of the selected anti-MHC coding gene is permitted to give rise to the anti-X repertoire; according to another hypothesis (Langman 1978) all germline anti-MHC genes except the selected anti-MHC gene are permitted to give rise to anti-X. These models depend upon the idea that the repertoire of anti-X receptors expressed by a T-cell population will have gaps with respect to a particular X antigen, depending upon the nature of the anti-MHC receptor expressed. However, considerable experimental evidence is inconsistent with these preclusion models (Müllbacher 1981a). Also, the models can account only for simple Ir-gene phenomena, i.e. low or high responses to a given X antigen associated with a single MHC allele.

In contrast, the cross-tolerance model (Müllbacher 1981a) can accommodate more complex phenomena in which certain alleles of one MHC locus can influence the strength of T-cell responses to a foreign antigen that are restricted to the antigenic product of a second locus. According to this theory, when T cells reactive to one self-MHC antigen plus X are cross-reactive to another self-MHC antigen alone, then self-tolerance to the second MHC antigen will result in suppression and

Table 10.1. Influence of different H-2 genes on susceptibility to infection with virulent Ectromelia virus. Mice of either sex between 2 and 4 months of age were inoculated subcutaneously in the hind footpad with 40μl of Moscow strain Ecromelia virus stock (Blanden 1970). Groups of 5 or 6 mice were given different doses of virus ranging from 10^6 down to 1 pfu in 10-fold steps, and deaths were recorded daily for 14 days. The LD_{50} values were calculated by the method of Reed & Muench (1938) and significances of differences in total mortality were determined by the X^2 test.

Mouse strain	H-2 genes								Dose of virus required to produce 50% mortality in plaque-forming units (PFU)
	K	A	Aβ	Eβ	J	E	S	D	
C57B/10(B10)	b	b	b	b	b	b	b		$>3 \times 10^5$
B10.A(4R)	k	k	k	k	b	b	b	b	$>3 \times 10^5$
B10.A(2R)	k	k	k	k	k	k	d	b	1.3×10^5
B10.A[a]	k	k	k	k	k	k	d	d	3.1×10^3
B10.A(5R)[b]	b	b	b	b	k	k	d	d	30

[a] Significantly more susceptible than B10, B10.A (4R) ($p < 0.02$), and B10.A (2R) ($p < 0.05$).
[b] Significantly more susceptible than B10, B10.A (4R), B10.A (2R) ($p < 0.001$), and B10.A ($p < 0.02$).

impairment of the response to the first. There is extensive experimental evidence consistent with this model (Müllbacher 1981b; Müllbacher et al. 1981; Müllbacher, Blanden & Brenan 1983).

The second theory, concerning the action of Ir genes through antigen presentation (Blanden 1980), depends on the key feature of MHC restriction, i.e. MHC and X antigens must be on the same cell in order to trigger T cells that recognize them. The unique operational feature of combinations of MHC and X antigen molecules in or on the same cell surface may be an intimate physical association. Such associations may be impossible to achieve when MHC and X are in or on separate cell surfaces. Thus low T-cell responses could occur when certain combinations of MHC and X antigens in or on an antigen-presenting cell fail to achieve the optimal physical association required for T-cell stimulation. There is no definitive evidence for or against such an idea but it can explain only simple Ir-gene effects, not the complex phenomena encompassed by the cross-tolerance model.

MHC and susceptibility to disease

MHC genes could influence susceptibility to disease by their effects upon protective or harmful T-cell responses. In the case of infectious diseases, inefficient induction of protective T-cell responses could result from:

1. Quantitatively insufficient MHC gene expression on transformed or infected cells or antigen-presenting cells.

2. Qualitative resemblance of one self-MHC antigen to a second MHC antigen plus X (from the infectious agent), so that self-tolerance suppresses the response to MHC + X.

3. Failure of physical association of self-MHC antigens and antigens of infectious agents.

Inefficient effector function of protective T-cell responses could also result from quantitatively insufficient MHC gene expression on subsets of infected or transformed cells (in the case of viral infection or cell transformation). Finally, immunopathological diseases can result from T-cell responses against infection or autoimmune T cells can appear during certain infections, though the regulation of this phenomenon is still poorly understood. Only careful analysis of individual diseases can determine which of these possibilities applies. The remainder of this chapter will briefly review current analysis of ectromelia virus infection of mice.

Mechanisms of recovery from ectromelia virus infection in mice

Recovery from ectromelia infection in mice depends on virus-immune effector T cells that are class I MHC-restricted, as demonstrated in cell-transfer experiments (Blanden 1970; 1971a; 1971b; Kees & Blanden 1976). Such T cells are detectable in the spleen 4 to 5 days after subcutaneous infection in the foot with virulent virus, and they reach peak activity 6 to 8 days after infection and then decline (O'Neill, Blanden & O'Neill 1983). Effector T cells can migrate through the bloodstream to sites of infection, such as in the liver, where they trigger an influx of neutrophil polymorphonuclear granulocytes lasting several hours and a more protracted accumulation of mononuclear phagocytes (Blanden 1971b; 1974).

Once effector T cells and mononuclear phagocytes are present in virus-infected lesions and continue to accumulate via blood, the complete clearance of infection takes place (Blanden 1971a; 1971b). The precise mechanisms by which clearance is achieved are difficult to define in vivo, but evidence from experiments in vitro point to three possibilities. First, ectromelia-infected cells are vulnerable to lysis by cytotoxic T cells within 1 h after infection (Ada et al. 1976), whereas completion of the viral replication process takes many hours. Hypothetically, cytotoxic T cells acting alone could prevent further spread of infection, and this is indeed observed when immune T cells are injected into infected recipient mice that have been irradiated to abrogate monocyte production (Blanden 1971a). Second, infection of mononuclear phagocytes by ectromelia virus is relatively unproductive (Blanden 1971b; Roberts 1964). Thus, phagocytosis of virions by macrophages can contribute to elimination of infection. Third, γ-interferon produced locally by activated T cells may also limit viral spread and enhance the efficiency of the clearance of infection.

Genetics of susceptibility to ectromelia virus infection

A nonMHC gene causes very large differences between inbred mouse strains in their susceptibility to infection with virulent ectromelia virus via the 'natural' route of infection subcutaneously in the foot (Fenner 1949; R.V. Blanden, B.D. Deak & H.O. McDevitt unpublished). This gene operates by controlling the rate at which infection spreads from the inoculation site to the visceral target organs — the liver and spleen (O'Neill & Blanden 1983). In susceptible mouse strains such as A/J or BALB/c, spread of infection is rapid and the liver is destroyed by cytopathic infection as early as 5 days after infection before the protective T-cell response can reach an effective level. MHC genes have no discernible effect on the outcome.

However, in resistant mouse strains which have the C57Bl/6 or C57BL/10 (B10) background, infection spreads slowly enough for the T-cell response to protect the animal from death mediated by cytopathic liver infection. In these cases, MHC genes have a considerable influence on the outcome of infection as indicated by differences in the 50% lethal dose between strains of congeneic B10 mice which differ within the H-2 gene complex, the murine MHC (Table 1).

The cause of death in susceptible B10 background strains such as B10.A (5R) is unknown but apparently it is not massive liver damage. Death in B10.A (5R), which occurs 9 to 14 days after infection, is associated with a variety of clinical features such as gut haemorrhages, subcutaneous oedema and persisting infection in the spleen, and to a lesser extent the liver. These persist for at least 11 days after infection, by which time virus has been eliminated from resistant mouse strains. Histological examination shows that the liver lesions are resolving at this time, whereas large necrotic foci persist in the spleen (O'Neill, Blanden & O'Neill 1983).

The first hypothesis examined was that the class I MHC-restricted T-cell response in susceptible strains such as B10.A (5R) is smaller or slower than in resistant strains such as B10, B10.A (2R) or B10.A (4R). However, all strains generated strong cytotoxic T-cell responses, as assayed on virus-infected target cells *in vitro* (O'Neill, Blanden & O'Neill 1983). There was no obvious defect in responsiveness that might account for susceptibility. In this context, it is worth noting that a complex Ir-gene effect which has been studied comprehensively (Müllbacher, Blanden & Brenan 1983; Doherty et al. 1978; Zinkernagel et al. 1978) occurs in the cytotoxic T-cell responses of B10.A (2R) or B10.A (4R) mice to poxviruses (vaccinia and ectromelia). The presence of H-2K region genes of the k haplotype (H-$2K^k$) in these strains causes a very weak cytotoxic T-cell response to poxviruses in association with the H-$2D^b$ region genes (see Table 1 for H-2 haplotypes). This is apparently because D^b in combination with poxvirus cross-reacts with K^k, so that self-tolerance to K^k suppresses most of the cytotoxic T-cell repertoire capable of anti-D^6-poxvirus responses (Müllbacher, Blanden & Brenan 1983). Clearly a weak cytotoxic T-cell response restricted to D^b is not essential for resistance to virulent ectromelia infection because B10.A (2R) and B10.A (4R) are as resistant as B10, though B10 has a much stronger D^b-restricted anti-poxvirus cytotoxic T-cell response.

The second hypothesis tested was that the defect in susceptible strains of mice was at the level of effector mechanisms in tissues in which viral infection persisted, such as the spleen, rather than at the level of induction of T-cell responses. To test this, virus-immune spleen cells from (BALB/c x B6) F_1 hybrid donors were transferred intravenously into B6, (BALB/c x B6) F_1 hybrids and B10.A (5R) recipient mice that had been infected intravenously with virulent ectromelia virus 24 h previously (Table 2). The impact of the immune cells on the amount of infectious virus in the spleens of recipients was quantified by titrating virus plaque-forming units persisting in the spleens at 24 h after cell transfer. Reduction of virus titres in the spleens and livers of recipients in this protocol is known to be triggered by class I MHC-restricted T cells (Blanden 1971a; Kees & Blanden 1976). The data in Table 2 indicate that (BALB/c x B6) F_1 hybrid mice containing the K^b-D^d gene combination, which allows viral persistence and death in B10.A (5R) mice, can generate ectromelia virus-immune T cells that clear infection efficiently from spleens of B6 recipients. However, no significant clearance of infection occurred in (BALB/c x B6) F_1 or B10.A (5R) recipients, thus validating the hypothesis that persistent infection in the spleens of susceptible mouse strains is due to a failure of effector mechanisms rather than a failure to generate the T cells that trigger those mechanisms. The data in Table 2 are representative of several experiments in which susceptible mouse strains such as B10.A (5R), B10.G, (B10 x B10.A 5R) F_1 hybrids or (B10.G x

Table 10.2. Antiviral effects of Ectromelia-immune T cells from BALB c × B6 F_1 donors in different strains of recipients. Donors were immunized intravenously with 2×10^4 pfu of attenuated Hampstead egg strain Ectromelia virus (Blanden 1971*a*) 5 days before their spleen cells were transferred intravenously to recipients (5×10^7 viable cells per recipient). Groups of 4 or 5 recipients were infected intravenously with 1.5×10^3 pfu of virulent Moscow strain Ectromelia virus 24 h before receiving immune spleen cells. Their spleens were removed for virus titration (Blanden 1940) 24 h after immune cell transfer.

Recipient strain	Dose of immune cells	H-2 class I genes		Mean $\log_{10}$ virus titre in recipient spleen ± S.E.
		K	D	
B6	5×10^7	b	b	<2.0
	0	b	b	5.16 ± 0.27
(BALB/c × B6)F_1	5×10^7	b/d	b/d	5.89 ± 0.15
	0	b	d	6.35 ± 0.07
B10.A (5R0)	5×10^7	b	d	6.01 ± 0.18
	0	b	d	6.19 ± 0.18

B10) F_1 hybrids provided virus-immune T cells that cleared infection efficiently from the spleens of B10 or B6 (resistant) recipients, but failed to reduce significantly infection in the spleens of syngeneic (susceptible) recipients (data not shown).

In some of the these experiments it was shown that infection was efficiently cleared from the livers of infected susceptible mice which received immune T cells, while at the same time infection was not cleared from the spleen of the same animals, indicating that the failure of effector mechanisms was not generalized.

In this model, class I MHC genes influence the process of viral clearance that is triggered by class I MHC antigen-restricted virus-immune T cells. The simplest hypothesis to account for this effect would involve class I MHC gene expression on the surface membranes of productively infected subsets of splenocytes. Thus, a splenocyte subpopulation (possibly a small proportion of the total spleen cell population) which allows efficient viral replication while displaying insufficient class I MHC antigen on the cell surface for efficient T-cell recognition could account for viral persistence in the spleen. This hypothesis is currently under investigation.

References

Ada, G.L., Jackson, D.C., Blanden, R.V., Tha Hla, R. & Bowern, N.A. (1976). Changes in the surface of virus-infected cells recognized by cytotoxic T cells. 1. Minimal requirement for lysis. *Scandinavian Journal of Immunology*, **5**, 23-30.

Benacerraf, B. & McDevitt, H.O. (1972). Histocompatibility-linked immune response genes. *Science*, **175**, 273-79.

Blanden, R.V. (1970). Mechanisms of recovery from a generalized viral infection: mousepox. 1. The effects of anti-thymocyte serum. *Journal of Experimental Medicine*, **132**, 1035-54.

Blanden, R.V. (1971a). Mechanisms of recovery from a generalized viral infection: mousepox. 2. Passive transfer of recovery mechanisms with immune lymphoid cells. *Journal of Experimental Medicine*, **133**, 1074-89.

Blanden, R.V. (1971b). Mechanisms of recovery from a generalized viral infection: mousepox. 3. Regression of infectious foci. *Journal of Experimental Medicine*, **133**, 1090-1104.

Blanden, R.V. (1974). T-cell response to viral and bacterial infection. *Transplantation Reviews*, **19**, 56-88.

Blanden, R.V. (1980). How do immune response genes work? *Immunology Today*, **1**, 33-36.

Doherty, P.C., Blanden, R.V. & Zinkernagel, R.M. (1976). Specificity of virus-immune effector T cells for H-2K or H-2D compatible interactions: implications for H-antigen diversity. *Transplantation Reviews*, **29**, 89-124.

Doherty, P.C., Biddison, W.E., Bennink, J.R. & Knowles, B.B. (1978). Cytotoxic T-cell responses in mice infected with influenza and vaccinia viruses may vary in magnitude with genotype. *Journal of Experimental Medicine*, **148**, 534-43.

Fenner, F. (1949). Mouse-pox (infectious ectromelia of mice): a review. *Journal of Immunology*, **63**, 341-73.

Kees, U. & Blanden, R.V. (1976). A single genetic element in H-2K affects mouse T-cell anti-viral function in poxvirus infection. *Journal of Experimental Medicine*, **143**, 450-76.

Langman, R.E. (1978). Cell-mediated immunity and the major histocompatibility complex. *Reviews of Physiology, Biochemistry & Pharmacology*, **81**, 1-34.

Müllbacher, A. (1981a). Natural tolerance: a model for Ir-gene effects in the cytotoxic T-cell response to H-Y. *Transplantation*, **32**, 58-60.

Müllbacher, A. (1981b). Neonatal tolerance to alloantigens alters major histocompatibility complex-restricted patterns. *Proceedings of the National Academy of Sciences of the USA*, **78**, 7689-91.

Müllbacher, A. & Blanden, R.V. (1979). H-2 linked control of cytotoxic T-cell responsiveness to alphavirus infection: presence of H-2D^k during differentiation and stimulation converts stem cells of low responder genotype to T cells of responder phenotype. *Journal of Experimental Medicine*, **149**, 786-90.

Müllbacher, A., Blanden, R.V. & Brenan, M. (1983). Neonatal tolerance of major histocompatibility complex antigens alters Ir-gene control of the cytotoxic T-cell response to vaccinia virus. *Journal of Experimental Medicine*, **157**, 1324-38.

Müllbacher, A., Sheena, J.H., Fierz, W. & Brenan, M. (1981). Specific haplotype preference in congeneic F1 hybrid mice in the cytotoxic T-cell response to the male specific antigen H-Y. *Journal of Immunology*, **127**, 686-89.

O'Neill, H.C. & Blanden, R.V. (1983). Mechanisms determining innate resistance to ectromelia virus in C57Bl mice. *Infection & Immunity*, **41**, 1391-94.

O'Neill, H.C., Blanden, R.V. & O'Neill, T.J. (1983). H-2-linked control of resistance to ectromelia virus infection in B10 congeneic mice. *Immunogenetics*, **18**, 225-66.

Reed, L.J. & Muench, H. (1938). A simple method of estimating fifty per cent end points. *American Journal of Hygiene*, **27**, 493-98.

Roberts, J.A. (1964). Growth of virulent and attenuated ectromelia virus in cultured macrophages from normal and ectromelia-immune mice. *Journal of Immunology*, **92**, 837-42.

von Boehmer, H., Haas, W. & Jerne, N.K. (1978). MHC-linked immune-responsiveness is acquired by lymphocytes of low responder mice differentiating in thymus of high-responder mice. *Proceedings of the National Academy of Sciences of the USA*, **75**, 2439-42.

Zinkernagel, R.M. & Doherty, P.C. (1974a). Restriction of *in vitro* T cell-mediated cytotoxicity in lymphocytic choriomeningitis within a syngeneic or semiallogeneic system. *Nature*, **248**, 701-2.

Zinkernagel, R.M. & Doherty, P.C. (1974b). Immunological surveillance against altered self-components by sensitized T lymphocytes in lymphocytic choriomeningitis. *Nature*, **251**, 547-48.

Zinkernagel, R.M. & Doherty, P.C. (1979). MHC-restricted cytotoxic T cells: studies on the biological role of polymorphic major transplantation antigens determining T-cell restriction specificity, function and responsiveness. *Advances in Immunology*, **27**, 51-177.

Zinkernagel, R.M., Althage, A., Cooper, S., Kreeb, G., Klein, P.A., Sefton, B., Flaherty, L., Stimpfling, J., Shreffler, D. & Klein, J. (1978). Ir genes in H-2 regulate generation of anti-viral cytotoxic T cells: mapping to K or D and dominance of unresponsiveness. *Journal of Experimental Medicine*, **148**, 592-612.

11

The bovine major histocompatibility complex and disease resistance

T.E. ADAMS and M.R. BRANDON

Genes in the major histocompatibility complex (MHC) have a profound effect on the ability to respond to specific antigens. This suggests that MHC genes may have a major effect on resistance to diseases in which the immune system is protective or in autoimmune diseases in which the immune system itself may initiate the disease process. The most important question in relation to ruminants is whether MHC products can be used as a means of selecting for disease resistance. At present, studies in ruminants are directed to finding associations between resistance to particular diseases or parasites and class I antigens. Little work has been done on defining the class II antigens of the ruminant MHC; this still requires substantial effort. A novel approach which is now feasible, due to recent advances in molecular biology and gene manipulation, involves the cloning of the MHC genes of ruminants and the subsequent insertion of these genes into the germline of ruminants and other species. The insertion of MHC genes from other animals into the ruminant germline is also a possibility. This approach will allow a better understanding of the role of the MHC in resistance and susceptibility to bacterial, viral and parasitic infections.

The MHC, immune response genes and disease resistance

The concept that hereditary factors may influence an individual's resistance or susceptibility to toxins, infectious agents and allergic diseases is not new. Early studies on animals demonstrated that genetic factors could influence the immune response to complex natural antigens, the natural level of certain antibodies and natural resistance to a number of pathogenic organisms. For example, Wright and Lewis (1921) found that the susceptibility of different strains of guinea pigs to tuberculosis was genetically determined. In 1943, Schiebel showed that by selective inbreeding of guinea pigs it was possible to develop strains of animals which were either good or poor producers of diphtheria antitoxin.

Studies in inbred strains of mice yielded similar results. Gorer and Schutze (1938) found that differences in the susceptibility to *Salmonella* infection in strains of mice could be correlated to differences in antibody levels, and Carlinfanti (1948) found hereditary influences on the levels of spontaneous red-cell alloagglutination. Other investigators found that strains of inbred mice differ in their responses to complex antigens, such as egg albumin and type I pneumococcal polysaccharide (Fink & Quinn 1953), tetanus toxoid (Ipsen 1959) and sheep red cells (Dineen 1964).

In 1963, the first specific immune response (Ir) genes were recognized, as a result of work by Kantor, Ojeda and Benacerraf with guinea pigs immunized with different 2-4 dinitrophenyl copolymers. These researchers found they could divide guinea pigs into two groups as measured by their response to these antigens. The high and low response characteristics of the two groups were genetically determined. In 1965, McDevitt and Sela reported similar genetic control of immune responses to synthetic copolymers in mice. Subsequent work established that in both the guinea pig (Ellman et al. 1970; Bluestein, Green & Benacerraf 1971) and the mouse (McDevitt & Chinitz 1969; McDevitt et al. 1972) autosomal dominant genes control immune responsiveness: these Ir genes are located within a particular region of the chromosome, the I region, near or within the MHC. Ir genes were found to exist, not only to synthetic antigens in these species, but also to complex natural antigens including ovomucoid, serum albumin, mouse myeloma IgA, cell-surface alloantigens such as Thy 1 and the male mouse histocompatibility antigen (Green 1974).

Other studies established that the ability to respond to some antigens is under polygenic control linked to the MHC. For example, the response to Thy 1.1 in mice is under the control of four loci, two within the H-2 complex, one loosely linked to H-2 and one independent of H-2 (Zaleski & Klein 1974). Examples of dual H-2-linked Ir-gene control (gene complementation) have also been found (Stimpfling & Durham 1972; Zaleski & Milgrom 1973), in particular control in mice of the immune response to a terpolymer of glutamic acid, lysine and phenylalanine (GLO). The observation that the F_1 hybrids from two nonresponder strains were able to mount an immune response to GLO was resolved by the identification of two complementing Ir genes, one located in the I-C and the other located in the I-A subregion (Dorf & Benacerraf 1975). The immune response to a number of other synthetic peptide antigens is also under the control of complementary Ir genes (Benacerraf & Dorf 1976; Benacerraf & Germain 1978).

Much attention has been given to identifying the Ir-gene product and the level at which it is expressed. Originally it was thought that this was a receptor present on T cells (Katz & Benacerraf 1975) and the demonstration of T cell-derived, Ia-positive, antigen-specific helper and suppressor factors (Taussig et al. 1975; Tada, Taniguchi & David 1976; Thze et al. 1977) gave this idea some credence. However, it was demonstrated that the Ir defect, resulting in an inability to mount a humoral response when challenged with antigen, could also be expressed at the level of the B cell (Katz et al. 1973; Kapp, Pierce & Benacerraf 1975) and the macrophage (Rosenthal & Shevach 1973; Benacerraf & Germain 1978) independent of T-cell function. These observations suggested that Ir genes can manifest their control at various cellular levels involved in generating the immune response. However, a number of investigators have suggested that the class II MHC antigens themselves may be the Ir-gene products (Miller 1978; Zaleski & Klein 1978; Klein et al. 1981).

The existence of specific Ir genes has been demonstrated in other species. MHC-linked gene control of antibody production has been shown in rats to a variety of natural and synthetic antigens (Wurzburg 1971; Gunther, Rude & Stark 1972; Gunther et al. 1973; Wurzburg, Schutt-Gerowitt & Rajewsky 1973; Amerding, Katz & Benacerraf 1974). The antibody response to at least 15 antigens has been shown to be regulated by Ir genes mapping to the RT1 complex (Gunther & Stark 1977), but distinct from the major class I locus (RT1.A) as shown by the use of genetic recombinants (Stark et al. 1977). Mechanisms regulating responder/nonresponder status appear to be the same as those operating in mice and guinea pigs (Wurzburg, Schutt-Gerowitt & Rajewsky 1973; Rude & Gunther 1974; Gunther & Rude 1975). Recent studies have highlighted the close association between Ia specificities, MLR phenotypes and Ir genes in wild and inbred rats. This close association between alleles at different genetic loci suggests a restricted polymorphism at the RT.1 loci due to strong selective pressures on some genes (Gunther 1979; Shonnard et al. 1979).

Ir-gene control of immune responses, linked to the RhLA, has been demonstrated in rhesus monkeys, most notably for the synthetic antigens GA and DNP-GL (Balner et al. 1973; Dorf et al. 1974a; Dorf, Balner & Benacerraf 1975). The unavailability of antisera identifying class II Rh LA specificities did not allow analysis of associations with particular classes of MHC antigens.

DLA-linked genetic control of antibody responses to the synthetic antigens GA and GT has been demonstrated in dogs (Vriesendorp, Grosse-Wilde & Dorf 1977) and apparent Ir-gene control over the humoral response to chicken lysozyme has been reported in pigs, linked to the SLA complex (Vaiman et al. 1975). Detailed studies in chickens have determined the existence of genetic influences, linked to the B complex, on antibody responses to DNP-IgG (Balcarova et al. 1974), (T,G)-A--L (Balcarova et al. 1975), GAT (Benedict et al. 1975) and tuberculin (Karakoz et al. 1974). The discovery of a recombination between Ir genes coding for GAT responsiveness and the class I antigens of the B complex indicates that these particular genes are closely linked to, but separate from, the class I genes of the B complex (Pevzner, Trowbridge & Nordskog 1978). Other studies in chickens have highlighted the polygenic nature of genetic control of antibody responses to simple and complex antigens, by genes linked to or independent of the B complex (Koch, Hala & Srup 1977; Palladino et al. 1977).

HLA-linked Ir genes in man have yet to be directly identified, and the indirect epidemiological evidence is equivocal (van Rood, de Vries & Munro 1977). Associations of differences in humoral antibody responses with HLA antigens resembling Ir-gene regulation have been reported for exposure to tetanus toxoid (Sasazuki et al. 1978), malaria (Osoba et al. 1979), diptheria and measles (Haverkorn et al. 1975) and pollen allergens (Marsh & Bias 1977). The development of suitable *in vitro* assays of immunological responsiveness, as established for synthetic antigens in mice (Kapp, Pierce & Benacerraf 1973), is of critical importance in overcoming inherent problems of human experimentation, if the existence of HLA-linked Ir genes is to be unequivocally confirmed.

Evidence for the existence of human Ir genes has come from another direction. Over the past decade, with the establishment of a more precise genetic map of the HLA system, and with the increase in the number and degree of standardization of specificities defined by tissue typing, there has been a rapid accumulation of data associating disease susceptibility to the genes of the MHC. These associations of disease states with HLA antigens have been extensively reviewed (Albert & Gotze 1977). From the first reported association between Hodgkins disease and HLA specificities (Amiel 1967), definite associations between certain HLA antigens and a variety of disease states have been found, including rheumatic, autoimmune, neurologic, endocrine, oncogenic gastrointestinal and dermatologic conditions. The strongest associations include ankylosing spondylitis, where the B27 antigen is found in over 90% of clinical cases, coeliac disease with B8 and psoriasis with B13. Albert & Gotze (1977) highlighted a number of features associated with

1. There is a high familial incidence of the disease, although the mode of inheritance is complex.

2. Immunological mechanisms are suspected in the pathogenesis of most HLA-associated disease.

3. Generally the strongest HLA associations are found to involve antigens

Until now, evidence relating to the existence of Ir genes and their role in disease resistance has been based both on epidemiological studies (particularly in man) and experiments in which immune responses to synthetic and natural antigens were studied. However, there is striking evidence available in mice and chickens for a direct role of the MHC in conferring disease resistance/susceptibility to an individual against a biological pathogen.

Lilly (1966; 1971) demonstrated that resistance/susceptibility to Gross virus leukaemogenesis (GvL) in mice is under the control of several genes. One of these, the Rgv-1 gene, is linked to the H-2 type. Thus, H-2^k mice are very susceptible to GvL, but H-2^b are resistant. Resistance is dominant, so that a H-2^b/H-2^k hybrid is resistant. Rgv-1 has subsequently been mapped to the I region of the H-2 complex, along with other genes influencing resistance to a number of radiation-induced viral leukaemias (Sato et al. 1973). Resistance to Friend virus leukaemogenesis (FvL) (Chesebro, Wehrly & Stimpfling 1974), AKR thymoma cells (Meruelo, Deak & McDevitt 1977) and Moloney leukaemia virus (Debre et al. 1979) is also under I-region control. H-2-associated susceptibility to a number of other oncogenic viruses has been documented (Tennant & Snell 1968), while the polygenic nature of resistance/susceptibility is well established, involving H-2 and nonH-2 loci to such pathogens (Debre et al. 1979).

Recently the H-2 complex has been found to influence susceptibility to infectious agents other than viruses. The pronounced role of the H-2 complex was observed in mice infected with *Trichinella spiralis* (Wassom, David & Gleich 1979) and *Shistosoma mansoni* (Claas & Deedler 1979), while in mice susceptible to *Leishmania donovani* infection, long-term recovery is linked to H-2 haplotype (Blackwell, Freeman & Bradley 1980). In all of these studies, the polygenic nature of mechanisms playing a role in recovery is emphasized and again found to encompass both H-2 and nonH-2 genes.

MHC-associated resistance to disease has been studied extensively in chickens. Chickens homozygous for the B^2 or B^{21} alleles are resistant to the contagious lymphoma of Marek's disease (Pazderka et al. 1975), while other studies have established B^{21}-associated allele(s) as dominant resistance genes (Stone, Briles & McGibbon 1977). Paradoxically researchers after Pazderka et al. (1975) have found B^2/B^2 birds to be highly susceptible to Marek's disease.

The B locus has also been found to influence susceptibility of birds to Rous sarcoma virus-induced tumours. Schierman, Watanabe & McBride (1977) and Collins et al. (1977) have demonstrated that progression or regression of the virally induced tumours is dependent on alleles mapping close to or within the MHC. The capacity to cause regression of the tumour is dominant and in both groups there is an association with the B^2 allele.

Bacon, Kite and Rose (1974) have also found an association between B-locus alleles and the development of autoimmune thyroiditis. Pevzner, Nordskog & Kaeberle (1975) hypothesized that the higher mortality found in B1 homozygous female chickens injected with *Salmonella pullorum* is associated with an Ir gene. B^1 homozygotes produce less antibody to *S. pullorum* than either B^1B^2 or B^1B^{19} heterozygotes.

Pevzner, Trowbridge and Nordskog (1978) have presented evidence for a crossover between the genes coding for class I antigens and a gene (Ir-GAT) controlling immune response to the synthetic polypeptide GAT within the B complex, the MHC of chickens. The Ir-GAT^1 and Ir-GAT^{19} alleles control low and high immune response to GAT, respectively. Both low and high responders were recovered as recombinants from B^1 and B^{19} birds. The low-responder haplotypes are homozygous for the Ir-GAT^1 allele and the high-responder haplotyes carry the Ir-GAT^{19} allele. Mortality during the first 6 months after hatch for B^1B^1 nonresponder birds was 39%, compared with 19% for the B^1B^1 high responders. This agreed with earlier observations of mortality differences associated with B-locus alleles. Additionally, B^1 homozygote low-responders have a low response to *S. pullorum* and are susceptible to Marek's disease (Nordskog et al. 1977). In effect, the B^1 allele, or a gene close to it, is associated with general low immune competence in chickens. The polygenic control of the responses to disease in chickens has also been emphasized, in particular in the studies of Crittendon et al. (1972), who found birds identical serologically for the B locus that differed in their susceptibility to Marek's disease. Whether this was manifested by closely linked Ir genes which underwent recombination is not known.

The functional and biological role of the MHC

Although the MHC was first identified on the basis of its role in regulating rejection of allografts between individuals of a species, its principal biological function is unlikely to be in clinical transplantation. That the MHC has an important biological role may be inferred from the fact that it is perpetuated in phylogeny as a gene cluster whose products retain remarkable structural and functional homology, an observation indicative of selective forces at work. In the mammalian and avian species so far studied, the MHC is characterized by genes coding for major cell-surface antigens, for the ability to mount an immune response to a variety of antigens and for components of the complement system. Certainly the demonstration of Ir genes and of genes linked to the MHC influencing the susceptibility or resistance of an individual to disease is of immense biological importance.

A number of pioneering studies (Katz, Hamaoka & Benacerraf 1973; Shevach & Rosenthal 1973; Zinkernagel & Doherty 1974) implicated a biological role for the MHC in two immunologically important areas, protection of the host against infectious viral agents and regulation of immune cell-cell interaction in the generation of immune responses. As a result of these early observations, a great deal of attention has been directed towards elucidating the functional role of the MHC and its gene products in these areas. This subject has been extensively reviewed (see Zinkernagel & Doherty 1979).

Insight into the function and biological role of the MHC has been forthcoming only recently and often conflicting results have been reported. It is hoped that further research will resolve such differences so that a better understanding may be reached of how and why the MHC functions. This knowledge will be particularly valuable, not only in understanding immune functions, but also in clinical terms relating to HLA disease associations and to the role the MHC could play in domestic livestock immunogenetics.

The bovine MHC
Current research on the bovine MHC
So far only a limited number of serologically defined antigens (class I) have been described accurately for the bovine MHC. Evidence suggests that most of the presently defined specificities belong to the same genetic system, which may be single-locus or multilocus in nature (Amorena & Stone 1978; Spooner et al. 1978; Caldwell 1979; Adams 1980; Stear, Newman & Nicholas 1982).

Aspects of disease resistance and immune responsiveness in cattle
Differences in disease resistance are reported between various breeds of cattle and amongst individuals within breeds, though these are poorly documented. Spooner, Bradley and Young (1975), in reviewing this topic with relevance to dairy cattle, cited a number of cases where there were definite genetic and breed influences on the occurrence and heritability of disease states. Such diseases include mastitis, milk fever, ketosis and hypomagnesaemia. Two calfhood diseases mentioned are of particular interest, neonatal diarrhoea and pneumonia, because susceptibility to both (as with other calfhood diseases) depends on the amount of protective colostral antibody obtained from the mother, a factor itself that may be genetically influenced.

Studies in West Africa on the susceptibility of different breeds of cattle to trypanosomiasis, a disease of major importance in animal production (Murray, Morrison & Whitelaw 1982), showed that Zebu cattle suffer high mortality in response to experimental or natural trypanosome infection, whereas N'Dama cattle, a so-called 'trypanotolerant' breed, are able to survive. Differences in the susceptibility of the two breeds are reflected in differences in the levels of parasitaemia and severity of anaemia. A concurrent study revealed both trypanosome-resistant and trypanosome-susceptible strains of inbred mice.

Studies in Australia also indicate genetic control of disease resistance in cattle. For example, Dodt (1977) has shown that infectious kerato-conjunctivitis (pink-eye) occurs far less in *Bos taurus* x *Bos indicus* crosses than in *Bos taurus* cattle, and that the disease is of shorter duration and less destructive in the crossbred animals. This does not appear to be a result of any physical characteristic associated with *Bos indicus* strain. Studies in India and Japan have suggested different levels of resistance to brucellosis, babesiosis and tuberculosis in various breeds of cattle (K. Kondo personal communication).

Other studies indicate genetic differences with respect to antibody production. Dimmock (1973) found marked breed differences in the antibody response of cattle immunized with red blood cells. Similarly, Sellei and Rendel (1968) have observed considerable variation in antibody titres within dizygotic twin pairs immunized with cattle red blood cells, apparently as a result of genetic factors. Lie (1979) has described genetic influences regulating the antibody response of young bulls inoculated with human serum albumin.

Differences in immune responses and susceptibility to disease could result from immune response genes operating within the bovine MHC, particularly in view of the results of studies in other animals. If so, this leaves the opportunity open for the possible exploitation of genetic markers in the selective breeding of animals for disease resistance. A detailed study of the bovine transplantation antigens, with the long-term aim of analysing bovine Ir genes is obviously of great potential and importance for animal production.

Allograft survival
Evidence that polymorphic lymphocyte surface antigens, defined either by serological or cellular techniques, represent gene products of the MHC has been provided by their role in influencing allograft survival. Matching donor-recipient pairs on the basis of shared specificities leads to significant prolongation of allografts in humans, rhesus monkeys, pigs, mice and rats (Gotze 1977; Morris, Batchelor & Festenstein 1978; Ivanyi

186 T. E. Adams and M. R. Brandon

Table 1. The relationship between histoincompatibility for BoLa class I antigens and skin allograft survival in cattle. Lymphocyte antigen typing was performed by a microlympho-cytotoxicity test using a large panel of BoLA class I typing sera (Adams 1980). Skin grafts were examined 5 days after operation and subsequently at 2-day intervals. All grafts were read by a third party who had no knowledge of the BoLa status between heifer pairs. All grafts were performed in duplicate.

Donor	Recipient	Shared antigens	Allograft survival (days)
159	171	none	10, 10
171	159		10, 10
141	112	none	10, 10
112	141		12, 12
144	167	none	12, 12
167	144		10, 10
163	116	3	17, 17
116	163		17, 17
98	72	2	17, 17
72	98		15, 15
71	136	2	17, 17
136	71		14, 14

1979). These observations in outbred, heterogeneous populations of animals have been facilitated by using siblings or littermates for donor-recipient combinations for grafting, thereby reducing the influence of undetected (null) MHC antigens and presumably the cumulative effects of minor H loci.

Preliminary evidence from Amorena and Stone (1978) showed significant enhancement of skin allograft survival between cattle matched for class I antigens. To confirm this observation, skin grafts were performed between pairs of cattle which did or did not share class I specificities.

A highly significant enhancement of skin allograft survival was obtained when heifer donor-recipient pairs were matched on the basis of shared class I antigens (Tables 1 and 2). Allografts performed between unrelated heifer pairs sharing two or more antigens survived on average 6 days longer than allografts performed between class I mismatched, unrelated heifers. Skin grafts between class I mismatched heifers were rejected 10 to 12 days after grafting (mean of 10.67 ± 0.98 SD), while those exchanged between class I matched pairs survived 14 to 17 days after operating (mean 16.17 ± 1.27 SD), a highly significant ($p > 0.001$) prolongation of graft survival (see Figure 1).

Table 2. Comparison of the mean survival time of skin allografts between heifers compatible and incompatible for BoLA class I antigens. Survival times of allografts for individual heifers are shown in Table 1.

Donor-recipient relationship	Number of grafts	Mean survival time (days) ± SD
Shared class I antigens	12	16.17 ± 1.27
Incompatible for class I antigens	12	10.67 ± 0.98

t = 12.14; 22 d.f.; *p* <0.001

The results of these experiments suggest that the serologically defined BoLA specificities represent major transplantation antigens of cattle, and thus may be considered products of a bovine MHC analogous to that described in other species. The significant enhancement of allograft survival obtained by matching donor-recipient heifer pairs for two or more class I antigens, an average of 6 days, parallels observations made by others using related individuals in other species when defining MHC antigens. Grafts between pig siblings matched for class I antigens survived an average of 10 days, while those between mismatched siblings were rejected on average 7 days after grafting (Vaiman et al. 1970). Similarly, skin allografts performed between siblings matched for lymphocyte antigens survive up to 10 days longer than those between mismatched siblings in humans (Dausset et al. 1970), dogs (Cohen & Kozaki 1969) and rhesus monkeys (Balner et al. 1971).

Bovine B-lymphocyte alloantigens

Until recently, the existence of an MHC region coding for class II antigens in cattle was only speculated (Adams, Brandon & Morris 1977). However, Usinger, Curie-Cohen and Stone (1977), using the results of all paired one-way mixed lymphocyte culture tests on families of half siblings, have suggested that the lymphocyte-defined system in cattle contains a minimum of two loci. Recently, Newman, Adams and Brandon (1982) published evidence of a B-lymphocyte alloantigen system in cattle. The high degree of polymorphism, the restricted distribution on blood cells and the evidence for linkage to bovine class I antigens suggest that this system represents the bovine equivalent of the class II antigens of other species.

A two-colour fluorescence micro-lymphocytotoxicity test was performed as described by van Rood, van Leeuwen and Ploem (1976) with minor modifications. This technique allows the identification of B lymphocytes (having cell-surface immunoglobulin) by the presence of a green fluorescent FITC 'tag' on their surface and simultaneously the detection of cytotoxicity by the uptake of ethidium bromide (red

Figure 1. An autograft (AT) and a rejected allograft (AL) seen from two different angles 10 days after grafting was performed.

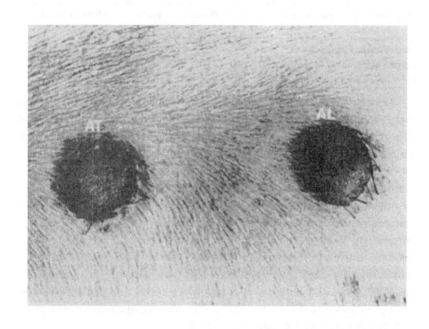

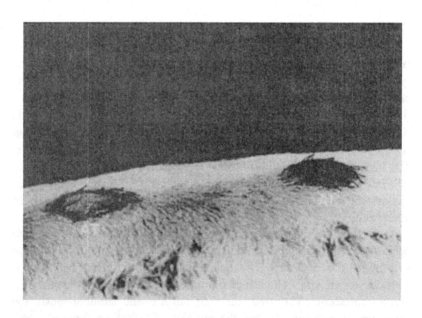

fluorescence) stain by dead cells. In this way, cytotoxic antibodies against alloantigens specific to B lymphocytes can be detected.

Typing sera were prepared by removing antibody to class I antigens from alloantisera produced by immunization of steers with peripheral blood lymphocytes. Antisera were absorbed with appropriate cells, (platelets which express only class I antigens and leucocytes or lymphocytes which express relevant class I antigens, but not the class II antigens under study) at room temperature for 1 h and care was taken to limit any nonspecific absorption of the antibody by using as few cells as possible. The absorbed antisera were used to type 110 cattle from 10 paternal half-sibling families, as well as a panel of 100 unrelated cows. The data obtained from the families were used to define ten B-lymphocyte antigen specificities genetically. Confirmation was obtained from the absorption data and statistical analysis of the typing results from the panel of 100 unrelated cattle. The CHI and CLUSTER computer programs were used (Pickbourne et al. 1977). Within the limits of the analysis, antisera were found that recognized one, two, three or four different B-lymphocyte specificities. Table 3 summarizes the results of the absorption analysis of the anti-B-lymphocyte sera.

The ten genetically defined specificities were present in 41 of 110 cattle from the families, while 58 individuals (53%) did not express any antigens detectable by the typing sera. Although not all of the cattle in the families expressed antigens identified by the antisera, there was a sufficient number to show that the antigens were inherited in an autosomal codominant fashion (Table 4). The family studies and population data suggest that the bovine B-lymphocyte antigen system is highly polymorphic. It can be assumed that because of the high percentage of blank individuals (53%) and the presence of 10 partially defined specificities, the degree of polymorphism within this system will be shown to be much greater as more typing antisera become available.

Ir genes

Evidence suggesting the genetic control of immune responsiveness and disease resistance has been found in cattle (Sellei & Rendel 1968; Dimmock 1973; Spooner, Bradley & Young 1975; Dodt 1977; Lie 1979). However, the genetic definition of the bovine MHC is still in its infancy and no experimental studies are known in which an attempt has been made to confirm the existence of Ir genes in cattle.

The use of antigens with few components or simple structures (e.g. example synthetic amino-acid polymers), a limiting dose of complex antigens, or alloantigens differing only slightly from self-components has

Table 3. Complexity of B lymphocyte-specific antisera as determined by absorption.

Minimum number of specificities recognized by an antiserum	Number of antisera
1	15
2	5
3	4
4	3

Table 4. Inheritance of bovine B-lymphocyte antigens in four informative families.

ID	Sire genotype	ID	Dam genotype	ID	Calf genotype	Breed
J–1	1/2	C–1	2/3	C–1A	2/3	Jersey
		C–2	2/3	C–2A	2/3	
		C–3	1/ [a]	C–3A	1/ [a]	
		C–4	1/4	C–4A	1/2	
		C–5	1/4	C–5A	1/4	
A117	7/ –	A148	–/–	463	–/–	Angus
		A90	6/ –	464	7/ –	
		A43	–/–	465	7/ –	
		A119	–/–	466	–/–	
		A52	3/ –	467	7/ –	
A110	1/ –	A8	–/–	202	–/–	Angus/ Holstein
		F10	6/ -	208	–/–	Friesian
		90	5/ [a]	206	5/ –	
		140	–/–	205	1/ –	
		137	–/–	201	–/–	
MG-2	8/ –	B52	8/ [a]	211	8/ [a]	Murray Grey
		147	–/–	217	8/ –	
		151	9/ [a]	221	9/ –	
		155	–/–	223	8/ –	
		152	–/–	220	–/–	

[a] Homozygosity cannot be excluded.

been critical in the identification of Ir genes. These antigens possess a small number of distinct determinants that trigger an immune response, and thus probe the immune system with the selectivity necessary to single out the function of highly specific Ir genes. A preliminary study has been done to identify bovine Ir genes to the synthetic polypeptides GA, GT,

Table 5. Cattle numbers and antigen doses used to study the existence of bovine Ir genes. Initial dose with Freunds complete adjuvant administered intramuscularly. Second dose administered intradermally without adjuvant.

Antigen	Immunizing dose (μg)	Secondary challenge (μg)	No. of cows	No. of calves	Bull
GA	100	10	10	10	1
GT	100	10	9	9	1
GAT	100	10	21	21	—
GAT	100	10	16	16	—
(T,G)-A--L	250	10	14	14	1

Table 6. Humoral antibody responses of cattle immunized with synthetic peptide antigens at 1/10 serum dilution.

Antigen	Number of cattle with a humoral response		
	<5% antigen bound	5-30% antigen bound	>30% antigen bound
GA	21	0	0
GT	19	0	0
GAT	74	0	0
(T,G)-A--L	2	23	4

GAT and (T,G)-A--L, and to establish the existence of genetic control over immune responsiveness. These antigens were chosen because in other species the immune response to these antigens is subject to Ir-gene control linked to the MHC.

The experimental approach adopted involved the study of the immune response to an antigen in selected, unrelated family groups of cattle, thus allowing the identification of any genetic control over responsiveness. Individual families were chosen where a number of unrelated cows were available along with their calves from a single sire. A total of 143 cattle were used, comprising five families. Each family was immunized with one antigen only. The number of cattle in each family, the antigen used and the immunizing dose are given in Table 5. Cattle received a deep intramuscular inoculation of antigen emulsified in Freund's complete adjuvant (final volume 0.5 ml), followed by a secondary immunization 21 days later with 10 µg of antigen in 0.1 ml of PBS (pH 7.8), administered intradermally. The presence of delayed-type hypersensitivity (DTH) reactions as a result of antigen challenge was noted 24 h later. Blood samples were collected before primary immunization and at 21 and 28 days afterwards. Serum was obtained and frozen at -20°C until assayed. Antibody titres were determined using a modified Farr assay (Dunham et al. 1973). A number of sera from cattle displaying DTH to GAT were also assayed by solid phase radioimmunoassay.

No humoral antibody response was detected in sera from bulls, cows or calves after primary or secondary immunization with either GA, GT or GAT (Table 6). In contrast, 27 of 29 animals immunized with the multichain polymer (T,G)-A--L mounted intermediate or high antibody responses by day 28 (Table 6). Two animals (one cow and one calf) were low- or nonresponders.

A positive skin-test response was observed in only 1 out of 21 cattle immunized with GA and in 2 out of 19 cattle immunized with GT. In direct contrast to the results obtained for antibody titres, 26 out of 74 cattle immunized with GAT showed positive or strongly positive skin responses. The opposite situation was found in cattle immunized with (T,G)-A--L, where despite demonstrable antibody responses, only 3 out of 29 were positive on skin testing. Results are presented in Table 7.

The variable results obtained in this preliminary study of the genetic control of the immune response in cattle to a number of synthetic peptide immunogens probably reflect the ambitious nature of the approach used. When studying a population of animals for the existence of Ir-gene control of the antibody response to an antigen, one would expect to encounter both responder and nonresponder individuals. This is supported by the observations made here with respect to the humoral response to (T,G)-A--L. It is, therefore, surprising to note that in 114 cattle immunized with GA, GT or GAT, not one showed a humoral response of any kind. The development of DTH in response to secondary challenge with antigen in a number of animals suggested that some stimulation of the immune response had occurred.

Immunization with 250 µg of (T,G)-A--L did invoke antibody responses, with evidence of responder and nonresponder cattle. These observations were compounded by the predominance of equivocal or intermediate responders (23 of the 27 cattle regarded as producing significant levels of antibody), which highlights the purely arbitrary choice of cut-off values for the classification of responder, intermediate or nonresponder cattle. Ideally an immunization regime is required which will maximize the differences between individual cattle, if they exist. The large number of intermediate responders found when dogs were immunized with GA (Vriesendorp, Grosse-Wilde & Dorf 1977) prompted the authors to suggest this was indicative of multigene control of the

Table 7. Cellular responses of cattle immunized with synthetic peptide antigens, measured in terms of delayed skin reaction 24 h after challenge and judged according to the extent of the induration.

Antigen	Sensitizing dose (μg)	no reaction	Skin test positive[a]	strongly positive[b]
GA	10	20	–	1
GT	10	17	1	1
GAT	10	48	11	15
(T,G)-A--L	10	26	2	1

[a] 75–100% increase in control skin thickness.
[b] > 100% increase in control skin thickness.

immune response. In chickens, antibodies have been established which recognize three different parts of the (T,G)-A--L polymer, controlled by three different loci, one of which is not MHC-linked. In mice, multilocus control of antibody production in response to GAT immunization has been reported (Dorf et al. 1974b).

There is no evidence for simple inheritance of parental responder/nonresponder status by calves in our study. This may be interpreted as further evidence for the complexity of the control of antibody responses to (T,G)-A--L. Out of 74 cattle immunized with GAT, 26 showed DTH when given a secondary challenge of 10 µg of antigen. DTH was much more prevalent in the calves than in their dams. The two bulls which sired the calves were not available, so they could not be evaluated with respect to cellular reactivity to antigen.

An apparent breed difference emerged between the Charolais and Brahman cattle in mounting a cellular response. Only seven Brahmans out of a total of 32 tested (including calves) displayed DTH, compared to 18 out of 42 Charolais. This difference was seen in both the dams and the calves and was not attributable to inherent skin thickness differences between the two breeds. Breed differences in cattle also appear to influence the animals' ability to mount humoral responses to red-cell immunization; Brahman crossbreeds produced far higher titres and maintained antibody levels much longer than a number of British breeds (Dimmock 1973), while the antibody response to keyhole limpet haemocyanin has been observed to be significantly different between *Bos taurus* and *Bos indicus* cattle systemically and locally challenged (Banyard 1980).

This study shows clearly that the definition of MHC-linked Ir genes in cattle will not only rely on the extensive analysis of the determinants of the bovine MHC, but will also require rigorous evaluation of experimental conditions (e.g. antigen type and dose and type of adjuvant), under which the function of these genes may be clearly observed.

Genetic manipulation of the sheep MHC

Sheep MHC genes

Genes of the sheep MHC have been cloned by using cloned human and mouse DNA for the detection of the homologous sheep DNA sequences. Sheep genomic libraries were screened with a human cDNA probe specific for human MHC class I genes. The ^{32}P-labeled probe was hybridized with phage and immobilized on nitrocellulose filters, and the hybridizing clones were isolated and purified. At present, seven clones containing sheep MHC class I genes have been isolated. These genomic clones are currently being characterized (U. Novak unpublished). Clones have been expanded and the sheep DNA inserts of each clone investigated using restriction endonucleases. Transformation of thymidine kinase-negative C3H mouse L cells with selected clones, together with herpes simplex virus tk gene, is in progress to produce transformants expressing sheep MHC class I molecules (U. Novak unpublished). Transformants expressing sheep MHC class I molecules are being screened by radioimmunoassay with monoclonal antibodies to sheep class I antigens (M.R. Brandon, Mackay & Gogolin unpublished) and sheep alloantisera.

Gene transfer of sheep MHC genes

Research has begun on methods to transfer mammalian genes, and in particular MHC genes, to the germline of sheep. At present the only successful approach for transferring genes into the mammalian germline is by microinjection of the male pronucleus into fertilized mouse eggs. This requires both the visualization of the male pronucleus and a precise knowledge of the development of the mouse egg following fertilization.

In vitro fertilization has been reported for a number of species including mouse and man and this approach facilitates gene transfer. No successful *in vitro* fertilization method has been described for sheep, though experiments are being undertaken with sheep oocytes and sperm (Brackett, Trounson, Gianaroli & M.R. Brandon unpublished) to achieve *in vitro* fertilization and to devise methods for visualizing the pronuclei of fertilized sheep eggs.

Concluding remarks

The association of MHC class I and class II antigens, as defined by use of alloantisera, with disease resistance and susceptibility is an imprecise and indirect approach to understanding the biological significance of the relationship between the MHC and disease. Work on cloning the MHC genes began very recently, but investigators have already cloned and determined the complete nucleotide sequences of several MHC genes in man and mouse (Lee, Trowsdale & Bodmer 1982; Steinmetz and colleagues 1982) and are on the verge of cloning many more. The genetic map of the MHC obtained by a serological description of MHC products is now being complemented with a molecular map which localizes precisely the genes encoding the various types of MHC polypeptides. For humans, this work has revealed a complexity of class II genes (Watson et al. 1983) not shown by serology.

It is now feasible to examine the importance of class I and class II MHC genes for disease resistance and susceptibility directly in experimental animals. Several significant advances have occurred in biology in the last 3 years that allow the synthesis of new hybrid animals, i.e. animals with a complete genotype to which is added new genetic information (Constantini & Lacy 1981; Wagner et al. 1981; Palmiter et al. 1982). This approach will offer an alternative to the animal geneticists' established and slow technique of breeding selectively for a particular characteristic. If a gene or genes in the MHC confers disease resistance to one or a number of bacterial, viral or parasitic infections, then the technical possibility now exists to transfer the desirable gene(s) to the germline of highly productive animals selected by conventional breeding techniques.

Three recent publications have illustrated how the biological importance of the MHC will be determined (Goodenow et al. 1982; Mellor et al. 1982; Rabourdin-Combe & Mach 1983). Goodenow and colleagues (1982) transformed thymidine kinase-negative C3H mouse L cells with the cloned $H-2L^d$ gene together with the herpes simplex virus tk gene to produce transformants expressing L^d molecules. The foreign L^d gene products expressed by cloned mouse L cells were shown to be virtually indistinguishable from Balb/c spleen L^d molecules. The ability to place cell-surface recognition molecules on the surfaces of foreign cells makes it possible to study the functional importance of these molecules. This approach was subsequently used by Mellor and colleagues (1982) who introduced an $H-2k^b$ gene into mouse L cells and showed that the transformed cells can be killed by allospecific anti-$H-2k^b$ cytotoxic T cells. Moreover, when the $H-2k^b$ transformed L cells are infected with influenza virus, they can be killed by an $H-2k^b$-restricted, influenza virus-specific cytotoxic T-cell line. The latter experiment offers an exquisite proof of the phenomenon known as H-2 restriction (Zinkernagel & Doherty 1974).

Rabourdin-Combe and Mach (1983) isolated cDNA clones of the HLA-DR β chains, HLA-DR α chains and the HLA-Dr invariant chain and used these clones to identify the corresponding genes from human genomic libraries. Mouse L cells were simultaneously transfected with all three genes and normal expression of HLA-DR antigens was shown at the surface of the transfected cells. This represents the first example of DNA-mediated gene transfer with the involvement of several genes necessary for the expression of a multimeric protein. Although it is not possible to extrapolate directly the transfection of a cell line with multiple genes to the transfer of multiple genes to the germline of an animal, it suggests that manipulation of mammalian genes is extremely flexible and open to exploitation in domestic livestock.

Acknowledgements

The authors express their thanks to Miss S. Cooke for typing the manuscript. This work was supported in part by a grant from the Australian Meat Research Committee.

References

Adams, T.E., (1980). A study of the bovine MHC. Ph.D. thesis. Canberra, Australian National University.

Adams, T.E., Brandon, M.R. & Morris, B. (1979). Genetic aspects of disease resistance in cattle. *Animal Blood Groups & Biochemical Genetics*, **10**, 155-63.

Albert, E. & Gotze, D. (1977). The major histocompatibility system in man. In *The major histocompatibility system in man and animals* (ed. D. Gotze), pp. 7-78. Berlin: Springer Verlag.

Amerding, D., Katz, D.H. & Benacerraf, B. (1974). Immune response genes in rats. 1. Analysis of responder status to synthetic polypeptides and low doses of serum albumin. *Immunogenetics*, **1**, 329-39.

Amiel, J.L. (1967). Study of the leucocyte phenotypes in Hodgkins disease. In *Histocompatibility testing* (eds. E.S. Curtoni, P.L. Mahinz & M.R. Tosi), pp. 79-81. Copenhagen: Munksgaard.

Amorena, B. & Stone, W.H. (1978). Serologically defined (SD) locus in cattle. *Science*, **201**, 159-60.

Bacon, L.D., Kite, J.H. & Rose, N.L. (1974). Relationship between the major histocompatibility (B) locus and autoimmune thyroiditis in obese chickens. *Science*, **186**, 274-75.

Balcarova, J., Derka, J., Hala, K. & Hraba, T. (1974). Genetic control of immune response to dinitrophenol group in inbred chickens. *Folia Biologica –Praha*, **20**, 346-49.

Balcarova, J., Gunther, E., Hala, K., Rude, E. & Hraba, T. (1975). Further evidence for the genetic control of immune responsiveness to (T,G)-A--L by the B system in chickens. *Folia Biologica –Praha*, **21**, 406-8.

Balner, H., Dorf, M.E., de Groot, M.L. & Benacerraf, B. (1973). The histocompatibility complex of rhesus monkeys. 3. Evidence for a major MLR locus and histocompatibility linked Ir genes. *Transplantation Reviews*, **5**, 1555-60.

Balner, H., Gabb, B.W., Dersjant, H., van Vreesijk, V.W. & van Rood, J.J. (1971). Major histocompatibility locus of rhesus monkeys (RGL-A). *Nature*, **230**, 177-80.

Banyard, M.R.C. (1980). Immune response and disease resistance of the bovine eye. Ph.D. thesis. Canberra, Australian National Unversity.

Benacerraf, B. & Dorf, M.E. (1976). Genetic control of immune responses and immune suppressions by I-region genes. *Cold Spring Harbor Symposia on Quantitative Biology*, **41**, 465-76.

Benacerraf, B. & Germain, R.N. (1978). The immune response genes of the major histocompatibility complex. *Immunological Reviews*, **38**, 70-119.

Benedict, A.A., Pollard, L.D., Morrow, P.R., Abplanalp, M.A., Maurer, P.A. & Briles, W.E. (1975). Genetic control of immune responses in chickens. 1. Responses to a terpolymer of poly (Glu60Ala30Tyr10) associated with the major histocompatibility complex. *Immunogenetics*, **2**, 313-24.

Blackwell, J., Freeman, J. & Bradley, D. (1980). Influence of H-2 complex on acquired resistance to *Leishmania donovani* infection in mice. *Nature*, **283**, 72-74.

Bluestein, H.G., Green, I. & Benacerraf, B. (1971). Specific immune response genes of the guinea pig. 1. Dominant genetic control of immune responsiveness to copolymers of L-glutamic acid and L-alanine and L-glutamic acid and L-tyrosine in random bred Hartley guinea pigs. *Journal of Experimental Medicine*, **134**, 471-81.

Caldwell, J. (1979). Polymorphism of the BoLA system. *Tissue Antigens*, **13**, 319-26.

Carlinfanti, E. (1948). The predisposition for immunity. *Journal of Immunology*, **59**, 1-7.

Chesebro, B., Wehrly, K. & Stimpfling, J. (1974). Host genetic control of recovery from Friend leukaemia virus-induced splenomegaly: mapping of a gene within the major histocompatibility complex. *Journal of Experimental Medicine*, **140**, 1457-67.

Claas, F.H.J. & Deedler, A.N. (1979). H-2 linked immune response to murine experimental *Shistosoma mansoni* infections. *Journal of Immunogenetics*, **6**, 167-76.

Cohen, I. & Kozaki, M. (1969). The production of isoantibodies in littermate dogs after allogeneic skin grafting. *Transplantation*, **7**, 468-74.

Collins, W.M., Briles, W.E., Zsigray, R.M., Dunlop, W.R., Corbett, A.C., Clark, K.K., Marks, J.L. & McGrail, T.P. (1977). The B locus (MHC) in the chicken: association with the fate of RSV-induced tumours. *Immunogenetics*, **5**, 333-43.

Constantini, F. & Lacy, E. (1981). Introduction of a rabbit betaglobin gene into the mouse germline. *Nature*, **294**, 9294.

Crittendon, L.B., Purchase, H.G., Solomon, J.J., Okazaki, W. & Burmester, B.R. (1972). Genetic control of susceptibility to the avian leucosis complex. 1. The leucosis-sarcoma virus group. *Poultry Science*, **51**, 242-61.

Dausset, J., Rapapat, F.T., Lagrand, L., Colombani, J. & Marcelli-Barge A. (1970). Skin allograft survival in 238 human subjects: role of specific relationships at four gene sites of first and second HLA loci. In *Histocompatibility testing* (ed. P. Terasaki), pp. 381-97. Copenhagen: Munksgaard.

Debre, P., Gisselbrecht, S., Poza, F. & Levy, J.P. (1979). Genetic control of sensitivity to Maloney leukaemia virus in mice. 2. Mapping of three resistant genes within the H-2 complex. *Journal of Immunology*, **123**, 1806-12.

198   T. E. Adams and M. R. Brandon

Dimmock, C.R. (1973). Blood group antibody production in cattle by a vaccine against Babesia argentinia. Research in Veterinary Science, 15. 305-9.

Dineen, J.K. (1964). Sources of immunological variation. Nature, 202, 101-2.

Dodt, R.M. (1977). The prevalence of bovine kerato-conjunctivitis in a beef cattle herd in North Queensland. Australian Veterinary Journal, 53, 128-31.

Dorf, M.E. & Benacerraf, B. (1975). Complementation of H-2 linked Ir genes in the mouse. Proceedings of the National Academy of Sciences of the USA, 72, 3671-75.

Dorf, M.E., Balner. H. & Benacerraf, B. (1975). Mapping of the immune response genes in the major histocompatibility complex of the rhesus monkey. Journal of Experimental Medicine, 142, 673-93.

Dorf, M.E., Balner, M., de Groot, M.L. & Benacerraf. B. (1974a). Histocompatibility-linked immune response genes in the rhesus monkey. Transplantation Proceedings, 6, 119-23.

Dorf, M.E., Dunham, E.K., Johnson, J.P. & Benacerraf, B. (1974b). Genetic control of immune response: the effect of nonH-2 linked genes on antibody production. Journal of Immunology, 112, 1329-36.

Dunham, E.R., Dorf, M.E., Shreffler, D.C. & Benacerraf, B. (1973). Mapping the H-2 linked genes governing, respectively, the immune responses to a glutamic acid-alanine-tyrosine copolymer and to limiting doses of ovalbumin. Journal of Immunology, 111, 1621-25.

Ellman, L., Green, I., Martin, W.J. & Benacerraf, B. (1970). Linkage between the poly-L-lysine gene and the locus controlling the major histocompatibility antigens in strain 2 guinea pigs. Proceedings of the National Academy of Sciences of the USA, 66, 322-28.

Fink, M.A. & Quinn, U.A. (1953). Antibody production in inbred strains of mice. Journal of Immunology, 70, 61-67.

Goodenow, R.S., McMillan, M., Nicholson, M., Sher, B.T., Eakle, K., Davidson, N. & Hood, L. (1982). Identification of the class I genes of mouse major histocompatibility complex by DNA-mediated gene transfer. Nature, 300, 231-37.

Gorer, P.A. & Schutze, H.S. (1938). Genetic studies on immunity in mice. 2. Correlation between antibody formation and resistance. Journal of Hygiene, 38, 647-62.

Gotze, D. (1977). Major histocompatibility system. In The major histocompatibility system in man and animals (ed. D. Gotze), pp. 1-6. Berlin: Springer Verlag.

Green, J. (1974). Genetic control of immune responses. Immunogenetics, 1, 4-21.

Gunther, E. (1979). Close association between particular I region-determined cell-surface antigens and Ir gene-controlled immune responsiveness to synthetic polypeptides in wild rats. European Journal of Immunology, 9, 391-401.

Gunther, E. & Rude, E. (1975). Genetic complementation of histocompatibility-linked Ir genes in the rat. Journal of Immunology, 115, 1387-93.

Gunther, E. & Stark, O. (1977). The major histocompatibility system of the rat (Ag-B or H-1 system). In The major histocompatibility system in man and animals (ed. D. Gotze), pp. 207-53. Berlin: Springer Verlag.

Gunther, E., Rude, E. & Stark, O. (1972). Antibody response in rats to the synthetic polypeptide (T,G)-A--L genetically linked to the major histocompatibility system. European Journal of Immunology, 2, 151-55.

Gunther, E., Rude, E., Meyer-Delius, M. & Stark, O. (1973). Immune
response genes linked to the major histocompatibility system in the rat.
Transplantation Proceedings, **7** (suppl. 1), 147-50.

Haverkorn, M.J., Hofman, B., Masurel, N. & van Rood, J.J. (1975). HL-A-
linked genetic control of immune response in man. *Transplantation Reviews*,
22, 120-24.

Ipsen, J. (1959). Differences in primary and secondary immunizability in
inbred strains of mice. *Journal of Immunology*, **83**, 448-57.

Ivanyi, P. (1979). The major histocompatibility antigens in various species.
Current Topics in Microbiology & Immunology, **61**, 1-90.

Kantor, F.S., Ojeda, A. & Benacerraf, B. (1963). Studies on artificial antigens.
1. Antigenicity of DNP-poly-lysine and DNP copolymers of lysine and
glutamic acid in guinea pigs. *Journal of Experimental Medicine*, **117**, 55-69.

Kapp, J.A., Pierce, C.W. & Benacerraf, B. (1973). Genetic control of immune
responses *in vitro*. 2. Cellular requirements for the development of primary
plaque-forming cell responses to random terpolymer L-glutamic acid 60-L-
alanine 30-L-tyrosine 10 (GAT) by mouse spleen cells. *Journal of
Experimental Medicine*, **138**, 1121-32.

Kapp. J.A., Pierce C.W. & Benacerraf, B. (1975). Genetic control of immune
responses *in vitro*. 6. Experimental conditions for the development of helper
T-cell activity specific for the terpolymer L-glutamic acid 60-L-alanine 30-L-
tyrosine 10 (GAT). *Journal of Experimental Medicine*, **142**, 50-60.

Karakoz, I., Krejci, J., Hala, K., Blaszchizk, B., Hraba, T. & Pekarek, J.
(1974). Genetic determination of tuberculin sensitivity in chicken inbred
lines. *European Journal of Immunology*, **4**, 545-48.

Katz, D.H. & Benacerraf, B. (1975). The function and interrelationships of T-
cell receptors, Ir genes and other histocompatibility gene products.
Transplantation Reviews, **22**, 175-95.

Katz, D.H., Hamaoka, T. & Benacerraf, B. (1973). Cell interactions between
histoincompatible T and B lymphocytes. 2. Failure of physiologic
cooperative interactions between T and B lymphocytes from allogeneic
donor strains in humoral response to hapten protein conjugates. *Journal of
Experimental Medicine*, **137**, 1405-18.

Katz, D.H., Hamaoka, T., Dorf, M.E., Maurer, P.H. & Benacerraf, B.
(1973). Cell interactions between histoincompatible T and B lymphocytes. 3.
Involvement of the immune response (Ir) genes in the control of
lymphocyte interactions in responses controlled by the gene. *Journal of
Experimental Medicine*, **138**, 734-39.

Klein, J., Juretic, A., Baxevanis, C.N. & Wagy, Z.A. (1981). The traditional
and new version of the mouse H-2 complex. *Nature*, **292**, 455-60.

Koch, C., Hala, K. & Srup, P. (1977). Immune response genes in chickens:
the multiarous responsiveness to (T,G)-A--L. In *Avian immunology* (ed.
A.A. Benedict), pp. 233-43. New York: Plenum Press.

Lee, J.S., Trowsdale, J. & Bodmer, W.F. (1982). cDNA clones coding for the
heavy chain of human HLA-DR antigen. *Proceedings of the National
Academy of Sciences of the USA*, **79**, 545-49.

Lie, O. (1979). Genetic analysis of some immunological traits in young bulls.
Acta Veterinaria Scandinavica, **20, 372-86.**

Lilly, F. (1966). The inheritance of susceptibility to the Gross leukaemia virus
in mice. *Genetics*, **53**, 529-39.

Lilly, F. (1971). Influence of H-2 type on Gross virus leukaemogenesis in
mice. *Transplantation Proceedings*, **3**, 1239-41.

Marsh, D.G. & Bias, W.B. (1977). Basal serum IgE levels and HLA antigen frequencies in allergic subjects. 2. Studies in people sensitive to rye grass groups I and ragweed antigen E and a postulated immune response (Ir) loci in the HLA region. *Immunogenetics,* **5**, 235-51.

McDevitt, H.O. & Sela, M. (1965). Genetic control of the antibody response. *Journal of Experimental Medicine,* **122**, 517-31.

McDevitt, H.D. & Chinitz, A. (1969). Genetic control of antibody response: relationship between immune response and histocompatibility (H-2) type. *Science,* **163**, 1207-8.

McDevitt, H.D., Deak, B.D., Shreffler, D.C., Klein, J., Stimpfling, J.H. & Snell, G.D. (1972). Genetic control of the immune response: mapping of the Ir locus. *Journal of Experimental Medicine,* **135**, 1259-78.

Mellor, A.L., Golden, L., Weiss, E., Bullman, H., Hurst, J., Simpson, E., James, R.F., Townsend, A.R., Taylor, P.M., Schmidt, W., Ferluga, J., Leban, L., Santamaria, M., Atfield, G., Festenstein, H. & Flavell, R.A. (1982). Expression of murine H-2Kb histocompatibility antigen in cells transformed with cloned H-2 genes. *Nature,* **298**, 529-34.

Meruelo, D., Deak, B. & McDevitt, H.O. (1977). Genetic control of cell-mediated responsiveness to an AKR tumour-associated antigen. *Journal of Experimental Medicine,* **146**, 1367-79.

Miller, J.F.A.P. (1978). Influence of genes of the major histocompatibility complex on the reactivity of thymus derived lymphocytes. *Contemporary Topics in Immunobiology,* **8**, 1-17.

Morris, B.J., Batchelor, J.R. & Festenstein, H. (1978). Matching for HLA in transplantation. *British Medical Bulletin,* **34**, 259-62.

Murray, M., Morrison, W.I. & Whitelaw, D.D. (1982). Host susceptibility to African trypanosomiasis: trypanotolerance. *Advances in Parasitology,* **21**, 1-68.

Newman, M.J., Adams, T.E. & Brandon, M.R. (1982). Serological and genetic identification of a bovine B-lymphocyte alloantigen system. *Animal Blood Groups & Biochemical Genetics,* **13**, 123-39.

Nordskog, A.W., Pevzner, I.Y., Trowbridge, C.L. & Benedict, A.A. (1977). Immune response and adult mortality associated with the B locus in chickens. *Advances in Experimental Medicine & Biology,* **38**, 245-56.

Osoba, D., Dick, H.M., Voller, A., Goosen, T.J., Goosen, T., Draper, C.C. & de the, G. (1979). Role of the HLA complex in the antibody response to malaria under natural conditions. *Immunogenetics,* **8**, 323-38.

Palladino, M.A., Gilmour, D.G., Scafuri, A.R., Stone, H.A. & Thorbecke, G.J. (1977). Immune response differences between two inbred chicken lines identical at the major histocompatibility complex. *Immunogenetics,* **5**, 253-59.

Palmiter, R.D., Brinster, R.L., Hammer, R.E., Trumbauer, M.E., Rosenfeld, M.G., Brinberg, N.C. & Evans, R.M. (1982). Dramatic growth of mice that develop from eggs microinjected with metallothionein-growth hormone fusion genes. *Nature,* **300**, 611-15.

Pazderka, R., Longenecker, B.M., Law, G.R.J., Stone, H.A. & Ruth, R.F. (1975). Histocompatibility of chickens selected for resistance to Marek's disease. *Immunogenetics,* **2**, 93-100.

Pevzner, I.Y., Nordskog, A.W. & Kaeberle, M.I. (1975). Immune response and the B blood group locus in chickens. *Genetics,* **80**, 753-59.

Pevzner, I.Y., Trowbridge, C.L. & Nordskog, A.W. (1978). Recombination between genes coding for immune response and the serologically determined antigens in the chicken's B system. *Immunogenetics,* **7**, 25-33.

Pickbourne, P., Richards, S., Bodmer, J.G. & Bodmer, W.F. (1977). Data organization and methods of analysis. In *Histocompatibility testing* (ed. W.F. Bodmer), pp. 295-324. Copenhagen: Munksgaard.

Rabourdin-Combe, C. & Mach, B. (1983). Expression of HLA-DR antigens at the surface of mouse L-cells cotransfected with cloned human gene. *Nature,* **303,** 670-74.

Rosenthal, A.S. & Shevach, E.M. (1973). Function of macrophages in antigen recognition by guinea pig T lymphocytes. 1. Requirement for histocompatible macrophages and lymphocytes. *Journal of Experimental Medicine,* **138,** 1194-1212.

Rude, E. & Gunther, E. (1974). Genetic control of the immune response to synthetic polypeptides in rats and mice. *Progress in Immunology,* **2,** 223-33.

Sasazuki, T., Kohno, Y., Iwamoto, I., Tanimura, M. & Naito, S. (1978). Association between a HLA haplotype and low responsiveness to tetanus toxoid in man. *Nature,* **272,** 359-61.

Sato, H., Boyse, E.A., Aoki, T., Iritani, C. & Old, L.T. (1973). Leukaemia-associated transplantation antigens related to murine leukaemia virus. The xol system: immune response controlled by a locus linked to H-2. *Journal of Experimental Medicine,* **138,** 593-606.

Schiebel, I.F. (1943). Hereditary differences in the capacity of guinea pigs for the production of diphtheria anti-toxin. *Acta Pathologica et Microbiologica Scandinavica,* **20,** 464-84.

Schierman, L.W., Watanabe, D.H. & McBride, R.A. (1977). Genetic control of Rous sarcoma regression in chickens: linkage with the major histocompatibility complex. *Immunogenetics,* **5,** 325-32.

Sellei, J. & Rendel, J. (1968). The genetic control of antibody production: a study of isoimmune antibodies in cattle twins. *Genetic Research,* **11,** 271-87.

Shevach, E.M. & Rosenthal, A.A. (1973). Function of macrophages in antigen recognition by guinea pig T lymphocytes. 2. Role of the macrophage in the regulation of genetic control of the immune system. *Journal of Experimental Medicine,* **138,** 1213-19.

Shonnard, J.W., Davis, B.K., Cramer, D.V., Radka, S.F. & Gill, T.J. (1979). The association of immune responsiveness, mixed lymphocyte responses, and Ia antigens in natural populations of Norway rats. *Journal of Immunology,* **124,** 778-83.

Spooner, R.L., Bradley, J.S. & Young, G.B. (1975). Genetics and disease in domestic animals with particular reference to dairy cattle. *Veterinary Record,* **96,** 125-30.

Spooner, R.L., Leveziel, H., Grosclaude, F., Oliver, R.A. & Vaiman, M. (1978). Evidence for a possible major histocompatibility complex (BLA) in cattle. *Journal of Immunogenetics,* **5,** 335-46.

Stark, O., Gunther, E., Kohoustova, M. & Vojeik, L. (1977). Genetic recombination in the major histocompatibility complex (H-1, Ag-B) of the rat. *Immunogenetics,* **5,** 183-87.

Stear, M.J., Newman, M.J. & Nicholas, F.W. (1982). Two closely linked loci and one apparently independent locus code for bovine lymphocyte antigens. *Tissue Antigens,* **20,** 289-99.

Steinmetz, M., Minard, K., Horvath, S., McNicholas, J., Srelinger, J., Wate, C., Long, E., Mach, B. & Hood, L. (1982). A molecular map of the immune response from the major histocompatibility complex of the mouse. *Nature,* **300,** 35-42.

Stimpfling, J.H. & Durham, T. (1972). Genetic control by the H-2 gene complex of the alloantibody response to an H-2 antigen. *Journal of Immunology,* **108,** 947-51.

Stone, H.A., Briles, W.E. & McGibbon, W.H. (1977). The influence of the major histocompatibility complex on Marek's disease in the chicken. *Advances in Experimental Medicine & Biology*, **88**, 299-308.

Tada, T., Taniguchi, M. & David, C.S. (1976). Properties of the antigen-specific suppressive T-cell factor in the regulation of antibody response in the mouse. 4. Special subregion assignment of the gene(s) that code for the suppressive T-cell factor in the H-2 histocompatibility complex. *Journal of Experimental Medicine*, **144**, 713-25.

Taussig, M.J., Munro, A.J., Campbell, R., David, C.S. & Staines, N. (1975). Antigen-specific T-cell factor in cell cooperation: mapping within the I region of H-2 complex and ability to cooperate across allogeneic barriers. *Journal of Experimental Medicine*, **142**, 694-700.

Tennant, J.R. & Snell, G.D. (1968). The H-2 locus and viral leukaemogenesis as studied in congeneic strains of mice. *Journal of the National Cancer Institute*, **41**, 597-604.

Thze, J., Waltenbaugh, C., Dorf, M.E. & Benacerraf, B. (1977). Immunosuppressive factor(s) specific for L-glutamic acid^{50}-L-tyrosine^{50} (GT). 2. Presence of I-J determinants on the GT suppressive factor. *Journal of Experimental Medicine*, **146**, 287-92.

Usinger, R., Curie-Cohen, M. & Stone, W.H. (1977). Lymphocyte-defined loci in cattle. *Science*, **196**, 1017-18.

Vaiman, M., Renard, C., La Fage, P., Ameteau, J. & Wizza, P. (1970). Evidence for a histocompatibility system in swine (SL-A). *Transplantation*, **10**, 155-64.

Vaiman, M., Renard, C., Ponceau, M., Lecontre, J. & Villiers, P.A. (1975). Alloantigens sous la dépendance de la région SL-A controlant la raction lymphocytaire chez le porc. *Comptes Rendus Hebdomadaires des Séance de l'Academie des Sciences: Série D Sciences Naturelles*, **280**, 2809-12.

van Rood, J.J., van Leeuwen, A. & Ploem, J.S. (1976). Simultaneous detection of two cell populations by two-colour fluorescence and application to the recognition of B-cell determinants. *Nature*, **262**, 795-97.

van Rood, J.J., de Vries, R.R.P. & Munro, A. (1977). The biological meaning of transplantation antigens. *Progress in Immunology*, **3**, 338-50.

Vriesendorp, H.M., Grosse-Wilde, H. & Dorf, M.E. (1977). The major histocompatibility system of the dog. In *The major histocompatibility system in man and animals* (ed. D. Gotze), pp. 129-63. Berlin: Springer Verlag.

Wagner, T.E., Hoppe, P.C., Jollick, J.D., Scholl, D.R., Modinka, R.L. & Gault, J.B. (1981). Microinjection of rabbit betaglobin gene into zygotes and its subsequent expression in adult mice and their offspring. *Proceedings of the National Academy of Sciences of the USA*, **78**, 6376-80.

Wassom D.L., David, C.S. & Gleich G.J. (1979). Genes within the major histocompatibility complex influence susceptibility to *Trichinella spiralis* in the mouse. *Immunogenetics*, **9**, 491-96.

Watson, A.J., De Mars, R., Trowbridge, I.S. & Bach, F.H. (1983). Detection of a novel human class II HLA antigen. *Nature*, **304**, 358-61.

Wright, S. & Lewis, P.A. (1921). Factors in the resistance of guinea pigs to tuberculosis with special regard to inbreeding and heredity. *American Naturalist*, **55**, 636-42.

Wurzburg, U. (1971). Correlation between the immune response to an enzyme and histocompatibility type in rats. *European Journal of Immunology*, **1**, 496-97.

Wurzburg, U., Schutt-Gerowitt, H. & Rajewsky, K. (1973). Characterization of an immune response gene in rats. *European Journal of Immunology*, **3**, 762-66.

Zaleski, M. & Milgrom, F. (1973). Complementary genes controlling immune response to AKR antigen in mice. *Journal of Immunology,* **110**, 1238-44.

Zaleski, M. & Klein, J. (1974). Immune response of mice to Thy 1.1 antigen, genetic control by alleles at the Ir-5 locus loosely linked to the H-2 complex. *Journal of Immunology,* **113**, 1170-77.

Zaleski, M. & Klein, J. (1978). Genetic control of immune responses to Thy 1 antigen. *Immunological Reviews,* **38**, 120-62.

Zinkernagel, R.M. & Doherty, P.C. (1974). Restriction of *in vitro* T cell-mediated cytotoxicity in lymphocytic choriomeningitis within a synseneic or semi-allogeneic system. *Nature,* **248**, 701-2.

Zinkernagel, R.M. & Doherty, P.C. (1979). MHC-restricted cytotoxic T cells: studies on the biological role of polymorphic major transplantation antigens determining T-cell restriction-specificity, function and responsiveness. *Advances in Immunology,* **27**, 51-177.

Structural organization and ontogeny

12

The lymphoid apparatus of the sheep and the recirculation of lymphocytes

B. MORRIS, N.C. PEDERSEN and
W. TREVELLA

The various lymphoid organs of the sheep develop at different times throughout intrauterine and postnatal life and the lymphoid apparatus does not reach its full anatomical, physiological and functional maturity until some months after birth. However, the foetal lamb acquires an extensive range of immunological reactivities before birth and well before the lymphoid apparatus is fully developed. Most nonimmunoglobulin-bearing lymphocytes come from the thymus in the foetal lamb while the immunoglobulin-bearing lymphocyte population comes principally from the Peyer's patches. An extensive recirculation of lymphocytes between the blood and lymph is established in the foetus before the lymphoid apparatus is fully developed and is mostly directed through peripheral lymph nodes. This occurs in the absence of immunoglobulins and in the absence of foreign antigens. After birth much larger numbers of lymphocytes migrate through the gut tissues and this intestinal cell traffic is the predominant component of the lymphocyte recirculation in postnatal and adult life. This large-scale traffic of cells through the gut probably occurs in response to antigenic stimulation. There are variations in the cell populations in lymph coming from different tissues due to the nonrandom migration of different classes of lymphocytes. For example, cells with surface immunoglobulins are less prevalent in peripheral lymph than in central lymph. Cells derived from the gut appear to recirculate from the blood to the intestinal lymph while cells from the lymph derived from peripheral lymph nodes tend to recirculate through tissues drained by these nodes. The extent of lymphocyte migration through lymph nodes challenged with antigen and through tissues where cell-mediated immune responses are occurring is greatly enhanced and reactive. Sensitized cells are continually being added to and removed from the lymph as the localized immune response is propagated systemically throughout the lymphatic system. These processes lead to the reassortment and restructuring of both the free-floating and fixed-cell populations in the lymphoid apparatus changing its physiological and biochemical potentialities.

During organogenesis, cells develop special structural relationships with one another which give rise to tissues of particular form and function. The cells within an organ differentiate into a stable homotypic population with a predictable life history and a fixed genetic expression. Liver cells join with other liver cells to form the liver, epidermal cells with epidermal cells to form the skin, and so on. The fixed structural relationships of the cells in most tissues ensure that their metabolic activities are regulated by the activities of neighbouring cells, and this in turn establishes the biochemical anatomy of the internal milieu in which the cells exist. The tissue fluid of the liver is modified by the activities of the hepatic parenchymal cells and the Kupffer cells; the tissue fluid of the ovary is modified by the activities of the luteal cells and the granulosa cells. In these circumstances, fixed cells have few reactive options open to them and few opportunities for encountering regulatory stimuli other than those originating from within the tissue itself: thus the life history of the cells within an organ is preordained by mutuality.

The lymphoid apparatus develops quite differently from most other tissues. Some components of the lymphoid tissue are quite unstructured in their final form and the constituent cells retain a high degree of mobility, moving in and out of the tissue continuously. These cells retain a range of reactive and differentiative options denied the fixed cells of most other tissues.

In the sheep, the earliest circulating white cells are macrophages, which are present in the circulation of foetuses at 24 days gestation. These cells represent a primitive discriminatory system which can distinguish alien materials in the absence of opsinins and immunoglobulins (Ig) (M.Al Salamin personal communication). The cells which eventually comprise the lymphoid apparatus originate principally in the thymus and the Peyer's patches. The thymus first contains lymphocytes at about 40 days gestation and becomes the principal organ of lymphopoiesis by 60 days. Lymphocytes devoid of any surface Ig are spawned from the thymus and migrate by way of the lymph and blood to the lymph nodes, the spleen, the gut and other tissues. These cells are first found in the circulation at around 50 days gestation. Thymectomy done at 55 to 60 days gestation reduces the total number of recirculating lymphocytes in the lamb to around 20% of normal levels. The Peyer's patches become a principal site of lymphopoiesis from 100 days gestation onwards and are the source of the Ig-positive lymphocyte population. Removal of the Peyer's patches from the foetal lamb prevents the development of this population of circulating lymphocytes (Gerber 1979). The thymus and the Peyer's patches have reached their maximum development at around the third month after birth and then undergo a more or less synchronous involution (Reynolds & Morris 1983).

A large proportion of the lymphocyte population becomes associated caducously with lymph nodes and other aggregations of lymphoid tissue throughout the body, moving into and out of these tissues by way of the blood and lymphstream. This metastatic behaviour provides the lymphocytes with opportunities for experiencing a range of internally and externally derived stimuli in a variety of situations. In developmental terms, this diffuse structure and distribution of the lymphoid tissue can be seen as a strategy for avoiding the constraints imposed on the life history of cells by formal organogenesis.

The central role played by the lymphocyte in the immune response has led to attempts to associate all aspects of the life history of this cell with immunological functions. In this regard the metastatic behaviour of lymphocytes has generally been interpreted in terms of immune surveillance, as a mechanism for recruiting, selecting and stimulating specific immunocompetent cells by exposing them to antigens and for distributing immune effector cells to sites of antigen deposition and focuses of infection. It would be most unlikely if the peregrinations of lymphocytes throughout the body were unrelated to these activities, but this is not to say that all features of the life history of lymphocytes have an immunological basis.

Patterns of lymphocyte migration in the sheep

A large proportion of the lymphocyte population in the sheep exists as free-floating cells in the lymph and blood and as migrating cells in the tissues and tissue fluid. Lymphocytes are carried passively in the blood and lymphstream but they move out of the circulation and into and through the tissues by virtue of their inherent mobility. The mobility of these cells is a function of their state of activation. The degree of movement and the rate of locomotion of individual cells varies in any population.

Lymphocytes leave the blood principally in the lymphoid tissues by way of the postcapillary venules. They traverse the endothelium of the venules by passing between cells and penetrating the basement membrane (Schoefl 1972). In sheep, unlike rodents, vessels with high endothelium are rarely found and most lymphocyte traffic occurs through capillaries with an attenuated endothelial membrane.

Depending on the nature of the tissue into which the lymphocytes migrate and their mode of arrival, the transit time of the migrant cells between the blood and the lymph may vary from a few minutes to days or weeks. Lymphocytes that enter a lymph node from the blood have different domains of migration and a different transit time out of the node, compared to lymphocytes entering a node by way of the lymph. (Fahy et al. 1980b; Trevella & Morris 1980).

Lymphocytes normally enter nonlymphoid tissues in relatively small numbers. In certain pathological circumstances, however, this traffic can be very extensive and may be of the same order or greater than that occurring in lymph nodes (Smith, McIntosh & Morris 1970). The histological picture of mononuclear cell infiltration is the manifestation of a traffic of lymphocytes into a tissue which exceeds the traffic out of the tissue. Lymphocytes leave the interstitial spaces by passing into the lymphatic capillaries between the junctions of lymphatic endothelial cells. Once in the lymph, they are carried centripetally through the lymph nodes to the bloodstream.

The number of cells leaving the lymph nodes of sheep varies with the size and location of the node. The cell output from the popliteal node which weighs around 1 g is of the order of 2×10^7 cells per hour; from the prescapular node which may be 5 times as large, the output is of the order of 5 to 10×10^7 cells per hour. Most of these are recirculating cells migrating through the lymph nodes from the blood (Hall & Morris 1965). There is a very large traffic of cells from the gut of sheep, particularly in young lambs up until about 1 year of age (Heath, Lascelles & Morris 1962). The output in the intestinal lymph of lambs may be of the order of 2 to 5×10^8 cells per hour; most of these cells come from the general gut tissue and the *lamina propria* of the small intestine; removal of the mesenteric lymph nodes and the Peyer's patches does not diminish significantly the cell output in the intestinal lymph (Gerber 1979; Trevella & Morris 1980). Antigenic stimulation arising from the gut may be responsible for at least part of the dramatic increase in cell output in the intestinal lymph which occurs during the first months of postnatal life. Many of the blast cells appearing in the intestinal lymph at this time have the characteristics of antibody-forming cells. The traffic of cells through most other somatic tissues, with the exception of the liver, is similar and of the order of a few million cells per hour. The lymph from the liver has a cell concentration between 2 and 5×10^6 cells per ml and a cell output of 1 to 2×10^7 cells per hour (Smith, McIntosh & Morris 1970).

While the number of cells coming from lymphoid and nonlymphoid tissues varies greatly, there are also differences in the constitution of these populations of cells. The migrant cell population in lymph coming from nonlymphoid tissues, except the gut, has a significantly smaller proportion of cells with surface Ig than the population of cells leaving lymph nodes. If it is assumed that these cell populations reflect the migrant populations entering the tissues from the blood, then it seems that cells with surface Ig are restricted from entering nonlymphoid tissues relative to their thymus-derived colleagues (Scollay, Hall & Orlans 1976; Miller & Adams 1977). Peripheral lymph from nonlymphoid tissue also contains from 5 to 10% macrophages, a cell type that is absent in lymph draining from lymph nodes. These observations highlight the reassortment that occurs in lymphocyte populations during their migration between blood and lymph (Fahy et al. 1980a).

Lymphocyte recirculation

The recirculation of lymphocytes between the blood and lymph is the manifestation of some biotactic influence which directs the migratory habits of these cells. This migration occurs largely through fixed lymphoid tissue and, as a consequence, some means must exist to enable lymphocytes to recognize where they are within the blood vascular compartment. The nature of the recognitive elements on lymphocytes and endothelial cells is unknown, although there is evidence in some species of a specialized metabolism in the endothelial cells of postcapillary venules which suggests that immunoglobulins or sulphated glycoproteins on the endothelium may play a role in directing lymphocyte migration (Sordat, Hess & Cottier 1971; Andrews, Ford & Stoddart 1980). The migration pathways of lymphocytes can change in response to general pathological or specific immunological stimuli and cause extensive alterations in the number of lymphocytes entering and leaving a tissue. In the case of the hydronephrotic kidney at least, this occurs without any histological conversion of the normal endothelium of the renal vasculature to the high endothelium characteristic of the postcapillary venules of rodents (Smith, McIntosh & Morris 1970).

Lymphocyte recirculation in the foetal lamb

Lymphocytes are present in the lymph of foetal lambs from at least 70 days gestation. Their appearance in the blood at around 50 days precedes the development of the lymph nodes and their associated lymphatics so that the pattern of recirculation from blood to lymph is probably not established until about 65 days gestation. At this time the foetus is first able to respond to antigenic challenge, but the lymphoid apparatus has not yet developed its mature histological structure. The number of lymphocytes involved in this recirculation increases exponentially throughout gestation. The size of the recirculating pool of lymphocytes has been calculated to be around 5.7×10^9 cells in foetuses 130 to 135 days of age, increasing to 1.2×10^{10} cells in foetuses near to term (Pearson, Simpson-Morgan & Morris 1976).

A large-scale traffic of lymphocytes also takes place through some of the peripheral lymph nodes of the foetus. The output from the prescapular lymph node of the 120-day-old foetus is around 2.5×10^7 cells per hour. This is of the same order as the number of cells migrating through the foetal gut (Cahill et al. 1979; Morris & Simpson-Morgan in press; M.W. Simpson-Morgan, W. Trevella, A.R. Hugh, S.J. McClure & B. Morris unpublished).

Most of the recirculating cells in the foetus are thymus-derived and are devoid of surface Ig; only about 2 to 4% of cells in the blood have surface Ig before birth. Both Ig-positive and Ig-negative populations of cells are present in the central lymph and blood in similar proportions (Gerber 1979).

The proportion of Ig-positive cells in thymectomized lambs is significantly higher (12 to 50%) than in normal lambs, reflecting the substantial reduction in the thymus-derived Ig-negative cell population. The residual cells in thymectomized lambs appear to recirculate more slowly between the blood and the lymph (Cole & Morris 1971; 1973; Pearson, Simpson-Morgan & Morris 1976; Fahey 1976).

The observation of an extensive recirculation of lymphocytes at an early stage of foetal development demonstrates that the migration of lymphocytes cannot be seen as conditioned by antigenic contact. The foetal lamb is agammaglobulinaemic and develops in an environment free of any foreign antigenic stimulation. As a consequence, the metastatic behaviour of lymphocytes must be an intrinsic property of these cells.

Lymphocyte recirculation in the adult sheep

The extensive traffic of lymphocytes from blood to lymph established early in foetal development is maintained throughout postnatal life. When cells are taken from the thoracic duct lymph or from the efferent lymph of individual lymph nodes, labeled with a radioactive marker and reinjected intravenously into the animal, they reappear promptly in the lymph (Figure 1). This migration occurs through all lymph nodes and through all tissues, but the magnitude of the cell traffic varies in different sites. By far the most extensive recirculation of cells occurs through the gut. This is in contrast to what happens in the foetus.

The kinetics of recirculation of lymphocytes are surprisingly similar in all tissues, as shown by the general form of the specific activity-time curves of lymph coming from different tissues of one animal (Figure 1). As the migrant cells are entering tissues which have vastly different numbers of lymphoid cells, it seems that there must be some close relationship between the number of lymphocytes in a particular tissue and the rate at which lymphocytes enter and leave that tissue.

Although lymphocytes migrate from the blood to the lymph in all tissues, there is now some evidence that these patterns of migration are not strictly random. This is apparent from the fact that the proportion of Ig-positive and Ig-negative cells in peripheral lymph is lower than in central lymph (Scollay, Hall & Orlans 1976; Miller & Adams 1977). There is also evidence of a bias in lymphocyte traffic when the migration patterns of lymphocytes collected from the popliteal or prescapular lymph are

Figure 1. The reappearance of isologous normal lymphocytes in lymph from various tissues in the sheep. The results of an experiment, in which 1.86×10^9 lymphocytes labeled with 6.99×10^6 cpm ^{51}Cr were injected into the bloodstream, are shown in the upper figure and the results of an experiment with 1.24×10^9 lymphocytes labeled with 13.8×10.6^6 cpm ^{51}Cr in the lower figure. The specific activity is given as the number of labeled cells per 10^3 cells. Upper figure: ○renal afferent lymph; ■right popliteal efferent lymph; □left popliteal efferent lymph. Lower figure: □left popliteal afferent lymph; ◇left prefemoral efferent lymph; ◆right prefemoral efferent lymph.

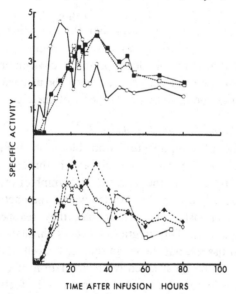

compared with the migration of cells collected from the intestinal lymph (Scollay, Hopkins & Hall 1976; Cahill et al. 1977). It seems that after birth cells collected from a particular site in the lymphoid apparatus tend to migrate back to that general site and reappear in the lymph from that region in greater numbers than in lymph from elsewhere. The general inference from this is that antigenic or other experiences encountered by lymphocytes in some way alter their subsequent life history.

Lymphocyte migration following antigenic stimulation
Humoral antibody responses

Studies on the response of single lymph nodes to antigenic stimulation have revealed that dramatic changes occur in the cell content

Figure 2. The primary cellular and antibody response of the popliteal node to an injection of 1 x 10^9 chicken red blood cells, and the effect of infusing popliteal efferent lymph collected during the period 48 to 72 h after challenge into the lumbar complex of lymph nodes. Total cells; _____IgM antibody; _____ IgG antibody.

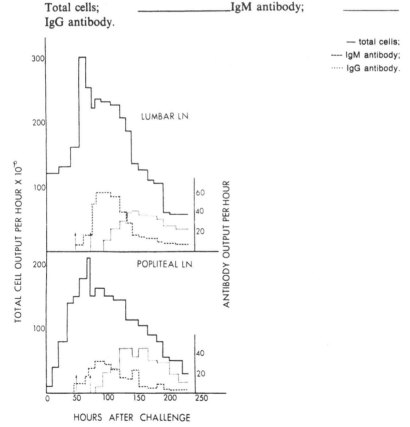

of the efferent lymph coming from a node challenged with antigen (Figures 2 and 3). In the first 24 h after an antigenic challenge, the cell traffic leaving the node is greatly reduced. Subsequently, cells are recruited from the blood into the node, where they undergo processes of selection, proliferation and differentiation into antibody-synthesizing cells. At 72 to 96 h after challenge, the cell output in the lymph increases to a maximum of 5 to 10 times above the prestimulation level, and large blast cells and antibody-forming cells appear. The cells leaving the node in the lymph propagate the immune response to more centrally disposed nodes. In this way, a localized immune response becomes systemic and a widespread immunological memory is established (Fahy et al. 1980b).

Cell-mediated responses

Lymphocyte migration is an obvious aspect of cell-mediated response. In fact, delayed-type hypersensitivity (DTH) reactions have been characterized histopathologically by the extent to which the lesions become infiltrated with mononuclear cells. However, the histopathology of a DTH reaction gives no indication of the dynamic nature of the lesion or of the extent to which lymphocytes are entering and leaving

Figure 3. The result of the same experiment shown in Figure 2 recorded in terms of the specific plaque-forming cell and blast-cell responses in the popliteal and lumbar lymph. _____
Blast cells;   _____plaque-forming cells.

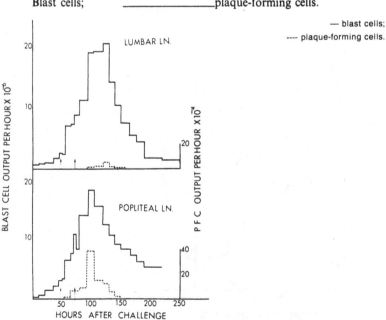

the area. The magnitude of this traffic is vividly illustrated in the reaction that takes place in a renal allograft (Pedersen & Morris 1970).

When a primary kidney allograft is placed in a recipient, the circulation of lymphocytes throughout the lymphoid apparatus in general and the allograft in particular is profoundly modified. It was proposed originally by Medawar (1958) that lymphoid cells reactive to donor histocompatibility antigens become sensitized peripherally in the graft by contact with the grafted donor cells. These sensitized cells then travel via afferent lymphatics to the regional lymph node where they multiply to give rise to immune effector cells which leave the node in the lymph, enter the blood and home back to the graft where they bring about its destruction. For this concept to have physiological significance, the sensitized cells must exhibit some specific homing propensity directed towards the allograft so that they are able to accumulate preferentially in the tissue and destroy it.

Various experiments have been done to decide whether lymphocytes accumulating in an allograft arrive there by some specifically directed migration, but the interpretation of the results is far from clear. While the notion of specific migration and homing is accepted as part of the life history of lymphocytes in regard to their general physiological behaviour, the extension of this concept to explain certain immunological and pathological phenomena is difficult to justify on the basis of most experimental data (Morris 1980). Studies on the migrations of cells collected from the lymph draining a renal allograft illustrate some aspects of this problem.

Once a renal allograft is installed in a recipient, there is an exponential increase in the cell traffic from the blood into the lymph coming from the graft over the first 4 to 5 days. At the end of this time, the migrant cell population has changed from a population of predominantly small lymphocytes with around 2% macrophages to one that contains 35 to 60% blast cells and 10 to 25% macrophages. These striking cellular changes make it obvious that the graft is acting to modify the migrant cell population; inevitably the cells leaving the graft in the lymph differ from those entering the graft from the blood. Many of the cells in the lymph are 'sensitized' against the graft. When these cells are labeled with a radioactive marker and reintroduced intravenously into the host, the specific activity of the cell population in lymph from the allograft reaches significantly higher levels than the specific activity of the cell population in lymph from other regions of the body (Figure 4). This result is consistent with the idea that sensitized cells can home back to the graft, although it does not prove that this actually happens.

The population of cells recovered from the allograft lymph contains cells which had escaped retention in the graft when they first migrated through it, as well as newly sensitized and unsensitized cells. The experiment shows that many of the sensitized cells that were not retained in the graft during their previous transit from the blood were passed through the graft on their second migration. As the relative likelihood of sensitized or unsensitized cells being arrested in the graft is unknown, the specific activity time curve shown in Figure 4 cannot be assumed to be due to the preferential migration of sensitized cells into the graft.

The immunological specificity of the migration of sensitized lymphocytes

In order to test the specificity of the migration of sensitized lymphocytes into a renal allograft, a series of experiments was done in which cells were collected from renal allograft lymph, labeled and injected back into the host after various manoeuvres had been performed to perturb the lymphocyte traffic through other tissues.

Renal allografts were placed (1) into a sheep that had had one of its own kidneys grafted into its neck, (2) into a sheep that had had one of

Figure 4. The reappearance of isologous lymphocytes in lymph from a renal allograft, in the renal lymph from the recipient's own kidney and in the lymph from the popliteal node: 2.09 x 10⁹ allograft lymph cells were collected between 22 and 48 h after grafting, labeled with 16.9 x 10⁹ cpm ^{51}Cr and injected intravenously. The specific activity is given as the number of labeled cells per 10^3 cells. ○Cells in allograft lymph; ●cells in lymph from recipient's own kidney; △cells in popliteal efferent lymph.

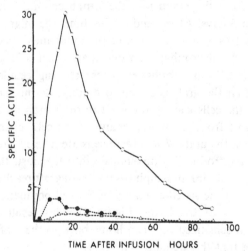

Figure 5. The reappearance of isologous lymphocytes in lymph from a renal allograft, in lymph from a hydronephrotic kidney and in lymph from the cervical lymph duct: 9.36 x 10⁸ allograft lymph cells were collected between 31 and 52 h after grafting, labeled with 5.46 x 10⁶ cpm ⁵¹Cr and injected intravenously. The specific activity is given as the number of labeled cells per 10⁷ cells. ○Cells in allograft lymph; ■cells in lymph from the hydronephrotic kidney □cells in cervical duct lymph.

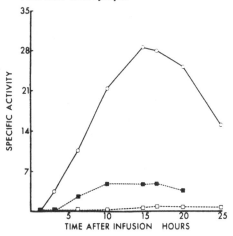

Figure 6. The reappearance of isologous lymphocytes in lymph from a renal allograft, in the peripheral leg lymph draining a tuberculin reaction in the lower hind limb and in efferent lymph from the prefemoral and popliteal lymph nodes: 7.5 x 10⁸ cells from allograft lymph were collected between 30 and 53 h after grafting, labeled with 2.8 x 10⁶ cpm ⁵¹Cr and injected intravenously. The specific activity is given as the number of labeled cells per 10⁷ cells. ○Cells in allograft lymph; ■cells in peripheral lymph from the tuberculin reaction; □cells in prefemoral lymph; ●cells in popliteal lymph.

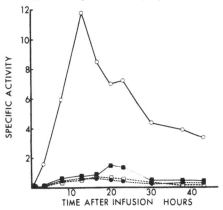

its kidneys made hydronephrotic by ligation of the ureter, and (3) into a sheep that had had a tuberculin reaction established in the lower leg some 3 days previously. In each case the intervention led to a greatly enhanced circulation of lymphocytes into the particular tissues. Comparisons between the kinetics of appearance of the labeled cells collected originally from the allograft lymph showed that, even when the cell traffic through other tissues of the body was deranged by some pathological or immunological process, the cells sensitized in the allograft reappeared in the lymph draining from the graft in a significantly higher proportion than in the lymph coming from the other tissues in which the lymphocyte traffic was increased (Figures 5, 6 and 7). However, the question of whether the redirection of the lymphocyte traffic into the allograft occurred by some specific alteration to the cells has not been answered by these experiments, and the results do not demonstrate that the process of sensitization of lymphocytes intravascularly alters the probability of a cell being extracted from the circulation as it passes through the allograft. The only way this question could be answered would be by determining the specific activities of the migrating cell populations on each side of the blood capillary membrane through which the cells are migrating.

The specificity of the accumulation of cells in areas where cell-mediated

Figure 7. The reappearance of isologous lymphocytes in lymph from a renal allograft, in lymph from an autografted kidney and in prefemoral lymph: 1.39 x 10⁹ cells from allograft lymph were collected between 45 and 66 h after grafting, labeled with 5.47 x 10⁶ cpm ⁵¹Cr and injected intravenously. □Cells in allograft lymph; ■cells in autograft lymph; ○cells in prefemoral lymph.

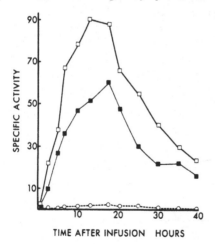

reactions are taking place bears on the relationship between lymphoid cells generated in the regional lymph node and cells accumulating in the reaction site. Najarian and Feldman (1961) claimed that cells from the lymph node regional to a tuberculin reaction had a specific pattern of migration back into the tissue where the tuberculin reaction was occurring. Turk (1962), McCluskey, Benacerraf and McCluskey (1963) and Turk and Oort (1963) failed to demonstrate this specificity. Similarly, Kay and Rieke (1963) failed to demonstrate any specificity in the migration of tuberculin-sensitized cells. They proposed that because of this lack of any demonstrable specificity, only small numbers of sensitized cells may be required to produce the characteristic pathology of cell-mediated reactions. Similar negative conclusions regarding the specificity of migration of sensitized cells derived from immunized lymph nodes were reached by Werdelin and McCluskey (1970), by Prendergast (1964) and by Hall (1967).

It seems physiologically consistent to ascribe a homing capability to lymphocytes to explain the metastatic aspect of their life history and the role they play in immune reactions. The homing concept suggests that lymphocytes have some means of establishing their position in the intravascular compartment and a memory, akin to an immunological memory, for past places and past experiences which compels them to return to certain tissues. The homing concept requires some recognitive mechanism between lymphocytes and endothelial cells that is tissue specific and unrelated to the phenomenon of chemotaxis (Morris 1980).

Conclusions

There is indisputable evidence that lymphocytes possess the capacity to migrate in a nonrandom fashion. Lymphocytes enter certain tissues more readily than other tissues and more readily than other cells; lymphocytes migrate for the most part through lymphoid tissues in preference to nonlymphoid tissues. It has also been established that within the lymphocyte population there are subsets of cells which show patterns of migration which are nonrandom. However, it is not known whether the migration patterns of lymphocytes are consistent for any given cell. It seems likely that the migratory life history of a cell may change in response to a variety of environmental experiences. Thus any concept that a lymphocyte possesses a home has to encompass the idea that its home address changes frequently, possibly each time it passes from the blood to the lymph (Morris 1980).

What is certain in regard to the migration of lymphocytes is that this process involves the continual reassortment of cells from within the migrant population to bring certain categories of cells together and to disassociate others. Some cells are withdrawn temporarily or eliminated permanently from the system, while others proliferate and differentiate into new classes of cells. Some become incorporated into the fixed lymphoid tissue while others retain an ephemeral association with their colleagues as free-floating constituents of the lymph. In this way the lymphoid apparatus is restructured continually by processes similar to those involved in organogenesis in the foetus. However, this restructuring is never completed during the life of the animal. As a consequence, the outcome is not a stable, differentiated, homotypically structured tissue with a single function, but one in a state of disorganized order, consistent with its retaining a wide range of physiological and biochemical capabilities.

References

Andrews, P., Ford, W.L. & Stoddart (1980). Metabolic studies of high-walled endothelium of postcapillary venules in rat lymph nodes. *Ciba Foundation Symposium,* **71**, 211-30.

Cahill, R.N.P., Poskitt, D.C., Frost, H. & Trnka, Z. (1977). Two distinct pools of recirculating T lymphocytes. *Journal of Experimental Medicine,* **145**, 420-28.

Cahill, R.N.P., Poskitt, D.C., Heron, I. & Trnka, Z. (1979). The collection of lymph from single lymph nodes and the intestines of foetal lambs *in utero. International Archives of Allergy & Applied Immunology,* **59**, 117-20.

Cole, G.J. & Morris, B. (1971). The growth and development of lambs thymectomized *in utero. Australian Journal of Experimental Biology & Medical Science,* **49**, 33-53.

Cole, G.J. & Morris, B. (1973). The lymphoid apparatus of the sheep: its growth, development and significance in immunologic reactions. *Advances in Veterinary Science & Comparative Medicine,* **17**, 225-63.

Fahey, K.J. (1976). Humoral immune responses in foetal sheep. Ph.D. thesis. Canberra, Australian National University.

Fahy, V.A., Morris, B., Trevella, W. & Zukoski, C.F. (1980a). The physical and functional heterogeneity of circulating lymphocyte populations. *Blood Cells,* **6**, 11-18.

Fahy, V.A., Gerber, H.A., Morris, B., Trevella, W. & Zukoski, C.F. (1980b). The function of lymph nodes in the formulation of lymph. *Monographs in Allergy,* **16**, 82-99.

Gerber, H.A. (1979). Functional studies in gut-associated lymphoid tissue. Ph.D. thesis. Canberra, Australian National University.

Hall, J.G. (1967). Studies of the cells in the afferent and efferent lymph of lymph nodes draining the site of skin homografts. *Journal of Experimental Medicine,* **125**, 737-54.

Hall, J.G. & Morris, B. (1965). The origin of the cells in the efferent lymph from a single lymph node. *Journal of Experimental Medicine,* **121**, 901-10.

Heath, T.J., Lascelles, A.K. & Morris, B. (1962). The cells of sheep lymph. *Journal of Anatomy,* **96**, 397-408.

Kay, K. & Rieke, W.O. (1963). Tuberculin hypersensitivity: studies with radioactive antigens and mononuclear cells. *Science,* **139**, 487-90.

McCluskey, R.T., Benacerraf, B. & McCluskey, J.W. (1963). Studies on the specificity of the cellular infiltrate in delayed hypersensitivity reactions. *Journal of Immunology*, **90**, 466-77.

Medawar, P.B. (1958). The homograft reaction. *Proceedings of the Royal Society of Biology*, **174**, 155-72.

Miller, H.R.P. & Adams, E.P. (1977). Reassortment of lymphocytes in lymph from normal and allografted sheep. *American Journal of Pathology*, **87**, 59-80.

Morris, B. (1980). The homing of lymphocytes. *Blood Cells*, **6**, 3-7.

Morris, B. & Simpson-Morgan, M.W. (in press). The development of immunological reactivity in foetal lambs. *Annals of the New York Academy of Sciences*.

Najarian, J.S. & Feldman, J.D. (1961). Passive transfer of tuberculin sensitivity by tritiated thymidine-labeled lymphoid cells. *Journal of Experimental Medicine*, **114**, 779-89.

Pearson, L.D., Simpson-Morgan, M.W. & Morris, B. (1976). Lymphopoiesis and lymphocyte recirculation in the sheep foetus. *Journal of Experimental Medicine*, **143**, 167-75.

Pedersen, N.C. & Morris, B. (1970). The role of the lymphatic system in the rejection of homografts: a study of lymph from renal transplants. *Journal of Experimental Medicine*, **131**, 936-69.

Prendergast, R.A. (1964). Cellular specificity in the homograft reaction. *Journal of Experimental Medicine*, **119**, 377-88.

Reynolds, J.D. & Morris, B. (1983). The evolution and involution of Peyer's patches in foetal and postnatal sheep. *European Journal of Immunology*, **13**, 627-35.

Schoefl, G.I. (1972). The migration of lymphocytes across the vascular endothelium in lymphoid tissues. *Journal of Experimental Medicine*, **136**, 568-88.

Scollay, R.G., Hall, J.G. & Orlans, E. (1976). Studies on the lymphocytes of sheep. 2. Some properties of cells in various compartments of the recirculation lymphocyte pool. *European Journal of Immunology*, **6**, 121-25.

Scollay, R.G., Hopkins, J. & Hall, J.G. (1976). Possible role of surface Ig in the nonrandom recirculation of small lymphocytes. *Nature*, **260**, 528-29.

Smith, J.B., McIntosh, G.H. & Morris, B. (1970). The traffic of cells through tissues: a study of peripheral lymph in sheep. *Journal of Anatomy*, **107**, 87-100.

Sordat, B., Hess, M.W. & Cottier, H. (1971). IgG immunoglobulin in the wall of postcapillary venules: possible relationship to lymphocyte recirculation. *Immunology*, **20**, 115-18.

Trevella, W. & Morris, B. (1980). Reassortment of cell populations within the lymphoid apparatus of the sheep. *Ciba Foundation Symposium*, **71**, 127-44.

Turk, J.L. (1962). The passive transfer of delayed hypersensitivity in guinea pigs by the transfusion of isotopically labeled lymphoid cells. *Immunology*, **5**, 478-88.

Turk, J.L. & Oort, J. (1963). A histological study of early stages of the development of the tuberculin reaction after passive transfer of cells labeled with (^{3}H) thymidine. *Immunology*, **6**, 140-47.

Werdelin, O. & McCluskey, R. T. (1971). The nature and the specificity of mononuclear cells in autoimmune inflammations and the mechanisms leading to their accumulation. *Journal of Experimental Medicine*, **133**, 1242-43.

13

Cellular constituents and structural organization of the bovine thymus and lymph node

W.I. MORRISON, P.A. LALOR,
A.K. CHRISTENSEN and P. WEBSTER

The thymus is concerned with the production of functionally mature T cells. In cattle, it consists of two lobes, each composed of multiple lobules with distinct cortical and medullary regions. The thymus is supported by a network of epithelial cells; in the cortex, these are delicate spindle-shaped cells which express high levels of class II MHC antigens but little or no class I antigen, whereas, in the medulla they are more pleomorphic and are rich in both class I and class II MHC antigens. The medulla also contains a population of interdigitating cells which express high levels of class II MHC antigens. The cortical lymphocytes, which make up about 85% of thymic lymphocytes, are largely negative for class I and class II MHC antigens and exhibit high affinity for the lectin, peanut agglutinin (PNA), whereas the medullary lymphocytes express detectable levels of class I MHC antigens, are negative for class II antigens and show low affinity binding of PNA. In cattle, as in other species, the structure of the lymph node is adapted to trapping and responding to foreign material which gains access to the lymphatic system. The lymph nodes possess a structural framework of reticulum cells and reticulin fibres permeated by sinuses which are also lined by reticulum cells. The solid lymphoid tissue is segregated into follicular areas, populated predominantly with B cells, and paracortical areas containing mainly T cells. Each compartment contains specialized accessory cells, namely, the follicular dendritic cells and the interdigitating cells of the paracortex. The latter, along with B cells express high levels of class II MHC antigens, whereas class I antigens are expressed on all of the cellular constituents, although they appear to be present at much lower concentrations on germinal centre lymphoblasts.

Introduction

The life history of a lymphocyte encompasses two distinct phases of development. The first of these phases occurs in primary lymphoid tissues, namely, the bonemarrow, the thymus and possibly the Peyer's patches. Undifferentiated stem cells develop into lymphocytes which mature and proliferate to produce different populations of immunocompetent lymphocytes, each comprising clones of cells capable of recognizing and responding to individual antigenic determinants. This process is largely independent of exposure to foreign antigens. The second phase of development, which occurs in secondary diffuse or solid lymphoid tissues, involves the proliferation and differentiation of immunocompetent lymphocytes, in response to foreign antigen, to produce specific immune effector cells. The lymphoid tissues in which the different phases of lymphocyte development occur have distinct structural arrangements, with microenvironments adapted to the specific functions of each tissue.

Although studies of the phenotype and behaviour of lymphocytes *in vitro* have provided a great deal of information on their ontogeny and the way in which they participate in immune responses, much less is known about these events as they occur *in situ*. There are a number of potential pitfalls in extrapolation of results obtained from *in vitro* experiments, since the preparation of cell suspensions from lymphoid tissues results in loss of the spatial relationships of different cell types to one another, and certain nonlymphocytic cell types are not readily obtained in cell suspension. It is, therefore, important in considering the results of manipulations *in vitro*, to have detailed knowledge of the environments in which the cells reside in the animal.

This chapter will consider the cellular components and structural arrangement of the thymus and lymph node of cattle, as examples, respectively, of a primary and secondary lymphoid organ, in relation to the specialized functions of these tissues. Information derived from studies in other species will also be included, particularly in areas which have not been studied in any detail in cattle.

The thymus

General features

Before describing the cellular constituents of the bovine thymus, it is worth considering briefly some general aspects of thymic development, cellular kinetics and structure. Although much of this information has been derived from studies in laboratory animals, the high degree of conservation in the structure and development of the thymus between different mammalian species suggests that the observations will in general also be valid for ruminants.

Experimental evidence indicates that the thymus is concerned with the production of functionally mature lymphocytes capable of participating in cell-mediated immune responses (Miller 1962; Parrott, DeSouza &

East 1966; Davies et al. 1969a, 1969b). Laboratory animals which are congenitally athymic or have been thymectomized at birth are unable to mount cell-mediated immune responses or antibody responses to T-dependent antigens. Such animals exhibit profound lymphocytic depletion of the paracortical areas of the lymph nodes and periarteriolar regions of the spleen, and are highly susceptible to many infectious diseases. Congenital aplasia of the thymus in man, referred to as the Di George syndrome, is associated with a similar immunodeficiency (Lischner & Huff 1975).

There are no reports of primary congenital thymic aplasia in ruminants. However, profound thymic atrophy has been described in association with a congenital defect in the intestinal absorption of zinc in Black Pied Danish cattle of Friesian descent (Andersen et al. 1970; Flagstad 1976). Such animals appear to be highly susceptible to infections and have deficient cell-mediated immune responses, as evaluated by delayed-type hypersensitivity skin reactions (Brummerstedt et al. 1974). However they have normal numbers of blood lymphocytes and treatment with zinc results in repopulation of the thymus and marked clinical improvement (Brummerstedt et al. 1971). Experiments involving thymectomy in ruminants have failed to confirm an absolute dependence on the thymus for cell-mediated immune responses, although some effects have been observed (Cole & Morris 1971a; 1971b; 1971c; Fahey, Outteridge & Burrells 1980; Snider, Adams & Pierce 1981; Snider & Pierce 1981). Thus, calves or lambs thymectomized at birth or during foetal life subsequently show marked reduction in the population of surface immunoglobulin negative (sIg-negative) lymphocytes and depletion of periarteriolar areas and paracortical areas in spleen and lymph nodes, respectively. Such animals exhibit deficiency in certain but not all cell-mediated immune responses and show some increase in susceptibility to infections. However, they are able to survive and show relatively normal growth rates and indeed, in some instances, after a time exhibit regeneration of the sIg-negative lymphocyte population. It is possible that the results of these studies reflect the technical difficulties in completely removing the thymus, as in some instances, small remnants of thymic tissue were found in the thymectomized animals (Fahey, Outteridge & Burrells 1980; Snider & Pierce 1981). Alternatively, in ruminants, other lymphoid tissues, either in the absence of the thymus or under normal circumstances, may be capable of supporting maturation of T lymphocytes.

In mammals, the thymus is highly developed at birth, although there is a period of rapid growth in the immediate postnatal period. Development and growth of the organ are independent of exposure to antigen. In mice, the thymus consists of two lobes each with a distinct demarcation into cortex and medulla. Evidence has been obtained that prethymic lymphocytes produced in the bonemarrow migrate to the thymus (Ford et al. 1966; Moore & Owen 1967); at least a proportion of these immigrant cells enter the cortex and, as they mature, move inwards towards the medulla (Weissman 1973). The finding that medullary thymocytes are phenotypically similar to thymic emigrant lymphocytes and to peripheral T cells (van Ewijk, van Soest & van der Engh 1981; Scollay, Butcher & Weissman 1980; Scollay 1982) also tends to support this as the main pathway of maturation. However, there is evidence that some cells enter the medulla directly (Brumby & Metcalf 1967). Thus, the precise pathway(s) followed by lymphocytes during maturation in the thymus are far from clear (see reviews by Weissman et al. 1982; Scollay 1983).

In the mouse, there is considerable proliferation of lymphocytes in the thymus, particularly in the outer cortex, and it has been estimated that the entire lymphocyte population turns over every 3 to 4 days (Metcalf 1967). However, since only about 1% of lymphocytes produced in the thymus ever leave the organ (Metcalf 1967; Scollay 1983), there is obviously a high rate of cell death. It has been suggested that this may represent the outcome of a process of selection whereby lymphocytes with inappropriate antigenic receptors are eliminated in the thymus. Mature T cells generally possess receptors for either class I or class II major histocompatibility (MHC) antigens plus foreign antigens. Studies using chimaeric mice have shown that the specificity of some T-cell populations for self-MHC antigens together with foreign antigen is determined by the thymic microenvironment, i.e. the T cells in such animals preferentially recognize foreign antigen when presented on cells with the same MHC antigens as expressed on the thymic epithelium (Zinkernagel 1978). However, other studies along similar lines have produced conflicting results (Wagner et al. 1981). Thus, the precise role of the thymus in selection of recognition specificities of T cells and the mechanism by which the putative selection might be achieved remain unclear.

Another notable feature of the thymus in laboratory rodents is the existence of a blood-thymus barrier which precludes entry of macromolecules into the thymic cortex. The arterial blood supply of the thymus enters the peripheral medulla (Raviola & Karnovsky 1972). From here, arterioles extend into the cortex and give rise to an anastomosing arcade of capillaries in the outer cortex. The capillaries flow back into the medulla where they join postcapillary venules. The endothelial junctions of the cortical capillaries have been shown to be impermeable to macromolecules (Raviola & Karnovsky 1972). Although a small amount of material traverses the endothelial cells by vesicular transport, this material appears to be phagocytosed by cells which form a more or less continuous layer around the capillaries. Raviola and Karnovsky (1972) described these cells as macrophages, but in subsequent studies, they have been identified as epithelial cells linked by desmosomal junctional complexes (Clark 1973; Duijvestijn & Hoefsmit 1981). Unlike the cortical capillaries, the blood vessels in the medulla are highly permeable to macromolecules (Raviola & Karnovsky 1972). Thus, antigens present in the bloodstream can gain access to the medulla, although they apparently cannot enter the cortex from the medulla. The lymphocytes in the cortex, therefore, appear to be protected from exposure to foreign antigens.

The bovine thymus
General structure

The following results are based on studies of over 30 bovine thymuses. The animals were mainly of the Boran breed (*Bos indicus*), ranging in age from newborn to 3 years old and were maintained indoors free from the protozoan and helminth infections which are prevalent in East Africa. Thymic tissue was examined by conventional histological and enzyme histochemical techniques, and by immunofluorescence or immunoperoxidase to detect MHC antigenic determinants and receptors for the lectin, peanut agglutinin (PNA). Immunofluorescence was carried out both on suspensions of thymocytes, using a fluorescence-activated cell sorter (FACS II, Becton Dickinson, Mountain View, California), and on frozen tissue sections. A small number of samples was also examined by electron microscopy. Six monoclonal antibodies were used, three of which are thought to recognize determinants on bovine class I MHC molecules and three which are thought to recognize determinants on bovine class II MHC molecules (Spooner & Pinder 1983; Lalor, Morrison & Black this volume) (Table 1). Biochemical analysis of the target molecules has not yet been completed, although preliminary results confirm that one of the antibodies (P3) precipitates a molecule composed of two polypeptides with molecular masses characteristic of class I MHC antigens (A. Bensaid personal communication).

In cattle the thymus is composed of two lobes, one located in the anterior mediastinum and the other extrathoracically in the lower neck. In newborn calves, the two lobes together weigh from 25 to 75 g. In the

Table 1. Monoclonal antibodies used to stain tissues and cell suspensions.

Monoclonal antibody	Host restricted	Frequency of positive animals	Putative specificity
P12	no	100%	class I MHC
B4/18	yes	30%	class I MHC
P3	yes	12%	class I MHC
R1	no	100%	class II MHC
P1	yes	30%	class II MHC
P2	yes	37%	class II MHC

first 3 to 4 months of life, the thymus undergoes a marked increase in size reaching weights of up to 350 g. Thereafter, the organ grows more slowly until the animal reaches sexual maturity, after which the thymus undergoes a slow progressive involution. As in other species, the bovine thymus is highly susceptible to stress as a result of either poor nutrition or infectious disease and under such circumstances may undergo marked atrophy.

The lobes of the bovine thymus are made up of multiple lobules each of which consists of cortex and medulla, the latter being contiguous between all lobules (Figure 1A). The arterioles and venules conveying the blood supply to and from the thymus are located in the outer medulla of each lobule. The perivascular areas contain a network of reticular fibres and prominent bundles of collagen (Figure 1B). However, outside of these perivascular areas, the main structural framework of the thymus comprises epithelial cells. In laboratory rodents the epithelial cells of the thymus have been shown to be derived in the embryo from both endoderm and ectoderm in the regions of the third branchial pouch and third branchial cleft respectively (Hair 1974; Cordier & Heremans 1975). The ectoderm-derived component is thought to be restricted to the medulla.

Cell suspensions prepared from bovine thymus contain 98% or more lymphocytes. A large proportion of these lymphocytes are significantly smaller, have a lower cytoplasm to nucleus ratio and are significantly denser than lymphocytes in peripheral blood and lymph nodes. In contrast to mononuclear cells obtained from peripheral blood or lymph nodes, of which the majority stain for putative class I MHC antigens, only 10 to 15% of the thymic lymphocytes express detectable levels of the determinants recognized by P12, P3 and B4/18 (Figure 2). Staining of frozen sections reveals that the positive cells are found exclusively in the medulla (Figure 3A). Approximately 85% of thymocytes stain strongly

Figure 1. Normal bovine thymus showing (A) demarcation of lobules into cortex and medulla (HE x 65) and (B) localization of reticulin predominantly in the perivascular areas in the peripheral medulla (reticulin stain x 70).

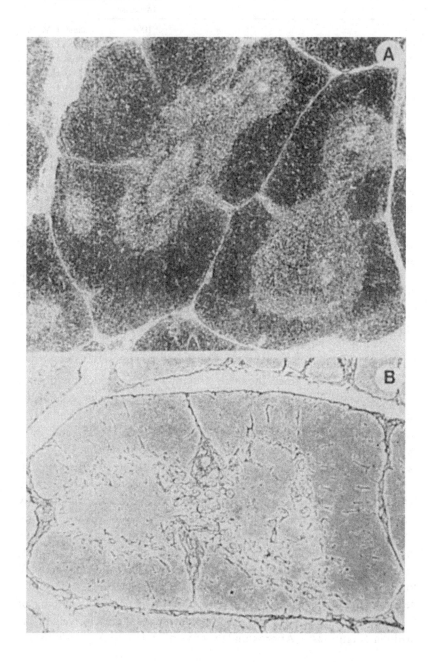

Figure 2. FACS fluorescence profiles of cell suspensions prepared from normal bovine lymph node and thymus, showing staining of lymph node cells with monoclonal antibodies B4/18 (100% positive) and R1 (11% positive) and thymic cells with monoclonal antibody B4/18 (16% positive) and PNA (84% positive). Staining with PNA was carried out using directly conjugated lectin (Vector Laboratories Inc., California), whereas staining with monoclonal antibodies utilized an indirect method using FITC-labeled sheep anti-mouse immunoglobulin. Controls were incubated with the anti-mouse conjugate alone.

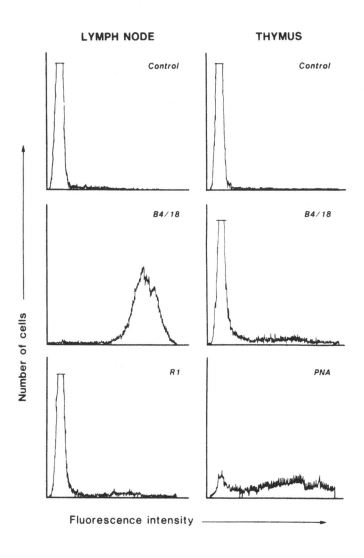

with fluorescein-labeled PNA (Figure 2), the remaining negative cells also stain weakly when high concentrations of the labeled lectin are used. In frozen sections, it can be seen that strongly staining cells are confined to the cortex. Only a very small number of cells (less than 2%) stain for the putative class II MHC determinants.

The cortex

The lymphocytes in the thymic cortex are densely packed within a fine network of epithelial cells. The latter are difficult to distinguish in histological sections but at the ultrastructural level can be recognized by their euchromatic nucleus, relatively sparse perinuclear cytoplasm and delicate cytoplasmic processes extending between the surrounding lymphocytes. The cytoplasm of these cells is markedly vacuolated and contains moderate numbers of small electron-dense granules. The cortical epithelial cells stain strongly for putative class II MHC antigens but very weakly or not at all for the putative class I MHC antigens (Figure 3A and B).

The other cell type found in the thymic cortex is the macrophage. Small numbers of macrophages, which stain strongly for nonspecific esterase activity, are found scattered through the cortex, being most numerous in the inner cortical areas. Many of these macrophages have voluminous cytoplasm containing phagocytosed lymphocytes in varying stages of degeneration.

The medulla

The medulla has a more complex structure and contains a greater variety of cell types than the cortex. Epithelial cells are more numerous and more pleomorphic than in the cortex and, in addition, there is a population of interdigitating cells which is not found in the cortex.

In the outer medulla, surrounding the arterioles and venules, there are distinct perivascular regions which vary in width from about 3 to 10 cell diameters (Figure 3C). The outer boundary of these areas is delineated by a basement membrane-like structure on which lies a layer of epithelial cells linked by desmosomal junctions. These epithelial cells in some areas are cuboidal whereas at other sites they are extremely flattened. At intervals there are large gaps in this epithelial cell layer which theoretically would permit free movement of cells between the medulla and the perivascular areas. Epithelial cells do not extend into the perivascular

Figure 3. Normal bovine thymus: (A) stained by immunofluorescence with monoclonal antibody P3 which is specific for bovine class I MHC antigen showing diffuse staining of the medulla but no detectable staining in the cortex (x 235); (B) stained by immunofluorescence with monoclonal antibody R1 which is thought

(continued)

to be specific for bovine class II MHC antigen, showing a distinctive reticular pattern of staining of the epithelial cells in the cortex and more diffuse staining in the medulla (x 235); (C) an area of peripheral medulla, showing distinct perivascular areas (P) can be seen containing numerous cells and with a discrete peripheral boundary (arrows) (1 μm section, methylene blue x 500).

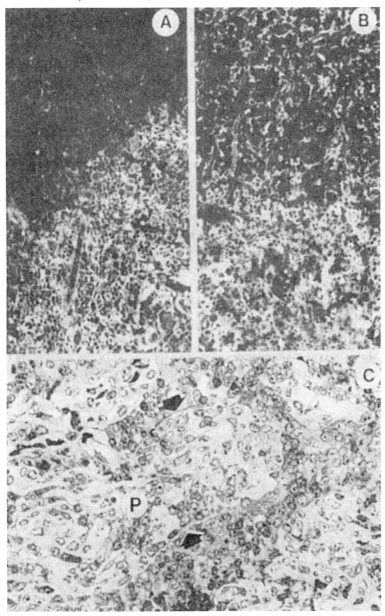

areas. Instead, these areas contain moderate numbers of fibroblasts and extensive bundles of collagen. Apart from this difference, the cell types found in the perivascular areas are apparently similar to those in the surrounding medulla. Small numbers of lymphocytes can be observed traversing the endothelium of the postcapillary venules in the medulla. However, it is not possible to ascertain the direction in which these lymphocytes are migrating. Whether the perivascular areas represent a discrete compartment of the thymus in which maturation of thymocytes is completed or whether it is merely a holding area for cells in the process of leaving or entering the organ is not known.

The epithelial cells in the medulla are very pleomorphic; they range from large cells (up to 20 μm in diameter) with a euchromatic nucleus and abundant discrete cytoplasm to smaller spindle- or stellate-shaped cells which more closely resemble the cortical epithelial cells, but have more abundant cytoplasm, less delicate cytoplasmic processes and in some instances, more condensed chromatin. There are numerous desmosomal junctions between the cytoplasmic processes of adjacent epithelial cells, and prominent bundles of intermediate filaments are present in the cytoplasm (Figure 4). In studies of the thymus in other species (Duijvestijn & Hoefsmit 1981), these have usually been referred to as tonofilaments, because of their resemblance to the keratin filaments found in other epithelial cells. In the bovine thymus, these tonofilaments are particularly abundant in some of the large medullary epithelial cells. The Hassal's corpuscles found in the central medulla are composed of accumulations of these cells with a central area of keratin deposition.

The other main nonlymphocytic cell type found in the medulla is the interdigitating cell. This is a large cell with a characteristically irregular shaped nucleus and voluminous pale staining cytoplasm containing relatively few organelles and with numerous peripheral cytoplasmic processes. Similar cells have been described in the medulla of human and rat thymuses (Kaiserling, Stein & Muller-Hermelink 1974; Duijvestijn & Hoefsmit 1981) and, in the latter species, it has been shown that these cells are derived from the bonemarrow (Barclay & Mayrhofer 1981). Small numbers of macrophages and mast cells and occasional neutrophils are also present in the medulla. Mast cells tend to be most numerous in the perivascular areas.

In sections of bovine thymus stained by immunofluorescence for the putative class I MHC antigens, diffuse staining of the medullary areas is observed, suggesting that most of the nonlymphocytic cells, as well as the lymphocytes, in this location express class I MHC antigens (Figure 3A). There is also extensive staining of nonlymphocytic cells in the

Figure 4. Electron micrograph of thymic medulla showing (A) the network of epithelial cells (e) (x 4,725), (B) a desmosomal junction between two epithelial cells (x 28,000), and (C) keratin filaments within the cytoplasm of an epithelial cell (x 14,000).

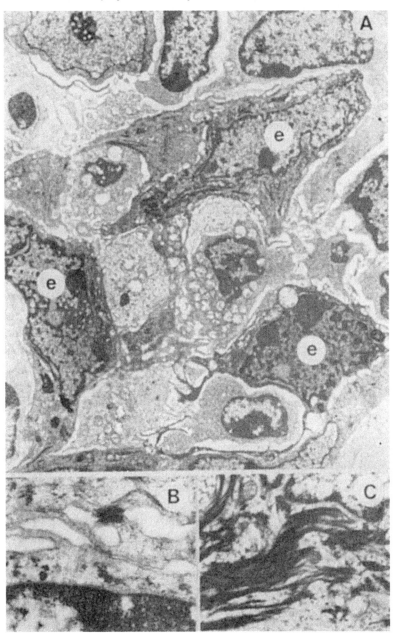

medulla for putative class II MHC determinants (Figure 3B). These positive cells almost certainly include both epithelial cells and interdigitating cells, as positive cells are found not only throughout the medulla but also in the perivascular areas. A similar distribution of MHC antigens has also been reported in mouse, rat and human thymus (Rouse et al. 1979; Bhan et al., 1980; Janossy et al. 1980; van Ewijk, Rouse & Weissman 1980; Barclay & Mayrhofer 1981).

The lymph node
General features
The lymph nodes evolved relatively late in phylogeny, compared with the thymus, and thus represent a later refinement of the immune system (Jonsson & Christensen 1978). Lymph nodes are found throughout the body, superimposed on the lymphatic system. They act as filters for foreign materials which enter lymph and provide a favourable environment for the interaction of different cell types in the generation of immune responses to lymphborne antigens. The structure of the lymph node is in a highly dynamic state, with large numbers of lymphocytes continuously entering the node from the blood and afferent lymph and leaving in the efferent lymph (reviewed by Morris 1972). For example, the output of lymphocytes in lymph draining from the prefemoral lymph node (weighing approximately 3 to 5 g) of a 6-month-old calf is in the region of 2 to 5 x 10^9 cells per day (Emery 1981). The majority of these cells are recirculating lymphocytes. In sheep, it has been estimated that peripheral lymph nodes receive approximately 90% of the cellular input from the blood, the remaining 10% of cells arriving in afferent lymph (Hall & Morris 1962). The latter contains not only lymphocytes but also up to 20% large macrophage-like cells (Smith, McIntosh & Morris 1970; Hall, Scollay & Smith 1976; Miller & Adams 1977), which are thought to be important in transportation of antigen into the lymph nodes and possibly also replenishing the accessory cell pool in the node. These large nonlymphocytic cells are absent, or present in extremely small numbers, in efferent lymph. The arrangement of a dual circulation of cells through the lymph nodes, from blood and lymph, provides maximum opportunity for specifically reactive lymphocytes to come into contact with antigen or antigen-bearing cells arriving in the lymph node from the tissues.

The lymph node is permeated by a system of sinuses. Multiple afferent lymphatics enter through the lymph node capsule and empty into the subcapsular sinus. Lymph may then pass via cortical or intermediate sinuses into an anastomosing system of sinuses in the medulla of the node. These medullary sinuses converge at the hilus where they are continuous with one or more efferent lymphatic vessels.

The remaining 'solid' lymphoid tissue consists of the medullary cords and the cortex. The cortical areas of lymph nodes are clearly demarcated into regions populated predominantly by sIg-positive lymphocytes, the B-dependent follicular areas, and regions populated mainly by sIg-negative lymphocytes, the T-dependent paracortical areas. It is within the paracortical areas that lymphocytes enter the lymph node from the blood through specialized postcapillary venules (Gowans & Knight 1963). Each of these compartments also possesses its own specialized accessory cell population. This arrangement of the node dictates that different lymphocyte populations, once they enter the lymph node, have distinct preferences for where they localize. Moreover, different types of immune responses (e.g. humoral or cell-mediated) involve different compartments of the lymph node to a greater or lesser extent.

The bovine lymph node

The following discussion is based on detailed studies on lymph nodes from over 40 cattle using histological, enzyme histochemical and immunofluorescence or immunoperoxidase techniques, similar to those outlined for the thymus. A few samples of lymph nodes were also

Figure 5. Distribution of reticulin in normal bovine lymph node. The B-dependent follicles, most of which contain germinal centres that are relatively devoid of reticulin fibres, can be seen in the superficial cortex and around the connective tissue trabeculae in the deep cortex. Elsewhere, the reticulin fibres are more prominent in the perifollicular areas than in the central paracortex (reticulin stain x 60).

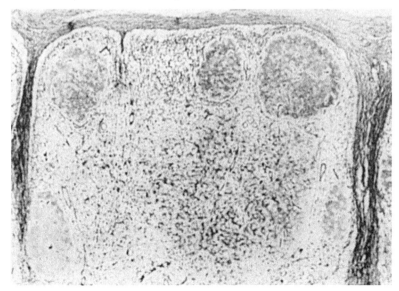

examined by electron microscopy. The studies concentrated on peripheral lymph nodes, namely, the prescapular, prefemoral and popliteal nodes.

The reticulin and reticulum cell network

The structural framework of the lymph node is provided by a network of reticulum cells and reticulin. Reticulin is a rather ill-defined material

Figure 6. Electron micrograph of normal bovine lymph node showing a reticulum cell (R) in the paracortex. The cell has a moderately well developed endoplasmic reticulum and its cytoplasm surrounds an area of reticulin (arrow) (x 5,350).

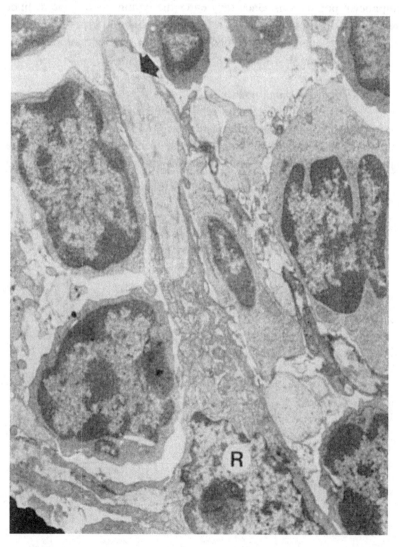

which is identified histologically by its affinity for silver stains (Figure 5). The precise basis of this staining reaction is unknown. At the ultrastructural level, reticulin is seen to consist of amorphous or fine fibrillar material containing variable numbers of fibres showing distinct banding similar to collagen fibres (Figure 6). Within the lymph nodes there is a network of reticulin enveloped by reticulum cells. These are elongated or spindle-shaped, electron-dense cells which are relatively poorly differentiated (Figure 6). It is generally assumed that they produce the reticulin ground substance although this has not been formally proven. Certainly, in bovine lymph nodes many of the reticulum cells contain rough-surfaced endoplasmic reticulum and have a prominent Golgi apparatus, suggestive of active protein synthesis. The reticulin network is present throughout the lymph node except within the germinal centres. The superficial cortex around the follicular areas and adjacent to the subcapsular sinus contains much coarser bundles of reticulin than does the paracortex (Figure 5). The reticulum cells also form a continuous layer around blood capillaries and the perivascular connective tissue of larger blood vessels. The reticulum cells abutt and overlap with each other and at some points exhibit desmosomal junctions.

The nature of the cells which line the lymph node sinuses has been a matter of some confusion over the years, as they have been variously referred to as littoral cells, reticuloendothelial cells and endothelial lining cells. We have not examined the structure of the bovine lymph node sinuses at the ultrastructural level. However, from studies in other species, it is clear that these sinus lining cells do not lie on a basement membrane but are associated with an underlying layer of reticulin (Clark 1962; Moe 1963; Farr, Cho & de Bruyn 1980). It has been proposed that they are reticulum cells because of this and their morphological similarity to the reticulum cells elsewhere in the node (Hoefsmit 1975). The lymph node sinus wall consists of two layers of reticulum cells with an intervening layer of reticulin. Bundles of reticular fibres which traverse the lumena of the sinuses are also enveloped in reticulum cells. The reticulum cells and reticulin within areas of solid lymphoid tissue are attached to the outer aspect of the sinus walls, thus providing continuity of the network throughout the entire node. At all levels within the lymph node the reticulin and areas of connective tissue are separated from the resident lymphoid cells by a layer of reticulum cells.

In laboratory animals, it has been shown that the sinus lining reticulum cells form a continuous layer at all levels except in the inner lining of the subcapsular sinus, where there are small gaps in the reticulum cell layer which enable cells to cross the sinus wall (Clark 1962). This is thought to be the major route by which cells and antigens in afferent lymph enter the substance of the lymph node. The sinus reticulum cells, unlike those elsewhere in the node, possess a well-developed lysosomal apparatus and have considerable phagocytic capacity. It has been demonstrated in sheep that, following infusion of various particulate antigens into afferent lymph, 75% or more of the antigen can be trapped in the drainage lymph node (Fahy et al. 1980). By contrast, only about 1% of a soluble antigen was retained in the node. Phagocytosis of particulate material such as carbon can be visualized not only in the sinuses but also in the superficial cortex immediately beneath the subcapsular sinus. It is believed that the filtering capacity of the lymph node is facilitated by a slowing of lymph flow in the medullary sinuses and creation of turbulence in lymph flow by the reticular trabeculae in the sinuses. Whether or not reticulum cells have other functions related more directly to induction of immune responses in the lymph node is not known.

The follicular areas

Histologically, bovine lymph nodes have a segmented appearance due to thick connective tissue trabeculae which extend inwards from the capsule into the deep cortex (Figure 5). The follicular areas are found in the superficial cortex and in the deeper cortex adjacent to the sinuses which extend inwards around the connective tissue trabeculae (Figures 5 and 7A). Surface Ig-positive cells are also found more diffusely in the medullary cords and in the cortico-medullary transitional areas. It is possible that the formation of discrete follicular accumulations of B cells in the peripheral cortex is associated with the entry of lymphocytes and antigens from the lymph in these areas. In other words, follicular formation may occur principally in areas adjacent to sinuses which have gaps in the reticulum cell lining. In adult cattle, a proportion of the follicles are primary follicles composed entirely of small sIg-positive lymphocytes, whereas others, termed secondary follicles, comprise a germinal centre surrounded by a mantle of small sIg-positive lymphocytes. In the peripheral lymph nodes of calves, germinal centres appear in the first week of life, and in healthy 6 to 12 month old animals 30 to 70% of follicles within a lymph node may contain germinal centres. Thus, the lymph nodes of normal healthy cattle are extremely active. The germinal centres contain dense granular deposits of immunoglobulins (Figure 7A). By analogy with findings in other species (Mandel et al. 1980), these deposits probably represent immune complexes on the surface of follicular dendritic cells (FDC). Active germinal centres are usually demarcated into two distinct zones; the inner pole distal from the adjacent sinus

Figure 7. Normal bovine lymph node: (A) stained by immunofluorescence with FITC-labeled sheep anti-bovine immunoglobulin, showing staining of B cells in two follicles lying on either side of a connective tissue trabeculum, the follicle on the right contains a germinal centre in which there are coarse granular deposits of immunoglobulin (x 135); (B) stained by immunoperoxidase with monoclonal antibody R1, showing staining of lymphocytes in the follicular (areas) and large nonlymphocytic cells in the paracortex (P) (x 105).

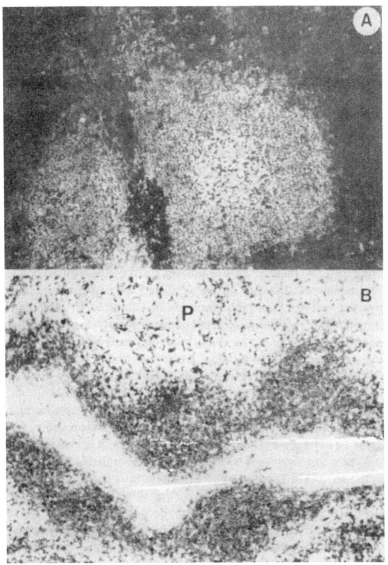

contains mainly lymphoblasts with high mitotic activity and the outer less active cap region contains fewer lymphoblasts and more numerous nonlymphocytic cells. The granular deposits of Ig in the germinal centres are usually much more abundant in the outer cap region suggesting that this part is rich in FDC. The significance of this zonal arrangement of the germinal centre is not known. The high proliferative activity in germinal centres is associated with lymphocyte death and the presence of macrophages which contain phagocytosed dead and dying lymphocytes (tingible body macrophages).

Many of the blast cells in germinal centres appear to be negative for surface IgM, as detected in tissue sections, at least in the inner part of the centre where interpretation is not masked by Ig on the FDC. This is in agreement with studies in other species in which the germinal centre cells have been shown to be mainly sIgM-negative but have low levels of sIgG or sIgA (Butcher et al. 1982; Rose et al. 1982). In mouse and human lymph nodes, small numbers of T cells (mainly expressing helper cell phenotypes) are found within both primary and secondary follicles (Rouse et al. 1982). We have not yet examined the distribution of cells bearing T-cell markers in cattle. Approximately 10 to 18% of cells in suspensions of bovine lymph node cells stain for putative class II MHC determinants (Figure 2). Staining of tissue sections has confirmed that these are the B lymphocytes, i.e. there is staining of lymphocytes in the follicular areas but not in the paracortex (Figure 7B). The question of whether the FDC stain for class II antigens could not be resolved because of the staining of other cells in the germinal centre. When sections are stained for putative class I MHC antigens, the cells of the germinal centres are found to express much lower levels of these antigens than other cellular components of the node, in many instances there being virtually no staining (Figure 8A). The germinal centre cells also stain strongly with PNA (Figure 8B) (J. Ellis, N. McHugh & W.I. Morrison unpublished), as has been observed in other species (Rose et al. 1980; 1982).

Although the structure of germinal centres has been known for many years and, indeed, they have been recognized as prominent components of secondary lymphoid tissues since the last century, it is only in the last 5 to 6 years that significant information on their probable functions has been obtained. The function of germinal centres is thought to be based on the properties of the FDC (Klaus et al. 1980; Mandel et al. 1980. These cells form a network within the germinal centre. At the ultrastructural level, they can be seen to have relatively small amounts of perinuclear cytoplasm but have long slender cytoplasmic processes, with irregular invaginations of the plasma membrane, extending between

Figure 8. Normal bovine lymph node stained: (A) by immunofluorescence with monoclonal antibody P12 showing virtually no detectable staining of cells within a germinal centre (G) (x 250); (B) stained with peroxidase-labeled peanut agglutinin, showing staining of the germinal centre cells (x 180).

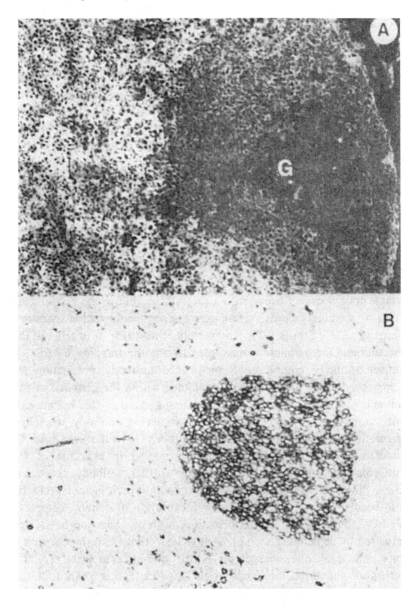

the surrounding lymphocytes. Whether the FDC is a modified reticulum cell or is derived from the bonemarrow is not known. Studies in mice have shown that it is only possible to isolate FDC by subjecting lymphoid tissues to rigorous enzymatic treatment (Humphrey & Grennan 1982). Such isolated cells were found to have receptors for the Fc region of IgG and for C3; they did not adhere to glass, were not phagocytic, contained only weak lysosomal enzyme activity and did not express detectable levels of class II MHC antigens. These cells are, therefore, distinct from macrophages and from the interdigitating or dendritic cells found in the paracortical areas. Previous studies in mice showed that antigen is trapped on the surface of FDC in the form of antigen-antibody complexes and that this is most effectively achieved through fixation of complement, presumably by binding to the C3 receptors (reviewed by Klaus et al. 1980; Mandel et al. 1980). It is still unclear whether the immune complexes are transported into the follicle by a population of Fc-bearing lymphocytes (Brown et al. 1970) or by macrophages or by the precursors of the FDC, if these cells originate outside of the node.

The two main functions which have been proposed for germinal centres are the generation of memory B cells and the regulation of level and duration of an antibody response. There is considerable evidence that initial production of antibody during a humoral immune response in the lymph node does not occur within the follicles, but in the perifollicular areas or medullary cords, before germinal centre formation (Movat & Fernando 1965; Straus 1977; Kamperdijk et al. 1978). It is believed that the stimulus for germinal centre formation is the trapping by FDC of antigen-antibody complexes formed once antibody production has commenced. Such antigen may be retained within the germinal centres for many months. There is evidence to suggest that this retention of antigen may serve to maintain levels of circulating antibody, whereby if circulating levels of antibody fall, antibody is withdrawn into the circulation from the germinal centre, resulting in exposure of free antigenic determinants which stimulates further antibody production (Tew, Phipps & Mandel 1980). The production of B-memory cells has also been shown to be dependent on localization of immune complexes on FDC and germinal centre formation, both of which can be severely impaired by depletion of complement (Klaus et al. 1980). Inoculation of mice with small quantities of preformed immune complexes results in generation of B-memory cells about a week earlier than following inoculation with antigen alone. The finding that localization of immune complexes in germinal centres can also result in generation of memory cells with specificity for the idiotype of the antibody in the complexes has led to the suggestion that the germinal centres may be involved in negative regulation of antibody responses by anti-idiotype antibodies (Klaus 1978).

The paracortical areas

The paracortical areas are populated by sIg-negative lymphocytes and distinct populations of nonlymphocytic cells. The latter comprise small numbers of macrophages and numerous interdigitating cells. In contrast to macrophages, the interdigitating cells do not appear to be actively phagocytic and contain only a few or no lysosomal granules; they have a characteristic irregularly shaped nucleus and voluminous translucent cytoplasm, relatively devoid of organelles and showing numerous peripheral cytoplasmic processes (Figure 9). Similar interdigitating cells have been described in mouse and rat secondary lymphoid tissues (Veerman, 1974; Veerman & van Ewijk 1975; Friess 1976).

In bovine lymph nodes stained for putative class II MHC antigens, although the lymphocytes in the paracortex are negative, numerous brightly staining large stellate-shaped cells are observed (Figure 7A). These are much too numerous to be accounted for by the macrophage population, and almost certainly represent the interdigitating cells. A similar pattern of staining for class II MHC antigens has also been detected in mouse, rat and human lymph nodes (Hoffmann-Fezer et al. 1978; Barclay 1981; Janossy et al. 1981; Poppema et al. 1981). The interdigitating cell is probably the same cell as the dendritic cell originally described by Steinman and Cohn (1973) in cell suspensions prepared from mouse lymph node, spleen and Peyer's patch. These dendritic cells were shown to be derived from bonemarrow but could be distinguished from macrophages by their poor phagocytic activity and by the absence of Fc and C3 receptors; they also lacked various lymphocyte surface markers, were rich in class II MHC antigens and were potent stimulators of both syngeneic and allogeneic mixed leucocyte reactions (Steinman & Nussenzweig 1980). Present evidence suggests that the dendritic cell is an important accessory cell in T cell-mediated immune responses (van Voorhis et al. 1983).

It is believed that the majority of macrophages and dendritic/interdigitating cells in the lymph nodes enter the nodes in the afferent lymph, rather than from the blood. Rat lymph nodes surgically deprived of afferent lymphatics become severely depleted for these cell types (Hendriks, Eestermans & Hoefsmit 1980). Studies of afferent lymph in rabbits have demonstrated the presence of numerous large frilly cells, only some of which are phagocytic and have Fc and C3 receptors (Kelly et al. 1978). A large percentage of these frilly cells were found to enter the paracortical areas of the lymph node. From these and other studies (Kamperdijk et al. 1978), it is suggested that the population of interdigitating cells in the paracortical areas of the lymph nodes are continually turning over and being replenished by cells arriving in the afferent lymph. It is also likely, as suggested by Balfour and colleagues (1982), that these cells are important in transportation of antigen into the paracortical areas. The afferent lymph cells and the interdigitating cells have many properties in common with the epidermal Langerhans cells in the skin (Silberberg-Sinakin et al. 1980), although, unlike the interdigitating cells, Langerhans cells possess Fc and C3 receptors (Stingl et al. 1977). Furthermore, in guinea pigs, uptake of antigen by epidermal Langerhans cells has been demonstrated and antigen-containing cells with morphological characteristics of Langerhans cells have been observed in afferent lymph and in lymph nodes regional to the site of inoculation of antigen (Silberberg-Sinakin et al. 1980).

Another prominent feature of the paracortical areas is the presence of specialized postcapillary venules which are the portals for entry of lymphocytes from the bloodstream. In cattle, many of these vessels have a distinct cuboidal endothelium and, as in other species, they contain numerous lymphocytes in the process of crossing the endothelium. Such vessels can be seen in the lymph nodes of neonatal calves, although they are less prominent than in older animals. These specialized venules are not static structures but can fluctuate in the degree of development, depending on the activity of the lymph node (Herman 1980). Thus, under conditions of heavy antigenic challenge to the node, there is an increase in the number of high endothelial vessels due to activation of normally quiescent segments of venule. Studies on rat lymph nodes suggest that afferent lymph is required for development of the modified postcapillary venules, as in lymph nodes surgically deprived of afferent lymphatics the high endothelial venules disappeared (Hendriks, Eestermans & Hoefsmit 1980). Evidence was obtained which suggested that macrophages or soluble products of macrophage-lymphocyte interactions may be involved in stimulating the vascular changes (Hendriks & Eestermans 1983). Nevertheless, since there is considerable recirculation of lymphocytes during foetal life (Morris, Pederson & Trevella this volume), it is clear that the high endothelial venules are functional in circumstances where there is no antigenic stimulation.

Although there is considerable information on the routes by which circulating cells enter the lymph node, relatively little is known about the way in which they exit from the node and how the outflow of cells is regulated. Superficially, on histological examination of a bovine lymph node, the paracortical areas appear to be composed of continuous sheets of lymphocytes and accessory cells in a matrix of reticulum cells. However, on closer scrutiny, it is often possible to distinguish longitudinal channels with a more or less continuous lining of flattened cells and containing dense accumulations of lymphocytes. Such channels can more readily be seen in the lymph nodes of cattle which have died from theileriosis (Figure 10); in some of these lymph nodes there is profound lymphocytic depletion but the structural framework of the node is retained intact. Indeed, studies by Kelly (1975) in rabbits indicate that the paracortex is made up of a system of cords surrounding the venules, with intervening channels which he termed paracortical sinuses. These paracortical sinuses emptied into the medullary sinuses through narrow apertures at the cortico-medullary

Figure 9. Electron micrograph of normal bovine lymph node showing two interdigitating cells in the paracortex (x 14,500).

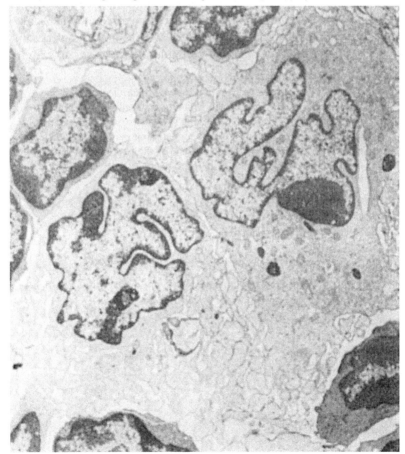

junction. Similar channels have also been described in the paracortex of mouse and rat lymph nodes (Soderstrom & Stenstrom 1969). Furthermore, on the basis of morphological observations on resting and antigen-stimulated lymph nodes, Kelly suggested that the outflow of cells from the cortex may be regulated by clumping of lymphocytes and plugging of the sinus outlets. These changes could be induced by infusion into afferent lymphatics of supernatants of lymph node cells which had been sensitized *in vivo* and stimulated *in vitro* with BCG (Kelly et al. 1972). Thus, these authors suggested that soluble products generated either at the site of antigen inoculation or in the lymph node could produce changes in the node which result in retention of lymphocytes.

It has been known for many years that antigenic stimulation of sheep lymph nodes results in a transient period of decreased cell output in efferent lymph (termed 'shutdown'), within a few hours of administration of antigen (Hall & Morris 1965). A similar phenomenon can also be induced by inoculation of antigen-antibody complexes or substances which activate complement (McConnell & Hopkins 1981). Evidence that prostaglandin E_2 (PGE_2) may be the molecule involved in effecting

Figure 10. Lymph node from an animal which died of theileriosis showing an area of paracortex which is markedly depleted of lymphocytes. A postcapillary venule (P) can be seen with several dilated sinus-like channels in the surrounding paracortex. These structures are probably the equivalent of the paracortical sinuses described by Kelly (1975) in the rabbit.

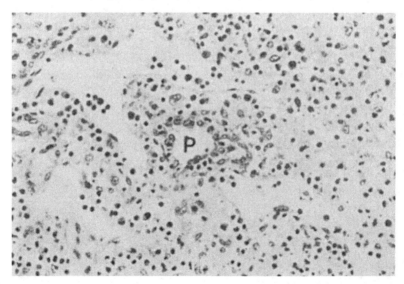

retention of lymphocytes in the lymph node was obtained by Hopkins, McConnell and Pearson (1981) who showed that the period of shutdown correlated with a peak of PGE_2 in efferent lymph and that administration of inhibitors of PGE_2 synthesis abrogated the shutdown phenomenon.

Concluding remarks

The observations on the bovine thymus illustrate a close resemblance in the constituent cell populations and structure, with the thymus of other species. In summary, the lymphocytes in the cortex, which make up about 85% of the thymocyte population are relatively devoid of both class I and class II MHC antigens and express high levels of the PNA receptor on their surface. They are closely associated with a fine network of epithelial cells which exhibit high levels of class II MHC antigens but little or no detectable class I antigen. By contrast, the medullary lymphocytes express detectable levels of class I MHC antigen, have only low levels of PNA receptors and are associated with two main types of accessory cell, a heterogeneous population of epithelial cells and the interdigitating cells, both of which are rich in class I and also class II MHC antigens. Most of the questions regarding how this structural arrangement exerts its influence on maturation and selection of thymic lymphocytes and at what levels within the thymus these processes occur remain unanswered.

As in other species, the structure of the bovine lymph node is adapted to monitoring and responding to foreign materials which gain access to the lymphatic system. Lymph nodes are highly efficient at trapping foreign antigens by virtue of their marked phagocytic capacity and the tendency for antigen-bearing cells in afferent lymph to be retained within the nodes. Moreover, continuous recirculation of lymphocytes from blood to lymph through the lymph nodes, via specialized postcapillary venules, maximizes recruitment of specifically antigen reactive lymphocytes. Both the input of lymphocytes from the blood and the output in the lymph can be modulated by subtle structural changes in the node, probably mediated by molecules released during initiation of immune responses. These functional features of the lymph node enable efficient induction of immune responses with the production of effector cells which can recirculate to the original site of tissue insult.

The lymphocyte populations in bovine lymph nodes are segregated into distinguishable compartments. The follicular areas are populated predominantly by B lymphocytes which express surface Ig and class II MHC antigens; the blast cells within the germinal centres of the follicles are characterized by their high affinity for PNA and expression of much lower level of class I MHC antigens than on cells elsewhere in the lymph node. The paracortical areas are populated mainly by T lymphocytes which are negative for surface Ig and class II MHC antigens. Each of these areas has its own specialized accessory cell, namely, the follicular dendritic cell and the interdigitating cell, the latter being rich in class II MHC antigens.

While useful information on the definition of lymphocyte and accessory cell populations in cattle is beginning to emerge, there is much scope for further investigation. More markers are required for identification and isolation of functionally significant populations of cells. Furthermore, information on the efficacy and type of immune response generated, in relation to such factors as the molecular nature of the antigen, the accessory cells which take up a given antigen, and the sites of localization of antigen or antigen-laden cells in the lymph node is required. Their size gives domestic ruminants certain advantages for studies of such problems, since surgical cannulation techniques can readily be performed to monitor the events afferent and efferent to a stimulated lymph node.

Acknowledgements

We wish to thank Dr Max Murray for encouragement and advice during the course of these studies. We are also grateful to Mr Niall MacHugh and Mr Douglas Bovell for expert technical assistance.

References

Andersen, E., Flagstad, T., Basse, A. & Brummerstedt, E. (1970) Evidence of a lethal trait, A46, in Black Pied Danish cattle of Friesian descent. *Nordisk Veterinarmedicin,* **22,** 473-85.

Balfour, B., O'Brien, J.A., Perera, M., Clarke, J., Sumerska, T. & Knight, S.C. (1982). The effect of veiled cells on lymphocyte function. In *In vivo immunology: histophysiology of the lymphoid system* (eds. P. Nieuwenhuis, A.A. van den Broek & M.G. Hanna), Advances in Experimental Medicine & Biology 149, pp. 447-54. New York: Plenum Press.

Barclay, A.N. (1981). Different reticular elements in rat lymphoid tissue identified by localization of Ia, Thy 1 and MRC OX2 antigens. *Immunology,* **44,** 727-36.

Barclay, A.N. & Mayrhofer, G. (1981). Bonemarrow origin of Ia-positive cells in the medulla of rat thymus. *Journal of Experimental Medicine,* **153,** 1666-71.

Bhan, A.K., Reinherz, E.L., Poppema, S., McCluskey, R.T. & Schlossman, S.F. (1980). Localization of T-cell and major histocompatibility complex antigens in the human thymus. *Journal of Experimental Medicine,* **152,** 771-82.

Brown, J.C., de Jesus, D.G., Holborow, E.J. & Harris, G. (1970). Lymphocyte-mediated transport of aggregated human γ-globulin into germinal centre areas in normal mouse spleen. *Nature,* **228,** 367-69.

Brumby, M. & Metcalf, D. (1967). Migration of cells to the thymus:

demonstrated by parabiosis. *Proceedings of the Society for Experimental Biology & Medicine*, **124**, 99-103.

Brummerstedt, E., Flagstad, T., Basse, A. & Andersen, E. (1971). The effect of zinc on calves with hereditary thymus hypoplasia (lethal trait A46). *Acta Pathologica Microbiologica Scandinavica*, **79**, 686-87.

Brummerstedt, E., Andersen, E., Basse, A. & Flagstad, T. (1974). Lethal trait A46 in cattle: immunological investigations. *Nordisk Veterinarmedicin*, **26**, 279-93.

Butcher, E.C., Reichert, R.A., Coffman, R.L., Nottenburg, C. & Weissman, I.L. (1982). Surface phenotype and migratory capability of Peyer's patch germinal centre cells. In In vivo *immunology: histophysiology of the lymphoid system* (eds. P. Nieuwenhuis, A.A. van den Broek & M.G. Hanna), Advances in Experimental Medicine & Biology 149, pp. 765-72. New York: Plenum Press.

Clark, S.L. (1962). The reticulum of lymph nodes in mice studied with the electron microscope. *American Journal of Anatomy*, **110**, 217-57.

Clark, S.L. (1973). The intrathymic environment. In *Thymus dependency* (eds. A.J.S. Davies & R.L. Carter), Contemporary Topics in Immunobiology 2, pp. 77-99. New York: Plenum Press.

Cole, G.J. & Morris, B. (1971a). The growth and development of lambs thymectomized *in utero*. *Australian Journal of Experimental Biology & Medical Science*, **49**, 33-53.

Cole, G.J. & Morris, B. (1971b) The cellular and humoral response to antigens in lambs thymectomized *in utero*. *Australian Journal of Experimental Biology & Medical Science*, **49**, 54-74.

Cole, G.J. & Morris, B. (1971c) Homograft rejection and hypersensitivity reactions in lambs thymectomized *in utero*. *Australian Journal of Experimental Biology & Medical Science*, **49**, 75-88.

Cordier, A.C. & Heremans, J.F. (1975). Nude mouse embryo: ectodermal nature of the primordial thymic defect. *Scandinavian Journal of Immunology*, **4**, 193-96.

Davies, A.J.S., Carter, R.L., Leuchars, E. & Wallis, V. (1969a). The morphology of immune reactions in normal, thymectomized and reconstituted mice. 2. The response to oxazolone. *Immunology*, **17**, 111-26.

Davies, A.J.S., Carter, R.L., Leuchars, E., Wallis, V. & Koller, P.C. (1969b). The morphology of immune reactions in normal, thymectomized and reconstituted mice. 1. The response to sheep erythrocytes. *Immunology*, **16**, 57-69.

Duijvestijn, A.M. & Hoefsmit, E.C.M. (1981). Ultrastructure of the rat thymus: the microenvironment of T-lymphocyte maturation. *Cell & Tissue Research*, **218**, 279-92.

Emery, D.L. (1981). Kinetics of infection with *Theileria parva* (East Coast fever) in the central lymph of cattle. *Veterinary Parasitology*, **9**, 1-16.

Fahey, K.J., Outteridge, P.M. & Burrells, C. (1980). The effect of prenatal thymectomy on lymphocyte subpopulations in the sheep. *Australian Journal of Experimental Biology & Medical Science*, **58**, 571-83.

Fahy, V.A., Gerber, H.A., Morris, B., Trevella, W. & Zukoski, C.F. (1980). The function of lymph nodes in the formulation of lymph. In *Essays on the anatomy and physiology of lymphoid tissues* (eds. Z. Trnka & R.N.P. Cahill), Monographs in Allergy 16, pp. 28-99. Basel: S. Karger.

Farr, A. G., Cho, Y. & de Bruyn, P.P.H. (1980). The structure of the sinus wall of the lymph node relative to its endocytic properties and transmural cell passage. *American Journal of Anatomy*, **157**, 265-84.

Flagstad, T. (1976). Lethal trait A46 in cattle: intestinal zinc absorption. *Nordisk Veterinarmedicin*, **28**, 160-69.

Ford, W.L., Micklem, H.S., Evans, E.P., Gray, J.G. & Ogden, D.A. (1966). The inflow of bonemarrow cells to the thymus: studies with part body irradiated mice injected with chromosome-marked bonemarrow and subjected to antigenic stimulation. *Annals of the New York Academy of Sciences*, **129**, 283-96.

Friess, A. (1976). Interdigitating reticulum cells in the popliteal lymph node of the rat: an ultrastructural and cytochemical study. *Cell & Tissue Research*, **170**, 43-60.

Gowans, J.L. & Knight, E.J. (1963). The route of recirculation of lymphocytes in the rat. *Proceedings of the Royal Society of London*, **159**, 257-82.

Hair, J. (1974). The morphogenesis of thymus in nude and normal mice. In *Proceedings of the first international workshop on nude mice* (eds. J. Rygaard & C.O. Poulsen), pp. 23-30. Stuttgart: Gustav Fischer Verlag.

Hall, J.G. & Morris, B. (1962). The output of cells in lymph from the popliteal node of the sheep. *Quarterly Journal of Experimental Physiology*, **47**, 360-69.

Hall, J.G. & Morris, B. (1965). The immediate effect of antigens on the cell output of a lymph node. *British Journal of Experimental Pathology*, **46**, 450-54.

Hall, J.G., Scollay, R. & Smith, M. (1976). Studies on the lymphocytes of sheep. 1. Recirculation of lymphocytes through peripheral lymph nodes and tissues. *European Journal of Immunology*, **6**, 117-20.

Hendriks, H.R. & Eestermans, I.L. (1983). Disappearance and reappearance of high endothelial venules and immigrating lymphocytes in lymph nodes deprived of afferent lymphatic vessels: a possible regulatory role of macrophages in lymphocyte migration. *European Journal of Immunology*, **13**, 663-69.

Hendriks, H.R., Eestermans, I.L. & Hoefsmit, E.C.M. (1980). Depletion of macrophages and disappearance of postcapillary high endothelial venules in lymph nodes deprived of afferent lymphatic vessels. *Cell & Tissue Research*, **211**, 375-89.

Herman, P.G. (1980). Microcirculation of organized lymphoid tissues. In *Essays on anatomy and physiology of lymphoid tissues* (eds. Z. Trnka & R.N.P. Cahill), Monographs in Allergy 16, pp. 26-142. Basel: S. Karger.

Hoefsmit, E.C.M. (1975). Mononuclear phagocytes, reticulum cells and dendritic cells in lymphoid tissues. In *Mononuclear phagocytes in immunity, infection and pathology* (ed. R. van Furth), pp. 129-46. London: Blackwell Scientific Publications.

Hoffmann-Fezer, G., Gotze, D., Rodt, H. & Thierfelder, S. (1978). Immunohistochemical localization of xenogeneic antibodies against Ia^k lymphocytes on B cells and reticular cells. *Immunogenetics*, **6**, 367-77.

Hopkins, J., McConnell, I. & Pearson, J.D. (1981). Lymphocyte traffic through antigen-stimulated lymph nodes. 2. Role of prostaglandin E$_2$ as a mediator of cell shutdown. *Immunology*, **42**, 225-31.

Humphrey, J.H. & Grennan, D. (1982). Isolation and properties of spleen follicular dendritic cells. In *In vivo immunology: histophysiology of the lymphoid system* (eds. P. Nieuwenhuis, A.A. van den Broek & M.G. Hanna), Advances in Experimental Medicine & Biology 149, pp. 823-27. New York: Plenum Press.

Janossy, G., Tidman, N., Papageorgion, E.G., Kung, P.C. & Goldstein, G.

(1981). Distribution of T-lymphocyte subsets in the human bonemarrow and thymus: an analysis with monoclonal antibodies. *Journal of Immunology*, **126**, 1608-13.

Janossy, G., Thomas, J.A., Bollum, F.J., Granger, S., Pizzolo, G., Bradstock, K.F., Wong, L., McMichael, A., Ganeshauru, K. & Hoffbrand, A.V. (1980). The human thymic microenvironment: an immunohistologic study. *Journal of Immunology*, **125**, 202-12.

Jonsson, V. & Christensen, B.E. (1978). The phylogenetic evolution of the lymph node. *Scandinavian Journal of Haematology*, **20**, 5-12.

Kaiserling, E., Stein, H. & Muller-Hermelink, H.K. (1974). Interdigitating reticulum cells in the human thymus. *Cell & Tissue Research*, **155**, 47-55.

Kamperdijk, E.W.A., Raaymakers E.M., de Leeuw, J.H.S. & Hoefsmit, E.C.M. (1978). Lymph node macrophages and reticulum cells in the immune response. 1. The primary response to paratyphoid vaccine. *Cell & Tissue Research*, **192**, 1-23.

Kelly, R.H. (1975). Functional anatomy of lymph nodes. 1. The paracortical cords. *International Archives of Allergy & Applied Immunology*, **48**, 836-49.

Kelly, R.H., Wolstencroft, R.A., Dumonde, D.C. & Balfour, B.M. (1972). Role of lymphocyte activation products (LAP) in cell-mediated immunity. 2. Effectors of lymphocyte activation products on lymph node architecture and evidence for peripheral release of LAP following antigenic stimulation. *Clinical & Experimental Immunology*, **10**, 49-65.

Kelly, R.H., Balfour, B.M., Armstrong, J.A. & Griffiths, S. (1978). Functional anatomy of lymph nodes. 2. Peripheral lymphborne mononuclear cells. *Anatomical Record*, **190**, 5-21.

Klaus, G.G.B. (1978). Antigen-antibody complexes elicit anti-idiotypic antibodies to self-idiotypes. *Nature*, **272**, 265-66.

Klaus, G.G.B., Humphrey, J.H., Kunkl, A. & Dungworth, D.W. (1980). The follicular dendritic cell: its role in antigen presentation in the generation of immunological memory. *Immunological Reviews*, **53**, 3-28.

Lischner, H.W. & Huff, D.S. (1975). T-cell deficiency in Di George syndrome. In *Immunodeficiency in man and animals* (eds. D. Bergsing, R.A. Good, J. Finstad & N.W. Paul), pp. 16-21. Sunderland, Massachusetts: Sinauer Associates.

Mandel, T.E., Phipps, R.P., Abbot, A. & Tew, J.G. (1980). The follicular dendritic cell: long-term antigen retention during immunity. *Immunological Reviews*, **53**, 29-59.

McConnell, I. & Hopkins, J. (1981). Lymphocyte traffic through antigen-stimulated lymph nodes. 1. Complement activation within lymph nodes initiates cell shutdown. *Immunology*, **42**, 217-31.

Metcalf, D. (1967). Lymphocyte kinetics in the thymus. In *The lymphocyte in immunology and haemopoiesis* (ed. J.M. Yoffey), pp. 333-41. London: Edward Arnold.

Miller, H.R.P. & Adams, E.P. (1977). Reassortment of lymphocytes in lymph from normal and allografted sheep. *American Journal of Pathology*, **87**, 59-76.

Miller, J.F.A.P. (1962). Immunological significance of the thymus of the adult mouse. *Nature*, **194**, 1318-19.

Moe, R.E. (1963). Fine structure of the reticulum and sinuses of lymph nodes. *American Journal of Anatomy*, **112**, 311-35.

Moore, M.A.S. & Owen, J.J.J. (1967). Experimental studies on the development of the thymus. *Journal of Experimental Medicine*, **126**, 715-25.

Morris, B. (1972). The cells of lymph and their role in immunological reactions. In *Handbuch der allgemeinen pathologie*, pp. 405-56. Berlin: Springer Verlag.

Movat, H.Z. & Fernando, N.V.P. (1965). The fine structure of the lymphoid tissue during antibody formation. *Experimental & Molecular Pathology*, **4**, 155-88.

Parrott, D.M.V., DeSouza, M.A.B. & East, J. (1966). Thymus-dependent areas in the lymphoid organs of neonatally thymectomized mice. *Journal of Experimental Medicine*, **123**, 191-203.

Poppema, S., Bhan, A.K., Reinherz, E.L., McCluskey R.T. & Schlossman, S.F. (1981). Distribution of T-cell subsets in human lymph nodes. *Journal of Experimental Medicine*, **153**, 30-41.

Raviola, E. & Karnovsky, M.J. (1972). Evidence for a blood-thymus barrier using electron-opaque tracers. *Journal of Experimental Medicine*, **136**, 466-98.

Rose, M.L., Booth, R.J., Habeshaw, J.A. & Robertson, D. (1982). Separation and characterization of tonsillar germinal centre lymphocytes using peanut lectin. In *In vivo immunology: histophysiology of the lymphoid system* (eds. P. Nieuwenhuis, A.A. van den Broek & M.G. Hanna), Advances in Experimental Medicine & Biology 149, pp. 773-80. New York: Plenum Press.

Rose, M.L., Birbeck, M.S.C., Wallis, V.J., Forrester, J.A. & Davis, A.J.S. (1980) Peanut lectin-binding properties of germinal centres of mouse lymphoid tissue. *Nature*, **286**, 364-66.

Rouse, R.V., van Ewijk, W., Jones, P.P. & Weissman, I.L. (1979). Expression of MHC antigens by mouse thymic dendritic cells. *Journal of Immunology*, **122**, 2508-15.

Rouse, R.V., Weissman, I.L., Ledbetter, J.A. & Warnke, R.A. (1982). Expression of T-cell antigens by cells in mouse and human primary and secondary follicles. In *In vivo immunology: histophysiology of the lymphoid system* (eds. P. Nieuwenhuis, A.A. van den Broek & M.G. Hanna), Advances in Experimental Medicine & Biology 149, pp. 751-56. New York: Plenum Press.

Scollay R. (1982). Thymus cell migration: cells migrating from thymus to peripheral lymphoid organs have a mature phenotype. *Journal of Immunology*, **128**, 1566-70.

Scollay, R. (1983). Intrathymic events in the differentiation of T lymphocytes: a continuing enigma. *Immunology Today*, **4**, 282-86.

Scollay, R., Butcher, E.C. & Weissman, I.L. (1980). Thymus cell migration. Quantitative aspects of cellular traffic from the thymus to the periphery in mice. *European Journal of Immunology*, **10**, 210-18.

Silberberg-Sinakin, I., Gigli, I., Baer, R.L. & Thorbecke, G.J. (1980). Langerhans cells: role in contact hypersensitivity and relationship to lymphoid dendritic cells and to macrophages. *Immunological Reviews*, **53**, 203-32.

Smith, J.B., McIntosh, G.H. & Morris, B. (1970). The traffic of cells through tissues: a study of peripheral lymph in sheep. *Journal of Anatomy*, **107**, 87-100.

Snider, T.G. & Pierce, K.R. (1981). Lymphoreticular alterations in neonatally thymectomized and anti-lymphocyte globulin treated Holstein-Friesian calves. *Veterinary Immunology & Immunopathology*, **2**, 331-42.

Snider, T.G., Adams, L.G. & Pierce, K.R. (1981). Effect of neonatal thymectomy and anti-lymphocyte globulin on nonspecific mitogen stimulation in Holstein-Friesian calves. *Veterinary Immunology & Immunopathology*, **2**, 175-88.

Soderstrom, N. & Stenstrom, A. (1969). Outflow paths of cells from the lymph node parenchyma to the efferent lymphatics. *Scandinavian Journal of Immunology*, **6**, 186-96.

Spooner, R.L. & Pinder, M. (1983). Monoclonal antibodies potentially detecting bovine MHC products. *Veterinary Immunology & Immunopathology*, **4**, 453-58.

Steinman, R.M. & Cohn, Z.A. (1973). Identification of a normal cell type in peripheral lymphoid organs of mice. 1. Morphology, quantitation and tissue distribution. *Journal of Experimental Medicine*, **137**, 1142-62.

Steinman, R.M. & Nussenzweig, M.C. (1980). Dendritic cells: features and functions. *Immunological Reviews*, **53**, 127-47.

Stingl, G., Wolff-Schreiner, E.C., Pichler, W.J., Gschnait, F., Knapp, W. & Wolff, K. (1977). Epidermal Langerhans cells bear Fc and C3 receptors. *Nature*, **268**, 246.

Straus, W. (1977). Cytochemical observations on the development of antibody-forming cells in popliteal lymph nodes by an improved immunoperoxidase method. *Histochemistry*, **53**, 273-83.

Tew, J.G., Phipps, R. P. & Mandel, T.E. (1980). The maintenance and regulation of the humoral immune response: persisting antigen and the role of follicular antigen-binding dendritic cells as accessory cells. *Immunological Reviews*, **53**, 175-201.

van Ewijk, W., Rouse, R.V. & Weissman, I.L. (1980). Distribution of H-2 microenvironments in the mouse thymus. *Journal of Histochemistry & Cytochemistry*, **28**, 1089-99.

van Ewijk, W., van Soest, P.L. & van der Engh, G.H. (1981). Fluorescence analysis and anatomic distribution of mouse T-lymphocyte subsets defined by monoclonal antibodies to the antigens Thy 1, Lyt 1, Lyt 2 and T 200. *Journal of Immunology*, **127**, 2594-604.

van Voorhis, W.C., Valinsky, J., Hoffman, E., Luban, J., Hair, L.S. & Steinman, R.M. (1983). Relative efficacy of human monocytes and dendritic cells as accessory cells for T-cell replication. *Journal of Experimental Medicine*, **158**, 174-91.

Veerman, A.J.P. (1974). On the interdigitating cells in the thymus-dependent area of the rat spleen: a relation between the mononuclear phagocytic system and T lymphocytes. *Cell & Tissue Research*, **148**, 247-57.

Veerman, A.J.P. & van Ewijk, W. (1975). White pulp compartments in the spleen of rats and mice. A light and electron microscopic study of lymphoid and nonlymphoid cell types in T and B areas. *Cell & Tissue Research*, **156**, 417-41.

Wagner, H., Hardt, C., Stockinger, H., Pfizenmaier, K., Bartlett, R. & Rollinghoff, M. (1981). Impact of thymus on the generation of immunocompetence and diversity of antigen-specific MHC-restricted cytotoxic T-lymphocyte precursors. *Immunological Reviews*, **58**, 95-129.

Weissman, I.L. (1973). Thymus-cell maturation. Studies on the origin of cortisone-resistant thymic lymphocytes. *Journal of Experimental Medicine*, **137**, 504-10.

Weissman, I.L., Rouse, R.V., Kyewski, B.A., Lepault, F., Butcher, E.C., Kaplan, H.S. & Scollay, R.G. (1982). Thymic lymphocyte maturation in the thymic microenvironment. In *The influence of the thymus on the generation of the T-cell repertoire* (eds. H. Wagner, M. Rollinghoff & K. Pfizenmaier), Behring Institute Research Communications 70, pp. 242-51.

Zinkernagel, R.M. (1978). Thymus and lymphohaemopoeitic cells: their role in T-cell maturation in selection of T cells' H-2-restriction specificity and in H-2 linked Ir-gene control. *Immunological Reviews*, **42**, 224-70.

14

Ontogeny of immune responses in cattle

B.I. OSBURN

An understanding of the ontogeny of immune responses in the foetal calf
helps explain the high prevalence of abortion and congenital infections in
cattle. There is a sequential development of structural, cellular and functional
aspects of the immune system. Functional studies with infectious organisms
indicate that antibody to parainfluenza virus may be produced as early as 120
days of gestation, whereas antibody to bovine viral diarrhoea virus is not
produced until 190 days of gestation and to *Trichomonas foetus* not until 30
days after birth. The presence of specific antibodies and elevated
immunoglobulin levels in infected foetuses provides a means of diagnosing
congenital infections.

Introduction

An understanding of developmental immunology and maternal-
foetal placentation helps explain the unusual preponderance of congenital
infections and abortions in ruminants. Abortion or congenital disease
may occur in 20 to 50% of pregnant animals in any one herd or flock.
This rate of congenital infection is considerably higher than observed in
human populations. Ungulate foetuses are essentially
agammaglobulinaemic since the syndesmochorial placenta does not
permit transfer of maternal immunoglobulins to the foetal circulation.
This placentation differs from the haemochorial system of primates where
certain classes of immunoglobulins cross the placenta (Brambell 1970).

Ruminant foetal infections may result in foetal death with subsequent
mummification and/or abortion, foetal deformities, abortion of live but
weak calves, premature delivery of apparently normal calves or
immunologically compromised (tolerant) calves. The varied patterns of
congenital disease are dependent on the infectious agent, the stage of
gestation and organ development when infection occurs, and the ontogeny
of the foetal host defence system at the time of infection. Since the foetal

immune system is capable of responding to a wide variety of antigens including those associated with many pathogenic microbes, foetal serum has been used for diagnostic purposes.

The ruminant is ideal for studying ontogeny of immune responsiveness because maternal immunoglobulins are not present to interfere with interpretation of foetal responses. The relatively slow maturation and the size of the foetus makes it possible to sample adequate amounts of blood and serum for analyses. Since the foetuses are in a sterile environment, inoculation of antigens leads to responses limited to the inciting stimulus. In this chapter, the technical approaches and information gained from studies on the ruminant foetus will be considered.

Technical approaches

A variety of procedures have been used to study ontogeny of immunity in the bovine foetus. Infection of bovine foetuses comes about naturally, in many instances, by infectious agents which cross the placenta from the maternal circulation. Such infections often result in foetal death and abortion. Infectious agents which multiply once they cross the placenta, but do not cause abortion, potentially provide a natural method for immunizing the foetus. One of the approaches to studying foetal responses is to inoculate pregnant cows at different stages of gestation and then collect foetuses by caesarean section at varying periods after exposure or at birth to evaluate the foetal immune response. This procedure has the advantage of reducing the surgical manipulation of the cow and it stimulates natural routes of foetal infection. The disadvantages include the chance that the incidence of foetal infection is reduced, as placental infection does not always occur and the exact time of foetal infection is not known. It also requires the inoculation of replicating organisms or antigens capable of crossing the placenta.

Direct inoculation of the foetus has been used in a number of experimental studies. This approach usually requires surgical intervention through a laparotomy incision in the right or left paralumbar fossa. The pregnant uterus is located and the inoculum inserted transplacentally into foetal tissues. It is also possible to open the uterine wall and expose the foetus before inoculation. Catheters can be inserted in foetal vessels, allowing periodic inoculation and/or sampling. The major advantages of these approaches are that they assure inoculation of antigens into the foetus and that the exact time of inoculation or infection of the foetus is known. The disadvantages are the need for surgical procedures and the increased risk of introducing saprophytic infections which may cause rapid foetal and maternal septicaemias and death.

Organ development

In cattle as in other species, there is a sequential seeding of central and peripheral lymphoid tissue with lymphocytes (Schultz, Dunne & Heist 1973; Osburn, MacLachlan & Terrell 1982). Blood islands are first recognized in the vascular lumen of the yolk sac. Haemopoiesis and lymphopoiesis occur in the embryonic and foetal liver. Lymphocytes are evident in the thymus at 42 days of gestation. This occurs before the first observation of lymphocytes in the blood at 45 days of gestation. Undoubtedly the lymphocytes in the thymus migrate from the liver. By 55 days of gestation, lymphocytes are present in the spleen. Peripheral lymph nodes are populated by 60 days, mesenteric lymph nodes by 100 days and the lymphoid tissue of the gastrointestinal tract at 175 days (see Table 1).

During ontogeny, there is a sequential appearance of immunoglobulins, even in the absence of obvious exogenous antigenic stimuli. IgM, although in minimal quantities, is evident in serum by 130 days and IgG by 145 days of gestation (Schultz, Confer & Dunne 1971; Schultz, Dunne & Heist 1973). To date, specific antibody activity has not been associated with these immunoglobulins. Similarly, cell markers such as receptors for the Fc portion of immunoglobulin and C3 receptors appear before mid-gestation (Renshaw et al. 1977; MacLachlan, Schore & Osburn 1984a). The significant degree of development of the lymphoid organs and their cellular constituents suggests that a degree of immunological maturation has occurred by mid-gestation.

Maturation of humoral immune responsiveness

Humoral immune responses have been evaluated in the foetal calf primarily by quantitation of total immunoglobulin or specific antibody activity in serum. Table 1 summarizes the sequential maturation of the antibody responses in foetal and neonatal life. Even though there is morphological evidence of a lymphoid system by mid-gestation, the ability of the bovine foetus to respond to antigens with detectable antibodies is not acquired until some time after birth. It is evident that the bovine foetus can make antibodies to parainfluenza-3 virus as early as 120 days of gestation (Swift & Kennedy 1972) (Table 1). However, antibody to bovine viral diarrhoea virus is not detected until 190 days of gestation (Braun, Osburn & Kendrick 1973) and antibody to *Trichomonas foetus* is not detected until 30 days after birth (Kerr & Robertson 1954).

In bovine foetuses infected with bluetongue virus at 120 days of gestation, precipitating antibody and virus-neutralizing antibodies were analysed in samples collected from 128 days of gestation until birth. The first antibody activity detected was precipitating antibodies to group-specific antigens by 145 days of gestation. Virus neutralizing activity to serotype specific antigen was not evident until sometime between 175 days of gestation and birth. The ability of the foetus to make antibodies to one particular viral subunit antigen does not necessarily mean that it is capable of responding to other subunit viral antigens (MacLachlan, Schore & Osburn 1984b). Cytotoxic antibodies have been detected at 243 days of gestation in foetal calves that had been inoculated with mouse myeloma cells at 220 days (Osburn, MacLachlan & Terrell 1982).

Maturation of cell-mediated immune responsiveness

The foetal calf appears to acquire the ability to mount cell-mediated immune responses at around 280 days of gestation (Osburn, MacLachlan & Terrell 1982). Collection and analyses of foetal lymphocytes for cell-mediated activity is difficult, primarily because of accessibility to the foetal vascular system. Most current data are confined to responses to phytomitogens (Renshaw et al. 1977; MacLachlan, Schore & Osburn 1984a). Lymphocyte stimulation assays performed on lymphocytes from calves before mid-gestation indicate that these cells are functionally responsive. Attempts to demonstrate lymphocyte proliferative responses to purified protein derivate (PPD) and bluetongue virus in sensitized foetal calves 20 to 50 days after inoculation at 125 days of gestation were not successful (MacLachlan, Schore & Osburn 1984a). In other studies, immunization of foetal calves with *Mycobacterium bovis*, tetanus toxoid or *Brucella abortus* between 168 and 248 days of gestation were sufficient to stimulate cell-mediated responses detectable at birth by delayed hypersensitivity skin tests and lymphocyte stimulation (Rossi et al. 1978).

Development of accessory systems

The immune system requires accessory components, such as complement, neutrophils, macrophages and basophils, to be fully competent. Hence, an understanding of the ontogeny of these systems is important in evaluating the factors involved in acquisition of immune competence. There is evidence for complement activity in the bovine foetus as early as 70 to 90 days of gestation (Barta et al. 1972; Culbertson 1978). Although modest haemolytic activity is present this early in foetal life, the levels remain low (less than 50% of adult activity) during the remainder of gestation (Culbertson 1978). Furthermore, C3 activity remains less than 50% of the adult level during the foetal and early

Table 1. Maturation of the bovine immune system during ontogeny.

Sequential maturation of lymphoid and accessary systems	Gestation day	Defined immune responses
Thymus	42[a]	–
Lymphocytes in blood	45[a]	–
Spleen	55[a]	–
IgM-bearing lymphocytes	59[a]	–
Peripheral lymph nodes	60[a]	–
Bacteriocidal activity in serum	75[b]	–
Phytohaemagglutinin lymphocyte blastogenesis	78[c]	–
Haemolytic complement activity and measurable C3 in serum	90[d]	–
Mesenteric lymph node	100[a]	–
	120[e]	Antibody to parainfluenza-3 virus
IgG in serum	130	–
Blood granulocytes	130[f]	
	140[g]	Neutralizing antibody to parvovirus
	141[h]	Antibody to *A. marginale*
IgG-bearing lymphocytes	145[i]	Agar-gel precipitating antibody to bluetongue virus
IgM in serum	145[ag]	Increased IgM to parvovirus
	145[j]	Interferon produced to bovine viral diarrhoea virus
	155[a]	Anti-maternal RBC antibody
	162[k]	Antibody to *Leptospira saxkoebing*
	165[a]	Anti-infectious bovine rhinotracheitis antibody
	170[e]	Antibody to inactivated infectious bovine rhinotracheitis virus
Lymphoid tissue of gastrointestinal tract	175	–
	190[m]	Neutralizing antibody to bovine viral diarrhoea virus
	210	Cytotoxic antibody to mouse cells
	243[n]	Complement-fixing antibodies to *Chlamydia*
IgA plasma cells in intestine	Birth[iop]	Agglutinins to *Brucella abortus* Neutralizing antibodies to bluetongue virus
	Birth[q]	Cell-mediated response to *M. bovis*, tetanus toxoid, and *B. abortus*
	7[r]	Antibody to egg albumin

7[a]	Antibody to human albumin
7[l]	Antibody to chicken RBC
7[l]	Antibody to OX174 bacteriophage
slow until 14[u]	Antibody to parainfluenza-3 virus
none until 14[j]	Antibody to *S. pullorum*
some at 1 month[e]	Antibody to *B. abortus*
none to small amounts slow appearance, no memory rapid loss of antibody	
none until 24[s]	Antibody to *S. dublin*
none until 30[j]	Antibody to *K. pneumoniae* polysaccharide
none until after 30	Antibody to *T. foetus*

References: [a]Schultz, Dunne & Heist 1973; [b]Barta et al. 1972; [c]Renshaw et al. 1977; [d]Culbertson 1978; [e]Swift & Kennedy 1972; [f]Schultz, Confer & Dunne 1971; [g]Liggett, DeMartini & Pearson 1982; [h]Trueblood, Swift & Bear 1971; [i]MacLachlan, Shore & Osburn 1984b; [j]Rinaldo et al. 1976; [k]Fennestad & Borg-Peterson 1962; [l]Osburn 1975; [m]Braun, Osburn & Kendrick 1973; [n]Osburn 1973; [o]Rossi et al. 1978; [p]Yamin & Sleight 1980; [q]Little & Orcutt 1922; [r]Husband & Lascelles 1975; [s]Smith & Ingram 1965; [t]Schultz 1973; [u]Thorsen, Sanderson & Bittle 1969; [v]Kerr 1956; [w]Kerr & Robertson 1956.

neonatal period (Culbertson 1978; Schwartz & Osburn 1974). Moreover, the functional capacity of the system is questionable, since it has been observed that direct intravenous inoculation of endotoxin into foetal calves does not lead to complement consumption (Culbertson 1978).

Although monocytes and granulocytes are present in the foetal circulation, essentially no basic studies on the functional ontogeny of neutrophils, macrophages, eosinophils or basophils in the bovine foetus have been reported. In other species, these cells mature slowly and their functional capacity does not compare to that in the postnatal period. Interestingly, neutrophils are rarely associated with inflammatory responses in foetal animals.

Immune tolerance

Immune tolerance is a state of antigen-specific nonreactivity in which an immune response to the antigen cannot be detected by conventional serological or cell-mediated techniques. In cattle, nonidentical twin calves usually exhibit haemopoietic chimaerism (Owen

1945; Marcum, Lasley & Day 1972). The chimaerism occurs through the fusion of placental vessels which allows a free exchange of haemapoietic stem cells between calves. The calves do not respond *in vitro* to the cotwins' leucocytes (Emery & McCullagh 1980b) and skin grafts from the cotwin are rejected only after prolonged periods (Emery & McCullagh 1980a). These findings suggest that the twins develop tolerance to each others haemopoietic cells during foetal life. It has also been suggested that immune tolerance occurs in congenital bluetongue virus infection and bovine viral diarrhoea virus infection in cattle (Johnson & Muscoplat 1973; Luedke, Jochim & Jones 1977; Coria & McClerkin 1978). These viruses have been associated with a variety of clinical manifestations, including congenital infection and disease and various forms of disease in postnatal life. Bluetongue virus infection during foetal life can cause foetal death with subsequent abortion, hydranencephaly, porencephaly and weak calves; postnatally it causes subclinical infection or oral ulceration, dermatitis and coronitis. Infection of foetuses with bovine viral diarrhoea virus causes alopecia or cerebellar hypoplasia and postnatally subclinical infection or dermatitis, diarrhoea and ulceration of the oral mucosa. The possibility of tolerance has been based on the isolation of these viruses from newborn calves that lack specific antibody activity for the virus in the serum. In the case of bovine viral diarrhoea virus, this unresponsive state appears to persist for the life of the animal. One explanation for this effect is that bovine viral diarrhoea virus may impair B-lymphocyte maturation leading to a failure of these cells to synthesize immunoglobulin (Alturu et al. 1979). All cases of tolerance appear to be from natural infections. There is no experimental evidence that immune tolerance can be produced by inoculation of the agents into foetal calves.

Concluding remarks

Foetal infections are relatively common in cattle. The usual result of infection is foetal death and/or abortion. Foetuses are susceptible to infection because transplacental transfer of antibodies does not occur in cattle. The immaturity of the immune and nonimmune effector systems further jeopardizes the foetus's ability to overcome infection.

Foetal immune responsiveness can help in diagnosing congenital infections. Elevated immunoglobulin values in aborted foetuses provide a means of diagnosing infection (Sawyer et al. 1973). Specific serological tests of antibody activity are useful in identifying the cause of infection. In a number of viral infections, clearance of virus may occur and the only evidence of infection at the time of examination is the serological response. The collection of serum and body fluids from an aborted foetus is a useful diagnostic aid in determining the cause of abortions.

References

Alturu, D., Notowidjojo, W., Johnson, W.Z. & Muscoplat, C.C. (1979). Suppression of *in vitro* immunoglobulin biosynthesis in bovine spleen cells by bovine viral diarrhoea virus. *Clinical Immunology & Immunopathology*, 13, 254-60.

Barta, O., Barta, V., Ingram, E. & Hubbert W. (1972). Bactericidal activity of bovine foetal serum against smooth and rough strains of *Escherichia coli*. *American Journal of Veterinary Research*, 33, 731-40.

Brambell, F.W.R. (1970). *Transmission of passive immunity from mother to young*. New York: Elsevier.

Braun, R.K., Osburn, B.I. & Kendrick, J.W. (1973). Immunologic response of bovine foetus to bovine viral diarrhoea virus. *American Journal of Veterinary Research*, 34, 1127-32.

Coria, M.F. & McClerkin, A.W. (1978). Specific immune tolerance in an apparently healthy bull persistently infected with bovine viral diarrhoea virus. *Journal of the American Veterinary Medical Association*, 172, 449-51.

Culbertson, M.R. (1978). Bovine haemostatic and complement systems. 1. Ontogenic studies. 2. Foetal response to bacterial endotoxin. Ph.D. thesis. Davis, University of California.

Emery, D.L. & McCullagh, P. (1980a). Immunological reactivity between chimaeric cattle twins. 1. Homograft reaction. *Transplantation*, 29, 4-9.

Emery, D.L. & McCullagh, P. (1980b). Immunological reactivity between chimaeric cattle twins. 3. Mixed leucocyte reaction. *Transplantation*, 29, 17-22.

Fennestad, K. & Borg-Peterson, C. (1962). Antibody and plasma cells in bovine foetuses infected with *Leptospira saxkoebing*. *Journal of Infectious Diseases*, 110, 63-69.

Husband, A.J. & Lascelles, A.K. (1975). Antibody responses to neonatal immunization in calves. *Research in Veterinary Science*, 18, 201-7.

Johnson, D.W. & Muscoplat, C.C. (1973). Immunologic abnormalities in calves with chronic bovine viral diarrhoea. *American Journal of Veterinary Research*, 34, 1139-41.

Kerr, W.R. (1956). Active immunity experiments in very young calves. *Veterinary Record*, 68, 476-77.

Kerr, W.R. & Robertson, M.R. (1954). Passively and actively acquired antibodies for *Trichomonas foetus* in very young calves. *Journal of Hygiene*, 52, 253-57.

Liggett, H.D., DeMartini, J.C. & Pearson, L.D. (1982). Immunologic responses of the bovine foetus to parvovirus infection. *American Journal of Veterinary Research*, 43, 1355-59.

Little, R. & Orcutt, M. (1922). The transmission of agglutinins of *Bacillus abortus* from cow to calf in the colostrum. *Journal of Experimental Medicine*, 35, 161-71.

Luedke, A.J., Jochim, M.M. & Jones, R.H. (1977). Bluetongue in cattle: effects of *Culicoides variipennis*-transmitted bluetongue virus on pregnant heifers and their calves. *American Journal of Veterinary Research*, 38, 1687-95.

MacLachlan, N.J., Schore, C. & Osburn, B.I. (1984a). Lymphocyte blastogenesis in bluetongue virus or *Mycobacterium bovis*-inoculated bovine foetuses. *Veterinary Immunology & Immunopathology*, 7, 11-18.

MacLachlan, N.J., Schore, C.E. & Osburn, B.I. (1984b). Anti-viral responses of bluetongue virus-inoculated bovine foetuses and their dams. *American Journal of Veterinary Research*, **45**, 1469-73.

Marcum, J.B., Lasley, J.F. & Day, B.N. (1972). Variability of sex chromosome chimaerism from heterosexual multiple births. *Cytogenetics*, **11**, 388-99.

Osburn, B.I. (1973). Immune responsiveness of the foetus and neonate. *Journal of the American Veterinary Medical Association*, **163**, 801-3.

Osburn, B.I. (1975). Pathogenesis of foetal infection in ruminants. *Bovine Practice*, **10**, 7-10.

Osburn, B.I., MacLachlan, N.J. & Terrell, T.G. (1982). Ontogeny of the immune system. *Journal of the American Veterinary Medical Association*, **181**, 1049-52.

Owen, R.D. (1945). Immunogenetic consequences of vascular anastomoses between bovine twins. *Science*, **102**, 400-1.

Renshaw, H.W., Eckblad, W.P., Everson, D.O., Tassinari, P.D. & Amos, D. (1977). Ontogeny of immunocompetency in cattle: evaluation of phytomitogen-induced *in vitro* bovine foetal lymphocyte blastogenesis using a whole blood culture technique. *American Journal of Veterinary Research*, **38**, 1141-50.

Rinaldo, C.R., Isackson, D.W., Overall, J.C., Glasgow, L.A., Brown, T.T., Bistner, S.I., Gillespie, J.H. & Scott, F.W. (1976). Foetal and adult bovine interferon production during bovine viral diarrhoea virus infection. *Infection & Immunity*, **14**, 660-66.

Rossi, C.R., Kiesel, G.K., Kramer, T.T. & Hudson, R.S. (1978). Cell-mediated and humoral immune responses of cattle to *Brucella abortus*, *Mycobacterium bovis* and tetanus toxoid: immunization of the foetus. *American Journal of Veterinary Research*, **39**, 1742-47.

Sawyer, M.M., Osburn, B.I., Knight, H.D. & Kendrick, J.W. (1973). A quantitative serologic assay for diagnosing congenital infections of cattle. *American Journal of Veterinary Research*, **34**, 1281-84.

Schultz, R.D. (1973). Comments on the immune response of the foetus and neonate. *Journal of the American Veterinary Medical Association*, **163**, 804-9.

Schultz, R.D., Confer, F. & Dunne, H.W. (1971). Occurrence of blood and serum proteins in bovine foetuses and calves. *Canadian Journal of Comparative Medicine*, **35**, 93-98.

Schultz, R.D., Dunne, H.W. & Heist, C.W. (1973). Ontogen of the bovine immune response. *Infection & Immunity*, **7**, 981-91.

Schwartz, L.W. & Osburn, B.I. (1974). An ontogenic study of the acute inflammatory reaction in the foetal rhesus monkey. 1. Cellular response to bacterial and nonbacterial irritants. *Laboratory Investigation*, **31**, 441-53.

Smith, A.N. & Ingram, D.G. (1965). Immunological responses of young animals. 2. Antibody production in calves. *Canadian Veterinary Journal*, **6**, 226-32.

Swift, B.L. & Kennedy, P.C. (1972). Experimentally induced infection *in utero* of bovine foetuses with bovine parainfluenza-3 virus. *American Journal of Veterinary Research*, **33**, 57-63.

Thorsen, J., Sanderson, R. & Bittle, J. (1969). Bovine parainfluenza-3 vaccine studies. *Canadian Journal of Comparative Medicine*, **33**, 105-7.

Trueblood, M., Swift, B.L. & Bear, P. (1971). Bovine foetal response to *Anaplasma marginale*. *American Journal of Veterinary Research*, **32**, 1089-90.

Yamini, B. & Sleight, S.D. (1980). Immunoglobulin A response of the bovine foetus and neonate to *Escherichia coli*. *American Journal of Veterinary Research*, **41**, 1419-22.

Cellular immune responses

15

Regulatory and effector cell interactions in immune responses

R.N. GERMAIN

The effector activities of the immune system are the net results of the direct activation of T and B lymphocytes, responsible for cell-mediated and humoral immunity respectively, and a balance of amplifying and suppressive regulatory influences. This paper describes the role of several cell types and their products in this process, including Ia antigen-bearing macrophages, monokines and helper-inducer, suppressor and cytotoxic T lymphocytes. The various schemes of T-cell help for B-lymphocyte proliferation and differentiation are covered, as well as the differences in helper cell requirements of various B cells. The classic three-step T- suppressor pathway is described, and two recently reported I region-restricted suppressor pathways are reviewed which appear to be distinct from the I-J-dominated T-suppressor cell scheme. The role of T-helper cells in growth and differentiation is discussed, including recent data which indicate the existence of helper-independent cytotoxic T cells. The cell-surface molecules which play important roles as receptors for antigen, histocompatibility molecules and lymphokines are briefly described. Finally, an overall scheme is presented, depicting the dynamic events following antigen introduction into a naive animal.

Introduction: cells of the immune system

The immune system is composed of a diverse array of cell types found either circulating in blood and lymph, or organized into primary and secondary lymphoid tissues. The major cellular components are lymphocytes and reticuloendothelial cells (macrophages, dendritic cells), which in tissues such as bonemarrow, lymph node, spleen, Peyer's patches, and especially the thymus, may be associated with various stromal and functional epithelial elements.

Lymphocytes were originally thought to be a single cell type, based on their homogeneous morphological appearance in the resting state. Experiments on bursal resection in birds (Warner 1965) and thymectomy in birds (Cooper et al. 1966) and mammals (Miller, Marshall & White 1962; Aronson, Janovic & Waksman 1962) first revealed the split of lymphocytes into the B (for bursal or, in mammals, bonemarrow-derived) cells and T (for thymus-derived) cells. T cells were found to be critical for cell-mediated immunity (e.g. delayed-type hypersensitivity responses), while B cells were essential to the development of humoral immunity. This functional split was, however, found to be imprecise, as it was soon shown, that although B cells actually make antibodies, T cells are also required for optimal antibody production (Claman, Chaperon & Triplett 1966; Davies et al. 1964; Miller & Mitchell 1968). This introduced the concept of functional compartmentalization of lymphocytes into regulatory and effector categories. Subsequent research has revealed a totally unexpected functional heterogeneity of lymphocytes (Cantor & Weissman 1976). It is now believed that during differentiation, various lymphocytes acquire a specific functional program that limits their potential activities to a given regulatory or effector role. The overall pattern of immune responses results from the interaction between and among the large number of diverse lymphocyte subsets. Included in this array are regulatory T cells (helper, suppressor and contrasuppressor), effector T cells (cytolytic and delayed-type hypersensitivity-inducing cells), and B cells with different triggering requirements, as well as subsets of macrophages and dendritic cells with different accessory functions. The set of interactions of these elements will be reviewed here in general terms.

It is important to note that a hallmark of the immune system is its ability to discriminate self from nonself, and to respond to nonself in a specific manner. This specificity, as postulated by Burnet (1959), arises from the clonal distribution on lymphocytes of membrane-bound receptors capable of binding antigen. The progeny of a single lymphocyte (i.e. a clone) all possess the same receptor and it is the sum of these clonally distributed receptors which comprises the immune repertoire of an individual. The interactions among accessory, regulatory and effector cells are guided either by the antigen-binding capacity of these receptors, or the structure of the receptors themselves, preserving the specificity of the response.

Antigen presentation

Since the receptors of lymphocytes dictate the specificity of immune responses, antigen binding by these receptors must play a fundamental role in regulating which lymphocytes are activated at any given time. The form of antigen recognized by various lymphocytes and the role of specialized antigen-presenting cells in this process differ among lymphocyte subsets. B lymphocytes possess cell membrane-bound immunoglobulin (Raff 1970a; Melchers, von Boehmer & Phillips 1975) which, like free antibody molecules, is capable of binding unmodified (native) antigen in solution. Thus, B lymphocytes do not in principle require special 'processing' of antigen nor presentation of an array of antigen on a cell membrane. However, it appears that cross-linking of surface immunoglobulin receptors of B cells is important to the activation process. This is most easily demonstrated by the inability of anti-μ antibody (as an antigen analogue) in the Fab form to initiate B-cell triggering while bivalent F(ab')$_2$ is active (Monroe & Cambier 1983; De Franco et al. 1982). Thus, for most protein antigens, which are uni- or pauci-valent, display of multiple antigen molecules on a cell membrane may be important, even for B cells.

The situation is clearer for certain T lymphocytes. As first demonstrated for proliferating cells in guinea pigs (Rosenthal & Shevach 1973) and in mice (Yano, Schwartz & Paul 1977), and for cytolytic T cells in mice (Zinkernagel & Doherty 1974), many, if not all, T cells have a dual requirement for antigen-recognition (summarized in Figure 1). They require the nominal immunizing antigen, and in addition either a class

Figure 1. A summary of the recognition events and soluble mediators involved in the generation of cytotoxic T cells and induction of specific antibody production by Lyb 5-negative and Lyb 5-positive B cells

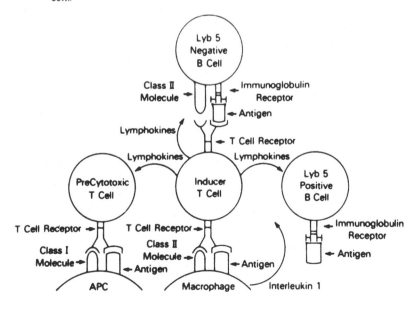

I (K,D,L) or class II (Ia) histocompatibility molecule. Cytotoxic T cells primarily respond to antigen in the context of class I molecules, while helper, and delayed-type hypersensitivity cells are usually restricted by class II molecules (Zinkernagel & Doherty 1979; Miller 1978).

Since class I molecules are present on virtually all somatic cells, it is not surprising that there is little evidence that cytotoxic T cells need to see antigen on specialized antigen-presenting cells. Rather, they seem capable of responding to antigen (possibly in its native form) present on the surface of most cell types. Optimal cytotoxic T-cell activity may nonetheless be achieved by recognition of antigen on antigen-presenting cells because of the role of these cells in triggering other T-helper cells which are important to cytotoxic T-cell generation.

Most investigations of the role of specialized antigen-presenting cells in immunity involve studies on responses of proliferating helper-type T cells *in vitro* (reviewed by Schwartz 1982). Many workers have shown that antigen alone cannot stimulate purified, primed L3T4-positive T cells, but rather they require the addition of viable Ia-positive cells (either macrophages, B cells or dendritic cells). Three major aspects of the function of these antigen-presenting cells have been explored: (1) the uptake and handling of protein antigen, (2) the role of surface class II major histocompatability complex (MHC) molecules (Ia) and (3) the production and release of antigen-nonspecific monokines (hormone-like mediators). It is now clear that Ia-positive macrophages and at least some subsets of B cells can present antigen (reviewed by Unanue 1981), and that both these cell types degrade the bulk of antigen they ingest into small peptides. A fraction (less than 1%) of the antigen retains its native structure, and a small amount is left as large proteolytic fragments (Schmidtke & Unanue 1971; Ellner & Rosenthal 1975; Grey, Colon & Chestnut 1982). Some antigen in each of these forms is recycled back to the cell surface. It has been shown that peptide derivatives of protein antigens are equal or superior to native antigen in eliciting T-cell responses *in vitro* (Thomas et al. 1981; Hansburg 1981), indicating that native antigen is not required for stimulation, and may not even be important. More recently, it was found that such peptides could be presented by cells that have been fixed with aldehydes, whereas the original proteins could not, suggesting that proteolysis is an important part of antigen processing and the function of antigen-presenting cells (Shimonkevitz 1983).

While such degradation of large proteins may often be required, it is not sufficient to cause T-cell activation, since free peptide or peptide bound to Ia-positive cells cannot trigger T cells. Rosenthal and Shevach

(1973) first demonstrated the requirement for histocompatibility matching between antigen-presenting cells and responding T cells for effective activation by antigen. This phenomenon has subsequently been shown to involve recognition of the polymorphic determinants possessed by Ia glycoproteins, and to involve a learning or selection process for self-Ia as the T cells mature in the thymus (Schwartz 1984). The requirement for Ia recognition can be demonstrated by varying the allelic form of Ia on the antigen-presenting cell or by blocking recognition of the appropriate self-Ia with anti-Ia antibody (Schwartz, Yano & Paul 1978). Such studies have shown that, just as the antigen-specific receptor of B cells is clonally distributed, recognition of Ia is restricted for each T-cell clone to a specific Ia molecule, i.e. to I-A or I-E of a given haplotype (Miller 1978; Schwartz 1984). This implies that Ia recognition is not mediated by a general class II-specific receptor, but rather, that antigen recognition and Ia restriction both reflect the specificity of clonally distributed single or dual receptors on the responding T cells. Our understanding of the structure of T-cell receptors is far from complete and so it is unclear if recognition of antigen and Ia occur without the two being in close approximation, or whether, as is frequently suggested, there exist complexes of Ia and antigen of finite stability. In particular, some researchers have published data supporting the existence of stable Ia-antigen complexes, shed by antigen-presenting cells exposed to antigen, and these complexes are said to provide the MHC-restricted signal to T cells (Erb, Feldmann & Hogg 1976; Puri & Lonai 1980).

The last component of the function of antigen-presenting cells involves monokine production. Although 'fixed' cells can present some antigen to immortalized T cells in the form of T-cell hybridomas (Shimonkevitz et al. 1983), many investigators have shown that disruption of physiological processes of antigen-presenting cells interferes with their activity, and that function can be restored to Ia-positive, antigen-bearing but inactivated antigen-presenting cells by adding supernatants of viable antigen-presenting cells (Mizel & Ben-Zvi 1980; Germain 1981) to the culture. Characterization of the active component of such supernatants suggests an important role for the 15,000-dalton protein monokine, interleukin-1 (IL-1) (Mizel 1980). The precise mode of action of IL-1 in the function of antigen-presenting cells is unknown, as is the signal for the release of IL-1. Some experiments indicate that T-cell recognition of Ia induces IL-1 release, which in turn is required to permit a T cell to proceed along the activation pathway upon recognition of Ia and antigen. A great deal of investigation is currently in progress in this area.

Serological and functional studies have shown that the antigen-presenting cells which stimulate helper and delayed-type hypersensitivity T cells all have both I-A and I-E subregion products on their surfaces (Cowing et al. 1978; Nepom, Benacerraf & Germain 1981), and no functionally distinct subsets of these cells have been convincingly demonstrated. However, there is disagreement as to the relative roles of macrophages and dendritic cells in these processes and growing support for the possible existence of a distinct subset of antigen-presenting cells, perhaps bearing distinct I-region products, which have special roles in triggering other T-cell subsets, such as suppressors and contrasuppressors.

Interaction of T-helper cells with B cells and cytotoxic T cells

Early experiments by Claman, Chaperon & Triplett (1966) and Miller and Mitchell (1968) established the importance of T lymphocytes in the generation of humoral responses by B cells. Since then, intensive investigation has revealed that T lymphocytes not only control whether or not a response is made by B cells, but the amount, nature (isotype, allotype, idiotype), affinity and, in some cases, specificity of the antibodies elicited by antigen (Gershon 1974). The T cells largely responsible for these effects are termed helper or inducer T cells and bear the Lyt 1-positive 2-negative, L3T4-positive phenotype in most circumstances. This is not to say that T-helper cells are homogeneous; subsets of these cells have been reported with a special ability to affect antibody production of specific isotype (Kishimoto & Ishizaka 1973), allotype (Herzenberg et al. 1976) or idiotype (Bottomly & Mosier 1979), to control class-switching between isotypes (Kawanishi, Saltzman & Strober 1983), or to amplify humoral responses only in the presence of 'true' T-helper cells, but not alone (Tada et al. 1978). This review will focus on the best characterized members of this family, the I region-restricted T-helper cells.

The previous section described the presentation of antigen to various T cells by antigen-presenting cells and pointed out the Ia-restricted recognition of antigen by Lyt 1-positive T-helper cells and delayed-type hypersensitivity cells. Such Ia-restricted helper cells, once activated by antigen plus Ia on antigen-presenting cells, have been found to interact with B lymphocytes by two pathways (Figure 1), which seem to correlate more with the nature of the target B cell than with a functional dichotomy in the T-helper cell pool itself (reviewed by Singer & Hodes 1983). The first pathway, initially reported by Kindred and Shreffler (1972) and extensively analysed by Katz, Hamaoka and Benacerraf (1973), involves an antigen-specific, Ia-restricted interaction of T-helper cells and B cells, in a manner analogous to the interaction of T-helper cells with antigen-presenting cells which bear antigen. Thus, the Ia-positive B cell acquires antigen in its surface immunoglobulin receptors, and 'presents' this antigen, probably after internalization and processing, on its membrane together with Ia. By a process involving cell-cell contact, the T cell recognizes this antigen/Ia display and provides an as yet unidentified signal to the B cell essential for initiating the process of activation.

This pathway provides an explanation for the requirements of physical linkage of haptens to carriers. As first shown by Mitchison (1971) and Raff (1970b), T-helper cells recognize the protein carrier. This is presented with Ia by B cells possessing Ig receptors able to acquire the complex by binding the hapten covalently linked to the protein carrier. Studies in several laboratories have shown that this 'cognate' recognition pathway of T-B interaction applies to a subset of B cells characterized as Lyb 5-negative, small resting cells. These are present in the CBA/N mouse strain, which possesses the xid mutation (Singer & Hodes 1983) and is deficient in Lyb 5-positive cells.

The second pathway does not acquire linked recognition of hapten and carrier. It involves Lyb 5-positive B cells or medium-to-large B cells and is the subset largely absent in CBA/N mice. In this case, direct T-B contact is not required. The absence of this requirement is well demonstrated by the ability of T-helper cells, once activated by histocompatible antigen-presenting cells and antigen, to help histoincompatible B cells, in contrast to the cognate pathway where I-region homology between T and B cells is necessary (Asano, Singer & Hodes 1981). In this case, the delivery of the helper signal appears to be mediated by soluble factors.

Analysis of B-cell clonal expansion and differentiation has shown that this is a complex, multistep process with T-helper cells playing several roles (reviewed by Howard & Paul 1983; Swain et al. 1982; Schimpl & Wecker 1975; Marrack et al. 1982). In model experiments, anti-μ antibody serves to replace specific antigen as the ligand for the surface Ig receptor. Anti-μ cross-linking of the receptors on B cells leads to a movement from the resting small lymphocyte in G_0 to an enlarged B cell in G_1. Such 'excited' B lymphocytes will undergo several rounds of division in the presence of the monokine IL-1, (perhaps released following recognition by T-helper cells of antigen plus Ia on antigen-presenting cells), if provided with, at least, B-cell growth factor derived from T-helper cells. B-cell growth factor has been characterized as a protein of 18,000 daltons. For the Lyb 5 subset, this growth factor acts only after cognate, Ia-restricted T-B interaction. For Lyb 5-positive cells, antigen recognition alone seems adequate to make the B cell receptive to these signals. Further differentiation of these dividing B cells into antibody secreting cells requires the action of two or more B-cell differentiation factors. These substances have not been well characterized, but include so-called T-cell replacing factor (Takatsu, Tominaga & Hamaoka 1980), and possibly γ-interferon (Roehm, Marrack & Kappler 1983). The cycling B cell has specific receptors, at least for T-cell replacing factor (Tominaga, Takatsu & Hamaoka 1980).

The growth and differentiation of cytotoxic T lymphocytes are also dependent on T-helper cells, as shown in Figure 1 (Cantor & Boyse 1975; Glasebrook & Fitch 1979). These lymphocytes, which lyse target cells bearing the appropriate antigen (e.g. virus) and class I MHC product, develop cell membrane receptors for the lymphokine interleukin-2 (IL-2) or T-cell growth factor, after recognition of antigen and class I MHC (Smith & Ruscetti 1981). T-helper cells, stimulated by antigen plus Ia on antigen-presenting cells, secrete IL-2 which binds to the IL-2 receptors of activated cytotoxic T cells, leading to clonal expansion. No direct proof is available showing that IL-2 induces differentiation as well as growth. To the contrary, some experiments indicate that IL-2, like B-cell growth factor for B cells, is insufficient to yield fully mature, active cytotoxic T cells, and that a cytotoxic cell-differentiation factor is required, which is also produced by Lyt 1-positive T-helper cells (Raulet & Bevan 1982; Wagner et al. 1982). This pathway of help for cytotoxic T cells thus closely resembles the help for Lyb 5-positive B cells, including the lack of any discernible requirement for contact or MHC-restricted interaction between the T-helper cells and the cytotoxic cells.

Recent work has provided evidence for a new type of T cell which combines properties of helper and cytotoxic cells (Glasebrook, Kelso & MacDonald 1983). These cells are restricted to class I MHC products, but nonetheless produce IL-2 when stimulated by antigen (possibly exclusively when on the membrane of antigen-presenting cells which are producing IL-1). The production of T-cell growth factor permits the maintenance of clonal expansion in the absence of conventional T-helper cells. The progeny of such cells have the functional lytic properties of classical cytotoxic T cells. It is not yet established if such cells produce their own cytotoxic cell-differentiation factor, nor the overall frequency of such cells in comparison to conventional T-helper cells and cytotoxic T cells.

Suppressor cell pathways and contrasuppressors

Gershon first suggested on logical grounds that the immune system should include a cell capable of shutting down an immune response once it began, in order to restore the system to homeostasis. His early experiments on 'infectious tolerance' first pointed to a negative regulation of immunity by T cells (Gershon & Kondo 1970; 1972), and subsequent investigation by Gershon and many others (reviewed by Germain 1980; Tada & Okumura 1979) has not only confirmed the existence of suppressor T cells, but revealed an almost baroque family of suppressor cells and suppressor factors which must interact with each other before suppression of T- or B-cell responses is seen. The best studied models will be briefly reviewed and their similarities and differences pointed out.

Germain and Benacerraf (1981) described a model based on studies in several laboratories of suppression of delayed-type hypersensitivity to the haptens, azobenzenearsonate (reviewed by Greene et al. 1982) and nitro hydroxyphenyl, and of the antibody response to the polypeptide L-glutamic acid ^{60}-L-alanine ^{30}-L-tyrosine[10]. This integrated model pathway involves three main T-suppressor cells, termed Ts_1, Ts_2 and Ts_3 and summarized in Figure 2. Introduction of antigen leads to the stimulation of Ts_1 cells, a process that to date has shown no apparent MHC restriction, but whose regulation is under control by I-region genes (Germain, Waltenbaugh & Benacerraf 1980). This may imply a role for I-region gene products in antigen presentation to Ts_1 cells. If so, the cell type involved may not be a classic macrophage-like, antigen-presenting cell since Ts_1 is induced more readily *in vivo* (Perry & Greene 1982) or

Figure 2. A summary of the induction pathways and effector functions of first-, second- and third-order T-suppressor cells in mice.

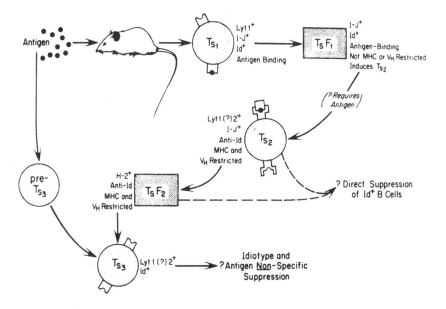

in vitro (Pierres and Germain 1978) when such cells are absent. Ts_1 cells are Lyt 1-positive 2-negative, I-J-positive and produce a soluble, antigen-specific factor of 30,000 to 50,000 daltons, bearing I-J determinants (Thze, Kapp & Benacerraf 1977). Krupi and colleagues (1982) have characterized putative Ts_1 factor from hybridomas as a single polypeptide chain of 25,000 daltons, but other laboratories have not confirmed this unexpected finding of a single polypeptide with both antigen-binding activity and MHC-linked determinants.

The only function known for Ts_1 factor is the stimulation of Ts_2, which are Lyt 1-positive 2-positive, I-J-positive second-order suppressor cells (Waltenbaugh et al. 1977; Germain et al. 1978). If this is done with Ts_1 factor in the absence of antigen, the Ts_2 cells are specific for the binding site of the Ts_1 factor (i.e. are anti-idiotypic). In those systems in which the Ts_1 factor shares serological cross-reactivity with major idiotypes of serum antibody, the Ts_2 cells bind to such Ig-borne idiotypes (Sy et al. 1980; Weinberger et al. 1980). If Ts_1 factor and antigen are used together, Ts_2 cells are often anti-antigen, instead of anti-idiotype (Kapp & Araneo 1982). In either case, Ts_2 cells have activity which is restricted by the I-J haplotype of the Ts_2 donor, and this same restriction is seen with the anti-idiotype or anti-antigen Ts_2 factor which can be derived from these cells (Kapp & Araneo 1982; Dietz et al. 1981). This Ts_2 factor is composed of two polypeptide chains, one binding antigen and the other bearing the I-J-associated determinants (Kapp et al. 1983). Ts_2 factor may have direct suppressive activity in some circumstances, but its best characterized function is the activation of a third cell, termed Ts_3.

Ts_3 cells are cyclophosphamide-sensitive, Lyt 2-positive, I-J-positive T cells generated by priming with antigen under immunogenic conditions (Sy et al. 1979; Sunday, Benacerraf & Dorf 1981; Sy et al. 1981). Ts_3 cells are antigen-binding, and in systems in which predominant idiotypes are induced, bear such determinants. This permits interaction either with antigen-specific Ts_2 cells via antigen bridges, or direct idiotype-anti-idiotype interaction with this other class of Ts_2 factor. Ts_2 factor causes primed Ts_3 cells to release Ts_3 factor, which is an antigen-binding, I-J-restricted, two-chain molecule of similar structure to Ts_2 factor (Okuda et al. 1981). This factor either directly suppresses certain cells, acts via additional T cells (Ts_4) to suppress B cells in an idiotype-specific manner, or acts via nonspecific mediators to suppress generally inducer-type T cells. These latter effector steps of suppression are poorly characterized at present.

In addition to I-J restriction of the T-suppressor factor-suppressor cell interaction, IgV_H-linked genes also impose restrictions on the ability of

T-suppressor factor to affect target T-suppressor cells (Minami et al. 1981). This has been shown for both Ts_2-Ts_3 interaction and the Ts_3-effector limb. The exact meaning of such restrictions is still unclear in cases where idiotype-anti-idiotype recognition is not involved.

Recent molecular biological studies of the MHC have raised questions about the nature of the I-J determinants expressed on T-suppressor cells which coincide with new findings on the involvement of some type of antigen-presenting cells in regulating the I-J restriction of T-suppressor cells. Thus, Hood and colleagues, using a series of nucleic acid probes covering the entire portion of chromosome 17 believed to encode I-J, could find no mRNA in active I-J-positive T-suppressor cells corresponding to this region of the MHC (Kronenberg et al. 1983). If this negative result is correct, and if the work of Minami, Honji and Dorf (1982) and Takaoki and colleagues (1982) on the importance of adherent cells in inducing I-J restrictions is supported by further investigation, then it is possible that the suggestion by Schrader (1979) may be correct — that I-J on T-suppressor factor is actually an anti-(I-J) receptor coded for outside the MHC and recognized by anti-(anti-I-J), anti-idiotypic antibodies in the alloantisera used. This view would look at T-suppressor factor as a model T-cell receptor from an antigen-specific, MHC-restricted T cell with dual receptors — one for antigen and one for I-J. A special subset of antigen-presenting cells would actually bear some form of class II MHC product (perhaps a variant of E_β) and would be recognized by this anti-I-J receptor. The constant region of the receptor-complex from T-suppressor cells would, when bound to antigen and I-J on such an antigen-presenting cell, transmit a signal to the next cell in the pathway.

The second major suppressor model is actually the first system to document the existence of antigen-specific suppressor factors and I-J subregion restriction between factor donor and target T cells. This model, developed and explored in detail by Tada, Taniguchi and colleagues (see Tada & Okumura 1979), involves priming mice twice with high doses of soluble carrier. Such mice have Lyt 2-positive, I-J-positive antigen-binding cells in spleen and thymus which produce a two-chain T-suppressor factor (Taniguchi et al. 1982). This factor has one peptide of 45,000 daltons which binds antigen and bears so-called C_T determinants (Tokuhisa & Taniguchi 1982), defined by Spurll and Owen (1981) and linked to the Ig-constant region coded for on mouse chromosome 12. The other chain of 25 to 30,000 daltons is linked by disulphide bonds to the antigen-binding chain and bears I-J determinants. Both chains are needed for the activity of this factor, which is to trigger antigen-primed Lyt 1-positive

2-positive, I-J-positive T cells to active suppressor status, or to aid maturation of a different Lyt 2-positive cell to such a state. This latter cell then produces an effector-suppressor factor which acts on T-helper cells. In some cases, the target cell for the T-suppressor factor appears to be anti-idiotypic rather than an antigen-specific (Taniguchi et al. 1981). Whether this anti-idiotypic cell then acts on another antigen-specific cell is not known at present.

The third suppressor pathway is the feedback suppressor system described by Cantor et al. (1978). These workers first demonstrated the involvement of Lyt 1-positive 2-negative cells in suppressor pathways, and distinguished these cells from Lyt 1-positive T-helper cells by virtue of additional cell-surface markers such as Qal and I-J. In its currently understood form (Cantor & Gershon 1979), this pathway involves the initial activation by antigen of Lyt 1-positive, Qal-positive, I-J-positive inducer T-suppressor cells. These cells release an antigen-binding polypeptide which has affinity for an I-J-bearing polypeptide from another cell (Yamauchi et al. 1981a). The I-J-bearing chain also possesses determinants apparently encoded by the IgV_H region, but distinct from immunoglobulin (Yamauchi et al. 1981b). The antigen-binding chain, in association with the I-J chain, form an active T-suppressor factor which is capable of acting in V_H-identical mice only to stimulate antigen-primed, Lyt 1-negative 2-positive, I-J-negative T-suppressor cells to release an I-J-negative, antigen-specific, MHC-restricted T-suppressor factor which inhibits T-helper cell function (Yamauchi et al. 1982; Flood, Yamauchi & Gershon 1982).

At present, the differences in detail among these three models cannot be fully reconciled. However, it is apparent that the general scheme is similar: there are inducer T-suppressor cells which elaborate soluble antigen-specific factors, whose target cells are antigen- or idiotype-specific second-order T-suppressor cells. These Ts_2 cells then elaborate a second T-suppressor factor which, either directly or via a third set of suppressor cells, evokes effector-suppressor function. The interaction between the Ts_1, Ts_2 and Ts_3 cells involves both idiotype- and MHC (particularly I-J)-restricted events. Further work, especially with cloned or immortalized suppressor cells, should help clarify the relationships of the various pathways and the mode of action of the suppressor factors.

More recently, two very distinct suppressor models have been described. The first, reported by Baxevanis et al. (1982), involves responses to the antigen lactate dehydrogenase. The response to this antigen is controlled by an H-2-linked immune response gene, and it was noted that responders lack expression of the I-E molecule (e.g. $H-2^b$). These workers have shown that the nonresponders possessing I-E have T-suppressor cells restricted to antigen plus I-E which prevents responses by T cells responding to I-A plus antigen. These Lyt 2-positive, I-J-positive suppressor cells require other 'inducing' suppressor cells for activity, but seem to function directly once activated, suppressing proliferating T cells via an antigen-specific, two-chain suppressor factor (Ikezawa et al. 1983). The interaction between T-suppressor factor and T-helper cells is restricted to combinations in which the helper cells are restricted to the I-A subregion haplotype of the suppressor cells producing the suppressor factor. The I-E restriction specificity of the cells producing the suppressor factor is not relevant to this interaction. Thus, in one sense, this may be considered a 'restriction specificity'-restricted T-suppressor cell pathway.

Another such pathway has been described by Asano and Hodes (1983). Working with cloned T-cell lines, they have recently demonstrated the existence of I-A-restricted T-suppressor cells which are only able to suppress T cells restricted to the same I-A haplotype. This restriction operates independently of the genotype of the T-helper cells being suppressed. Thus, T-helper cells from chimaeras which have donor cells of the 'a' haplotype, but which are restricted to the 'b' haplotype, are suppressed by 'c'-restricted supressor cells. No active supressor factor has been demonstrated yet in this model, nor is there any information which would relate this system to the factor-dependent models described above.

Just as Gershon reasoned that the immune system should have T-suppressor cells, he felt that another regulatory system would guard against excessive suppressor function. While this might seem like an entry into an endless cycle of positive and negative regulatory elements, Green, Flood and Gershon (1983) report evidence for a so-called 'contrasuppressor' pathway, involving cells distinct from either T-helper cells or T-suppressor cells. In this pathway, there is again a cascade of interacting cells, resulting in the generation of Lyt 1-positive cells bearing an I-J determinant distinct from that on effector-suppressor cells. These cells release a contrasuppressor factor which protects T-helper cells against the activity of T-suppressor factor. Whether 'anti-contrasuppressor' cells exist is unknown.

Conclusion

The variety and complexity of the interactions outlined in the preceding sections point out that we are only just beginning to gain some understanding of what happens between exposure to an antigen and the

274   R. N. Germain

final immune response generated. Much of the progress to date has come
from a combination of insight by individuals and technical advances that
have spread throughout the field. The recent application of advanced
technology in cell culture, somatic cell genetics, protein and membrane
biochemistry and molecular biology will permit a more precise analysis
of how antigen is seen by different T and B-lymphocyte subsets, the
structure and function of lymphokines controlling cell growth and
differentiation, the process of repertoire generation in T cells, and the
expression of genetic programs determining immunocyte function. We
can also look forward to practical applications of these new insights in
controlling disease.

Acknowledgements
I wish to thank Ms Shirley Starnes for her excellent and timely
preparation of this manuscript.

References
Aronson, B.G., Janovic, B.D. & Waksman, B.H. (1962). Effect of
    thymectomy on 'delayed' hypersensitive reaction. *Nature*, **194**, 99-100.
Asano, Y. & Hodes, R. (1983). T-cell regulation of B-cell activation: cloned
    Lyt 1-positive 2-negative T-suppressor cells inhibit the major
    histocompatibility complex-restricted interaction of T-helper cells with B
    cells and/or accessory cells. *Journal of Experimental Medicine*, **158**, 1178-90.
Asano, Y., Singer, A. & Hodes, R.J. (1981). Role of the major
    histocompatibility complex in T-cell activation of B-cell subpopulations:
    major histocompatibility complex-restricted and -unrestricted B-cell
    responses are mediated by distinct B-cell subpopulations. *Journal of
    Experimental Medicine*, **154**, 1100-15.
Baxevanis, C.N., Ishii, B., Nagy, Z.A. & Klein, J. (1982). H-2-controlled
    suppression of T-cell response to lactate dehydrogenase B. *Journal of
    Experimental Medicine*, **156**, 822-33.
Bottomly, K. & Mosier, D.E. (1979). Mice whose B cells cannot produce the
    T15 idiotype also lack an antigen-specific helper T cell required for T15
    expression. *Journal of Experimental Medicine*, **150**, 1399-1409.
Burnet, F.M. (1959). *The clonal selection theory of acquired immunity.*
    Cambridge University Press.
Cantor, H. & Boyse, E.A. (1975). Functional subclasses of T lymphocytes
    bearing different Lyt antigens. 2. Cooperation between subclasses of Lyt-
    positive cells in the generation of killer activity. *Journal of Experimental
    Medicine*, **141**, 1390-99.
Cantor, H. & Weissman, I. (1976). Development and function of
    subpopulations of thymocytes and T lymphocytes. *Progress in Allergy*, **20**,
    1-64.
Cantor, H. & Gershon, R.K. (1979). Immunological circuits: cellular
    composition. *Federation Proceedings*, **38**, 2058-64.
Cantor, H., Hugenberger, J., McVay-Bondreau, L., Eardley, D.D., Kemp, J.,
    Shen, F. & Gershon, R.K. (1978). Immunoregulatory circuits among T-cell
    sets: identification of a subpopulation of T-helper cells that induces
    feedback inhibition. *Journal of Experimental Medicine*, **148**, 871-77.

Claman, H.N., Chaperon, E.A. & Triplett, R.F. (1966). Thymus-marrow cell combinations: synergism in antibody production. *Proceedings of the Society for Experimental Biology & Medicine*, 122, 1167-71.

Cooper, M.D., Peterson, R.D., South, M.A. & Good, R.A. (1966). The functions of the thymus system and the bursa system in the chicken. *Journal of Experimental Medicine*, 123, 75-102.

Cowing, C., Pincus, S.H., Sachs, D.H. & Dickler, H.B. (1978). A subpopulation of adherent accessory cells bearing both I-A and I-E or C subregion antigens is required for antigen-specific murine T-lymphocyte proliferation. *Journal of Immunology*, 121, 1680-86.

Davies, A.J.S., Leuchars, E., Wallis, V., Marchant, R. & Elliot, E.V. (1964). The failure of thymus-derived cells to produce antibody. *Transplantation*, 5, 222-31.

De Franco, A.L., Raveche, E.S., Asofsky, R. & Paul, W.E. (1982). Frequency of B lymphocytes responsive to anti-immunoglobulin. *Journal of Experimental Medicine*, 155, 1523-36.

Dietz, M.H., Sy, M.-S., Nisonoff, A., Greene, M.I., Benacerraf, B. & Germain, R.N. (1981). Antigen- and receptor-driven regulatory mechanisms. 7. H-2-restricted anti-idiotypic suppressor factor from efferent suppressor T cells. *Journal of Experimental Medicine*, 153, 450-63.

Ellner, J.J. & Rosenthal, A.S. (1975). Quantitative and immunologic aspects of the handling of 2,4 dinitrophenyl guinea pig albumin by macrophages. *Journal of Immunology*, 114, 1563-69.

Erb, P., Feldmann, M. & Hogg, N. (1976). Role of macrophages in the generation of T-helper cells. 4. Nature of genetically related factor derived from macrophages incubated with soluble antigens. *European Journal of Immunology*, 6, 365-72.

Flood, P., Yamauchi, K. & Gershon, R.K. (1982). Analysis of the interactions between two molecules that are required for the expression of Lyt 2-suppressor cell activity: three different types of focusing events may be needed to deliver the suppressive signal. *Journal of Experimental Medicine*, 156, 361-71.

Germain, R.N. (1980). Antigen-specific T cell-suppressor factors: mode of action. In *Lymphokine reports* (eds. E. Pick & M. Landy), pp. 7-39. New York: Academic Press.

Germain, R.N. (1981). Accessory cell stimulation of T-cell proliferation requires active antigen processing, Ia-restricted antigen presentation, and a separate nonspecific second signal. *Journal of Immunology*, 127, 1964-66.

Germain, R.N. & Benacerraf, B. (1981). A single major pathway of T-lymphocyte interactions in antigen-specific immune suppression. *Scandinavian Journal of Immunology*, 13, 1-10.

Germain, R.N., Waltenbaugh, C. & Benacerraf, B. (1980). Antigen-specific T cell-mediated suppression. 5. H-2-linked genetic control of distinct antigen-specific defects in the production and activity of L-glutamic acid[50]-L-tyrosine[50] suppressor factor. *Journal of Experimental Medicine*, 151, 1245-59.

Germain, R.N., Thze, J., Kapp, J.A. & Benacerraf, B. (1978). Antigen-specific T cell-mediated suppression. 1. Induction of L-glutamic acid [60]-L-alanine[30]-L-tyrosine[10]-specific suppressor T cells *in vitro* requires both antigen-specific T cell-suppressor factor and antigen. *Journal of Experimental Medicine*, 147, 123-36.

Gershon, R.K. (1974). T-cell control of antibody production. In *Contemporary topics in immunobiology*, vol. 3, pp. 1-40.

Gershon, R.K. & Kondo, K. (1970). Cell interactions in the induction of tolerance: the role of thymic lymphocytes. *Immunology*, 18, 723-37.

Gershon, R.K. & Kondo, K. (1972). Infectious immunological tolerance. *Immunology*, **21**, 903-14.

Glasebrook, A.L. & Fitch, F.W. (1979). T-cell lines which cooperate in generation of specific cytolytic activity. *Nature*, **278**, 171-73.

Glasebrook, A.L., Kelso, A. & MacDonald, H.R. (1983). Cytolytic T-lymphocyte clones that proliferate autonomously to specific alloantigenic stimulation. 2. Relationship of the Lyt 2 molecular complex to cytolytic activity, proliferation and lymphokine secretion. *Journal of Immunology*, **130**, 1545-51.

Green, D.R., Flood, P.M. & Gershon, R.K. (1983). Immunoregulatory T-cell pathways. In *Annual review of immunology* (eds. W.E. Paul, C.G. Fathman & H. Metzger), pp. 439-63. Palo Alto, California: Annual Reviews Inc.

Greene, M.I., Nelles, M.J., Sy, M.-S. & Nisonoff, A. (1982). Regulation of immunity to the azybenzene-arsonate hapten. *Advances in Immunology*, **33**, 254-300.

Grey, H.M., Colon, S.M. & Chestnut, R.W. (1982). Requirements for the processing of antigen by antigen-presenting B cells. 2. Biochemical comparison of the fate of antigen in B-cell tumours and macrophages. *Journal of Immunology*, **129**, 2389-95.

Hansburg, D., Hannum, C., Inman, J.K., Appella, E., Margoliash, E. & Schwartz, R.H. (1981). Parallel cross-reactivity patterns of two sets of antigenically distinct cytochrome c peptides: possible evidence for a presentational model of Ir-gene function. *Journal of Immunology*, **127**, 1844-51.

Herzenberg, L.A., Okumura, K., Cantor, H., Sato, U., Shen, F.W., Boyse, E.A. & Herzenberg, L.A. (1976). T-cell regulation of antibody responses: demonstration of allotype-specific helper T cells and their specific removal by suppressor T cells. *Journal of Experimental Medicine*, **144**, 330-44.

Howard, M. & Paul, W.E. (1983). Regulation of B-cell growth and differentiation by soluble factors. In *Annual review of immunology* (eds. W.E. Paul, C.G. Fathman & H. Metzger), pp. 307-33. Palo Alto, California: Annual Reviews Inc.

Ikezawa, Z., Baxevanis, C., Nonaka, M., Abe, R., Tada, T., Nagy, Z. & Klein, J. (1983). Monoclonal suppressor factor specific for lactate dehydrogenase B. *Journal of Experimental Medicine*, **157**, 1855-66.

Kapp, J.A. & Araneo, B.A. (1982). Antigen-specific suppressor T-cell interactions. 1. Induction of an MHC-restricted suppressor factor specific for L-glutamic acid[50]-tyrosine[50]. *Journal of Immunology*, **128**, 2447-52.

Kapp, J.A., Araneo, B.A., Sorenson, C. & Pierce, C.W. (1983). Identification of H-2 restricted suppressor T-cell factors specific for lythmic acid L-tyrosine[50] (GT) and L-glutamic acid L-alanine[30]-L-tyrosine[10] (GAP). In *Ir genes: past, present and future* (eds. C. Pierce, S. Cullen, J. Kapp, B. Schwartz & D. Shreffler), pp. 553-63. Clifton, New Jersey: Humana Press.

Katz, D.H., Hamaoka, T. & Benacerraf, B. (1973). Cell interactions between histoincompatible T and B lymphocytes. *Journal of Experimental Medicine*, **137**, 1405-18.

Kawanishi, H., Saltzman, L. & Strober, W. (1983). Mechanism regulating IgA class-specific immunoglobulin production in murine gut-associated lymphoid tissues. *Journal of Experimental Medicine*, **158**, 649-69.

Kindred, B. & Shreffler, D.C. (1972). H-2 dependence of cooperation between T and B cells *in vivo*. *Journal of Immunology*, **109**, 940-43.

Kishimoto, T. & Ishizaka, K. (1973). Regulations of antibody response *in vitro*. 6. Carrier-specific helper cells for IgG and IgE antibody response. *Journal of Immunology*, **111**, 720-32.

Kronenberg, M., Steinmetz, M., Kobori, J., Kraig, E., Kapp, J.A., Pierce, C.W., Sorensen, C., Suzuki, G., Tada, T. & Hood, L. (1983). RNA transcripts for I-J polypeptides are apparently not encoded between the I-A and I-E subregions of the murine major histocompatibility complex. *Proceedings of the National Academy of Sciences of the USA,* **80,** 5704-8.

Krupi, K., Araneo, B.A., Kapp, J.A., Stein, S., Wilder, K.J. & Webb, D.R. (1982). Purification and characterization of a monoclonal T-cell suppressor factor specific for poly(LGlu^{60}LAla^{30}LTyr^{10}). *Proceedings of the National Academy of Sciences of the USA,* **79,** 1254-58.

Marrack, P., Graham, S., Kushnir, E., Leibson, J., Roehm, N. & Kappler, J.W. (1982). Nonspecific factors in B-cell responses. *Immunological Reviews,* **63,** 33-49.

Melchers, F., von Boehmer, H. & Phillips, R.A. (1975). B-lymphocyte subpopulations in the mouse: organ distribution and ontogeny of immunoglobulin-synthesizing and of mitogen-sensitive cells. *Transplantation Reviews,* **25,** 26-58.

Miller, J.F.A.P. (1978). Restrictions imposed on T-lymphocyte reactivities by the major histocompatibility complex: implications for T-cell repertoire selection. *Immunological Reviews,* **42,** 76-107.

Miller, J.F.A.P. & Mitchell, G.F. (1968). Immunological activity of thymus and thoracic duct lymphocytes. *Proceedings of the National Academy of Sciences of the USA,* **59,** 296-303.

Miller, J.F.A.P., Marshall, A.H.E. & White, R.G. (1962). The immunological significance of the thymus. *Advances in Immunology,* **2,** 111-62.

Minami, M., Honji, N. & Dorf, M.E. (1982). Mechanism responsible for the induction of I-J restrictions on Ts$_3$ suppressor cells. *Journal of Experimental Medicine,* **156,** 1502-15.

Minami, M., Okuda, K., Furusawa, S., Benacerraf, B. & Dorf, M.E. (1981). Analysis of T-cell hybridomas. 1. Characterization of H-2- and Igh-restricted monoclonal suppressor factors. *Journal of Experimental Medicine,* **154,** 1390-1402.

Mitchison, N.A. (1971). The carrier effect in the secondary response to hapten-protein conjugates. 2. Cellular cooperation. *European Journal of Immunology,* **1,** 18-27.

Mizel, S.B. (1980). Studies on the purification and structure-function relationships of murine lymphocyte activating factor (interleukin-1). *Molecular Immunology,* **17,** 571-77.

Mizel, S.B. & Ben-Zvi, A. (1980). Studies on the role of lymphocyte-activating factor (interleukin-1) in antigen-induced lymph node lymphocyte proliferation. *Cellular Immunology,* **54,** 382-89.

Monroe, J.G. & Cambier, J.C. (1983). B-cell activation. 1. Anti-immunoglobulin-induced receptor cross-linking results in a decrease in the plasma membrane potential of murine B lymphocytes. *Journal of Experimental Medicine,* **157,** 2073-86.

Nepom, J., Benacerraf, B. & Germain, R.N. (1981). Analysis of Ir-gene function using monoclonal antibodies: independent regulation of GAT and GLPhe T-cell responses by I-A and I-E subregion products on a single accessory cell population. *Journal of Immunology,* **127,** 31-34.

Okuda, K., Minami, M., Furasawa, S. & Dorf, M.E. (1981). Analysis of T-cell hybridomas. 2. Comparisons among three distinct types of monoclonal suppressor factors. *Journal of Experimental Medicine,* **154,** 1838-51.

Perry, L.L. & Greene, M.I. (1982). Conversion of immunity to suppression by *in vivo* administration of I-A subregion-specific antibodies. *Journal of Experimental Medicine,* **156,** 480-91.

Pierres, M. & Germain, R.N. (1978). Antigen-specific T cell-mediated suppression. 4. Role of macrophages in generation of L-glutamic acid^{60}-L-alanine^{30}-L-tyrosine^{10} (GAT)-specific suppressor T cells in responder mouse strains. *Journal of Immunology*, **121**, 1306-14.

Puri, J. & Lonai, P. (1980). Mechanism of antigen binding by T cells: H-2 (I-A)-restricted binding of antigen plus Ia by helper cells. *European Journal of Immunology*, **10**, 273-81.

Raff, M.C. (1970a). Two distinct populations of peripheral lymphocytes in mice distinguishable by immunofluorescence. *Immunology*, **19**, 637-50.

Raff, M.C. (1970b). Role of thymus-derived lymphocytes in the secondary humoral immune response in mice. *Nature*, **226**, 1257-58.

Raulet, D.H. & Bevan, M.J. (1982). A differentiation factor required for the expression of cytotoxic T-cell function. *Nature*, **296**, 754-57.

Roehm, N.W., Marrack, P. & Kappler, J. (1983). Helper signals in the plaque-forming cell response to protein-bound haptens. *Journal of Experimental Medicine*, **158**, 317-33.

Rosenthal, A.A. & Shevach, E.M. (1973). Function of macrophages in antigen recognition by guinea pig T lymphocytes. 1. Requirement for histocompatible macrophages and lymphocytes. *Journal of Experimental Medicine*, **138**, 1194-1212.

Schimpl, A. & Wecker, E. (1975). A third signal in B-cell activation given by TRF. *Transplantation Reviews*, **23**, 176-88.

Schmidtke, J.R. & Unanue, E.R. (1971). Macrophage-antigen interaction: uptake, metabolism and immunogenicity of foreign albumin. *Journal of Immunology*, **107**, 331-38.

Schrader, J.W. (1979). Nature of the T-cell receptor. *Scandinavian Journal of Immunology*, **10**, 387-93.

Schwartz, R.H. (1982). Functional properties of I-region gene products and theories of immune response (Ir)-gene function. In *Ia antigens: mice* (eds. S. Ferrone & C.S. David), vol. 1, pp. 161-218. Boco Raton, Florida: CRC Press.

Schwartz, R.H. (1984). The role of gene products of the major histocompatibility complex in T-cell activation and cellular interactions. In *Fundamental immunology* (ed. W.E. Paul), pp. 379-438. New York: Raven Press.

Schwartz, R.H., Yano, A. & Paul, W.E. (1978). Interaction between antigen-presenting cells and primed T lymphocytes: an assessment of Ir-gene expression in the antigen-presenting cell. *Immunological Reviews*, **40**, 153-80.

Shimonkevitz, R., Kappler, J., Marrack, P. & Grey, H. (1983). Antigen recognition by H-2-restricted T cells. 1. Cell-free antigen processing. *Journal of Experimental Medicine*, **158**, 303-16.

Singer, A. & Hodes, R.J. (1983). Mechanisms of T cell-B cell interaction. In *Annual review of immunology* (eds. W.E. Paul, C.G. Fathman & H. Metzger), pp. 211-41. Palo Alto, California: Annual Reviews Inc.

Smith, K.A. & Ruscetti, F.W. (1981). T-cell growth factor and the culture of cloned functional T cells. *Advances in Immunology*, **31**, 137-75.

Spurll, G.M. & Owen, F.L (1981). A family of T-cell alloantigens linked to IgH-1. *Nature*, **293**, 742-45.

Sunday, M.E., Benacerraf, R. & Dorf, M.E. (1981). Hapten-specific T-cell responses to 4-hydroxy-3-nitrophenyl acetyl. 8. Suppressor cell pathways in cutaneous sensitivity responses. *Journal of Experimental Medicine*, **153**, 811-22.

Swain, S., Wetzel, G., Soubiran, P. & Dutton, R. (1982). T-cell replacing factors in the B-cell response to antigen. *Immunological Reviews*, **63**, 111-28.

Sy, M.-S., Miller, S.D., Moorhead, J.W. & Claman, H.N. (1979). Active suppression of 1-fluoro-2,4-dinitrobenzene-immune T cells. *Journal of Experimental Medicine,* **149**, 1197-1207.

Sy, M.-S., Dietz, M., Germain, R.N., Benacerraf, B. & Greene, M.I. (1980). Antigen- and receptor-driven regulatory mechanisms. 4. Idiotype-bearing I-J-positive suppressor T-cell factors induce second-order suppressor T cells which express anti-idiotypic receptors. *Journal of Experimental Medicine,* **151**, 1183-95.

Sy, M.-S., Nisonoff, A., Germain, R.N., Benacerraf, B. & Greene, M.I. (1981). Antigen- and receptor-driven regulatory mechanisms. 8. Suppression of idiotype-negative, p-azobenzenearsonate-specific T cells results from the interaction of an anti-idiotypic second-order T-suppressor cell with a cross-reactive, idiotype-positive, p-azobenzenearsonate-primed T-cell target. *Journal of Experimental Medicine,* **153**, 1415-25.

Tada, T. & Okumura, K. (1979). The role of antigen-specific T-cell factors in the immune response. *Advances in Immunology,* **28**, 1-87.

Tada, T., Takemori, T., Okumura, K., Nonaka, M. & Tokuhisa, T. (1978). Two distinct types of helper T cells involved in the secondary antibody response: independent and synergistic effects of Ia-negative and Ia-positive helper T cells. *Journal of Experimental Medicine,* **147**, 446-58.

Takaoki, M., Sy, M.-S., Tominaga, A., Lowy, A., Tsurufuji, M., Finberg, R., Benacerraf, B. & Greene, M.I. (1982). I-J-restricted interactions in the generation of azobenzenearsonate-specific suppressor T cells. *Journal of Experimental Medicine,* **156**, 1325-34.

Takatsu, K., Tominaga, A. & Hamaoka, T. (1980). Antigen-induced T cell-replacing factor (TRF). 1. Functional characterization of a TRF-producing helper T-cell subset and genetic studies on TRF production. *Journal of Immunology,* **124**, 2414-22.

Taniguchi, M., Takei, I., Saito, T. & Tokuhisa, T. (1981). Activation of an acceptor T-cell hybridoma by AV_H-positive I-J-positive monoclonal suppressor factor. In *Immunoglobulin idiotypes,* (eds. C. Janeway, E. Sercarz & H. Wigzell), pp. 397-406. New York: Academic Press.

Taniguchi, M., Tokuhisa, T., Kanno, M., Kavita, Y., Shimizu, A. & Honjo, T. (1982). Reconstitution of antigen-specific suppressor activity with translation products of mRNA. *Nature,* **298**, 172-74.

Thze, J., Kapp, J.A. & Benacerraf, B. (1977). Immunosuppressive factor(s) extracted from lymphoid cells of nonresponder mice primed with L-glutamic acid[60]-L-alanine[30]-L-tyrosine[10] (GAT). 3. Immunochemical properties of the GAT-specific suppressive factor. *Journal of Experimental Medicine,* **145**, 839-56.

Thomas, D.W., Hsieh, K.-W., Schuster, J.L. & Wilner, G.D. (1981). Fine specificity of genetic regulation of guinea pig T-lymphocyte responses to angiotensin II and related peptides. *Journal of Experimental Medicine,* **153**, 583-94.

Tokuhisa, T. & Taniguchi, M. (1982). Two distinct allotypic determinants on the antigen-specific suppressor and enhancing T-cell factors that are encoded by genes linked to the immunoglobulin heavy chain locus. *Journal of Experimental Medicine,* **155**, 126-39.

Tominaga, A., Takatsu, K. & Hamaoka, T. (1980). Antigen-induced T cell-replacing factor (TRF). 2. X-linked gene control for the expression of TRF-acceptor site(s) on B lymphocytes and preparation of specific antiserum to that acceptor. *Journal of Immunology,* **124**, 2423-29.

Unanue, E.R. (1981). The regulatory role of macrophages in antigenic stimulation. 2. Symbiotic relationship between lymphocytes and macrophages. *Advances in Immunology,* **31**, 1-136.

Wagner, H., Hardt, C., Rouse, B.T., Rollinghoff, M., Scheurich, P. & Pfizenmaier, K. (1982). Dissection of the proliferation and differentiation signals controlling murine cytotoxic T-lymphocyte responses. *Journal of Experimental Medicine*, **155**, 1876-81.

Waltenbaugh, C., Thze, J., Kapp, J.A. & Benacerraf, B. (1977). Immunosuppressive factor(s) specific for L-glutamic acid[50]-L-tyrosine[50] (GT). 3. Generation of suppressor T cells by a suppressive extract derived from GT-primed lymphoid cells. *Journal of Experimental Medicine*, **146**, 970-85.

Warner, N.L. (1965). The immunological role of different lymphoid organs in the chicken. 4. Functional differences between thymic and bursal cells. *Australian Journal of Experimental Biology & Medical Science*, **42**, 439-50.

Weinberger, J.B., Germain, R.N., Benacerraf, B. & Dorf, M.E. (1980). Hapten-specific T-cell responses to 4-hydroxy-3-nitrophenyl acetyl. 5. Role of idiotypes in the suppressor pathway. *Journal of Experimental Medicine*, **152**, 161-69.

Yamauchi, K., Murphy, D., Cantor, H. & Gershon, R.K. (1981a). Analysis of antigen-specific, Ig-restricted cell-free material made by I-J-positive Lyt 1 cells (Lyt 1 TsiF) that induces Lyt 2-positive cells to express suppressive activity. *European Journal of Immunology*, **11**, 905-12.

Yamauchi, K., Murphy, D., Cantor, H. & Gershon, R.K. (1981b). Analysis of an antigen-specific, H-2-restricted cell-free product(s) made by I-J-negative Lyt 2 cells (Lyt 2 TsF) that suppresses Lyt 2 cell-depleted spleen cell activity. *European Journal of Immunology*, **11**, 913-18.

Yamauchi, K., Chao, N., Murphy, D. & Gershon, R.K. (1982). Molecular composition of an antigen-specific, Lyt 1 T-suppressor inducer factor: one molecule binds antigen and is I-J-negative, another is I-J-positive, does not bind antigen, and imparts an IgH-variable region-linked restriction. *Journal of Experimental Medicine*, **155**, 655-65.

Yano, A., Schwartz, R.H. & Paul, W.E. (1977). Antigen presentation in the murine T-lymphocyte proliferative response. 1. Requirement for genetic identity at the major histocompatibility complex. *Journal of Experimental Medicine*, **146**, 828-43.

Zinkernagel, R.M. & Doherty, P.C. (1974). Restriction of *in vitro* T cell-mediated cytotoxicity in lymphocytic choriomeningitis within a syngeneic or semi-allogeneic system. *Nature*, **248**, 701-2.

Zinkernagel, R.M. & Doherty, P.C. (1979). MHC-restricted cytotoxic T cells: studies on the biological role of polymorphic major transplantation antigens determining T-cell restriction-specificity, function and responsiveness. *Advances in Immunology*, **27**, 51-177.

16

The analysis of the effects of lymphokines on cytotoxic T-cell responses

M.A. SKINNER, N. CHRISTENSEN and

J. MARBROOK

A variety of humoral mediators stimulate the growth and/or differentiation of lymphoid cells. The mode of action of such lymphokines has been studied mainly by devising *in vitro* assays for biologically active molecules. If differentiating processes leading from stem cells to effector cells can be segregated into discrete steps, the progression of cells along specific lineages can be shown to depend on specific or nonspecific regulatory molecules. A description of the role of factors influencing the production of cytotoxic T cells can be used as an example of the effect of lymphokines on cells of one lineage. The *in vitro* generation of cytotoxic T lymphocytes, under conditions of limiting dilution, allows quantitative studies of the effect of lymphokines analogous to clonal assays for haemopoietic growth factors. The clonal analysis of antigen-stimulated and mitogen-stimulated cytotoxic T cells reveals the presence of other regulatory cells. Lymphokines also affect the overall specificity of effector cells. It is technically feasible to follow the growth of individual clones of cytotoxic T cells in primary cultures as a means of describing the basis of heterogeneity within precursor cell populations. The use of lymphokines in the clonal analysis of T cells allows the quantitation of such cells in normal lymphoid populations. This is a prerequisite for using *in vitro* assays to study the effect of biological and chemical agents on the immune system.

Introduction

One of the characteristics of the haemopoietic system is that it consists of distinct cell lineages. The precursor cells of each lineage are derived from a self-renewing pool of stem cells. The differentiation pathway of each cell lineage can be regarded as a series of discrete stages, the progression of cells through each differentiation stage being under regulatory control. The recognition of discrete stages depends on the facility with which migratory pathways and morphological and functional markers can be identified. Differentiation processes are closely linked to cell division and, therefore, the maintenance of subpopulations of cells requires that both proliferative and differentiative stages are 'controlled'.

The differentiation processes are generally regarded as being under the influence of mediators which are secreted by regulatory cells. Thus the flux of cells along one lineage can be controlled by the activity of cells of another lineage. The investigation of such processes consists of isolating discrete regulatory molecules and determining the cell of origin and the target cells, together with the role of such molecules in the progression of cells from stem cells through progenitor cells to effector cells.

The general strategies adopted to study the mode of action of regulatory factors involve devising routine procedures for the generation of factors followed by setting up sensitive assays for the biological activity of the particular lymphokine. In this way, factor preparations can be fractionated by standard biochemical procedures so that active molecules can be characterized and separated from each other. This traditional approach to studying the functional uniqueness of lymphokines will eventually be replaced by the use of recombinant DNA technology but this will not be reviewed here (see de Weck 1983).

Regulatory factors have been isolated from mitogen-stimulated primary cultures and from mitogen-stimulated tumour cells, and in early work a range of assays was used such that the same regulatory factors were measured in a variety of ways. A more rational approach was adopted by consensus with the introduction of the terminology of interleukins (Aarden et al. 1979). Representative examples of the assays used for interleukins are presented in Table 1. The assays in which primary cultures are used are separated from those utilizing factor-dependent tumour lines.

The regulatory network or cascade has been summarized diagrammatically by several authors (e.g. Wagner 1984) as a means of providing a conceptual framework in which to place the sequence of regulatory events involved in cytotoxic T-cell production. The detailed interactions have yet to be resolved fully but three broad regions can be identified. The first stage is the interleukin-1 (IL-1)-dependent secretion of interleukin-2 (IL-2) (Gillis & Mizel 1981). IL-1 is secreted by macrophages and promotes the differentiation of IL-2-producing T cells. The second stage involves the activation of cytotoxic lymphocyte precursor cells (CTLp). These cells acquire the capacity to react to IL-2 by the generation of the appropriate receptors following activation by mitogen or antigen (Wagner & Rollinghoff 1978). Thus the activation step may be seen as the sequential presentation of two signals, namely antigen and IL-2. Finally, the expansion of T-cell clones continues in the presence of IL-2 and forms the basis for the generation of T-cell lines (reviewed by Paul, Sredni & Schwartz 1981).

Within this lymphokine cascade (Farrar et al. 1982), it is becoming

Table 1. Examples of assay systems for interleukins.

	Assay systems	
	Primary cultures	Cell line assay
IL-1	thymus cells Mizel & Mizel (1981)	T-cell line Conlon (1983)
IL-2	co-stimulator assay Watson et al. (1979a)	T-cell line Gillis & Smith (1977)
IL-3	a comparison of assays Watson et al. (1979a)	

clear that other soluble factors contribute to effector-cell production. In addition to interleukins and interferon, at least one differentiation factor has been proposed (Reddehase et al. 1982; Raulet & Bevan 1982; Wagner et al. 1982).

The induction and expression of the enzyme 20 α-steroid dehydrogenase (20 αSDH) was thought to be closely associated with maturation stages of the T-cell lineage (Weinstein 1977). When Ihle and co-workers (Hapel et al. 1981) described a factor influencing the expression of 20 αSDH, it was described as interleukin-3, although it is now thought to be a colony-stimulating factor (Burgess 1984).

With the ability to set up reproducible assays for lymphokines, the question remains as to whether the functional significance of isolated factors in a normal *in vitro* or *in vivo* immune response can be delineated. The standard approach is to take what is assumed to be a nonstimulated lymphoid population and stimulate the cells in a way that leads to effector-cell (CTL) production. In following the effect of lymphokines on the immune response, any enhancement or inhibition of the response will present itself as an increase or decrease of effector-cell production. The basis for such fluctuations in effector-cell production can be interpreted most readily from an analysis of the clonal nature of the response. The studies discussed in this paper were carried out to examine the feasibility of setting up *in vitro* assays to analyse CTL responses on a clonal basis. The main questions which have been posed are: Are the CTL in a specific response derived from a single homogeneous pool of CTLp? Can one screen the population of clones contributing to a response for variations in specificity? Can the expansion of individual clones be followed in a 'primary' *in vitro* response?

The use of limiting dilution assays in analysing CTL production
It is generally assumed that in the differentiation of cells to

become CTL, the precursors are precommitted to a given antigen specificity. Within this conceptual framework, it follows that the role of lymphokines in CTL production is measured by the clonal analysis of responses at limiting dilution. The theoretical basis of this approach has been summarized by Miller (1982).

Briefly, a series of cultures is set up containing various numbers of responding cells, together with a constant and optimum number of stimulator cells. In investigating alloreactive precursors, the stimulator cells would be inactivated allogeneic lymphoid cells, nude spleen cells (Ceredig 1980) or F_1 spleen cells to achieve a one-way mixed leucocyte reaction (Skinner & Marbrook 1976). Similar limiting dilution cultures have been used to analyse anti-hapten and anti-virus responses in which the stimulator cells were mitomycin C-treated autologous cells, treated with trinitrobenzene sulphonate (Ching & Marbrook 1979) or infected with influenza virus (Komatsu et al. 1978). At appropriate times, the cultures are assayed for the presence of CTL and positive cultures are identified on the basis of cytotoxic activity. Using target cells labeled with (^{51}Cr) sodium chromate in a chromium-release assay (Brunner et al. 1970), an arbitrary level of chromium release is set, above which cytotoxic cells are judged to be present. This level is usually three standard deviations above the mean spontaneous release of chromium from the target cells.

Under optimal conditions, there should be a linear relationship between the number of responding cells and the number of CTL clones. The linear fit to the zero-order Poisson relationship is consistent with the clonal nature of the response, as shown in Figure 1. By appropriate treatment of the results of limiting dilution cultures, not only the frequency of CTLp can be calculated, but also the mean cytotoxicity per clone (Derry & Miller 1982). This allows the detailed analysis of cytotoxic T-cell responses, and any enhancement or inhibition of effector-cell production can be attributed to the effects on the number of clones or the clonal burst size.

The kinetics of an *in vitro* CTL response

The particular immune response which was used to investigate the production of CTL was an anti-fluorescein response of spleen cells from BALB/c mice (Christensen, Skinner & Marbrook 1983). Stimulator cells were treated with mitomycin C (20 μg/ml for 30 min at 37°C). The cells were then suspended in isotonic carbonate buffer (pH 9.0) containing 1 mM fluorescein isothiocyanate. The cells were washed in medium before being cultured. The source of lymphokines used in this work was the supernatant of cultures of rat spleen cells (RAFT) which had been stimulated with concanavalin A. Rat spleens were used as the most convenient source of cells for the production of large volumes of lymphokine-containing supernatants. The activity of the supernatants was determined by the antigen-stimulated thymocyte assays (Watson et al. 1979b) or by the ability to support growth of IL-2 dependent CTL lines (T-cell growth assay) (Gillis, Smith & Watson 1980).

Limiting dilution cultures were set up at various concentrations of responder cells in the presence of 10^5 stimulator cells in medium containing RAFT. The cultures were assayed for cytotoxic cells at various times using the chromium-release assay with fluoresceinated P815 mastocytoma cells as targets (FITC-P815). Frequency estimates of CTLp were derived from the results of the limiting dilution analysis. The kinetics of the appearance of anti-FITC-P185 activity is plotted in Figure 2. The maximum number of precursors (1 in 8000) was detectable on days 4, 5 and 6, after which the number of detectable positive cultures decreased.

In subsequent experiments, a different set of culture conditions was devized in which the limiting dilution cultures were set up as described above. However, on day 3, the cultures received additional and equal

Figure 1. A limiting dilution analysis of an anti-fluorescein response. The number of responder BALB/c spleen cells is plotted against the percent negative cultures. The data were fitted to the zero order of the Poisson equation by the least squares method. Correlation coefficients and Y intercepts were calculated. CTLp frequency estimates together with 95% confidence limits and chi-squared goodness-of-fit were also determined (see Christensen, Skinner & Marbrook 1983).

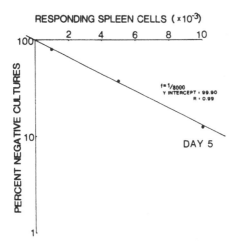

amounts of stimulator cells and RAFT. In a similar fashion, the replicate cultures were assayed at various times, from day 5 to day 13 after initial stimulation, to obtain frequency estimations at each time. The protocol of adding additional growth factors and stimulator cells on day 3 is referred to as a 'restimulation' culture. The results of these experiments are also plotted in Figure 2. It can be seen from these data that additional stimulation reveals a high frequency precursor (1 in 1000) which emerges later in the response, from day 7 onwards. The biphasic kinetics of cytotoxic responses in restimulated cultures might reflect the response from two discrete subsets (see below).

The cytotoxicity in single limiting dilution cultures

The kinetic data plotted in Figure 2 were derived from cultures which contained nonlimiting amounts of growth factors (RAFT)(Christensen, Skinner & Marbrook 1983). The frequency

Figure 2. The kinetics of anti-fluorescein CTL produced in limiting dilution cultures. Various concentrations of responding spleen cells from BALB/c mice were cultured with 10^5 fluoresceinated stimulator cells and RAFT as a source of growth factor. Each point represents the frequency estimate with the 95% confidence limits. ●represent estimates from primary cultures. ○represent data from experiments in which limiting dilution cultures received additional stimulator cells and RAFT on day 3. No confidence limits are indicated on the day 13 frequency value as it was derived from cultures containing a single concentration of responder cells.

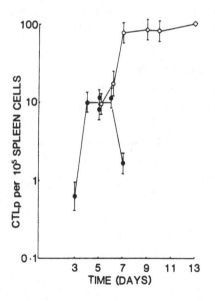

estimates were derived from linear dose-response curves, but it could not be determined from the experimental design whether the same clones detectable early were still detectable late in the 7-day response. To resolve this point, standard restimulated cultures were set up and half of each culture was assayed for specific anti-fluorescein CTL on day 5 and day 7. The cytotoxicity detected on day 5 was plotted against the cytotoxicity detected on day 7 in the same well, and the results are presented in **Figure 3.**

Figure 3. The effect of RAFT on the kinetics of cytotoxicity in limiting dilution cultures. Individual wells in a restimulated anti-FITC response were assayed for specific cytotoxicity on days 5 and 7. The plotted points represent individual wells. Cultures containing no cytotoxicity on both days are not plotted. In A microcultures contained 10^3 to 10^4 responder cells and 2 units of factor activity.Frequency on day 5 was 1:11,740. Frequency on day 7 was 1:7,572. In B microcultures contained 10^3 responder cells and 6.25 units of factor activity. Frequency on day 5 was 1:23,210. Frequency on day 7 was 1:1,235.

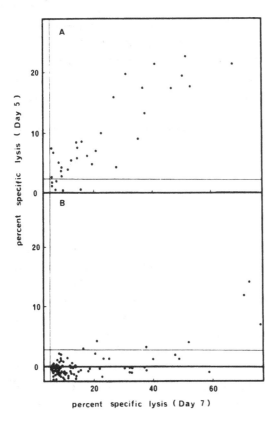

The results in Figure 3A were derived from an experiment set up with a lower concentration of RAFT than normally used (2 units) so that the high-frequency precursors were not detectable on day 7. It can be seen from the plotted data that the majority of cultures which were positive for cytotoxicity on day 5 were also positive on day 7. The culture conditions were devised so that CTL clones continued to expand after day 5.

The results in Figure 3B were plotted from the results of a similar restimulated *in vitro* response, except that a higher concentration of RAFT was used (6.25 units). Under these conditions, a high frequency of CTLp was detected on day 7. From Figure 3B, it can be seen that a relatively small number of cultures were positive on both days 5 and 7, while a large number of cultures contained specific CTL on day 7 alone. The majority of cultures, in which no CTL were detected on either day, are not plotted in Figure 3.

It may be concluded from these data that clones of cells which are detectably cytotoxic on day 5 continue to expand and produce higher amounts of cytotoxicity on day 7. It has been suggested by Christensen, Skinner and Marbrook (1983) that the high-frequency precursors detected on day 7 may represent a distinct stage of differentiation which requires a longer period to produce CTL than precursors detected in 5-day cultures. This suggestion requires data from further studies to eliminate other explanations for the apparent dual populations of CTLp.

The induction of anti-fluorescein cytotoxic responses under suboptimal conditions

The addition of exogenous lymphokine preparations to limiting dilution cultures ensures that the efficiency of precursor activation and clonal expansion leads to a linear dose-response curve. Under suboptimal conditions, which could be due to the addition of insufficient exogenous RAFT, the cloning efficiency of the cultures is impaired. This would be particularly apparent at low concentrations of responder cells, thus leading to nonlinear dose-response relationships. At higher cell concentrations, the presence of helper cells (IL-2 producers) would lead to the production of endogenous growth factors, and consequently the cloning efficiency of CTLp would increase. This phenomenon is illustrated in the results of the experiment plotted in Figure 4.

Limiting dilution cultures were set up at different concentrations of responder cells and restimulated on day 3. Acid-treated RAFT was added as a source of exogenous growth factors. This was obtained by dialysis against 0.1 M glycine-NaCl buffer (pH 2) followed by dialysis against fresh culture medium, as described by Raulet and Bevan (1982). The procedure was adopted originally to eliminate a differentiation factor which may be γ-interferon. One third of each culture was assayed for anti-fluorescein CTL on each of days 5, 6 and 7. The top panel of Figure 4 shows the semilog plot of responder cell number against the proportion

Figure 4. Cytotoxicity in limiting dilution cultures: the amount of specific anti-FITC lysis on days 5, 6 and 7. A range of concentrations of BALB/c spleen cells was set up in limiting dilution cultures with acid-treated RAFT. The cultures were restimulated on day 3. On days 5, 6 and 7 an aliquot was removed from each culture and assayed for cytotoxicity against FITC-P815 target cells. The top panel illustrates the dose-response relationships on days 5, 6 and 7. The four lower panels show data from cultures containing different numbers of responder cells. Cytotoxicity in each well is plotted against time of assay.

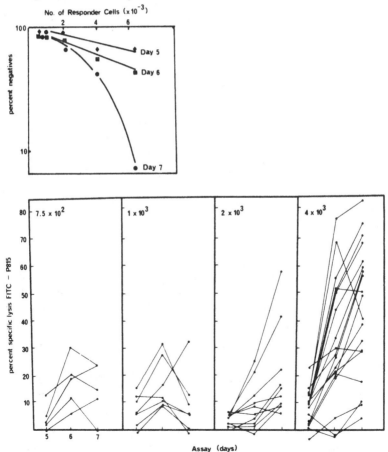

of negative cultures. The dose response on day 7 is clearly different from the results on days 5 and 6. The number of detectable CTL clones increases with time and, whereas the dose-response plots are linear on day 5 and 6, there is a lack of linearity on day 7, with an increase in the efficiency of CTL production at higher cell concentrations.

In the four lower panels of Figure 4, the cytotoxicity in individual wells is plotted against the time of assay. Each panel represents wells containing a different concentration of cells. The overall trend in the amount of cytotoxicity in each positive well can be followed. Thus, in cultures containing 7.5×10^2 and 10^3 cells, the detectable clones increase from day 5 to day 6 and decline thereafter. In contrast, there is a general expansion of clones in cultures containing 2×10^3 and 4×10^3 cells per culture throughout the 3 days of assay.

These data illustrate how endogenous 'helper factors' contribute to the efficiency of clonal activation and expansion. It should be noted that the results have been described in terms of clonal events, but even at the dilution used it cannot be assumed that single wells contain effector cells derived from a single clone (CTLp). However, the general conclusion can be made that suboptimal assays for CTLp affect both the number of CTLp and the degree of expansion of the clones.

The factor(s) in RAFT which contribute to the detection of high-frequency precursors in restimulated responses

The inclusion of RAFT as a source of lymphokines in the restimulation protocol is essential in order to detect the high-frequency CTLp on day 7. Further characterization of these factors was carried out in collaboration with Dr R. Prestidge. The limiting dilution assay was modified so that many fractions of growth factors could be assayed economically: 10^4 responding cells were cultured with 10^5 FITC-modified stimulator cells. According to the data in Figure 2, such cultures contained a mean of 10 high-frequency CTLp. The cultures were restimulated in the standard way and they received factor preparations at the beginning of the culture period and on day 3. With this assay, fractionated preparations of RAFT were analysed for their ability to generate specific anti-FITC CTL on day 7. An example of the results of this assay is summarized in Figure 5.

A RAFT sample was fractionated on a DEAE ion-exchange column and fractions were tested for IL-2 activity by the T-cell growth assay and the antigen-stimulated thymus assay (Gillis, Smith & Watson 1980). The activities measured by these assays were largely coincident, and the main contributory factor to the day 7 response was not separable from

IL-2 activity. Similar results have been obtained with RAFT fractionated by gel-exclusion chromatography.

At this stage, it cannot be assumed that an IL-2-like molecule is the only lymphokine contributing to the generation of high-frequency precursors. Lymphokines isolated from the murine lymphoma LBRM33.5A4 contained adequate IL-2, but were not efficient in supporting a specific day 7 high frequency response. If the cytotoxic response is dependent on a factor additional to IL-2, then either its size and charge properties are similar to IL-2 or a factor is generated endogenously which is dependent on IL-2.

The specificity of anti-fluorescein responses
Lymphokine preparations

The specific responses generated against FITC-syngeneic targets have been analysed at limiting dilution. At low concentrations of responder cells, there is a high probability that responses are clonal. Thus, the specificity or discriminating ability of 'clones' can be assessed by dividing each culture into two aliquots and assaying each half-culture against one of two different targets.

Figure 5. The profile of lymphokine activities in RAFT after fractionation by DEAE ion-exchange chromatography. Fractions from the column were assayed for IL-2 activity (□), the ability to induce CTL responses in antigen-stimulated thymocyte cultures (●) and the ability to generate a 7-day, restimulated anti-FITC response (▲).

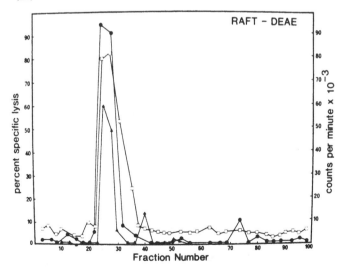

A series of limiting dilution cultures was set up with 10³ responder cells per culture (estimated 1 CTLp per culture) and restimulated on day 3. The individual wells were divided into two and each assayed against P815 or FITC-P815. The three panels in Figure 6 indicate the results of assaying the specificity of responses from cultures containing different batches of RAFT and semipurified IL-2.

The batch of lymphokine-containing culture supernatant used (RAFT A) supported CTL responses which did not appear to be highly specific. The use of RAFT which had been acid treated or purified by ion-exchange chromatography resulted in the generation of highly specific responses,

Figure 6. Effect of different lymphokine preparations on the specificity of CTL: 10³ responder cells were cultured in a standard restimulated culture. Various lymphokines were added as two doses, each containing 2.5 units of activity, on day 0 and day 3. Each culture was divided into two, and assayed against either FITC-labeled or unlabeled P815 target cells. Single points represent one culture; negative cultures are not plotted. Acid-treated RAFT was dialysed against glycine-NaCl buffer, pH 2, for 18 h followed by further dialysis against culture medium. SP IL-2 DEAE was the main fraction from a DEAE ion-exchange fractionation containing IL-2. IL-2 was assayed by the T-cell growth assay.

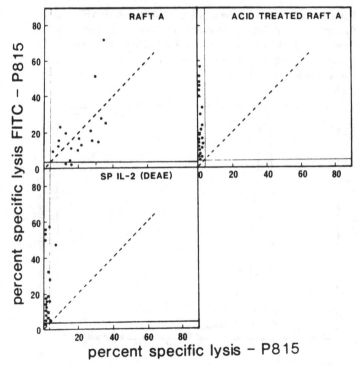

in terms of effector cells discriminating between modified and unmodified P815 targets. The lack of specificity in the results with RAFT A could not be attributable to residual concanavalin A, as similar results have been obtained with RAFT which has been passaged through a sephadex G25 column to remove any mitogen.

Figure 7. Effect of the concentration of lymphokines on the specificity of cytotoxic lymphocytes: 10^3 responder cells were set up in standard restimulated cultures. The units of RAFT added to the cultures represent the total amount added as two equal aliquots on days 0 and 3. Each microculture was assayed against both FITC-P815 and unmodified P815 target cells. RAFT A was passed through a sephadex G25 column to remove any residual concanavalin A. RAFT B had unusually high activity in the thymus assay. Both RAFT A and B gave a frequency estimate of 1:1000 in a restimulated, 7-day anti-FITC response with 5 units activity.

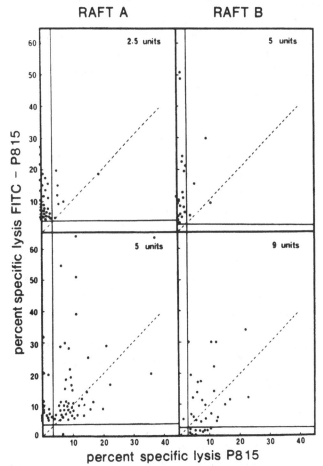

Different concentrations of lymphokines

A series of cultures was set up with 10^3 responder cells, 10^5 stimulator FITC-spleen cells and RAFT. The cultures were restimulated on day 3, and each culture was assayed against both P815 and FITC-P815 targets on day 7. The specific lysis against FITC-P815 is plotted against the specific lysis against P815 in Figure 7.

Two different batches of RAFT were used in these experiments and each batch was used at two concentrations. With batch A, the total RAFT added as two equal doses on day 0 and day 3 was either 2.5 or 5 units. With batch B, the total RAFT added was either 5 or 9 units. Lines are drawn to indicate the level of lysis above which a culture is judged to be positive for cytotoxicity. According to this criterion, 82% of the cultures with 2.5 units of RAFT A or 5 units of RAFT B were specific for FITC-P815 targets. This contrasts with only 26 and 3% of the cultures containing the higher concentrations of RAFT A and B respectively. As there is a mean number of 1 CTL per culture in these experiments, the specificity cannot be described in terms of individual clones (Miller 1982). It can be generally concluded, however, that the apparent specificity of CTL in limiting dilution culture can be influenced by the concentration of growth factors(s) added.

Conclusion

The main aim of this work was to measure various parameters of a CTL response so that the effect of added lymphokines could be assessed in relation to the clonal expansion of CTLp. The rationale was to take BALB/c mouse spleen cells as a representative sample of lymphoid cells, containing the full repertoire of T cells. From these cells anti-hapten CTLs were generated. Haptenated syngeneic spleen cells were treated with mitomycin C before being used as stimulator cells. RAFT was used as a source of T-cell growth factor(s) and it was recognized that a number of lymphokines, in addition to IL-2, may have been present in the culture supernatant. Any antigen-specific or H-2-restricted factors necessary for a CTL response would be generated indirectly from either responder or stimulator populations.

In vitro assays are essentially assays for discrete stages of the T-cell lineage. Two populations of CTLp were detected and, as summarized above, it has been suggested that these may represent precursors at different stages of differentiation, although this has not been demonstrated conclusively. A standard 'primary' *in vitro* response against FITC-P815 was derived from precursors occurring in the spleen at a frequency of approximately 1 in 8,000 cells. A restimulation protocol activated lymphoid cells with greater efficiency, such that analysis of a restimulated day-7 response gave an estimated CTLp frequency of 1 in 1,000 cells. A similar increase in precursor frequency estimations of cells secreting macrophage activation factor has been observed in restimulated cultures by Kelsoe and MacDonald (1982).

The effect of adding lymphokines on CTL production has been followed by assaying individual cultures on consecutive days. These experiments support the conclusion that there is a low-frequency, early-responding precursor and a high-frequency, late-responding CTLp. The late-responding population requires a higher concentration of added lymphokines to generate CTL on day 7. There are other reports of heterogeneity within the pool of CTLp in the spleen, particularly in relation to susceptibility to suppression (Goronzy et al. 1981; Hamann, Eichmann & Krammer 1983). Any relationship between the CTLp subpopulations reported in this work and those described by Goronzy and colleagues (1981) have yet to be established.

Finally the 'apparent' specificity of the response can be affected by the lymphokine concentration. It cannot be determined from the data whether the change of specificity (Figures 6 and 7) is due to the direct effect of lymphokine preparations on cytotoxic ability of the effector cells following the recognition of syngeneic or modified syngeneic target cells, or whether lymphokines are selecting different subsets of CTLp which are 'specific' or 'nonspecific'. Antigen-stimulated cultures also contain spontaneous or anomalous cytotoxic cells which contribute to the total cytotoxicity (Ching, Walker & Marbrook 1977; Roder, Karre & Kiessling 1981; Dorfman et al. 1982), and specific anti-viral responses appear to have an anti-self component both *in vitro* (Komatsu et al. 1978) and *in vivo* (Pfizenmaier et al. 1975). It may be significant that interferon produced in anti-viral responses may contribute to the specificity of anti-viral responses. In this regard, acid treatment of supernatants is useful, since γ interferon is labile under such conditions.

It has been a general finding that cloned specific T-cell lines growing in the presence of IL-2 do not always retain the fine specificity of the original cytotoxic T-cell clone (von Boehmer & Haas 1981). Similarly, specific T cells can acquire the phenotype and/or the specificity of natural killer (NK) cells (Brooks, Urdal & Henney 1983). The loss of fine specificity may depend on the lymphokine preparation, and apparent specificity could be affected by the culture condition without any change in the T-cell recognition mechanism. As suggested by Shortman et al. (1983), the cytolytic potential of a cell could be affected so that transient low-affinity interactions with a target cell are sufficient to allow a lethal hit.

Mitchison (1982) has stressed the importance of cloning in studying the specificity of T-cell responses and the object of this review has been to illustrate how limiting dilution cultures allow the clonal analysis of CTL responses. The ability to measure the number of clones and a combination of clone size and specificity has been adapted to investigate the effects of adding exogenous rat lymphokines. This approach is intended to demonstrate how the effect of lymphokines may be examined and to provide a basis for the further investigation of combinations of purified mouse factors.

Acknowledgements

This work was supported by the Medical Research Council of New Zealand and the Auckland Division of the Cancer Society of New Zealand. We wish to thank Prof. J.D. Watson and Dr R. Prestidge for their assistance in the purification of lymphokine preparations.

References

AAarden, L.A. et al. (1979). Revised nomenclature for antigen nonspecific T-cell proliferative and helper factors. *Journal of Immunology*, **132**, 2928-29.

Brooks, C.G., Urdal, D.L. & Henney, C.S. (1983). Lymphokine-driven 'differentiation' of cytotoxic T-cell clones into cells with NK-like specificity: correlations with display of membrane macromolecules. *Immunological Reviews*, **72**, 43-72.

Brunner, K.T., Mauel, J., Rudolf, H. & Chapins, B. (1970). Studies of allograft immunity in mice. 1. Induction, development and *in vitro* assay of cellular immunity. *Immunology*, **18**, 501-15.

Burgess, A.A. (1984). The complex mediators of cell growth and differentiation. *Immunology Today*, **5**, 155-58.

Ceredig, R. (1980). Frequencies of alloreactive cytotoxic T-lymphocyte precursors responding to IL-2 antigens alone. *Immunology*, **40**, 163-69.

Ching, L.-M. & Marbrook, J. (1979). The clonal analysis of cytotoxic lymphocytes against TNP-modified cells. *European Journal of Immunology*, **9**, 22-27.

Ching, L.-M., Walker, K.Z. & Marbrook, J. (1977). Spontaneous classes of cytotoxic T cells in culture. 1. Characteristics of the response. *Cellular Immunology*, **31**, 284-92.

Christensen, N., Skinner, M. & Marbrook, J. (1983). The analysis of an anti-fluorescein cytotoxic response. *European Journal of Immunology*, **13**, 701-7.

Conlon, P.J. (1983). A rapid biological assay for the detection of interleukin-1. *Journal of Immunology*, **131**, 1280-82.

Derry, H. & Miller, R.G. (1982). Measurement and calculation of CTLp frequencies. In *Isolation, characterization and utilization of T-lymphocyte clones* (eds. C.G. Fathman & F.W. Fitch), pp. 510-15. New York: Academic Press.

de Weck, A.L. (1983). The biology of lymphokines. In *Progress in immunology V* (eds. Y. Tamanura & T. Tada), pp. 307-14. New York: Academic Press.

Dorfman, N., Winkler, D., Burton, R.C., Kayassayda, N., Sabia, P. & Wunderlich, J. (1982). Broadly reactive murine cytotoxic cells induced *in vitro* under syngeneic conditions. *Journal of Immunology*, **129**, 1762-69.

Farrar, J.J., Benjamin, W.R., Hilfiker, M.L., Howard, M., Farrar, W.L. & Fuller-Farrar, J. (1982). The biochemistry, biology and role of interleukin-2 in the induction of cytotoxicity T-cell and antibody-forming B-cell responses. *Immunological Reviews, 63*, 129-66.

Gillis, S. & Mizel, S.B. (1981). T-cell lymphoma model for the analysis of interleukin-1-mediated T-cell activation. *Proceedings of the National Academy of Sciences of the USA, 78*, 1133-37.

Gillis, S. & Smith, K.A. (1977). Long-term culture of tumour-specific cytotoxic T cells. *Nature, 268*, 154-56.

Gillis, S., Smith, K.A. & Watson, J. (1980). Biochemical characterization of lymphocyte regulatory molecules. *Journal of Immunology, 124*, 1954-62.

Goronzy, J., Schaeffer, U., Eichmann, K. & Simon, M.M. (1981). Quantitative studies on T-cell diversity. 2. Determination of the frequencies and Lyt phenotypes of two types of precursor cells for alloreactive cytotoxic T cells in polyclonally and specifically activated splenic T cells. *Journal of Experimental Medicine, 153*, 857-70.

Hamann, U., Eichmann, K. & Krammer, P.H. (1983). Frequencies and regulation of trinitrophenyl-specific cytotoxic T-precursor cells: immunization results in release from suppression. *Journal of Immunology, 130*, 7-14.

Hapel, A.J., Lee, J.C., Farrar, W.L. & Ihle, J.N. (1981). Establishment of continuous cultures of Thy 1.2-positive Lyt 1-positive 2-negative T cells using purified interleukin-3. *Cell, 25*, 179.

Kelsoe, A. & MacDonald, H.R. (1982). Precursor frequency analysis of lymphokine-secreting alloreactive T lymphocytes. *Journal of Experimental Medicine, 156*, 1366-79.

Komatsu, Y., Nawa, Y., Bellamy, A.R. & Marbrook, J. (1978). Classes of cytotoxic lymphocytes can recognize uninfected cells in a primary response against influenza virus. *Nature, 274*, 802-4.

Miller, R.G. (1982). Clonal analysis by limiting dilution: an overview. In *Isolation, characterization and utilization of T-lymphocyte clones* (eds. C.G. Fathman & F.W. Fitch), pp. 219-31. New York: Academic Press.

Mitchison, N.A. (1982). Differentiation within the immune system: the importance of cloning. In *Isolation, characterization and utilization of T-lymphocyte clones* (eds. C.G. Fathman & F.W. Fitch), pp. 11-18. New York: Academic Press.

Mizel, S.B. & Mizel, D. (1981). Purification to apparent homogeneity of murine interleukin-1. *Journal of Immunology, 126*, 834-37.

Paul, W.E., Sredni, B. & Schwartz, R.H. (1981). Long-term growth and cloning of nontransformed lymphocytes. *Nature, 294*, 697-99.

Pfizenmaier, K., Trostmann, H., Rollinghoff, M. & Wagner, H. (1975). Temporary presence of self-reactive cytotoxic T lymphocytes during murine lymphocytic choriomeningitis. *Nature, 258*, 238-40.

Raulet, D.H. & Bevan, M.J. (1982). A differentiation factor required for the expression of cytotoxic T-cell function. *Nature, 296*, 754-56.

Reddehase, M., Suessmuth, W., Moyers, C., Falk, W. & Droege, W. (1982). Interleukin is not sufficient as helper component for the activation of cytotoxic T lymphocytes but synergizes with a late helper effect that is provided by irradiated I region-compatible stimulator cells. *Journal of Immunology, 128*, 61-68.

Roder, J.C., Karre, K. & Kiessling, R. (1981). Natural killer cells. *Progress in Allergy, 28*, 66-159.

Shortman, K., Wilson, A., Scolly, R. & Che, W.-F. (1983). Development of large granular lymphocytes with anomalous nonspecific cytotoxicity in

298   *M. A. Skinner and others*

clones derived from Lyt 2-positive T cells. *Proceedings of the National Academy of Sciences of the USA*, **80**, 2728-32.

Skinner, M.A. & Marbrook, J. (1976). An estimation of the frequency of precursor cells which generate cytotoxic lymphocytes. *Journal of Experimental Medicine*, **143**, 1562-67.

von Boehmer, H. & Haas, W. (1981). H-2 restricted cytotoxic and noncytotoxic T-cell clones: isolation, specificity and functional analysis. *Immunological Reviews*, **54**, 27-56.

Wagner, H. (1984). Where is MHC restriction determined? In *Progress in immunology V* (eds. Y. Tamanura & T. Tada), pp. 809-19. New York: Academic Press.

Wagner, H. & Rollinghoff, M. (1978). T-T interactions during *in vitro* cytotoxic allograft responses. *Journal of Experimental Medicine*, **148**, 1523.

Wagner, H., Hardt, C., Rouse, B.T., Rollinghoff, M., Scheurich, P. & Pfizenmaier, K. (1982). Dissection of the proliferative and differentiative signals controlling murine cytotoxic T-lymphocyte responses. *Journal of Experimental Medicine*, **155**, 1876-81.

Watson, J., Aarden, L., Shaw, J. & Paetkan, V. (1979a). Molecular and quantitative analysis of helper T cell-replacing factors on the induction of antigen-sensitive B and T lymphocytes. *Journal of Immunology*, **122**, 1633-38.

Watson, J., Gillis, S., Marbrook, J., Mochizuki, D. & Smith, K.A. (1979b). Biochemical and biological characterization of lymphocyte regulator molecules. 1. Purification of a class of murine lymphokines. *Journal of Experimental Medicine*, **150**, 849-61.

Weinstein, Y. (1977). 20 α-hydroxysteroid dehydrogenase: a T lymphocyte-associated enzyme. *Journal of Immunology*, **119**, 1223-29.

17

Bovine mixed leucocyte reactions and generation of cytotoxic cells

B.M. GODDEERIS, P.A. LALOR and

W.I. MORRISON

The inductive requirements for proliferation of bovine peripheral blood mononuclear cells (PBM) *in vitro*, in the allogeneic mixed leucocyte reaction (MLR), the autologous MLR and the autologous *Theileria* MLR were compared. In order to examine the role of monocytes in these reactions, methods were developed to deplete PBM of monocytes as well as to obtain purified monocytes. Depletion of monocytes (to less than 0·2%) could be achieved reproducibly by harvesting PBM from defibrinated blood followed by incubation on plastic for 2 h. Monocytes of greater than 95% purity were obtained by sorting with a cell sorter using a monoclonal antibody specific for monocytes within PBM. It was found that monocytes were required in the stimulator population for induction of both the allogeneic and autologous MLR and that purified monocytes were capable of eliciting both responses. However, an autologous MLR could only be induced if the responder population was depleted of monocytes. Stimulator cells fixed with glutaraldehyde did not induce either an autologous or allogeneic MLR. Genetically restricted cytotoxic cells were generated in the allogeneic MLR but not in the autologous MLR. Monocytes were not required for proliferation in the autologous *Theileria* MLR. Furthermore, the response could be elicited using *Theileria*-infected cells fixed with glutaraldehyde. However, stimulation with fixed cells was dependent on the presence of monocytes in the responder population. Cytotoxic cells were generated in the autologous *Theileria* MLR: using PBM from immune animals, at least a proportion of the effector population generated *in vitro* killed in a MHC-restricted way.

Introduction

Cell-mediated immune responses are important in immunity to many infectious diseases, notably those caused by intracellular microorganisms. *In vitro* assays of cell-mediated immune responses rely on detection of proliferation of lymphocytes to the antigen in question and/or generation of effector cells capable of mediating delayed-type hypersensitivity (DTH) reactions or cell-mediated cytotoxicity. There are specific requirements for recognition of antigen during induction of these responses. From studies in laboratory animals it is clear that molecules encoded by the major histocompatibility gene complex (MHC) play a central role in these recognition events. The MHC gene complex encodes two main types of cell-surface antigens: class I antigens expressed on the majority of haemopoietic cells and class II antigens expressed on resting B cells, a proportion of monocyte/macrophages and activated B and T cells. The responses to foreign antigens are characterized by proliferation of T-helper cells or cells mediating DTH reactions, in response to recognition of the antigen in conjunction with class II MHC molecules on the surface of an autologous antigen presenting cell; cytotoxic effector cells generated during these responses recognize antigen along with self-class I MHC antigen. Similarly, T lymphocytes can be induced to proliferate *in vitro* in response to allogeneic class II MHC antigens in mixed leucocyte reactions. Cytotoxic T lymphocytes generated in these reactions are specific predominantly for class I MHC antigens on the inducing allogeneic lymphocytes.

Although a number of investigators have used standard proliferative and cytotoxic assays to examine cell-mediated responses of bovine lymphocytes to infectious agents or to allogeneic leucocytes, there is relatively little information on the inductive requirements for, and the cell populations participating in, these responses in cattle. Our interest in this area stemmed from observations on the responses of cattle infected with the protozoon parasite *Theileria parva*. This tick-transmitted parasite infects cells of the lymphoid system in which it induces blast transformation and establishes a relationship whereby parasite division is synchronized with mitosis of the host cell, resulting in clonal expansion of parasitized cell populations. Infected cells can be established as continuously growing cell lines *in vitro*, either from cells isolated from an infected animal or by *in vitro* infection of cells from a noninfected animal with tick-derived sporozoites. Such cell lines have been useful in examining induction of cell-mediated immune responses *in vitro* and as target cells for detection of cytotoxic cells generated either *in vivo* or *in vitro*. In cattle undergoing immunization against *T. parva*, MHC-restricted, *Theileria*-specific cytotoxic cells are detected transiently in peripheral blood at the time of acquisition of immunity (Morrison et al. this volume). Parasitized cell lines have also been found to induce a potent proliferative response *in vitro* in autologous peripheral blood mononuclear cells (PBM): this autologous *Theileria* mixed leucocyte reaction (MLR) results in the generation of cytotoxic cells which, unlike those elicited *in vivo* during the acquisition of immunity to *T. parva*, are not always MHC-restricted or specific for *Theileria*.

In this chapter, we discuss the results of studies carried out to examine and compare the inductive requirements in the allogeneic MLR and the autologous *Theileria* MLR, in relation to generation and specificities of cytotoxic effector cells. Information is also presented on the autologous MLR which, during the course of these studies, was observed to occur between fractionated populations of normal leucocytes from the same animal.

The allogeneic mixed leucocyte reaction

The allogeneic MLR is considered an *in vitro* model for the recognition events and the generation of the destructive elements in the allograft rejection response (Bach et al. 1969; Hayry & Defendi 1970). It is an *in vitro* assay wherein proliferation of lymphocytes results from the coculture of leucocytes from two individuals of the same species; usually the proliferative reaction is made unidirectional by inhibiting proliferation of one cell population (e.g. by irradiation) while retaining its viability and stimulatory capacity (stimulator). The MLR differs from most lymphoproliferative responses to nonMHC antigens in that it does not require previous priming either *in vivo* or *in vitro* to obtain a measurable response and that presentation of foreign MHC antigens to T cells is not restricted by self-MHC molecules. Since the proliferative response in the MLR is induced principally by the allogeneic class II MHC antigens, this assay has been used in man as one of the methods of typing individuals for differences in class II MHC antigens.

In cattle, as in man, the leucocyte population most commonly tested in the MLR is PBM isolated by density gradient centrifugation from whole blood. Using such cells, Usinger, Curie-Cohen and Stone (1977) and Curie-Cohen, Usinger and Stone (1978) examined unidirectional MLRs in families of cattle in an attempt to determine the number of genetic loci which code for stimulatory determinants. On the basis of these studies they concluded that a minimum of two loci were involved. Furthermore, they demonstrated that the genes coding for stimulatory molecules are closely linked to those of the A locus coding for serologically defined class I MHC determinants (Usinger et al. 1981). In investigations on MLR reactivity between chimaeric and nonchimaeric twin cattle Emery and McCullagh (1980) showed that the leucocytes of chimaeric twins were mutually nonresponsive to each other, whereas those of nonchimaeric dizygotic twins responded to a degree similar to that observed with full siblings. This observation supported the contention that chimaerism is associated with mutual tolerance of the haemopoietic cells in such animals.

In a study on the various parameters affecting the induction of the MLR, Splitter, Everlith and Usinger (1981) showed that stimulation could be mediated by either lymphocytes or monocytes. This was based on the observation that PBM populations depleted or enriched for monocytes by adherence to plastic were equally capable of stimulating proliferation. Furthermore, Emery and McCullagh (1980) showed that efferent lymph cells were as effective, both as stimulators and responders, as PBM in the MLR. Since efferent lymph contains extremely small numbers of macrophages, these findings would suggest that macrophages are not required for stimulation in the bovine MLR.

Our studies of the bovine MLR have concentrated on the role of macrophage/monocytes in the induction and regulation of T-cell proliferation. We have adopted a standard technique using 5×10^5 PBM as responders and 2.5×10^5 PBM as stimulators (irradiated at 5,000 rads) in a final volume of 200 µl per well of a 96-well flat-bottom culture plate. Cultures were performed in RPMI 1640 medium supplemented with 2 mM glutamine, 5×10^{-5} M 2-mercaptoethanol (2ME), 50 µg/ml gentamycin and 10% heat-inactivated foetal bovine serum and were incubated in an atmosphere of 5% CO_2 in air. For optimal responses the most critical supplements in the culture medium were the addition of 2ME and the choice of a suitable batch of foetal bovine serum. Buffering of the medium with HEPES was found not to be essential. Proliferation, as evaluated by incorporation of ^{125}I iododeoxyuridine, was maximal on day 5 of culture. However, the magnitude of the responses varied enormously between individual cattle. This variation did not appear to be due entirely to the degree of genetic difference between responder and stimulator animals, since certain animals consistently gave poor responses to a number of different stimulator cells, whereas others gave potent responses to the majority of stimulators. Two background controls were included, one of responder cells alone and the other of responder cells plus autologous stimulators. As in the study of Emery and McCullagh (1980), two sets of chimaeric twins and one set of monozygotic twins were found to be mutually tolerant in the MLR and, in each instance, stimulator cells from the cotwin resulted in suppression rather than stimulation of proliferation. Several pairs of cattle which were fully matched at the A locus of the bovine MHC, but were of unrelated parentage, were all found to mount significant responses to each other.

A number of different methods for depletion and/or enrichment for monocytes in PBM were investigated. Enzyme histochemical staining for alpha-naphthyl acetate esterase (ANAE) activity was used as a means of identifying monocytes (Yang, Jantzen & Williams 1979). PBM obtained from jugular blood collected in an equal volume of Alsever's solution and separated on Hypaque-Ficoll (density of 1.077 g/ml; Pharmacia, Sweden) (referred to as intact PBM) were found to contain 5 to 20% of cells which exhibited diffuse cytoplasmic staining for ANAE. A variable proportion of the lymphocytes also contained one or two ANAE-positive granules. Incubation of PBM on plastic for 2 h caused a reduction in the number of diffuse ANAE-positive cells to approximately 1 to 3%. Incubation of PBM with sephadex G-10 (Jerrells et al. 1980) and subsequent separation of the unbound cells reduced the number of diffuse ANAE-positive cells to about 2 to 5%. Furthermore, a population of cells containing 70 to 80% ANAE-positive cells and representing 5% of the original PBM could be eluted from the sephadex G-10 with 1% lidocaine. During the course of studies on the influence of different methods of collecting unclotted blood on the yield of PBM, it was observed that the cells obtained from defibrinated blood were markedly depleted for ANAE-positive cells. The number of residual ANAE-positive cells varied somewhat from animal to animal but was always less than 2%. Furthermore, incubation of these PBM on plastic for 2 h consistently depleted monocytes to less than 0.2%. Unlike intact PBM, these cells when used below a certain concentration (2.5 x 10^5 cells per well) were incapable, in the absence of 2ME, of responding to the mitogen concanavalin A, indicating that depletion of monocytes was associated with loss of accessory cell function in this mitogen response. Adding autologous monocytes to the monocyte-depleted PBM restored the mitogenic resonse to levels observed in intact PBM (Goddeeris et al. in press). The depletion procedure did not appear to enrich preferentially for either B or T cells, as the proportion of surface immunoglobulin (sIg)-positive cells was similar to that in the starting population or to that of PBM obtained from blood collected in anticoagulant. Nevertheless, as the total yield of lymphocytes from defibrinated blood is lower than from blood collected in anticoagulant, it is possible that depletion of small subpopulations of lymphocytes may occur.

Two methods for enriching monocytes were tested. Firstly, hypotonic density gradient centrifugation (Feige, Overwien & Sorg 1982), using hypotonic (260 mosm/kg) medium and two layers of hypotonic percoll with densities of 1,060 and 1,054 g/ml, resulted in a band of leucocytes between the two layers of percoll which contained 30 to 50% ANAE-positive cells and represented about 6% of the original PBM population. Secondly, a 1-h incubation of PBM on a thin layer of gelatin precoated for 2 h with fresh bovine plasma and subsequent elution of the adherent cells with EDTA (Freundlich & Avdalovic 1983) gave a population of leucocytes containing 70 to 90% ANAE-positive cells which represented about 6% of the original PBM.

It was apparent from these studies that, while adequate depletion of PBM of monocytes could be achieved, the isolation of pure populations of monocytes was not possible. This became feasible with the production of a monoclonal antibody which reacted specifically with monocytes within populations of bovine PBM (Lalor et al. in press). Using this monoclonal antibody (MAb P8) with the fluorescence-activated cell sorter (FACS II, Becton Dickinson, Mountain View, CA), it was possible to obtain relatively pure populations of monocytes from PBM. When various PBM populations were separated into P8-positive and P8-negative cells using the FACS II, the P8-positive population was found to comprise 98% ANAE-positive cells, the majority of which adhered to plastic, whereas the P8-negative population contained less than 3% ANAE-positive cells. Later observations made on whole-blood leucocyte populations indicated that MAb P8 also recognizes neutrophilic granulocytes (J.Newson & J.Naessens personal communication).

When different combinations of intact and monocyte-depleted populations of PBM were tested in the MLR, the allogeneic monocyte-depleted PBM induced little or no proliferation either in intact or monocyte-depleted responder cells (Table 1). Both intact and monocyte-depleted populations proliferated strongly in response to allogeneic intact PBM. However, monocyte-depleted PBM also exhibited strong proliferative responses when stimulated by autologous intact PBM, so that this combination of responder and stimulator could not be used to evaluate allogeneic responses. These observations show that monocytes are required in the stimulator population to induce a significant proliferative response in the allogeneic MLR. This was supported by the finding that purified monocytes (95 to 98%), obtained from PBM enriched for monocytes by adherence to either plastic or plasma-coated gelatin and subsequent purification by the cell sorter with MAb P8, stimulated potent proliferative responses (Table 1), sometimes with numbers of monocytes as low as 1% of the responder PBM. The inability of the monocyte-depleted populations to act as stimulators was not due to a

Table 1. Comparison of different cell populations obtained from PBM as stimulators in the allogenic MLR. The allogenic MLR was performed in 96-well flat-bottom culture plates; 5×10^5 intact PBM were used as responders with 2.5×10^5 intact or monocyte-depleted PBM or 1.5×10^5 purified monocytes, irradiated at 5,000 rads, as stimulators. Cultures were pulsed on day 5 with 0.5 mCi ^{125}I iododeoxyuridine for 8 h. Intact PBM were isolated on Hypaque-Ficoll from blood collected in Alsever's solution. Monocyte-depleted PBM consisted of the plastic-nonadherent population of PBM harvested on Hypaque-Ficoll from defibrinated, blood; these contained less than 0.1% ANAE-positive cells. Monocytes were purified in a two-step procedure: PBM were first enriched for monocytes by gelatin-plasma adherence and then the monocytes were separated with the FACS using monoclonal antibody P8. This yielded a population comprising 95 to 98% ANAE-positive cells.

Responder	Allogeneic stimulator	Medium	Incorporation of ^{125}I iododeoxyuridine with different stimulators				
			Autologous		Allogeneic		
			Intact PBM	Monocyte-depleted PBM	Intact PBM	Monocyte-depleted PBM	Monocytes
B166	B487	879	302	374	30,743	1,540	—
B487	B470	1,656	313	1,806	75,664	5,517	—
B470	B487	331	81	179	81,719	972	—
C210	B166	705	519	—	37,179	—	70,599
B470	B166	319	331	—	16,982	—	23,444

loss of B cells, since the percentage of sIg-positive cells was similar to that in intact PBM. Since very small numbers of monocytes are required to induce proliferation, the finding by Splitter, Everlith and Usinger (1981) that monocyte-depleted populations were capable of stimulation probably related to inadequate depletion of monocytes in their studies. The conclusion that, in the allogeneic MLR, the monocyte is the principal stimulator cell type in resting PBM is also strengthened by our finding (Lalor et al. in press) that the stimulator cell is a large cell, identified by the monoclonal antibodies R1 and P5, with monocyte characteristics. These studies also show that, as in other species, the proliferating cells in the bovine MLR are T lymphocytes.

Fixation of stimulator cells with 0.05% glutaraldehyde for 2 min (Shimonkevitz et al. 1983) removed their capacity to induce a proliferative response in the bovine allogeneic MLR.

Our findings agree with studies in mice where only two cell types, macrophages and dendritic cells, have been identified as stimulators in the allogeneic MLR (Steinman & Witmer 1978; Ahmann et al. 1979; Minami, Shreffler & Cowing 1980). In the human MLR, besides these two cell types (Albrechtsen & Lied 1978; MacDermott & Stacey 1981; Van Voorhis et al. 1982), B cells (van Oers & Zeijlemaker 1977; Albrechtsen & Lied 1978; Gottlieb et al. 1979; MacDermott & Stacey 1981) and activated T cells (Engleman, Benike & Charron 1980) have also been implicated as stimulators, although it is unclear in these studies whether stimulation could have been induced by small numbers of contaminating monocytes or dendritic cells. In man, dendritic cells are present in low numbers (0.1 to 0.5%) in PBM (Van Voorhis et al. 1982) and these cells might be present at similar levels in bovine PBM and hence contribute to stimulation in the bovine MLR.

Studies on primary T-cell proliferative responses in mice indicate that, in addition to the stimulatory antigen, a second nonspecific signal, delivered by the factor interleukin-1 (IL-1), is also required (Germain 1981; Rock et al. 1983). A proportion of blood monocytes express class II MHC antigens and monocytes are one of the main producers of IL-1. Therefore, in the bovine MLR using intact PBM as responders and monocyte-depleted PBM as stimulators, IL-1 could potentially be produced by the responder population. Yet the B cells, present in the stimulator population and expressing class II MHC antigens, were unable to stimulate a response. It is possible that in resting PBM, the superior capacity of monocytes to act as stimulators, as compared to B cells, is associated with a quantitative or qualitative difference in the expression of class II MHC antigens. An alternative explanation is that there is a change in the expression of class II MHC antigens on the lymphocytes and/or monocytes of the stimulator population during the course of the MLR. The possibility that freshly harvested PBM do not express sufficient quantities of the relevant stimulatory molecules is suggested by the finding that glutaraldehyde-fixed stimulators can no longer induce proliferation. Glutaraldehyde fixation in other species has been shown to preserve the antigenicity of class II MHC molecules (Farr & Nakane 1981; Shimonkevitz et al. 1983).

Using a 4-h ^{51}Cr-release assay (Teale et al. 1985), it was possible to demonstrate that cytotoxic cells were generated during the MLR. Maximal cytotoxicity was detected on day 7 shortly after the peak of the proliferative response (Figure 1). The magnitude of the cytotoxic response varied markedly between different animal combinations and the level of cytoxicity generated did not always correlate with the degree of proliferation in the cultures. Preliminary observations suggest that prior priming of animals *in vivo* with allogeneic cells results in the generation of consistently higher levels of cytotoxicity in the MLR. The cytotoxic cells generated were always specific for the stimulator cells (Figure 1) or

Figure 1. Kinetics of generation of cytotoxic cells in a primary allogeneic mixed leucocyte reaction. Dead cells in the effector population were removed by centrifugation on Hypaque-Ficoll. Cytotoxicity was assayed in a standard 4-h ^{51}Cr-release assay using 5 x 10^4 target cells per well and effector-to-target ratios ranging from 80:1 to 5:1. Target cells consisted of *Theileria*-infected cells derived from the animal from which the stimulator cells were obtained (●) and from an animal with a different MHC A locus-encoded phenotype (○). Results obtained with an effector-to-target ratio of 40:1 are presented.

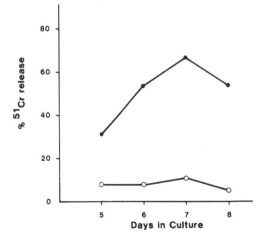

cells from other animals which shared class I MHC antigenic determinants with the stimultor animal (see Teale et al. this volume). The specificity of these effector cells for defined class I MHC antigens suggests that they are classical cytotoxic T cells, analogous to those described in other species (Eijsvoogel et al. 1973; Nahbolz et al. 1974).

The autologous mixed leucocyte reaction

In analysing the roles of different populations of PBM as responders in the allogeneic MLR, it was observed that monocyte-depleted PBM mounted strong proliferative responses to autologous stimulators (intact PBM). This reaction was termed the bovine autologous MLR because of its similarity to the syngeneic MLR in mouse (Howe 1973; von Boehmer & Adams 1973) and guinea pig (Yamashita & Shevach 1980) and to the autologous MLR in man (Opelz et al. 1975), in which purified T cells proliferate in response to autologous nonT cells. Such a proliferative response has not previously been described in cattle.

Based on a preliminary study of the kinetics of this response, a standard technique was adopted for the generation of an autologous MLR (Figure 2). This consisted of coculturing 5×10^5 monocyte-depleted PBM as responders with 1.25×10^5 autologous intact PBM as stimulators (5,000 rads) in a total volume of 200 µl culture medium (see allogeneic MLR)

Figure 2. The bovine autologous mixed leucocyte reactions of two cattle: (a) titration of stimulator cells and (b) time kinetics. Cultures were performed in 96-well flat-bottom culture plates using 5×10^5 monocyte-depleted PBM as responder cells. The stimulator cells were intact irradiated (5,000 rads) PBM which in the kinetics experiment (b) were added at a concentration of 1.25×10^5 per well. Proliferation was evaluated by measuring incorporation of ^{125}I iododeoxyuridine which was assayed on day 6 of the culture for the titration experiment (a).

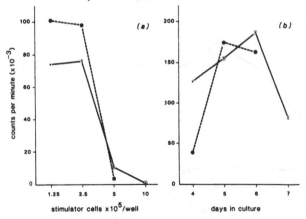

Table 2. Comparison of different cell populations obtained from PBM stimulators in the autologous MLR. The autologous MLR was performed in 96-well flat-bottom culture plates: 5×10^5 monocyte depleted PBM were used as responder cells with 1.25×10^5 autologous intact PBM or monocyte-depleted PBM or 1.25×10^4 monocytes, irradiated at 5,000 rads as stimulators. Cultures were pulsed on day 6 with 0.5 mCi of ^{125}I iododeoxyuridine for 8 h. The different cell populations were prepared as described in Table 1.

	Animal number	Incorporation of ^{125}I iododeoxyuridine with different stimulators (counts/min)			
		Medium	Intact PBM	Monocyte-depleted PBM	Monocytes
Experiment 1	B166	249	125,676	421	–
	B487	76	114,700	75	–
Experiment 2	B166	707	106,124	–	94,737
	C165	162	183,051	–	229,630

per well of a 96-well flat-bottom culture plate (Figure 2a). Occasionally smaller numbers of stimulators, sometimes as few as 2.5×10^4 cells, had to be used to elicit optimal responses. Maximum proliferation was obtained between days 5 and 6 of culture (Figure 2b). No autologous proliferation occurred when stimulators were rigorously depleted of monocytes (Table 2). Also, removal of class II MHC antigen-bearing cells, as defined by MaB R1 (Lalor et al. in press), from the stimulator population reduced or abolished proliferation. Conversely, strong proliferative reactions were obtained when plastic-adherent or sephadex G-10-adherent cells were used as stimulators. The stimulatory role of monocytes in the bovine autologous MLR was confirmed by using populations of monocytes (98% monocytes) purified with MAb P8. Numbers of irradiated monocytes ranging from 0.3 to 20% of the responder cell input induced strong proliferation in monocyte-depleted responders (Table 2). However, monocytes which were not irradiated did not induce proliferative responses when added at or above a concentration of 10% of the responder cell input, although they induced stronger proliferation than irradiated monocytes when used at or below a final concentration of 5% in the cultures. This absence of stimulation when nonirradiated monocytes were used at high concentrations, seemed to be caused by strong suppression of autologous proliferation since addition of 10% or more autologous purified monocytes to their autologous MLR (monocyte-depleted responders with autologous intact stimulators),

Table 3. Influence of the addition of different concentrations of autologous monocytes on the proliferative response in the autologous MLR. The autologous MLR consisted of a culture of 5×10^5 monocyte-depleted PBM as responder cells with 1.25×10^5 intact PBM as stimulator cells. Proliferation was assessed by incorporation of ^{125}I iododeoxyuridine on day 6 of culture. The results are expressed as counts per minute of radioactivity. The cell populations were obtained as described in Table 1.

Autologous cells added	Number of cells added per well ($\times 10^3$)				
	0	12.5	25	50	100
Monocytes	105,183	108,550	98,660	30,985	14,395
Irradiated monocytes (5,000 rads)	105,183	95,941	90,469	93,731	87,654

reduced strongly the proliferative response (Table 3). These results explained the absence of autologous proliferation in intact PBM, as intact PBM usually contain between 10 and 20% monocytes. From these findings it is now clear that monocytes can, in a dose-dependent fashion, either stimulate (concentrations less than 10%) or suppress (concentrations higher than 10%) autologous proliferation. This capacity of autologous monocytes to suppress the autologous MLR is radiosensitive, as irradiated autologous monocytes did not suppress proliferation. It is interesting to note that exposure of the stimulators to higher levels of radiation (10,000 or 30,000 rads) resulted in enhanced (10 to 80 times higher) proliferative responses in those autologous MLRs in which only moderate proliferation was achieved with stimulators given 5,000 rads. This further illustrates the sensitivity of the suppressive activity of monocytes to radiation dose.

The finding that stimulation of the bovine autologous MLR is mediated principally by monocytes differs from observations in the human autologous MLR and the murine syngeneic MLR. In both species, B cells, in addition to monocyte/macrophages, were found to have stimulatory activity (von Boehmer 1974; Opelz et al. 1975; Finke, Ponzio & Battisto 1976; Fernandez & Macsween 1981; Fournier & Charriere 1981; Brown et al. 1984). Furthermore, evidence has been obtained that dendritic cells are much more potent stimulators of the syngeneic and autologous MLR than either macrophages or B cells (Nussenzweig & Steinman 1980; Van Voorhis et al. 1982). We do not yet know whether bovine PBM contain significant numbers of dendritic cells, nor do we know whether the MAb P8 recognizes dendritic cells. The requirement for monocytes in the stimulatory population of the bovine autologous MLR may relate not only to the presence of the appropriate stimulatory antigen on these cells but also to the production of soluble mediators such as IL-1. It is of interest that intact PBM (including monocytes) fixed in 0.05% glutaraldehyde are no longer capable of stimulating the response. However, as discussed for the allogeneic MLR, it is possible that the stimulator cells undergo some quantitative or qualitative change in expression of the stimulatory molecule during the initial stages of the culture period. Studies of the autologous and syngeneic MLR in man and mouse indicate that class II MHC antigens are involved in stimulating the response (Gottlieb et al. 1979; Nussenzweig & Steinman 1980; Indiveri et al. 1983). Thus, it is possible that different cell types stimulate autologous proliferation as long as they express appropriate (quantity or quality) class II MHC antigens on their membranes.

Suppression of the autologous MLR by high concentrations of monocytes has been shown in man (Fernandez & Macsween 1981; Klajman, Drucker & Manor 1983; Steinberg et al. 1983; Twomey, Laughter & Brown 1983; Brown et al. 1984). In cattle, this suppressive activity appeared to be restricted to autologous interactions, as similar numbers of monocytes present in the allogeneic MLR did not lead to suppression. Thus, it would seem unlikely that the suppression was due to production of substances such as prostaglandins which are known to have general suppressive activity. It is possible that this negative regulation involves interaction of the monocytes with a population of lymphocytes. We have not been able to determine whether the stimulatory and suppressive activities of the monocytes reside in different subpopulations of monocytes or are exerted by the monocyte population as a whole.

When cells responding in the bovine autologous MLR were examined for cytotoxic activity, no killing of autologous lymphoblasts, either infected with *T. parva* or generated in cultures with concanavalin A, was observed although occasionally a degree of nonspecific cytotoxicity was encountered. Similar findings have been reported in the murine syngeneic MLR (Smith & Pasternak 1978; Lattime, Golub & Stutman 1980; Ponzio 1980) and the human autologous MLR (Vande Stouwe et al. 1977; Weksler et al. 1980; Pazderka et al. 1983), although in the latter nonspecific cytotoxicity was reported when autologous serum was used as a culture supplement (Tomonari 1980; Hausman et al. 1982; Hausman, Moody & Weksler 1983; Argov & Klein 1983; Goto & Zvaifler 1983). In both species the addition of heat-killed allogeneic leucocytes to the autologous or syngeneic MLR resulted in generation of cytotoxic cells specific for those allogeneic cells (Vande Stouwe et al. 1977; Ponzio, Matteoni & Sarley 1979). Since the heat-killed cells themselves are unable either to stimulate proliferation or induce cytotoxic cells, these findings suggest that the autologous reaction provides the necessary helper activity for generation of the alloreactive effectors.

The significance of the autologous MLR remains controversial. Several reports indicate that the proliferative reactions are responses to xenoproteins of the sera used to supplement the culture or separation media (Huber et al. 1982; Huber 1983; MacDermott & Bragdon 1983). However, in other studies in mice, the response could be generated with cells which had only been in contact with syngeneic serum (Hausman et al. 1983). In the bovine autologous MLR, using either medium supplemented with autologous serum (which was also used during depletion of responders of monocytes) or Iscove's serum-free medium, proliferative responses were obtained comparable to those achieved in media supplemented with foetal bovine serum. This strongly suggests that the reaction is not directed against alloantigens, which may be present in foetal bovine serum (Parish, Chilcott & McKenzie 1976), although the possibility that monocytes carry foreign antigens acquired *in vivo* cannot be excluded.

There is good evidence both in man and mouse that the response involves recognition of self-class II MHC antigenic determinants. Generation of suppressor (Smith & Knowlton 1979; Sakane & Green 1979; Pazderka et al. 1983) and helper (Lattime et al. 1981; Wolos & Smith 1982) functions in this reaction has been demonstrated. Furthermore, certain autoimmune and lymphoproliferative disorders are associated with impaired autologous MLR responsiveness (Weksler, Moody & Kozak 1981). These observations suggest that the response may represent an *in vitro* manifestation of a reaction which normally occurs during the response to foreign antigens and which involves MHC-restricted recognition of self-class II MHC antigens (Weksler, Moody & Kozak 1981).

The autologous *Theileria* mixed leucocyte reaction

PBM and cells from lymph nodes and spleen collected from cattle infected with *Theileria* induce proliferative responses in naive autologous PBM collected before infection (Emery & Morrison 1980; Emery et al. 1981). Cell lines transformed *in vitro* by *T. parva* also stimulate proliferation in autologous PBM from naive and immune animals in an undirectional MLR, termed the autologous *Theileria* MLR (Pearson et al. 1979; 1982; Emery & Kar 1983). These observations indicated the presence of parasite-induced antigenic changes on the surface of infected lymphoblasts.

PBM from most animals give an extremely potent proliferative response in the autologous *Theileria* MLR. This does not appear to differ significantly either in magnitude or kinetics between naive and immune cattle. In general, an optimal response is achieved by stimulating 5×10^5 responder PBM with 2.5×10^4 *Theileria*-infected autologous cells in a final volume of 200 µl per well of a 96-well flat-bottom culture plate. This gives maximal proliferation on day 5 or 6 of culture. However, the magnitude of the response varies between different cattle, as does the range of stimulator cell concentrations which will give a significant proliferative response. Some cell lines are stimulatory at high cell concentrations (2×10^5 per well) while others are not. This is due, at least in part, to differences between individual cell lines, as demonstrated by comparison of responses elicited by a series of phenotypically different cloned cell lines obtained from the same animal. When cloned cell lines derived by infection of T cell-enriched populations of PBM (Lalor et al. this volume) were compared with cloned cell lines obtained by infection of B cell-enriched populations of PBM, it was found that the former, but not the latter, failed to induce proliferation when added at high concentrations (i.e. 2×10^5 cells per well) to the autologous *Theileria* MLR (Figure 3). There is some indication that this inability to induce

Figure 3. The autologous *Theileria* mixed leucocyte reaction: stimulatory capacity of four autologous cloned cell lines at different cell concentrations. Cultures were performed in 96-well flat-bottom culture plates using 5×10^5 monocyte-depleted PBM per well as responder cells with irradiated (5,000 rads) stimulator cells. The latter were obtained from cloned *Theileria*-infected cell lines derived *in vitro* from preselected T lymphocytes (solid lines) or B lymphocytes (broken lines). Proliferation was evaluated on day 6 of culture by measuring the incorporation of ^{125}I iododeoxyuridine.

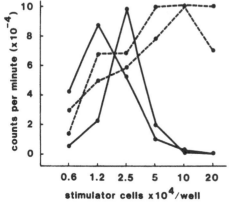

a proliferative response is caused by a suppressive effect of the parasitized T-cell line itself. Such a cell line abolished proliferation when added at 1 x 10⁵ cells per well to an autologous MLR with PBM from the same animal. Preliminary results indicate that this is related to the production of suppressor factors by the cell lines obtained from T cell-enriched populations when these cells are cultured at high cell concentrations (B.M. Goddeeris, C.L. Baldwin & W.I. Morrison unpublished). However, if such cell lines are subjected to high levels of irradiation (30,000 rads rather than 5,000 rads), they induce potent proliferative responses when used as stimulators at high cell concentrations, suggesting that they are no longer able to produce the suppressor factor(s).

There is no impairment of the proliferative response in the autologous *Theileria* MLR when responder PBM are depleted of monocytes and, indeed, with most animals a stronger response is obtained than with intact PBM. Monocyte-depleted PBM respond even in the absence of

Figure 4. The autologous *Theileria* mixed leucocyte culture: influence of fixation with glutaraldehyde of stimulator cells on the proliferative response of (a) intact PBM and (b) monocyte-depleted PBM. Cultures were performed in 96-well flat-bottom culture plates using 5 x 10⁵ responder cells per well. Stimultor cells were either irradiated (solid lines) or irradiated and fixed in 0.05 % glutaraldehyde (broken lines). Proliferation was evaluated on day 6 of culture by measuring the incorporation of ¹²⁵I iododeoxyuridine.

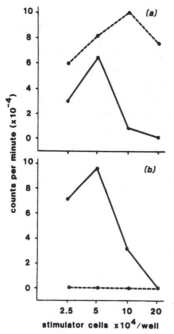

2ME in the culture medium. These findings indicate that monocytes are not required as accessory or presenting cells in the autologous *Theileria* MLR. In contrast to the finding that glutaraldehyde-fixed PBM are ineffective as stimulators in the autologous and allogeneic MLR, in the autologous *Theileria* MLR *Theileria*-infected cells, similarly fixed, induce potent proliferative responses in intact PBM (Figure 4a). However in naive animals when responder PBM are depleted of monocytes, the response to fixed *Theileria*-infected cells is abolished (Figure 4b). This finding indicates either that a metabolic product of the *Theileria*-infected cell replaces the requirement for monocytes in the reaction or that the proliferative responses of intact and monocyte-depleted PBM represent two different types of response. These possibilities are currently being explored.

The generation of cytotoxic cells in the autologous *Theileria* MLR has been demonstrated in a number of studies. However, in contrast to the effectors elicited *in vivo* during immunization which lyse only autologous parasitized cells (Eugui & Emery 1981; Emery et al. 1981; Emery & Kar 1983; Morrison et al. this volume), those generated in the autologous *Theileria* MLR were not restricted in specificity to the autologous parasitized cell. Pearson and colleagues (1979; 1982) detected cytotoxicity in autologous *Theileria* MLRs with PBM from immune but not naive cattle and observed a degree of specificity for the autologous parasitized cell lines as compared to allogeneic cell lines. However, significant levels of killing of uninfected lymphoblasts were detected. On the other hand, Emery and Kar (1983) generated in the autologous *Theileria* MLR similar levels of killing with PBM from either naive or immune cattle against both autologous and allogeneic targets. By pretreatment of target cells with inhibitors of carbohydrate synthesis (tunicamycin or 2-deoxyglucose), Emery and Kar (1983) also obtained evidence that at least a component of the cytotoxicity generated in the autologous *Theileria* MLR is directed against a different target antigen(s) than that recognized by the effectors induced *in vivo*. Target cells treated with these compounds were no longer susceptible to lysis by effectors generated *in vivo* but were still susceptible to lysis by those generated in the autologous *Theileria* MLR.

Our observations on the generation of cytotoxic cells in the autologous *Theileria* MLR indicate that there is a specific and a nonspecific component to the response. In general, the level of cytotoxicity and the specificity of the effectors generated in the reaction were found to vary considerably between individual animals. Cytotoxicity could be generated more consistently in cells from immune animals than from naive animals. In instances where cytotoxicity was generated with cells from naive animals, the effectors were predominantly nongenetically restricted in their killing of *Theileria*-infected lymphoblasts. Conversely, in immune animals, at least a proportion of the effector population generated *in vitro* was restricted to killing of the autologous *Theileria*-infected target. In some animals an additional component was nongenetically restricted. Preliminary findings indicate that the killing of uninfected lymphoblasts reported previously (Emery & Kar 1983) may be related to the presence of small residual amounts of concanavalin A on the target lymphoblast. Further studies are in progress to define the factors underlying the variability in specificity of the cytotoxic cells generated in the autologous *Theileria* MLR.

A potential complicating factor in studies of cell-mediated responses to *Theileria*-infected cells in cattle which received cultured cell lines is the generation of responses against antigens acquired in culture. Evidence that this can occur has been obtained in experiments using macrophages pulsed with plasma membrane antigen prepared from autologous *Theileria*-infected cell lines. PBM from cattle immunized with cultured cells, but not from cattle immunized by other methods, gave proliferative responses and generated genetically restricted cytotoxic cells in this assay (Morrison et al. in press). In some of the previous studies immune cattle had also come into contact with cultured cell lines.

Concluding remarks

The proliferative reactions considered in this chapter represent the *in vitro* manifestations of three different types of immunological response of bovine PBM to cell-surface antigens. The allogeneic MLR clearly is a specific response to foreign MHC antigens and, indeed, results in the generation of cytotoxic cells with a high degree of specificity for the stimulatory MHC molecules. Similar responses are thought to occur *in vivo* during graft rejection and graft *versus* host responses. The autologous MLR, on the other hand, would appear to be directed against normal self (possibly MHC) antigens and does not result in generation of specific cytotoxic cells. Since this reaction can only be elicited following manipulation of leucocyte populations *in vitro*, its relevance in immune responses *in vivo* is questionable. Nevertheless, an awareness that such a response can occur is important when isolated cell populations are used in studies of responses to foreign antigens. The autologous *Theileria* MLR seems to arise as a result of parasite-induced alterations in the parasitized stimulator cells. In immune animals at least a proportion of the cytotoxic cells generated in the reaction are genetically restricted, which suggests that a component of the response in PBM from immune animals is directed against specific antigenic changes induced on the surface of the parasitized cell. It is likely that this response is relevant to protective immunity *in vivo*. There also appears to be a nonspecific component of the response, as indicated by the similar high levels of proliferation obtained with cells from both naive and immune cattle and the detection of nongenetically restricted cytotoxic effectors. In cattle lethally infected with *T. parva*, cytotoxic effector cells with the capacity to kill a range of infected and uninfected target lymphoblasts are detected in the PBM during the later stages of infection (Emery et al. 1981). Thus, the nonspecific component of the autologous *Theileria* MLR may be relevant to the pathogenesis of infection with *T. parva* in cattle.

Acknowledgements

We thank A.J. Teale for phenotyping our cattle for MHC antigens and J.G. Magondu for guidance with the fluorescence-activated cell sorter. B.M. Goddeeris was supported by the Belgian Administration for Development Cooperation.

References

Ahmann, G.B., Nadler, P.I., Birnkrant, A. & Hodes, R.J. (1979). T-cell recognition in the mixed lymphocyte response. 1. NonT, radiation-resistant splenic adherent cells are the predominant stimulators in the murine mixed lymphocyte reaction. *Journal of Immunology*, **123**, 903-9.

Albrechtsen, D. & Lied, M. (1978). Stimulating capacity of human lymphoid cell subpopulations in mixed lymphocyte cultures. *Scandinavian Journal of Immunology*, **7**, 427-34.

Argov, S. & Klein, E. (1983). Effect of interferon on cell proliferation and generation of cytotoxic potential in mixed autologous and allogeneic lymphocyte cultures. *Scandinavian Journal of Immunology*, **17**, 211-23.

Bach, F.H., Bock, H., Graupner, K., Day, E. & Klostermann, H. (1969). Cell kinetic studies in mixed leucocyte cultures: an *in vitro* model of homograft reactivity. *Proceedings of the National Academy of Sciences of the USA*, **62**, 377-84.

Brown, M.F., Van, M., Abramson, S.I., Fox, E.J. & Rich, R.R. (1984). Cellular requirements for induction of human primary proliferative responses to trinitrophenyl-modified cells. *Journal of Immunology*, **132**, 19-24.

Curie-Cohen, M., Usinger, W.R. & Stone, W.H. (1978). Transivity of response in the mixed lymphocyte culture test. *Tissue Antigens*, **12**, 170-78.

Eijsvoogel, V.P., du Bois M.J.G.J., Meinesz A., Bierhorst-Eijlander, A., Zeylemaker, W.P. & Schellekens P.T.A. (1973). The specificity and the activation mechanism of cell-mediated lympholysis (CML) in man. *Transplantation Proceedings*, **5**, 1675-78.

Emery, D.L. & Kar, S.K. (1983). Immune responses of cattle to *Theileria parva* (East Coast fever): specificity of cytotoxic cells generated *in vivo* and *in vitro*. *Immunology*, **48**, 723-31.

Emery, D. & McCullagh, P. (1980). Immunological reactivity between chimaeric cattle twins. 3. Mixed leucocyte reaction. *Transplantation*, **29**, 17-22.

Emery, D.L. & Morrison, W.I. (1980). Generation of autologous mixed leucocyte reactions during the course of infection with *Theileria parva* (East Coast fever) in cattle. *Immunology*, **40**, 229-37.

Emery, D.L., Eugui, E.M., Nelson, R.T. & Tenywa, T. (1981). Cell-mediated immune responses to *Theileria parva* (East Coast fever) during immunization and lethal infections in cattle. *Immunology*, **43**, 323-36.

Engleman, E.G., Benike, C.J. & Charron, D.J. (1980). Ia antigen on peripheral blood mononuclear leucocytes in man. 2. Functional studies of HLA-DR-positive T cells activated in mixed lymphocyte reactions. *Journal of Experimental Medicine*, **152**, 114s-26s.

Eugui, E.M. & Emery, D.L. (1981). Genetically restricted cell-mediated cytotoxicity in cattle immune to *Theileria parva*. *Nature*, **290**, 251-54.

Farr, A.G. & Nakane, P.K. (1981). Immunohistochemistry with enzyme-labeled antibodies: a brief review. *Journal of Immunological Methods*, **4**, 129-44.

Feige, U., Overwien, B. & Sorg, C. (1982). Purification of human blood monocytes by hypotonic gradient centrifugation in percoll. *Journal of Immunological Methods*, **54**, 309-15.

Fernandez, L.A. & Macsween, J.M. (1981). The suppressive effects of monocytes in the autologous mixed lymphocyte reaction. *Immunology*, **44**, 653-59.

Finke, J.H., Ponzio, N.M. & Battisto, J.R. (1976). Isogeneic and allogeneic lymphocyte interactions may be controlled by cell-surface immunoglobulin tropism. *Cellular Immunology*, **26**, 284-94.

Fournier, C. & Charreire, J. (1981). Autologous mixed lymphocyte reaction in man. 1. Relationship with age and sex. *Cellular Immunology*, **60**, 212-19.

Freundlich, B. & Avdalovic, N. (1983). Use of gelatin/plasma coated flasks for isolating human peripheral blood monocytes. *Journal of Immunological Methods*, **62**, 31-37.

Germain, R.N. (1981). Accessory cell stimulation of T-cell proliferation requires active antigen processing, Ia-restricted antigen presentation and a separate nonspecific 2nd signal. *Journal of Immunology*, **127**, 1964-66.

Goddeeris, B.M., Baldwin, C.L., ole Moi Yoi, O. & Morrison, W.I. Improved methods for purification and depletion of monocytes from bovine peripheral blood mononuclear cells. *Journal of Immunological Methods*, in press.

Goto, M. & Zvaifler, N.J. (1983). Characterization of the killer cell generated in the autologous mixed leucocyte reaction. *Journal of Experimental Medicine*, **157**, 1309-23.

Gottlieb, A.B., Fu, S.M., Yu, D.T.Y., Wang, C.Y., Halper, J.M. & Kunkel, H.G. (1979). The nature of the stimulatory cell in the human allogeneic and autologous MLC reactions: role of isolated IgM-bearing B cells. *Journal of Immunology*, **123**, 1497-1503.

Hausman, P.B., Moody, C.E. & Weksler, M.E. (1983). Characterization of cytotoxic cells generated in the autologous mixed lymphocyte reaction. *Behring Institut Mitteilungen*, **72**, 78-86.

Hausman, P.B., Wrazel, L.J., Weksler, M.E. & Moody, C.E. (1982). The generation of nonspecific cytotoxic activity in the autologous mixed lymphocyte reaction. *Federation Proceedings*, **41**, 298.

Hausman, P.B., Moody, C.E., Innes, J.B., Gibbons, J.J. & Weksler, M.E. (1983). Studies on the syngeneic mixed lymphocyte reaction. 3. Development of a monoclonal antibody with specificity for autoreactive T cells. *Journal of Experimental Medicine*, **158**, 1307-18.

Hayry, P. & Defendi, V. (1970). Mixed leucocyte cultures produce effector cells: model *in vitro* for allograft rejection. *Science*, **168**, 133-35.

Howe, M.L. (1973). Isogeneic lymphocyte interaction: responsiveness of murine thymocytes to self-antigens. *Journal of Immunology*, **110**, 1090-96.

Huber, C. (1983). The role of foreign antigens in the induction of human AMLR. *Behring Institut Mitteilungen*, **72**, 117-21.

Huber, C., Merkenschlager, M., Gattringer, C., Royston, I., Fink, U. & Braunsteiner, H. (1982). Human autologous mixed lymphocyte reactivity is primarily specific for xenoprotein determinants adsorbed to antigen-presenting cells during rosette formation with sheep erythrocytes. *Journal of Experimental Medicine*, **155**, 1222-27.

Indiveri, F., Scudeletti, M., Pende, D., Barabino, A., Russo, C., Pellegrino, M.A. & Ferrone, S. (1983). Role of distinct domains of Ia antigens in autologous and allogeneic mixed lymphocyte reactions. *Behring Institut Mitteilungen*, **72**, 95-100.

Jerrells, T.R., Dean, J.H., Richardson, G.L. & Herberman, R.B. (1980). Depletion of monocytes from human peripheral blood mononuclear leucocytes: comparison of the sephadex G-10 column method with other commonly used techniques. *Journal of Immunological Methods*, **32**, 11-29.

Klajman, A., Drucker, I. & Manor, Y. (1983). Autologous mixed lymphocyte reaction in man: further characterization of responding cells. *Immunology Letters*, **6**, 13-19.

Lalor, P.A., Morrison, W.I., Goddeeris, B.M., Jack, R.M. & Black, S.J. Monoclonal antibodies identify phenotypically and functionally distinct cell types in the bovine lymphoid system. *Veterinary Immunology and Immunopathology*, in press.

Lattime, E.C., Golub, S.H. & Stutman, O. (1980). Lyt 1 cells respond to Ia-bearing macrophages in the murine syngeneic mixed lymphocyte reaction. *European Journal of Immunology*, **10**, 723-26.

Lattime, E.C., Gilles, S., David, C. & Stutman, O. (1981). Interleukin-2, production in the syngeneic mixed lymphocyte reaction. *European Journal of Immunology*, **11**, 67-69.

MacDermott, R.P. & Bragdon, M.J. (1983). Foetal calf serum augmentation during cell separation procedures accounts for the majority of human autologous mixed leucocyte reactivity. *Behring Institut Mitteilungen*, **72**, 122-28.

MacDermott, R.P. & Stacey, M.C. (1981). Further characterization of the human autologous mixed leucocyte reaction (MLR). *Journal of Immunology*, **126**, 729-34.

Minami, M., Shreffler, D.C. & Cowing, C. (1980). Characterization of the stimulator cells in the murine primary mixed leucocyte response. *Journal of Immunology*, **124**, 1314-21.

Morrison, W.I., Lalor, P.A., Goddeeris, B.M. & Teale A.J. (in press). Theileriosis: antigens and immunity. In *Parasite antigens, receptors and the immune response* (ed. T.W. Pearson). New York: Marcel Dekker.

Nabholz, M., Vives, J., Young, H.M., Meo, T., Miggiano, V., Rijnbeek, A. & Shreffler, D.C. (1974). Cell-mediated cell lysis *in vitro*: genetic control of killer cell production and target specificities in the mouse. *European Journal of Immunology*, **4**, 378-87.

Nussenzweig, M.C. & Steinman, R.M. (1980). Contribution of dendritic cells to stimulation of the murine syngeneic mixed leucocyte reaction. *Journal of Experimental Medicine*, **151**, 1196-212.

Opelz, G., Kiuchi, M., Takasugi, M. & Terasaki, P.I. (1975). Autologous stimulators of human lymphocyte subpopulations. *Journal of Experimental Medicine,* **142**, 1327-33.

Parish, C.R., Chilcott, A.B. & McKenzie, I.F.C. (1976). Low molecular weight Ia antigens in normal mouse serum. 1. Detection and production of a xenogeneic antiserum. *Immunogenetics,* **3**, 113-28.

Pazderka, F., Angeles, A., Kovithavongs, T. & Dossetor, J.B. (1983). Induction of suppressor cells in autologous mixed lymphocyte culture (AMLC) in humans. *Cellular Immunology,* **75**, 122-33.

Pearson, T.W., Lundin, L.B., Dolan, T.T. & Stagg, D.A. (1979). Cell-mediated immunity to *Theileria*-transformed cell lines. *Nature,* **281**, 678-80.

Pearson, T.W., Hewett, R.S., Roelants, G.E., Stagg, D.A. & Dolan, T.T. (1982). Studies on the induction and specificity of cytotoxicity to *Theileria*-transformed cell lines. *Journal of Immunology,* **128**, 2509-13.

Ponzio, N.M. (1980). Lymphocyte responses to syngeneic antigens. 1. Enhancement of the murine autologous mixed lymphocyte response by polyethylene glycol. *Cellular Immunology,* **49**, 266-82.

Ponzio, N.M., Matteoni, L.A. & Sarley, J.M. (1979). Potentiation of allospecific cytotoxic T-cell development by the autologous MLR. In *The molecular basis of immune cell function* (ed. J.G. Kaplan), pp. 525-28. Amsterdam: Elsevier.

Rock, K.L., Barnes, M.C., Germain, R.N. & Benacerraf, B. (1983). The role of Ia molecules in the activation of T lymphocytes. 2. Ia-restricted recognition of allo K/D antigens is required for class I MHC-stimulated mixed lymphocyte responses. *Journal of Immunology,* **130**, 457-62.

Sakane, T. & Green, I. (1979). Specificity and suppressor function of human T cells responsive to autologous nonT cells. *Journal of Immunology,* **123**, 584-89.

Shimonkevitz, R., Kappler, J., Marrack, P. & Grey, H. (1983). Antigen recognition by H-2-restricted T cells. 1. Cell-free antigen processing. *Journal of Experimental Medicine,* **158**, 303-16.

Smith, J.B. & Pasternak, R.D. (1978). Syngeneic mixed lymphocyte reaction in mice: strain distribution, kinetics, participating cells and absence in NZB mice. *Journal of Immunology,* **121**, 1889-92.

Smith, J.B. & Knowlton, R.P. (1979). Activation of suppressor T cells in human autologous mixed lymphocyte culture. *Journal of Immunology,* **123**, 419-21.

Splitter, G.A., Everlith, K.M. & Usinger, W.R. (1981). Bovine mixed leucocyte response: effect of selected parameters on cell responses. *Veterinary Immunology & Immunopathology,* **2**, 215-32.

Steinberg, A.D., Luger, T.A., Raveche, E.S., Siminovitch, K. & Smolen, J.S. (1983). The magnitude of the autologous mixed lymphocyte reaction itself is regulated. *Behring Institut Mitteilungen,* **72**, 153-62.

Steinman, R.M. & Witmer, M.D. (1978). Lymphoid dendritic cells are potent stimulators of the primary mixed leucocyte reaction in mice. *Proceedings of the National Academy of Sciences of the USA,* **75**, 5132-36.

Teale, A.J., Morrison, W.I., Goddeeris, B.M., Groocock, C.M., Stagg, D.A. & Spooner R.L. (1985). Bovine alloreactive cytotoxic cells generated *in vitro*: target specificity in relation to BoLA phenotype. *Immunology,* **55**, 355-62.

Tomonari, K. (1980). Cytotoxic T cells generated in the autologous mixed lymphocyte reaction. *Journal of Immunology,* **124**, 1111-21.

Twomey, J.J., Laughter, A. & Brown, M.F. (1983). A comparison of the regulatory effects of human monocytes, pulmonary alveolar macrophages

(PAMs) and spleen macrophages upon lymphocyte responses. *Clinical & Experimental Immunology*, **52**, 449-54.

Usinger, W.R., Curie-Cohen, M. & Stone, W.H. (1977). Lymphocyte-defined loci in cattle. *Science*, **196**, 1017-18.

Usinger, W.R., Curie-Cohen, M., Benforado, K., Pringnitz, D., Rowe, R., Splitter, G.A. & Stone, W.H. (1981). The bovine major histocompatibility complex (BoLA): close linkage of the genes controlling serologically defined antigens and mixed lymphocyte reactivity. *Immunogenetics*, **14**, 423-28.

van Oers, M.H.J. & Zeijlemaker, W.P. (1977). The mixed lymphocyte reaction (MLR): stimulatory capacity of human lymphocyte subpopulations. *Cellular Immunology*, **31**, 205-15.

van Voorhis, W.C., Hair, L.S., Steinman, R.M. & Kaplan, G. (1982). Human dendritic cell enrichment and characterization from peripheral blood. *Journal of Experimental Medicine*, **155**, 1172-87.

Vande Stouwe, R.A., Kunkel, H.G., Halper, J.P. & Weksler, M.E. (1977). Autologous mixed lymphocyte culture reactions and generation of cytotoxic T cells. *Journal of Experimental Medicine*, **146**, 1809-14.

von Boehmer, H. (1974). Selective stimulation by B lymphocytes in the syngeneic mixed lymphocyte reaction. *European Journal of Immunology*, **4**, 105-10.

von Boehmer, H. & Adams, P.B. (1973). Syngeneic mixed lymphocyte reaction between thymocytes and peripheral lymphoid cells in mice: strain specificity and nature of the target cell. *Journal of Immunology*, **110**, 376-83.

Weksler, M.E., Moody, C.E. & Kozak, R.W. (1981). The autologous mixed lymphocyte reaction. *Advances in Immunology*, **31**, 271-312.

Weksler, M.E., Moody, C.E., Ostry, R.F. & Casazza, B.A. (1980). Lymphocyte transformation induced by autologous cells. 10. Soluble factors that generate cytotoxic T lymphocytes. *Journal of Experimental Medicine*, **152**, 284s-91s.

Wolos, J.A. & Smith, J.B. (1982). Helper cells in the autologous mixed lymphocyte reaction. 3. Production of helper factor(s) distinct from interleukin-2. *Journal of Experimental Medicine*, **156**, 1807-20.

Yamashita, U. & Shevach, E.M. (1980). The syngeneic mixed leucocyte reaction: the genetic requirements for the recognition of self resemble the requirements for the recognition of antigen in association with self. *Journal of Immunology*, **124**, 1773-78.

Yang, T.J., Jantzen, P.A. & Williams, L.F. (1979). Acid alpha-naphthyl acetate esterase: presence of activity in bovine and human T and B lymphocytes. *Immunology*, **38**, 85-93.

18

Bovine alloreactive cytotoxic T cells

A.J. TEALE, W.I. MORRISON, R.L. SPOONER,

B.M. GODDEERIS, C.M. GROOCOCK and

D.A. STAGG

The target antigens recognized by bovine alloreactive cytotoxic T cells are discussed and compared with serologically defined bovine major histocompatibility complex (MHC) products which make up the bovine lymphocyte antigen (BoLA) system. The role of MHC products in T-cell function in other species is reviewed with particular emphasis on the use of cytotoxic T cells in the characterization of human and murine MHC products and in the development of cytotoxic T-cell technology in general. In the bovine system, serologically defined BoLA-A-locus products w8 and w11 are closely associated with, or identical to, the target determinants for cytotoxic T cells generated *in vitro* in peripheral blood mononuclear cells (PBM) of unprimed w8- and w11-negative cattle cocultured with w8/w11 PBM. There was little evidence of cytotoxicity directed against other products of the MHC. In similar studies of the specificity of cytotoxic T cells generated against PBM carrying the serologically defined specificity w6 and its subgroups, cytotoxicity could be detected which correlated with the distribution of the subgroups of w6 on target cells, but not with w6 itself. The generation and specificity of cytotoxic T cells in cocultures of w8/w11 responder PBM and w8/w11 stimulator PBM were also examined. In some cases, cytotoxic T cells were generated and their target specificity suggested that the products of a second cell-mediated lympholysis-controlling locus in cattle were being detected. The nature of these products is not known.

Introduction

The cell-surface glycoproteins recognized by alloreactive cytotoxic T cells are coded for by genes of the major histocompatibility complex (MHC) — a series of linked genes of elated function. This genetic complex was first discovered in mice, in which it is known for historical reasons as the H-2 system. The early work involved in its discovery has been extensively reviewed by Klein (1975) and Snell (1981). Early knowledge of the MHC came from studies of transplantation biology.

Little and Tyzzer (1916) elaborated on Little's hypothesis that susceptibility to transplanted tumours in mice was under the control of multiple genes (Little 1914). It is now clearly established that the most important of these genes reside in the MHC. Although the earliest work was carried out using tumours, Haldane (1933) proposed that graft rejection is a response to alloantigens, i.e. antigens which are present on normal tissues but which differ between individuals within a species. This is now firmly accepted.

Another important advance came with the demonstration that transplant rejection was an immune function (Gibson & Medawar 1943; Medawar 1944). Mitchison (1954) demonstrated that the principal effector agent in the allograft reaction could be adoptively transferred between individuals in the form of lymph node material but could not be transferred in serum. This showed that the effective response is primarily cell mediated rather than antibody mediated. Thus it is known that the cellular immune system of one individual is capable of responding to antigens characteristic of another unrelated individual and in so doing causes graft destruction. Moreover, the antigens triggering the response are the products of a discrete portion of the genome, the MHC.

Overview
Class I and II MHC products
The cell-surface glycoprotein products of the MHC are of two basic types, class I and class II. These differ in structure (Levy et al. Ch. 9), function (Klein et al. 1981) and with respect to cell and tissue distribution. Class I MHC antigens are expressed on the surface of most mammalian cells to varying degrees. Class II antigens are predominantly expressed on B lymphocytes, macrophages and dendritic cells.

MHC restriction
Although the ability to reject allografts has no value to higher organisms, the phenomenon serves to underline what is currently believed to be the true role of the MHC — the definition of self. One way of viewing this concept which accommodates most of what is known of the MHC is to regard immune surveillance as a homeostatic mechanism. Thus the immune system will respond to, and attempt to eliminate, any antigen which is not recognized as self. At least part of the immune system can therefore be conceptualized as scanning cell surfaces for self-aberration. Such recognized aberrations can be the result of additions of 'foreign' material such as virus antigens and simple chemical haptens. It

is now apparent that what is recognized are self-MHC antigens together with foreign antigens. This phenomenon, termed MHC restriction, was first described by Zinkernagel and Doherty (1974) and Doherty and Zinkernagel (1974) in studies of the mouse immune response *in vitro* to lymphocytic choriomeningitis virus-infected cells. They observed that immune cells derived from one mouse strain were able to lyse virus-infected cells from the original strain and from other strains sharing certain MHC antigens. Virus-infected cells of mouse strains differing with respect to these antigens were not lysed. The effector cells mediating the lytic response were also shown to be specific for virus antigens. Such MHC-restricted recognition was subsequently found to apply to cytotoxic cells of both human and mouse origin, specific for haptens (Shearer, Rehn & Garbarino 1975; Dickmeiss, Soeberg & Svejgaard 1977) and minor histocompatibility antigens (Bevan 1975; Goulmy et al. 1977). It is now known to apply also to the recognition of several viruses by human cytotoxic cells (McMichael et al. 1977; Kreth, ter Meulen & Eckert 1979; Moss et al. 1981).

In cattle, the recognition of parasite-induced alterations by cytotoxic T cells was strongly suggested to be MHC restricted by the results of studies of the immune response to lymphoblasts infected with the protozoan parasite *Theileria parva* (Eugui & Emery 1981). These workers demonstrated that cytotoxic T cells generated *in vivo* following challenge of *T. parva*-immune cattle were restricted to the autologous *Theileria*-infected targets. The relationship of the cattle used, with respect to MHC type, was not known. Subsequently, Preston, Brown and Spooner (1983) demonstrated a correlation between degree of lysis and degree of sharing of serologically defined MHC antigens between *in vivo*-generated effectors and *Theileria annulata*-infected target lymphoblasts. Our own studies have now conclusively shown that the cytotoxic T cell response to *T. parva* infection is MHC restricted (Morrison et al. this volume).

It appears to be a general rule that cytotoxic T cells are restricted by MHC class I products. This would be expected in view of their function which in most cases relies on an ability to recognize foreign antigen on a variety of cell types. There is no reason to suppose, however, that cytotoxic T cells which are restricted by class II products in the recognition of parasite antigens will not be found. Class II products appear to be almost universally involved in the restriction of recognition of foreign antigen on antigen-presenting cells by a second subset of T cells, T-helper cells. Both arms of the T cell-mediated immune response are therefore MHC restricted.

Alloreactivity of cytotoxic T cells

In addition to recognizing foreign antigens in some form of association with self, cytotoxic T cells are capable of reacting to and destroying allogeneic cells. Alloreactivity on the part of cytotoxic T cells may be viewed as a special case of MHC restriction. For some time it was regarded as an interesting phenomenon in its own right with its relevance limited to clinical transplantation efforts, but it may be regarded as a rather blunt manifestation of relevant and basic immune functions. Indeed, if it is accepted that the principal function of the MHC is the definition of self, alloreactivity is a readily acceptable, and arguably predictable, concept. The unifying view is that the recognition of self-MHC-coded antigens plus extrinsically derived antigen is not in itself fundamentally different from the recognition of foreign MHC antigen.

There are several theories (rapidly increasing in number) to explain the nature of the T-cell receptor-target antigen interaction (Doherty, Blanden & Zinkernagel 1976; Janeway, Wigzell & Binz 1976; Klein 1977; Reinherz, Meuer & Schlossman 1983; Siliciano, Brookmeyer & Shin 1983). The diversity of hypotheses is a reflection of the conceptual difficulties inherent in a dual-recognition system. Nevertheless, there are increasingly compelling reasons to suppose that, although the T cell is restricted by two factors (namely MHC and foreign antigen), a single receptor is involved in the recognition, and the MHC antigen and foreign antigen involved in some way cooperate in their interaction with that receptor. This view, unlike those invoking dual receptors for MHC and foreign antigen, readily accommodates the phenomenon of alloreactivity.

Any attempt to unify classical MHC restriction and T cell alloreactivity must explain the extremely potent nature of the alloreaction which involves the activation of numbers of T cells at least an order of magnitude greater than are activated in a conventional MHC-restricted response to foreign antigen (Fischer-Lindahl & Wilson 1977; Teh et al. 1977; Ryser & Robson-MacDonald 1979; Smith 1983). An increasingly favoured explanation was proposed by Burakoff and colleagues (1978) who found that both cytotoxic T cells generated against virus antigen on autologous cells and cytotoxic T cells generated against hapten-coupled autologous cells recognized unmodified allogeneic cells. The potent alloreactivity in the cytotoxic T-cell pool is suggested to be the result of cross-recognition of a variety of determinants on allogeneic cells, by a multiplicity of cytotoxic T cells specific for various foreign antigens on a self-MHC background. In support of this view, several reports were subsequently made of observations of such cross-reactivity in both mouse and human systems (von Boehmer et al. 1979; Sredni & Schwartz 1980; Braciale,

Andrew & Braciale 1981; Guimezanes, Davignon & Schmitt-Verhulst 1982; Gaston, Rickinson & Epstein 1983a). These findings, together with what is thought to be a single receptor/composite antigen mechanism of T cell function, strongly suggest that alloreactivity is not a special-case phenomenon but a predictable consequence of the basic MHC-restriction mechanism.

MHC definition by cytotoxic T cells

In addition to providing a relatively simple and powerful system in which to examine the function of immune T cells, T cell alloreactivity can be used to characterize the antigens coded for by the MHC. Serological techniques have been the most widely employed in MHC antigen definition (Sachs this volume; Spooner this volume). Indeed, defining alloantisera and monoclonal antibodies can show fine specificity, are convenient in use and in capable hands, are standardized reagents. However, there is some evidence to suggest that the alloreactive T cell repertoire and that of antibodies may not be identical.

For this reason serological typing may not always be expected to correlate with observed patterns of MHC restriction. The T cell system is capable of distinguishing the finer points of MHC polymorphism even between species and should be viewed as giving greater depth to the serological definition of MHC-coded antigens. Importantly, where inconsistencies arise in results obtained with antibodies and cytotoxic T cells used for MHC definition, it is the alloimmune cytotoxic T cell which detects the relevant details for MHC-restricted cytotoxic cells.

Cell-mediated lympholysis
Techniques

Cell-mediated lympholysis (CML) is the *in vitro* manifestation of alloimmune cytotoxic T cell function. The technical developments made in measuring CML have been reviewed by Klein (1978). Briefly, Govaerts (1960) and Rosenau and Moon (1961) observed that spleen or lymph node cells from graft recipients were able to kill cells derived from the relevant graft donors *in vitro*. These observations of target-cell destruction were made using monolayers of graft donor cells. Quantification of such effects was laborious and the need to generate cytotoxic (effector) cells *in vivo* added to the tedious nature of such work. The problem of measuring degree of target-cell killing was considerably alleviated by the adaptation of the ^{51}Cr-release assay to CML studies by Brunner and colleagues (1968). This technique, which has remained a standard technique, involves labeling of target cells with ^{51}Cr (usually as ^{51}Cr-sodium chromate) and measuring the ^{51}Cr released following exposure to cytotoxic effector cells.

Another major advance in studies of CML was made by Hayry and Defendi (1970) and Hodes and Svedmyr (1970) who demonstrated that effector cytotoxic T cells could be generated *in vitro* in cocultures of lymphoid cells derived from different donors, i.e. mixed leucocyte cultures (MLC). The responder mononuclear cell population, in which effectors are generated, are cultured with stimulator cells which have been either irradiated or treated with mitomycin C to prevent their proliferation but retain viability (one-way MLC).

Targets which take up relatively large quantities of ^{51}Cr and have low spontaneous release of the isotope over the course of the assay have obvious advantages in CML studies. Lightbody and colleagues (1971) reported that mitogen-stimulated lymphoblasts were superior to most previously used targets in this respect and they are readily producible. Stimulation of mononuclear leucocytes by concanavalin A (Con A) or phytohaemagglutinin (PHA) for 48 to 72 h is still a widely used technique for producing target cells for CML. Tumour cells have been used as target cells because the ^{51}Cr-labeling characteristics of such cells are generally extremely good. The use of the murine mastocytoma P815 in CML was first decribed by Brunner, Mauel & Schindler (1966) and this target has been widely used since then. However, Simon and colleagues (1984) have recently demonstrated that murine cytotoxic T cell clones may acquire specificity for P815 during culture whilst losing their original specificities. This observation introduces an important caveat into the use of tumour cell lines as target cells in assays of alloimmune cytotoxic T-cell function, at least in cases where effector cells have been maintained *in vitro* over periods of several months. The mechanisms responsible for the functional changes involved are not known. It is also not known whether other transformed (tumour type) targets are susceptible to this anomalous cytotoxic T cell activity.

The ^{51}Cr-release assay has undoubtedly been of enormous value in the detection of cytotoxic T cell function and indeed continues to be. However, it has limitations. First, it measures the combined results of the activity of various lytic mechanisms and not just those attributable to cytotoxic T cells. Second, target cells vary in the extent to which they can be lysed by a given effector. Our own observations of bovine cloned and bulk *T. parva*-infected targets have revealed this quite clearly. There are two principal components in this variation. The first involves determinant display and the second, inherent resistance to lysis. The latter complication can be overcome by use of the cold-target inhibition technique, whereby the ability of unlabeled cells to inhibit competitively lysis of ^{51}Cr-labeled targets is used to determine whether the target and inhibitor cells share the same target determinants. The third main limitation of the ^{51}Cr-release assay is that it does not measure the absolute number of effector cells present in a bulk effector population.

Finally, a significant technical development in studies of cell-mediated immune functions, including CML, has been the development of methods of cloning and long-term maintenance of T cells (Fathman & Fitch 1982; Fathman & Frelinger 1983). This was made possible by the finding that T-cell growth factors (TCGF) are produced by mitogen-stimulated leucocytes (Morgan, Ruscetti & Gallo 1976). It has made it possible to start elucidating the T cell repertoire and has provided a battery of monoclonal reagents of fine specificity to be added to alloantisera and monoclonal antibodies for use in dissection of MHC antigenic determinants. It is important to realize, however, that there is increasing evidence that culture clonality does not necessarily imply functional clonality. Specificity may be lost, particularly in cases where cultured cytotoxic T cells are maintained with high levels of growth factors and in the absence of feeder cells and/or specific antigenic stimulation (Acha-Orbea et al. 1983; Shortman, Wilson & Scollay 1984). Another factor contributing to loss of specificity appears to be time in culture, although it is also apparent that medium-term maintenance of cytotoxic T cells *in vitro* may lead to changes in, as distinct from loss of, specificity (Simon et al. 1984). It seems that where the maintenance of specificity is a requirement in the production of alloimmune cytotoxic T cells maintained *in vitro* over periods of months, a combination of allogeneic feeder/ stimulator cells and low to moderate levels of TCGF is the most successful approach (Fitch 1981). Our own observations of long-term cultured bovine alloimmune cytotoxic T cells support this generalization. Further, where uncloned populations of bovine alloreactive cells are concerned, periodic restimulation with allogeneic cells without the addition of exogenous TCGF appears to provide culture conditions in which specificity is often maintained (albeit, usually of a polyclonal nature) throughout the period during which the cultures are capable of responding to the stimulating antigens.

Revelations

A large number of individual reports of CML studies have appeared over the past decade. The majority have concerned work in murine and human systems and only a brief summary of the important points which have emerged will be given here.

The initial indications from work in both human and mouse systems were that class I, serologically defined MHC antigens were the principal targets for effector cells generated in primary MLC (Alter et al. 1973; Eijsvoogel et al. 1973; Nabholz et al. 1974). It should be noted that, as reviewed by Bach, Bach and Sondel (1976), the antigens responsible for induction of proliferation in MLC (so-called lymphocyte-defined antigens) were distinct from the serologically defined antigens recognized at that time. It is now clear that the lymphocyte-defined antigens referred to in these early reports are class II MHC antigens. Thus one set of MHC antigens is primarily responsible for stimulating the proliferative response in the MLC, whilst a different set is the target for cytotoxic effector cells arising as a result of that proliferation. Further, the proliferation occurring in MLC involves the production of growth and differentiation factors by T-helper cells, necessary for the maturation and proliferation of cytotoxic T-precursor cells (reviewed in detail by Germain this volume; Skinner, Christensen & Marbrook this volume). A recent report of the specificity of pig alloimmune cytotoxic T cells is consistent with the findings in mouse and man in that class I determinants on targets appear to direct CML in this species (Thistlethwaite et al. 1984).

Although numerous subsequent studies have confirmed the earlier reports that class I MHC antigens provide the principal target determinants for primary *in vitro*-generated cytotoxic T cells, it has proved possible to generate cytotoxic T cells directed to class II antigens in primary cultures (Wagner et al. 1975). However, as noted by Klein (1978), this generally occurs more readily after *in vivo* priming of the donors of responder cells.

Finally, it has become apparent that serological and CML definitions of MHC antigens, whilst overlapping to a large extent, do not do so completely. Further, the indications are that the variants recognized by classically MHC-restricted cytotoxic T cells correspond with those recognized by alloimmune cytotoxic T cells even in cases where serological definition fails to detect them. This is demonstrated by the ability of cytotoxic T cells to distinguish between the MHC antigens of inbred mouse strains and those of mutants arising within such strains differing in only a few amino-acid residues on a single MHC product (Klein 1978). An example in the human is provided by the HLA-A2 specificity. A proportion of serologically defined A2 individuals can be distinguished from the majority of A2 individuals on the basis of the restriction of influenza virus-specific (Biddison et al. 1980a) and Epstein-Barr virus-specific (Gaston, Rickinson & Epstein 1983b) cytotoxic T cells. Moreover, the A2 cellularly defined subtypes revealed in the influenza virus system are distinguishable from prototypic A2 by alloimmune cytotoxic T cells but are not distinguished serologically (Biddison et al. 1980b).

Heterogeneity within A2 has also been revealed using human cytotoxic T cells specific for the male H-Y antigen (Goulmy et al. 1982; Pfeffer & Thorsby 1982) and in studies of human cytotoxic T cells specific for alloantigens (Horai, van der Poel & Goulmy 1982; Spits et al. 1982). Further, variants of A2 detected by influenza virus-specific human cytotoxic T cells are distinguished from prototypic A2 by murine cytotoxic T cells specific for human MHC-coded antigens (Engelhard & Benjamin 1983).

Krangel, Biddison and Strominger (1983) carried out a comparative structural analysis of A2 serologically defined antigens and detected a limited number of amino-acid substitutions closely clustered in a region of the A2 antigen which they proposed was critical for recognition by cytotoxic T cells but not for recognition by alloantibodies. A recent report of studies of hybrid murine class I antigens as targets for alloimmune and MHC-restricted cytotoxic T cells is consistent with a close association of the relevant MHC determinants in these two systems (Allen et al. 1984).

Studies of bovine alloreactive cytotoxic cells

Studies were primarily undertaken to relate serological definition of bovine histocompatibility antigens to target specificity in CML. This objective has two components. The first concerns definition of the BoLA system itself and the second definition of restricting elements for cytotoxic T cell function as a prelude to studies of MHC restriction in cattle. Work has mainly been concentrated on the BoLA workshop specificities w6, w8 and w11.

Materials and methods
In vitro generation
Bovine alloreactive cytotoxic T cells were generated in MLC as described elsewhere (Goddeeris, Lalor & Morrison this volume; Teale et al. 1985). In brief, responder and irradiated stimulator peripheral blood mononuclear cells (PBM) derived from blood collected into Alsever's solution are cocultured for 6 or 7 days. At the end of this period, the cells are harvested, counted and assayed for cytotoxic function on ^{51}Cr-labeled target cells.
Cytotoxicity assay
In the studies reported here, cytotoxicity was measured in a standard

4-h ^{51}Cr-release assay carried out in flat-bottomed microculture wells using effector:target (E:T) ratios of 80:1 to 5:1. Maximum release was based on the counts in supernatants derived from rapid freezing/slow thawing of target-cell suspensions in water. Spontaneous release was measured in supernatants from target cells in the presence of assay medium only.

Target cells

Trial experiments were undertaken to compare the suitability of *Theileria parva*-infected and uninfected lymphoblasts as target cells in assays of cell-mediated lytic function. The establishment and maintenance of *Theileria*-infected cell lines was essentially performed as described elsewhere (Brown 1979; Stagg et al. 1981). Noninfected bovine lymphoblasts were prepared from PBM. Following separation, the cells were maintained *in vitro* in the presence of Con A and TCGF of bovine origin. After 10 to 14 days in culture, noninfected lymphoblasts possessed similar ^{51}Cr-labeling characteristics to *Theileria parva*-infected lymphoblasts. Further, no qualitative differences were observed in ^{51}Cr-release patterns from a variety of infected and noninfected targets when tested with various effector cells in cytotoxicity assays. The *Theileria parva*-infected lymphoblasts were routinely used in preference to noninfected lymphoblasts as targets because of their simple maintenance requirements.

Histocompatibility testing (BoLA typing)

So far, 17 serologically defined specificities believed to represent determinants on bovine class I histocompatibility antigens coded for at a single locus (BoLA-A locus) have been agreed by two international BoLA workshops (Spooner et al. 1979; Anon. 1982). The specificities are currently designated by Arabic numerals prefixed with a 'w' (e.g. w7). One specificity, w6, has been allocated two subgroups, w6.1 and w6.2 (Spooner this volume).

Donor cattle and target-cell lines were BoLA-typed as described by Teale and colleagues (1983), using a micro-lymphocytoxicity assay and a panel of typing antisera prepared at the Agriculture and Food Research Council's (AFRC) Animal Breeding Research Organization, Edinburgh. The panel contains reagents defining all of the workshop specificities. Target-cell lines were typed subsequent to the fourth *in-vitro* passage and all cell lines and their donor cattle were identical with regard to workshop specificities.

Results
The w8 and w11 specificities

Specificities w8 and w11 were studied concurrently in the same series of experiments (Teale et al. 1985). Thirteen Friesian cattle (castrated males or nonpregnant heifers between 1.5 and 3 years of age) with serologically defined specificities w8 and w11 were selected from a ranch in Kenya. They were the progeny of five sires and a larger number of dams. The group consisted of five cattle of w8/w11 phenotype, four cattle carrying w8 but not w11 at the A locus (w8 halfmatched) and four cattle carrying w11 but not w8 (w11 halfmatched).

In an initial series of experiments, two cattle from a different source (191 and 470 of A-locus phenotypes w7/w10 and w6/w7 respectively) were used as donors of responder cells for MLC generation of cytotoxic effectors. PBM from these two animals were stimulated with PBM of animals 811 and 821, both of which are of the w8/w11 phenotype.

The four effector populations generated were assayed on a panel of targets derived from all of the w8 and w11-bearing cattle and from five other animals which did not carry either w8 or w11 (mismatched). The pattern of cytotoxicity observed using the two effector populations derived from PBM of animal 470 is shown in Figure 1. The pattern and levels of cytotoxicity obtained with the two effectors derived from PBM of animal 191 were essentially the same except that these effectors achieved higher levels of cytotoxicity on w11 halfmatched targets than on w8 half-matched targets. The results using all four effectors can be summarized as follows:

1. Cytotoxicity was apparent with all targets possessing w8 or w11 or both with all four effector populations.
2. Low or undetected levels of cytotoxicity were found with targets not possessing either of these A-locus specificities (mismatched) (mean 8%).
3. Intermediate levels were observed with targets possessing either w8 or w11 (halfmatched) (56%).
4. There were no significant differences between the levels of cytotoxicity achieved on w8/w11 targets derived from stimulator cell donors and those achieved on other w8/w11 targets (matched) (81% and 72% respectively).
5. Both effectors derived from animal 191 achieved higher levels of lysis of w11 than w8 halfmatched targets (71% and 42% respectively); the former approaching the levels achieved on w8/w11 targets (79%).

The specificity of the cytotoxic cells for w8 and w11 was further investigated in cold-target inhibition assays. This involves addition to the assay of varying numbers of unlabeled (cold) targets which, if they share relevant determinants with the labeled targets, will competitively inhibit lysis. This technique offers a more definitive view of target specificity than the basic assay already discussed, principally because it addresses the question of whether the effectors are recognizing similar or different determinants on different target cells and negates any effects of target resistance to lysis. For a more detailed discussion of cold-target inhibition in CML see Nakamura and Cudkowicz (1982).

The results of the cold-target inhibition assays using effector cells generated in PBM of animal 191 during coculture with PBM of animal 821 are shown in Figure 2. It is clear that the matched w8/w11 targets are only slightly less effective as inhibitors of lysis than 821. Next in order of efficacy as inhibitors are the w11, w8 and mismatched cold targets. It is particularly significant that the combinations of two halfmatched cold targets were as effective at blocking as the 821 cold target alone. These data are entirely consistent with those obtained in the initial series of experiments.

It is apparent from the combined results of these two sets of experiments that where the responder cell does not share w8 or w11 with the stimulator cell in the generation of cytotoxic T cells, the cytotoxic response is overwhelmingly directed at determinants closely associated with the serologically defined w8 and w11 specificities. There is some suggestion from the slightly reduced ability of a matched w8/w11 cold inhibitor to block lysis of the 821 target, by comparison with the 821 target itself,

Figure 1. Specificity of two 470 effector populations (w6/w7) generated by stimulation with two different w8/w11 PBM. The solid bar indicates the target which is autologous with the stimulator. Effector:target ratio 80:1.

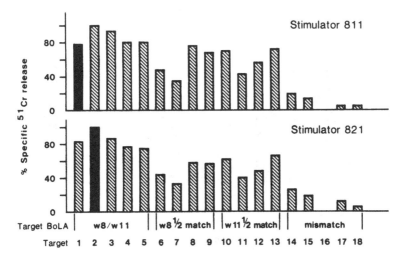

that a small component of cytotoxicity may be generated against determinants not associated with w8 or w11. However, if this was the case, such cytotoxicity was at a very low level as shown by the inability to distinguish w8/w11 targets in direct lysis assays and by the efficacy of cold-target combination blocking. In these latter cases, the cold-target inhibitors derived neither haplotype from the same sire as the 821 target donor, but the combinations proved to be extremely efficient in cold-target inhibition.

It is difficult to exclude entirely the possibility that lysis of w8 and w11-bearing targets was directed against the products of other putative MHC loci (which might be expected to be tightly linked to the A locus). However, although all of the w8- and w11-bearing cattle were obtained from a single source, they were the progeny of five sires (of diverse origins) and probably 13 dams. In view of this, it is concluded that in the cytotoxic T-cell generation combinations studied, the determinants recognized were closely associated on the same MHC class I molecules with the w8 and w11 serologically defined specificities.

The w6 specificity

Three types of experiments were carried out to investigate the w6 specificity and its subgroups. First, three responders (7092, 1862 and

Figure 2. Cold-target inhibition of lysis of the 821 target by the 191 effector using different cold targets. The effectors were generated in culture with 821 stimulators. The results presented were obtained with an effector:hot target ratio of 80:1 using a cold-target:hot-target ratio of 10:1. Where w8 and w11 halfmatched cold targets were used in combination(*), each was included at a cold:hot-target ratio of 5:1. The result with 821 cold targets is indicated by the solid bar.

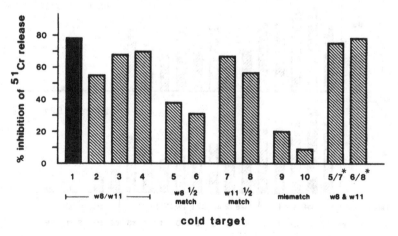

DS91) which did not carry the w6 specificity were used with two stimulators (7168 and 7995) which carried the serologically defined subtypes w6.1 and w6.2 respectively (Spooner et al. 1979; Anon. 1982) and which both carried the w10 specificity. Several separate experiments were undertaken to examine the specificity of cytotoxicity generated by these combinations and the results obtained were consistent. Target cells used for this work were derived from *Theileria annulata*-infected and *T. parva*-infected cell lines. The results of representative assays are presented in Figure 3.

It is apparent that, in those instances where responders did not share either workshop specificity with stimulator cells, cytotoxicity was consistently preferentially generated against w10-associated determinants rather than w6.1 or w6.2. However, although the number of halfmatched targets used was limited, there are indications that these MLC combinations did generate effectors capable of distinguishing the w6 subgroups and that very little, if any, cytotoxicity was associated with the w6 'broad' specificity. This result is particularly interesting in view

Figure 3. Specificity of three effector populations generated against the w6.1, wt.2 and w10 specificities. Effector:target ratio 40:1. NT = not tested.

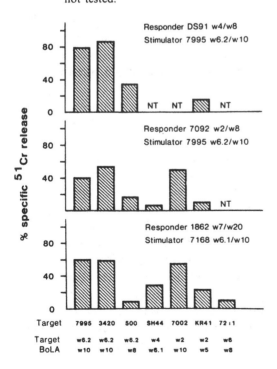

Table 1. *Cytotoxicity generated in responder PBM of animal 7168 (w6.1/w10) during coculture with stimulator PBM of 7995 (w6.2/w10). Effector:target ratio 72:1.*

Target	BoLA A-locus phenotype	% specific ^{51}Cr release
7995	w6.2/w10	43
500	w6.2/w8	42
DX52	w2/w6.ED100	0
7211	w6/w8	0
HF21	w9/w16	0
KR41	w2/w5	0

of the unknown relationship of the two main components of the w6.1 and w6.2 entities. There are two possibilities. First, the w6 broad. specificity may represent a 'public' component, i.e. a shared determinant(s) coded for by several class I alleles, each of which is individually definable by the subgroup specificity. A second possibility is that the 'broad' specificity and the subgroup specificities are products of separate loci in tight linkage disequilibrium. The CML results presented here support the former hypothesis in that should the broad specificity be a distinctive A-locus product in its own right, cytotoxicity against determinants on the w6 molecule would be expected to be readily generated. It is pertinent that in both mouse and human systems the predominant alloreactive and even xenoreactive cytotoxic T-cell response is to private rather than public (less polymorphic) determinants (Engelhard & Benjamin 1983).

The second series of experiments examined the cytotoxic response generated in MLC where responders and stimulators differed with respect to the w6 subgroup specificity and were matched for the second A-locus allele. In the case cited, the responder was w6.1/w10 and the stimulator w6.2/w10. Cytotoxicity was generated in the responder population (see Table 1) against the target which was autologous with respect to the stimulator used, and against the only other w6.2-bearing target in the panel. There was no evidence of cytolysis of the four remaining targets. Of these four, two did not share either A-locus specificity with the stimulator and two carried the w6 broad specificity. DX52, one of these w6 subgroup-mismatched targets, carries a w6-associated determinant defined by the typing serum ED100 which is distinct from w6.1 and w6.2 (Anon. 1982). The other w6-bearing target in this category (7211) carries no w6 subgroup specificity detectable with our panel of typing antisera.

The distinction between the w6.2 subgroup on the one hand and the

Figure 4. Specificity of 191 effector populations (w7/w10) generated in coculture with 820 PBM. Effector:target ratio 80:1.

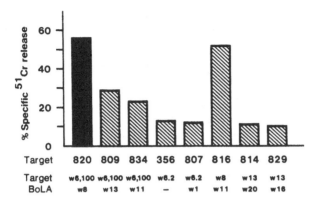

Target	820	809	834	356	807	816	814	829
Target	w6,100	w6,100	w6,100	w6.2	w6.2	w8	w13	w13
BoLA	w8	w13	w11	—	w1	w11	w20	w16

Table 2. Percentage of specific ^{51}Cr release obtained with effectors of w8/w11 phenotype generated in PBM in coculture with w8/w11 stimulators. Where figures are not given, assays were not performed.

Target	A-locus phenotype	Effector[a]					
		A	B	C	D	E	F
811	w8/w11	88	20	100	73	51	34
816	w8/w11					2	3
817	w8/w11	100	44	100	79	58	40
821	w8/w11	3	3	9	5	0	0
828	w8/w11	100			77	74	56
959	w8/w11				8		
809	w6/w13		7	2			
819	w11/w16		2	2			
820	w6/w8			12			
829	13/w16				7	0	0
834	w6/w11		2	6			
836	w8/w13		32	100			
436	w9/w20		33				
487	w5/w20	85		66	65	15	

[a] effector A: 816, stimulator 811; E:T 160:1
effector B: 816, stimulator 811; E:T 80:1
effector C: 816, stimulator 811; E:T 8:1 (maintained *in vitro* with periodic restimulation)
effector D: 816, stimulator 817; E:T 80:1
effector E: 995, stimulator 817; E:T 80:1
effector F: 995, stimulator 817; E:T 80:1

Table 3. Cattle used in studies of cytotoxicity generated in **PBM** of animals 816 and 995 during stimulation with **PBM** of 811, 817 and 821. Where not shown, information is not available.

Animal	A-locus phenotype	Sire	Inherited paternal haplotype
811	w8/w11	W	w11
816	w8/w11	E	
817	w8/w11	AD	w11
821	w8/w11	J	w11
828	w8/w11	N	w8
995	w8/w11		
809	w8/w13	N	w13
819	w11/w16	AD	w11
820	w6/w8	N	w8
829	w13/w16	AD	w13
834	w6/w11	W	w11
836	w8/w13	AD	w13
436	w9/w20		
487	w5/w20		

ED100 w6-associated specificity on the other, was confirmed in a third series of experiments using two stimulator PBM populations (820 and 834) of the w6 ED100/w8 and w6 ED100/w11 phenotypes respectively. The ED100 determinant(s) is believed to represent a further w6 subtype. The responder populations were of the w7/w10 (191) and w5/w20 (487) phenotypes. Results with the 191 effector generated in coculture with 820 PBM are shown in Figure 4. With a panel including targets carrying w6.2 and w6 ED100, together with other A-locus specificities not present on the relevant stimulator cells (i.e. w6 halfmatched targets), cytotoxicity was consistently greater with the w6 ED100 target cells than with targets carrying w6.2.

These studies, in addition to showing cytotoxic T-cell recognition of a third w6-associated specificity, again revealed a 'preference' on the part of effectors for the nonw6 A-locus determinants, with the degree of cytolysis consistently greater on w8 and w11 halfmatched targets. Also, there was again no evidence of cytotoxicity directed towards the w6 'broad' specificity.

NonA-locus specificities

MLCs have been established using responder and stimulator cells which both carry the w8 and w11 specificities. In some of these combinations, cytotoxicity was generated (Table 2). The cattle used for these studies are listed in Table 3 together with A-locus phenotype, sire and inherited paternal haplotype. Target lymphoblasts were infected with *T. parva parva* (Muguga).

Responder populations 816 and 995 in coculture with either 811 or 817 generated cytotoxicity against 811, 817 and 828 but not against 821 (and where appropriate 816). It did not prove possible to generate significant levels of cytotoxicity in PBM of 816 or 995 during stimulation with PBM of 821. Thus there appears to be at least two groups within our panel of w8/w11 cattle.

Consideration of the parentages of these cattle and the origins of haplotypes (Table 3) suggests that the determinants recognized by the cytotoxic T cells generated in these studies are not present on the w8 and w11 A-locus products. The haplotypes coding for the relevant antigen(s) in 811 and 817 are marked by the w8 phenotype and in animal 828 by the w11 phenotype. Further, as shown in Table 2, cytotoxicity was evident with both the 816 and 995 effectors on a nonw8/w11 target, 487. Lysis of a w8 halfmatched target (836) and a further nonw8/w11 target (436) was also observed.

The nature of the proposed second CML-controlling locus in cattle is subject for speculation. Stear, Newman and Nicholas (1982) presented serological evidence for a second class I locus in cattle, and similar results are being obtained by other workers (R.L. Spooner unpublished). Such a locus would indeed be a good candidate for the origin of the determinants detected by cytotoxic T cells generated in the w8/w11-matched cocultures.

A further consideration, however, is that the determinants recognized may be coded for by a class II locus. It is pertinent to point out that *Theileria parva*-infected lymphoblasts may be capable of expressing class II products (Black et al. 1981), although this remains to be confirmed.

In vitro propagation and cloning of bovine alloreactive T cells

It is possible to maintain bulk cultures of bovine MLC-generated, alloreactive T cells for periods of at least a year. This has been done by weekly restimulation with the relevant PBM preparation following, in the early stages, separation of viable responder cells from culture debris by centrifugation over Hypaque-Ficoll/diatrizoate sodium before restimulation. Such bulk cultures can be maintained without the addition of exogenous TCGF for varying periods of up to several months. In all cases we have found that the addition of TCGF is eventually required for the continued maintenance of the cultures.

Although the specificity of the cells as measured by cytotoxicity and the ability to proliferate when cocultured with various stimulator cells is usually maintained, it may be lost at any stage. Loss of specificity is likely to occur with the addition of TCGF in the form of supernatants from cultures of bovine PBM stimulated with Con A (Teale et al. 1985), especially when carried out early in the life of a culture. The problem is compounded by prolonged culture with added TCGF without restimulation with relevant PBM.

Preliminary results indicate that bovine alloreactive T cells can be cloned from early bulk cultures to which the addition of exogenous TCGF is not required. Clones have been established using seeding rates of 0.3 responder cells per round-bottomed microculture well together with 2 x 10^4 irradiated stimulator PBM and added TCGF. Further studies are in progress to define the effector functions and specificities of these clones.

Concluding remarks

It is apparent that the BoLA-A locus codes for target determinants in CML and thus conforms in this respect with class I loci of other species. Further, the results reported here give no indication of heterogeneity in class I products defined by our panel of serological typing reagents. However, studies of other serologically defined specificities are in progress and it remains to be seen whether or not this will hold true for all of them. The use of cloned cytotoxic T cells will provide a powerful means of investigating BoLA specificities for heterogeneity and it may be expected that the simplistic picture of the BoLA system which we have at present will not stand up to such rigorous scrutiny.

A relationship between w6 and its subgroups has been proposed on the basis of studies with alloreactive cytotoxic T cells. This relationship will require confirmation through studies using genetic and biochemical approaches. Finally, evidence has been obtained for a second CML-controlling locus in cattle. The nature of this locus is the subject of continuing studies.

Acknowledgements

Mr C.G.D. Brown of the Centre for Tropical Veterinary Medicine, Edinburgh kindly established the *T. annulata*-infected cell lines. Mr S. Kemp, Mrs P. Millar and Mrs A. Morgan of the AFRC's Animal Breeding Research Organization, Edinburgh were largely responsible for the tissue typing. During the period of the studies reported here, A.J.Teale and D.A. Stagg were supported by the Overseas Development Administration (UK) under Research Schemes 3555 and 3256/3791 respectively. C.M. Groocock was supported by the United States Department of Agriculture, Agricultural Research Service/International

Activities and B.M. Goddeeris by the Belgian Administration for Development Cooperation.

References

Acha-Orbea, M., Groscurth, P., Lang, R., Stutz, L. & Hengartner, M. (1983). Characterization of cloned cytotoxic lymphocytes with NK-like activity. *Journal of Immunology*, **130**, 2952-59.

Allen, M., Wraith, D., Pala, P., Askonas, B. & Flavell, R.A. (1984). Domain interactions of H-2 class I antigens alter cytotoxic T-cell recognition sites. *Nature*, **309**, 279-81.

Alter, B.J., Schendel, D.J., Bach, M.L., Bach, F.M., Klein, J. & Stimpfling, J.H. (1973). Cell-mediated lympholysis: importance of serologically defined H-2 regions. *Journal of Experimental Medicine*, **137**, 1303-9.

Anon. (1982). Proceedings of the second international bovine lymphocyte antigen (BoLA) workshop. *Animal Blood Groups & Biochemical Genetics*, **13**, 33-53.

Bach, F.M., Bach, M.L. & Sondel, P. (1976). Differential function of major histocompatibility complex antigens in T-lymphocyte activation. *Nature*, **259**, 273-81.

Bevan, M.J. (1975). The major histocompatibility complex determines susceptibility to cytotoxic T cells directed against minor histocompatibility antigens. *Journal of Experimental Medicine*, **142**, 1349-64.

Biddison, W.E., Ward, F.E., Shearer, G.M. & Shaw, S. (1980a). The self-determinants recognized by human virus-immune T cells can be distinguished from the serologically defined HLA antigens. *Journal of Immunology*, **124**, 548-52.

Biddison, W.E., Krangel, M.S., Strominger, J.L., Ward, F.E., Shearer, G.M. & Shaw, S. (1980b). Virus-immune cytotoxic T cells recognize structural differences between serologically indistinguishable HLA-A2 molecules. *Human Immunology*, **3**, 225-32.

Black, S.J., Jack, R., Lalor, P. & Newson, J. (1981). Analyses of *Theileria*-infected cell-surface antigens with monoclonal antibodies. In *Advances in the control of theileriosis* (eds. A.D. Irvin, M.P. Cunningham & A.S. Young), Current Topics in Veterinary Medicine & Animal Science 14, pp. 327-39. The Hague: Martinus Nijhoff.

Braciale, T.J., Andrew, M.E. & Braciale, V.L. (1981). Simultaneous expression of H-2 restricted and alloreactive recognition by a cloned line of influenza virus-specific cytotoxic T lymphocytes. *Journal of Experimental Medicine*, **153**, 1371-76.

Brown, C.G.D. (1979). Propagation of *Theileria*. In *Practical tissue culture applications* (eds. K. Maramorosch & H. Hirumi), pp 223-54. New York: Academic Press.

Brunner, K.T., Mauel, J. & Schindler, R. (1966). *In vitro* studies of cell-bound immunity: cloning assay of the cytotoxic action of sensitized lymphoid cells on allogeneic target cells. *Immunology*, **11**, 499-506.

Brunner, K.T., Mauel, J., Cerottini, J.-C. & Chapuis, B. (1968). Quantitative assay of the lytic action of immune lymphoid cells on ⁵¹Cr-labeled allogeneic target cells *in vitro*: inhibition by isoantibody and by drugs. *Immunology*, **14**, 181-96.

Burakoff, S.J., Finberg, R., Glimcher, L., Lemmonier, F., Benacerraf, B. & Cantor, H. (1978). The biological significance of alloreactivity: the ontogeny of T-cell sets specific for alloantigens or modified self-antigens. *Journal of Experimental Medicine*, **148**, 1414-22.

Dickmeiss, E., Soeberg, B. & Svejgaard, A. (1977). Human cell-mediated cytotoxicity against modified target cells is restricted by HLA. *Nature*, **270**, 526-28.

Doherty, P.C. & Zinkernagel, R.M. (1974). T cell-mediated immunopathology in viral infections. *Transplantation Reviews*, **19**, 89-120.

Doherty, P.C., Blanden, R.V. & Zinkernagel, R.M. (1976). Specificity of virus-immune effector T cells for H-2K or H-2D compatible interactions: implications for H-antigen diversity. *Transplantation Reviews*, **29**, 89-124.

Eijsvoogel, V.P., du Bois, M.J.V.J., Meinesz, A., Bierhorst-Eijlander, A., Zeylemaker, W.P. & Schellekens, P.Th.A. (1973). The specificity and activation mechanism of cell-mediated lympholysis (CML) in man. *Transplantation Proceedings*, **5**, 1675-78.

Engelhard, V.H. & Benjamin, C. (1983). Xenogeneic cytotoxic T-cell clones recognize alloantigeneic determinants on HLA-A2. *Immunogenetics*, **18**, 461-73.

Eugui, E.M. & Emery, D.L. (1981). Genetically restricted cell-mediated cytotoxicity in cattle immune to *Theileria parva*. *Nature*, **290**, 251-54.

Fathman, C.G. & Fitch, F.W. eds. (1982). *Isolation, characterization and utilization of T-lymphocyte clones*. New York: Academic Press.

Fathman, C.G. & Frelinger, J.G. (1983). T-lymphocyte clones. *Annual Review of Immunology*, **1**, 633-55.

Fischer-Lindahl, K. & Wilson, D.B. (1977). Histocompatibility antigen-activated cytotoxic T lymphocytes. 2. Estimates of the frequency and specificity of precursors. *Journal of Experimental Medicine*, **145**, 508-22.

Fitch, F.W. (1981). T-lymphocyte clones having defined immunological functions. *Transplantation*, **32**, 171-76.

Gaston, J.S.H., Rickinson, A.B. & Epstein, M.A. (1983a). Cross-reactivity of self-HLA-restricted Epstein-Barr virus-specific cytotoxic T lymphocytes for allo-HLA determinants. *Journal of Experimental Medicine*, **158**, 1804-21.

Gaston, J.S.H., Rickinson, A.B. & Epstein, M.A. (1983b). Epstein-Barr virus-specific cytotoxic T lymphocytes as probes of HLA polymorphism. *Journal of Experimental Medicine*, **158**, 280-93.

Gibson, T. & Medawar, P.B. (1943). The fate of skin homografts in man. *Journal of Anatomy*, **77**, 299-310.

Goulmy, E.A., Termitjelen, A., Bradley, B.A. & van Rood, J.J. (1977). Antigen killing by T cells of women is restricted by HLA. *Nature*, **266**, 544-45.

Goulmy, E.A., von Leeuwen, A., Blokland, E., van Rood, J.J. & Biddison, W.E. (1982). Major histocompatibility complex restricted H-Y-specific antibodies and cytotoxic T lymphocytes may recognize different self-determinants. *Journal of Experimental Medicine*, **155**, 1567-72.

Govaerts, A. (1960). Cellular antibodies in kidney homotransplantation. *Journal of Immunology*, **85**, 516-22.

Guimezanes, A., Davignon, J.L. & Schmitt-Verhulst, A.M. (1982). Multiple cytolytic T-cell clones with distinct cross-reactivity patterns coexist in anti-self and hapten cell lines. *Immunogenetics*, **16**, 37-46.

Haldane, J.B.S. (1933). The genetics of cancer. *Nature*, **132**, 265-67.

Hayry, P. & Defendi, V. (1970). Mixed lymphocyte cultures produce effector cells: model *in vitro* for allograft rejection. *Science*, **168**, 133-35.

Hodes, R.J. & Svedmyr, E.A.J. (1970). Specific cytotoxicity of H-2-

incompatible mouse lymphocytes following mixed culture *in vitro*. *Transplantation*, **9**, 470-77.

Horai, S., van der Poel, J.J. & Goulmy, E. (1982). Differential recognition of the serologically defined HLA-A2 antigen by allogeneic cytotoxic T cells. 1. Population studies. *Immunogenetics*, **16**, 135-42.

Janeway, C.A., Wigzell, H. & Binz, H. (1976). Two different VH gene products make up the T-cell receptors. *Scandinavian Journal of Immunology*, **5**, 993-1001.

Klein, J. (1975). *Biology of the mouse histocompatibility-2 complex*. New York: Springer Verlag.

Klein, J. (1977). Evolution and function of the major histocompatibility system: facts and speculations. In *The major histocompatibility system in man and animals* (ed. D.Gotze), pp. 339-78. New York: Springer-Verlag.

Klein, J. (1978). Genetics of cell-mediated lymphocytotoxicity in the mouse. *Springer Seminars in Immunopathology*, **1**, 31-49.

Klein, J., Juretic, A., Baxevanis C.N. & Nagy, Z.A. (1981). The traditional and a new version of the mouse H-2 complex. *Nature*, **291**, 455-60.

Krangel, M.S., Biddison, W.E. & Strominger, J.L. (1983). Comparative structural analysis of HLA-A2 antigens distinguishable by cytotoxic T lymphocytes. 2. Variant DK1: evidence for a discrete CTL recognition region. *Journal of Immunology*, **130**, 1856-62.

Kreth, H.W., ter Meulen, V. & Eckert, G. (1979). Demonstration of HLA-restricted killer cells in patients with acute measles. *Medical Microbiology & Immunology*, **165**, 203-14.

Lightbody, J., Bernoco, D., Miggiano, V.C. & Cepellini, R. (1971). Cell-mediated lymphocytes in man after sensitization of effector lymphocytes through mixed leucocyte cultures. *Giornale de Bacteriologia Virologia Immunologia Annales*, **64**, 243-54.

Little, C.C. (1914). A possible Mendelian explanation for a type of inheritance apparently nonMendelian in nature. *Science*, **40**, 904-6.

Little, C.C. & Tyzzer, E.E. (1916). Further experimental studies on the inheritance of susceptibility to a transplantable tumour carcinoma (JWA) of the Japanese waltzing mouse. *Journal of Medical Research*, **33**, 393-427.

McMichael, A.J., Ting, A., Zweerink, M.J. & Askonas, B.A. (1977). HLA restriction of cell-mediated lysis of influenza virus-infected human cells. *Nature*, **270**, 524-26.

Medawar, P.B. (1944). The behaviour and fate of skin autografts and skin homografts in rabbits. *Journal of Anatomy*, **78**, 176-99.

Mitchison, N.A. (1954). Passive transfer of transplantation immunity. *Proceedings of the Royal Society of London*, **142**, 72-87.

Morgan, D.A., Ruscetti, F.W. & Gallo, R.C. (1976). Selective *in vitro* growth of T lymphocytes from normal human bonemarrows. *Science*, **193**, 1007-8.

Moss, D.J., Wallace, L.E., Rickinson, A.B. & Epstein, M.A. (1981). Cytotoxic T-cell recognition of Epstein-Barr virus-infected B cells. 1. Specificity and HLA restriction of effector cells reactivated *in vitro*. *European Journal of Immunology*, **11**, 686-93.

Nabholz, M., Vives, J., Young, H.M., Mes, T., Miggiano, V., Rijnbeek, A. & Schreffler, D.C. (1974). Cell-mediated cell lysis *in vitro*: genetic control of killer cell production and target specificities in the mouse. *European Journal of Immunology*, **4**, 378-87.

Nakamura, I. & Cudkowicz, G. (1982). Fine specificity of auto- and alloreactive cytotoxic T lymphocytes: heteroclitic cross-reactions between mutant and original H-2 antigens. *Current Topics in Microbiology & Immunology*, **99**, 51-80.

Pfeffer, P.F. & Thorsby, E. (1982). HLA-restricted cytotoxicity against male-specific (H-Y) antigen: acute rejection of an HLA-identical sibling kidney. Clonal distribution of the cytotoxic cells. *Transplantation*, **33**, 52-56.

Preston, P.M., Brown, C.G.D. & Spooner, R.L. (1983). Cell-mediated cytotoxicity in *Theileria annulata* infection of cattle with evidence for BoLA restriction. *Clinical & Experimental Immunology*, **53**, 88-100.

Reinherz, E.L., Meuer, S.C. & Schlossman, S.F. (1983). The delineation of antigen receptors on human T lymphocytes. *Immunology Today*, **4**, 5-8.

Rosenau, W. & Moon, M.D. (1961). Lysis of homologous cells by sensitized lymphocytes in tissue culture. *Journal of the National Cancer Institute*, **27**, 471-77.

Ryser, J.E. & Robson-MacDonald, H. (1979). Limiting dilution analysis of alloantigen-reactive T lymphocytes. 1. Comparison of precursor frequencies for proliferative and cytolytic responses. *Journal of Immunology*, **122**, 1691-96.

Shearer, G.M., Rehn, T.G. & Garbarino, C.A. (1975). Cell-mediated lympholysis of trinitrophenyl-modified autologous lymphocytes: effector cell specificity to modified cell-surface components controlled by the H-2K and H-2D serological regions of the murine major histocompatibility complex. *Journal of Experimental Medicine*, **141**, 1348-64.

Shortman, K., Wilson, A. & Scollay, R. (1984). Loss of specificity in cytolytic T-lymphocyte clones obtained by limit dilution culture of Lyt 2 T cells. *Journal of Immunology*, **132**, 584-93.

Siliciano, R.F., Brookmeyer, R. & Shin, H.S. (1983). The diversity of T-cell receptors specific for self-MHC gene products. *Journal of Immunology*, **130**, 1512-20.

Simon, M.M., Weltzein, H.U., Buhring, H.J. & Eichmann, K. (1984). Aged murine killer T-cell clones acquire specific cytotoxicity for P815 mastocytoma cells. *Nature*, **308**, 367-70.

Smith, J.B. (1983). Frequency in human peripheral blood of T cells which respond to self, modified self and alloantigens. *Immunology*, **50**, 181-87.

Snell, G.D. (1981). Studies in histocompatibility. *Science*, **213**, 172-78.

Spits, H., Breuning, M.H., Ivanyi, P., Russo, C. & de Vries, J.E. (1982). *In vitro*-isolated human cytotoxic T-lymphocyte clones detect variation in serologically defined HLA antigens. *Immunogenetics*, **16**, 503-12.

Spooner, R.L., Oliver, R.A., Sales, D.I., McCoubrey, C.M., Millar, P., Morgan, A.G., Amorena, B., Bailey, E., Bernoco, D., Brandon, M., Bull, R.W., Caldwell, J., Cwik, S., van Dam, R.H., Dodd, J., Gahne, B., Grosclaude, F., Hall, J.G., Hines, H., Leveziel, H., Newman, M.J., Stear, M.J., Stone, W.H. & Vaiman, M. (1979). Analysis of alloantisera against bovine lymphocytes: joint report of the first international bovine lymphocyte antigen (BoLA) workshop. *Animal Blood Groups & Biochemical Genetics*, **10**, 63-68.

Sredni, B. & Schwartz, R.H. (1980). Alloreactivity of an antigen-specific T-cell clone. *Nature*, **287**, 855-57.

Stagg, D.A., Dolan, T.T., Leitch, B.L. & Young, A.S. (1981). The initial stages of infection of cattle cells with *Theileria parva* sporozoites *in vitro*. *Parasitology*, **83**, 191-97.

Stear, M.J., Newman, M.J. & Nicholas, F.W. (1982). Two closely linked loci and one apparently independent locus code for bovine lymphocyte antigens. *Tissue Antigens*, **20**, 289-99.

Teale, A.J., Kemp, S.J., Young, F. & Spooner, R.L. (1983). Selection by major histocompatibility type (BoLA) of lymphoid cells derived from a bovine chimaera and transformed by *Theileria* parasites. *Parasite Immunology*, **5**, 329-36.

Teale, A.J., Morrison, W.I., Goddeeris, B.M., Groocock, C.M., Stagg, D.A. & Spooner, R.L. (1985). Bovine alloreactive cytotoxic cells generated *in vitro*: target specificity in relation to BoLA phenotype. *Immunology*, **55**, 355-62.

Teh, H.-S., Marley, E., Phillips, R.A. & Miller, R.G. (1977). Quantitative studies on the precursors of cytotoxic lymphocytes. 1. Characterization of a clonal assay and determination of the size of clone derived from single precursors. *Journal of Immunology*, **118**, 1049-56.

Thistlethwaite, J.R., Auchincloss, M., Pescovitz, M.D. & Sachs, D.M. (1984). Immunologic characterization of MHC-recombinant swine: role of class I and II antigens in *in vitro* immune respones. *Journal of Immunogenetics*, **11**, 9-19.

von Boehmer, H., Hengartner, H., Nabholz, M., Lernhardt, W., Schreier, M.H. & Haas, W. (1979). Fine specificity of a continuously growing killer-cell clone specific for H-Y antigen. *European Journal of Immunology*, **9**, 592-97.

Wagner, H., Gotze, D., Ptschelinzew, W. & Rollinghoff, M. (1975). Induction of cytotoxic T lymphocytes against I region-coded determinants: *in vitro* evidence for a third histocompatibility locus in the mouse. *Journal of Experimental Medicine*, **142**, 1477-87.

Zinkernagel, R.M. & Doherty, P.C. (1974). Restriction of *in vitro* T cell-mediated cytotoxicity in lymphocytic choriomeningitis virus infection within a syngeneic or semi-allogeneic system. *Nature*, **248**, 701-2.

19

Antibody-dependent cellular cytotoxicity and natural killer cell activity in cattle: mechanisms of recovery from infectious diseases

M.P. LAWMAN, D. GAUNTLETT and

F. GALLERY

It has been postulated that while antibody is important in preventing infection with a variety of pathogenic microorganisms, cell-mediated immunity may play a more important role in recovery from infection. It is possible that some of the important mechanisms of recovery require the interaction of both humoral and cellular components of the immune system. The purpose of this chapter is to review both published and ongoing research on the functional role of two cellular immune mechanisms described as having importance in bovine immunity. The first of these is an antibody-dependent mechanism involving the interaction of Fc receptor-bearing effector cells and immunoglobulins, the specificity of which is governed by the antibody. This is termed antibody-dependent cellular cytotoxicity. The second mechanism is independent of antibody and is mediated by natural killer cells. Both antibody-dependent cellular cytoxicity (ADCC) and natural killer (NK) cells have been described in cattle. The polymorphonuclear leucocyte is the predominant mediator of ADCC in cattle; the lymphoid K cell is poorly defined in cattle and less effective in mediating lysis compared to the equivalent cell in humans, mice and rats. The effector cell mediating NK activity has been isolated in many species and characterized morphologically as a large granular lymphocyte. However, while NK-like activity has been clearly demonstrated in cattle, the effector cell has not been fully characterized. A variety of target cells has been demonstrated for both mechanisms in cattle and a number of other mammalian species; these include virus-infected cells, tumour cells, bacteria and parasites.

Introduction

Immunity against infectious diseases is a complex phenomenon, in many ways reflecting the complexity of the immune system. Exposure to an infectious agent may result in a humoral and/or a cell-mediated immune response. The induction of these responses involves interaction between different cell types, each having specific function(s). There is usually a requirement for antigen presentation on specialized accessory cells and for production of specific soluble mediators, i.e. lymphokines or monokines, to produce the appropriate immune response. The immune system is not an anatomically discrete organ but is disseminated throughout the body. It is a conglomeration of different cell types, either dispersed in various nonlymphoid organs or localized in specialized lymphoid tissues. The immune system of ruminants shares many of the characteristics of that of other mammals but also has unique and highly specialized immunological features (Butler 1982).

The subject of immunity to infectious agents has been reviewed previously (Notkins 1974; Allison 1974; Blanden 1974; Rouse & Babiuk 1978; Lawman 1982). The overall objective of studies on immunity to infectious organisms is to obtain reliable information on the mechanism(s) that are successful in controlling infection. It has been accepted that the humoral response (production of antibody) plays an important role in protection against primary infection. Of a more debatable nature, however, is the role of antibody in recovery from such infection, and it has been postulated that cellular and cell-mediated immunity may play a more important role in some instances. It is particularly important to point out that cellular functions may intimately involve the specific antibody response.

This chapter reviews both published work and current research in our laboratory on two cellular immune parameters described as having importance in bovine immunity. Most of the work described here focuses on anti-viral immunity, but where possible we shall refer to work on other etiological agents of bovine diseases. The two mechanisms covered in this review are antibody-dependent cellular cytotoxicity (ADCC), involving Fc receptor-bearing effector cells and sometimes facilitation by complement, and antibody-independent mechanisms mediated by natural killer cells (NK) and monocyte/macrophages. These antibody-independent mechanisms should not be confused with those mediated by cytotoxic T lymphocytes, which are highly specific for antigen plus self major histocompatibility (MHC) antigenic determinants, a phenomenon referred to as MHC restriction.

Antibody-dependent cellular cytotoxicity (ADCC)

Antibody undoubtedly plays an important role in mediating protection against many viruses and bacteria, either during primary infection or against reinfection (see reviews by Perlmann, Perlmann & Wigzell 1972; Pearson 1978). The role of antibody in recovery from virus infections is less certain, since considerations such as the manner or route of viral spread become important. Viruses that spread by the extracellular route (destruction of the host cell and liberation of infectious viruses) could be controlled by a specific humoral response. However, for viruses such as herpes viruses and pox viruses that can avoid the extracellular route by spreading between contiguous cells, control by antibody is likely to be less effective. It has been shown by Rouse, Wardley and Babiuk (1976a) that the bovine herpes virus — infectious bovine rhinotracheitis (IBR) virus — is capable of spreading intracellularly *in vitro* 2 h earlier than dissemination by the extracellular route.

As noted by Rouse and Babiuk (1978), further evidence, albeit circumstantial, can be derived for or against the importance of antibody during recovery from infections such as herpes viruses or picornaviruses by examining hosts in which the capacity to mount antibody responses is impaired. For example, the disease associated with infection with poliomyelitis virus can be much more severe in patients suffering from agammaglobulinaemia (Allison 1974). On the other hand, in the case of certain herpes virus infections, selective impairment of the humoral response does not result in any increase in the severity of disease (Allison 1974; Bloom & Rager-Zisman 1975). A further argument against the role of antibody during herpes virus infection is that recovery does not always correlate with the production of specific antibody.

The important factor in recovery from virus infections which spread intracellularly is the removal of the source of virus, i.e. the virus-infected cell. While antibody on its own can neutralize infectious virus particles, it can also facilitate the destruction of virus-infected cells by fixation of complement. Complement-dependent antibody killing of virus-infected target cells has been recorded, with the activation of complement occurring via either the classical or alternate pathways. An important factor is the kinetics of lysis of the virus-infected target cells. Babiuk, Wardley and Rouse (1975) have shown in cattle that the lysis of herpes virus-infected target cells by antibody plus complement occurs late in the viral replication cycle after the virus has spread via the intracellular route.

A third mechanism involving antibody in the destruction of virus-infected cells is through the interaction of antibody with normal leucocytes. This phenomenon, first described by Moller (1965) and later studied extensively by Perlmann and Holm (1969) and Perlmann, Perlmann and Wigzell (1972), has been termed antibody-dependent cellular cytotoxicity (ADCC). While the exact mechanism of lysis is unknown, it has been well established that the specificity of the system is determined by antibody (primarily IgG) which is bound to membrane-

Table 1. The effector cells reported to be involved in antibody-dependent cellular cytoxicity (ADCC).

Cell type	Species	Reference
Lymphoid K cell (nonadherent, nonphagocytic)		
nonT, nonB cell	mouse/human/rabbit/rat/bovine	1, 2, 3
T cell	mouse/human	1, 5
Myeloid K cell (adherent, nonphagocytic or phagocytic)		
immature granulocyte	mouse	6
polymorphonuclear leucocyte	mouse/human/bovine	7, 8
monocyte-macrophage	mouse/human/bovine	9, 10, 11
Eosinophil	rat	12
Platelet	rat	13

1. Greenberg et al. 1975.
2. Wisloff, Froland & Michaelsen 1974.
3. Grewal & Rouse 1979.
4. Lamon et al. 1977.
5. Saal et al. 1977.
6. Pearson 1978.
7. Gale & Zighelboim 1974.
8. Wardley, Babiuk & Rouse 1976.
9. Kohl et al. 1977.
10. Shore, Melewicz & Gordon 1977.
11. Rouse, Wardley & Babiuk 1976a.
12. Capron et al. 1981.
13. Joseph et al. 1983.

bound antigens, while the effector cells in the system bear receptors for the Fc portion of the immunoglobulin. The interaction between the antibody-coated target cells and the effector cell is important in initiating the cytolytic event.

Effector cell population
The cell types which mediate ADCC are listed in Table 1. The most important feature of these cells is the presence of Fc receptors. Greenberg, Shen and Roitt (1973) and Greenberg and colleagues (1973) indicated that in mice the effector cell in ADCC is a nonT (lacking Thy 1), nonB (lacking surface immunoglobulin) nonphagocytic adherent cell with Fc receptors. Greenberg referred to these cells as myeloid killer cells (K cells). Further studies by this group indicated, again in mice, that ADCC could be mediated by a nonadherent, nonT and nonB lymphocyte. These effector cells, which were termed lymphoid K cells (Greenberg et al. 1975),

lack Thy 1 antigen and surface immunoglobulin, have receptors for Fc and complement and morphologically are intermediate sized lymphocytes with a low nuclear to cytoplasmic ratio. The nucleus is ovoid and eccentrically placed with no prominent nucleoli, and the cytoplasm lacks peroxidase granules. Studies in man have also identified human ADCC-effector cells as nonT cells (Moller & Svehag 1972; Cerottini & Brunner 1974) and nonB cells, because they lack surface immunoglobulin. Passage of lymphocytes through nylon wool does not abrogate the ADCC activity, and patients who are hypogammaglobulinaemic (lacking detectable B cells) can still mediate ADCC (Wisloff & Froland 1973). Finally, normal human lymphocytes depleted of Fc-bearing cells are no longer capable of mediating ADCC (Wisloff, Froland & Michaelsen 1974). Wisloff and colleagues concluded that, as in the mouse, the human effector cell in ADCC is a nonT, nonB cell with lymphocyte morphology. However, there is evidence that in some instances human T cells (Wahlin, Perlmann & Perlmann 1976; Saal et al. 1977) and murine T cells (Lamon et al. 1977) are capable of mediating ADCC. The 'myeloid K cells' which mediate ADCC have been shown to include immature granulocytes (Pearson 1978), polymorphonuclear leucocytes (Gale & Zighelboim 1974) and monocytes and macrophages (Kohl et al. 1977; Shore, Melewicz & Gordon 1977). Evidence for the existence of two populations of effector cells capable of mediating ADCC has also been obtained in rats (Garovoy et al. 1976), rabbits (Gelfand, Resch & Prester 1972; Resch, Gelfand & Prester 1974) and cattle (Rouse, Wardley & Babiuk 1976b; Grewal & Rouse 1979).

Effector mechanism

The exact nature of target-cell lysis is a matter of conjecture, though the involvement of antibody is well established. The major immunoglobulin involved in ADCC is the IgG isotype (Cerottini & Brunner 1974). However, other immunoglobulin classes have occasionally been implicated: IgM (Lamon et al. 1975; Lamon et al. 1977), IgE (Capron et al. 1981) and IgA (Shen & Fanger 1981). The failure to induce ADCC with either Fab_1 or $F(ab)_2$ fragments of antibody (Perlmann, Perlmann & Wigzell 1972; Maclennan 1972; van Boxel et al. 1974) strongly supports the evidence that binding of the Fc region of immunoglobulins with Fc receptors of the effector cells is the primary interaction required. It has also been demonstrated that blocking of Fc receptors with unrelated antigen/antibody complexes (van Boxel et al. 1973; Greenberg, Shen & Roitt 1973; Greenberg et al. 1973; Scornick et al. 1974), anti- B-lymphocyte alloantisera (Mikulski & Billing 1977), Fc-reactive rheumatoid factor (Austin & Daniels 1976), protein A (Austin & Daniels 1976) or heat-aggregated IgG (Maclennan 1972; Jewel & Maclennan 1973) results in impairment or inhibition of the ADCC reaction. It is well established that exposure of effector cells to free specific antibody or to soluble antigen-antibody complexes (in antibody excess) related to the target-cell antigen primes these cells so they can subsequently kill the appropriate target cells (Saksela, Imir & Makela 1975). However, these methods of activating ADCC are not as efficient as the use of antibody-coated target cells, and they require higher levels of antibody.

Analysis of the kinetics of target-cell lysis during ADCC indicates that lysis follows the one-hit model, suggesting that target-cell lysis results from a single interaction with an effector cell (Ziegler & Henny 1975). These authors and others (Perlmann, Perlmann & Wigzell 1972) concluded that following the single hit, the effector cell becomes inactivated and unable to mediate the ADCC reaction against further target cells.

There are further requirements for target-cell lysis. In studies relating to early events in target-effector cell interactions (Scornick 1974), direct contact and binding between effector and target cell was found to be essential. This interaction is not mediated by nonspecific binding by unrelated receptors on the effector cell, but requires specific interaction between antibody bound to the target cell and the Fc receptor found on the effector cell. Throughout the ADCC reaction, the target and effector cells remain in contact. Scornick (1974) demonstrated that target cells undergoing lysis exhibit an increase in sensitivity to osmotic changes, leading to a temperature dependence in the permeability of intracellular constituents. Lysis of the target cell can be prevented if the reaction is carried out at 4°C. Scornick (1974) suggested that other steps in the cytotoxic reaction also depend on temperature.

The initiation of the lytic event has been shown not to involve complement. Sera used in ADCC are usually heat inactivated. Inhibitors of complement activation do not affect the ADCC reaction (Pollack & Nelson 1973). Also, mice deficient in certain complement components (C4 to C6) were shown to be capable of mediating ADCC. Depletion of complement by treating sera with cobra-venom factor also did not inhibit ADCC (van Boxel et al. 1974). On the other hand, complement is capable of enhancing the ADCC reaction.

For lysis to occur, the effector cell needs to be viable and metabolically active. Chemicals or agents capable of causing perturbation of the cell membrane, e.g. inhibition of microtubule formation and/or function, also inhibit ADCC, while inhibition of DNA and/or RNA synthesis does not

Table 2. Target cells lysed in ADCC assays.

Target	Selected reference
Virus-infected target cell	
Moloney sarcoma	Skurzak et al. 1972
Gross leukaemia virus	Delanduzuri, Kedar & Fahey 1974
lymphocytic choriomeningitis virus	Zinkernagel & Oldstone 1976
measles virus	Kreth & ter Meulen 1977
herpes simplex virus I/II	Shore et al. 1976
Epstein-Barr virus	Pearson & Orr 1976
vaccinia virus	Zinkernagel & Oldstone 1976
Tumour cell	
choriocarcinoma	Wunderlich et al. 1975
human leukaemia cells	Durantez & Zighelboim 1976
Wilm's tumour cells	Kumar, Waghe & Taylor 1977
Bacteria	
Escherichia coli	Hagberg, Ahlstedt & Hanson 1982
Salmonella typhimurium	Nencioni et al. 1983
Parasites	
Shistosoma mansoni	Chung, Asch & Bruce 1982
Brugia malayi	Sim, Kwa & Mak 1982
Trypanosoma cruzi	Kipnis et al. 1981
malaria	Brown & Smalley 1980
Transplantation antigens	
HLA antigen (class I)	O'Toole, Tiptaft & Stevens 1982

inhibit ADCC (Strom et al. 1975). The involvement of protein synthesis in ADCC is debatable, since Strom and colleagues (1975) reported that inhibition of protein synthesis inhibits ADCC, whereas Dickmeiss (1974) recorded no inhibitory effect on ADCC. Modulation of cAMP and cGMP also affects levels of ADCC (Garovoy et al. 1976): an increase in cAMP is accompanied by a decrease in the level of ADCC, whereas an increase in cGMP results in an increase in ADCC. Finally, ADCC has been shown to operate against a variety of targets, the most frequently recorded being virus-infected cells or tumour cells (Table 2).

Modulation of ADCC by other immune effector mechanisms

Other effector molecules such as complement and interferon have been implicated in the augmentation of ADCC. As stated earlier, there is evidence that complement is not involved in the initiation or in the terminal stage of ADCC. However, investigations into the effect on ADCC of target cell-bound C3b showed that — at concentrations of anti-target immunoglobulins too low to cause ADCC — C3b cooperated with the bound antibody to produce significant target-cell lysis (Ghebrehiwet, Medicus & Müller-Eberhard 1979). Ghebrehiwet and colleagues reported that the enhancement of ADCC by C3b is dose-dependent and can be increased by the attachment of properdin, which is polyvalent for C3b and may cause cross-linking of target cell-bound C3b. Further evidence for the involvement of complement in ADCC came from studies by Sundsmo and Müller-Eberhard (1979). They were able to detect neoantigenic determinants specific for the membrane-attack complex of complement on the surface of effector cells during ADCC. These neoantigenic determinants were detected at the zone of contact between the effector and target cell. Antiserum to the neoantigen inhibited ADCC by up to 79%.

It has also been reported that IgM antibody can potentiate ADCC mediated by IgG. Anti-bovine erythrocyte IgM, which on its own is not capable of mediating ADCC, was found to enhance IgG-mediated ADCC of bovine erythrocytes at suboptimal concentrations of IgG (Perlmann et al. 1981).

Herberman, Ortaldo and Bonnard (1979) showed that interferons (predominantly α and β types) produced by lymphoblastoid cells exposed to Sendai virus or by leucocytes or fibroblasts and activated with polyinosinic acid:polycytidylic acid (poly I:C) are capable of augmenting ADCC. The α- and β-interferon needed to be incubated with the effector cell for at least 18 to 20 h in order to obtain any augmentation. The ADCC was blocked if the effector cell was preincubated with antibody to interferon produced by lymphoblastoid or fibroblast cells. Merrill (1983) observed that at optimal concentrations of antibody, interferon prohibits K cells from binding to the target cell, resulting in engagement of fewer effectors. However, if 0.1 to 0.01 times the optimal concentration of antibody was used, the ADCC was enhanced. Recently it has been shown that protein A is capable of inducing interferon and that this interferon is capable of augmenting the ADCC phenomenon. This may explain previous conflicting results with protein A, as protein A itself has been found to inhibit ADCC. However, when protein A is added to ADDC assays in the presence of immune complexes, there is an observed increase in ADCC activity against chicken erythrocytes (Sulica et al. 1976). From this observation it was concluded that the increase in levels of ADCC may be due in part to the relation of protein A to the Fc region of the antibody molecule, inducing conformational changes that could lead to the stable attachment of protein A-antibody complexes to the cell surface.

Natural killer (NK) cells
In a recent paper on natural killer cells, Herberman (1982)
stated:

> ...Within the last two years, the pace, scope and even the
> character of research on effectors of natural immunity have
> undergone profound changes. That natural cell-mediated
> immunity plays an important role in protecting the host against
> threats from transformed cells and from the environment has
> attained wide acceptance....

Natural killer (NK) cells have been described as a population of
nonadherent mononuclear cells, morphologically defined as large granular
lymphocytes. A number of early studies showed that NK cells can lyse,
independently of antibody, a wide range of syngeneic, allogeneic and
xenogeneic target cells, from normal fibroblasts to nonlymphoid tumour
cells. In addition to the diverse range of potential targets the effector cell
is also heterogeneous (Djeu 1982). The presence of these cells in
nonimmune donors has led to the hypothesis that they may serve as a
primary immune surveillance mechanism against neoplastic and/or virus-
infected cells (Kiessling & Wigzell 1979; Djeu 1982).

Morphology and characteristics of NK cells
In studies in rats, humans and mice, NK cells have been described
as having a high cytoplasm-to-nucleus ratio, intracytoplasmic
azurophilic granules and generally a kidney shape nucleus (Reynolds et
al. 1982; Grossi & Ferrarini 1982; Tagliabue et al. 1982). NK cells have
been successfully isolated using a discontinuous percoll gradient adjusted
to 285 mosm/kg H_2O (Timonen, Ortaldo & Herberman 1981). Results on
characteristic cell-surface markers are confusing and inconclusive. Most
studies suggest that NK cells are nonadherent, surface Ig-negative and
Fc receptor-positive. Murine NK cells have also been shown to possess
the Lyt 5 antigen, which is present on T cells but not on B lymphocytes
or macrophages (Cantor et al. 1979). A further marker, HNK-1, has
been defined and shown to be present only on NK cells and K cells (Abo,
Cooper & Balch 1982). NK cells in man and other species constitute a
complex and heterogeneous system. Results of studies using monoclonal
antibodies to identify cell-surface markers on human NK cells indicate
that there may be subpopulations of NK cells belonging to different cell
lineages. There are arguments for and against NK cells arising from T-
cell (Habu & Okumura 1982; Koo, Cayre & Mittl 1982) or macrophage
lineages (Sun & Lohmann-Mathes 1982), but there is also evidence that

NK cells may represent a separate and new cell lineage (Ferrarini & Grossi 1982; Ortaldo 1982).

Effector mechanisms

As with effector cells in ADCC reactions, direct cell-to-cell contact between the NK cell and the target cell is required. Perussia and Trinchieri (1981) showed that NK cells are inactivated after direct interaction with target cells and are unable to kill further target cells which are susceptible to NK lysis. These authors also showed that the inactivation of the NK effector cell is completed within 3 to 4 h after exposure to target cells. The observed inactivation is not due to inhibitory factors or suppressor cells. These results do not exclude the possibility that the NK cell may lyse more than one target cell; nevertheless, after one or more lytic events, in the absence of α-interferon, the NK cell is no longer active. This inactivation is reversible, as cytotoxicity may be restored by incubating NK effector cells with α-interferon.

The biochemical mechanism or mechanisms of killing by NK cells are not fully understood. NK cells have been shown to contain several acid hydrolases and the pattern of staining with azo-dye techniques suggests that the acid hydrolases may be associated with cell granules. Electron microscopic studies have also localized acid hydrolases in the granules (Grossi & Ferrarini 1982). Serine-dependent proteases have been implicated in the lytic mechanism. This protease activity is either membrane bound or has to be secreted upon activation of the NK cell. Further studies have been undertaken to elucidate the role of lysosomal enzymes (Roder et al. 1980), cellular hydrolases (Zagury et al. 1981), phospholipases (Hoffman et al. 1981), oxidative metabolism (Goldfarb, Timonen & Herberman 1982) and neutral serine proteases (Goldfarb, Timonen & Herberman 1982). Wright and Bonavida (1981) reported that human peripheral blood lymphocytes, previously stimulated with lectin, release a cytotoxic factor which selectively lyses NK-sensitive targets. This soluble factor is sensitive to proteolytic digestion, is stable at -20°C and has a molecular mass of 10,000 daltons. Other soluble mediators such as lymphotoxins (Yamamoto et al. 1982) and active cellular products of unknown nature (Saksela, Carpen & Virtanen 1982) have also been postulated as being involved in the lytic mechanism, and there is evidence that the generation of oxygen radicals (O_2-, H_2O_2 and OH-) may be important in the lytic mechanism. Hefland, Werkmeister and Roder (1982) have shown that within seconds of NK-target cell interaction there is a burst of oxygen radical generation by the effector cell. However, the authors hypothesize that these oxygen radicals are not directly active in the lytic mechanism but activate other events that are cytolytic.

Distribution of NK cells and influence of age

Studies on human lymphoid tissues have indicated that the blood and spleen contain the majority of NK cells, with the lymph node, bonemarrow and thymus contributing less than 1% (Abo, Cooper & Balch 1982). A similar distribution has been observed in the mouse, with high NK-cell activity found in peripheral blood and spleen and, in addition, in the lung and intestinal mucosa, but low to no activity in lymph nodes, thymus and bonemarrow (Tagliabue et al. 1982). The effect of age has also been studied in mice, with NK activity primarily occurring in young animals between 4 and 8 weeks of age (Tagliabue et al. 1982).

Targets susceptible to NK-cell activity

NK cells have been shown to be capable of killing a variety of target cells. These include tumour cells — leukaemia (Zarling et al. 1979), carcinoma and sarcoma cells (Klein et al. 1980; Vose & Moore 1980) — parasites such as *Trypanosoma cruzi* (Hatcher & Kuhn 1982) and *Trichomonas vaginalis* (Landolfo, Martinotti & Martinetto 1982), fungi such as *Cryptococcus neoformans* (Murphy & McDaniel 1982), and cells infected with viruses such as lymphocytic choriomeningitis virus (Welsh, Zinkernagel & Hallenbeck 1979), cytomegalovirus (Quinnan & Manischewitz 1979), Moloney leukaemia virus (Kiessling, Klein & Wigzell 1975) and herpes simplex virus (Kohl et al. 1981).

Augmentation of NK-cell activity

The best studied agents that augment NK-cell activity are the interferon inducers such as poly I:C, lipopolysaccharide, pyran and viruses (Djeu 1982). Direct evidence for the involvement of interferon as the soluble factor augmenting NK-cell activity was obtained by Djeu, Heinbaugh and Holden (1979). They showed that treatment of murine splenic NK cells with interferon resulted in a rapid increase in NK-cell activity and that this augmentation was lost by addition of antiserum containing antibody to interferon. Significantly less time was required to augment NK activity by incubating NK cells with interferon than the time required for interferon inducers to be effective. Other agents known to induce interferon have been studied in relation to augmenting NK-cell activity. Significant increases in activity were seen with a variety of mitogens (phytohaemagglutinin, concanavalin A and *Corynebacterium parvum*) and *Mycoplasma* (*M. hyorhinis*, *M. hominis* and *M. orale*) (Djeu 1982). Another stimulant of NK activity is the lymphokine interleukin-2 (Kuraibayashi, Gilks & Kern 1981). The mechanism by which purified interleukin-2 potentiates NK activity is unknown, but is thought to be independent of interferon.

Common characteristics of effector cells mediating NK and ADCC

Cells mediating NK or ADCC activity appear to share some of the same characteristics and cell-surface markers. Both populations are nonadherent and lack phagocytic activity, they share the NK-1 and Lyt 5 surface markers and both respond to interferon. Biological assays such as cold-target competition or adsorption to monolayers also suggest that NK cells and the cells which mediate ADCC may be overlapping populations (Ferrarini & Grossi 1982). Further work is required to study the presence or absence of other NK-cell markers on cells mediating ADCC activity to elucidate further the common or divergent origin of these two effector cells.

ADCC and complement-augmented ADCC in bovine immunity

Both ADCC and complement-augmented ADCC have been shown to operate in cattle (Rouse, Wardley & Babiuk 1976a; Rouse et al. 1977). As in humans and other species, several types of cells can mediate ADCC in cattle, namely lymphoid K cells, adherent mononuclear cells and polymorphonuclear cells (Grewal & Rouse 1979; Rouse, Wardley & Babiuk 1976b; Wardley, Rouse & Babiuk 1976). Rouse, Wardley and Babiuk (1976b) undertook studies to determine the nature of the cell types mediating ADCC in cattle and also to evaluate the influence of different target cells on the ADCC mechanism. They used as effector cells bovine peripheral blood leucocytes isolated from Hypaque-Ficoll gradients and bovine mammary gland leucocytes. The target cells were antibody-sensitized chicken red blood cells (CRBC) and herpes virus (IBR)-infected bovine kidney primary cells. The two effector cell populations were capable of lysing the virus-infected target cells in the presence of specific antibody, but the mammary gland cells gave a higher level of lysis against both targets. It was also shown that peripheral blood leucocytes depleted of plastic-adherent cells or phagocytes were capable of lysing only the CRBC target. Anti-macrophage serum inhibited lysis against both CRBC and virus-infected targets, but anti-lymphocyte serum only inhibited lysis of the CRBC target. From these observations, Rouse, Wardley and Babiuk (1976b) hypothesized that there are at least two types of effector cell capable of mediating ADCC against antibody-sensitized target cells: one a nonadherent, nonphagocytic cell type and the other an adherent cell with characteristics of a monocyte-macrophage.

Grewal and Rouse (1979) undertook further characterization of bovine leucocytes involved in ADCC. They used cell-surface markers and physical separation techniques to ascertain the nature of the ADCC effector cells from peripheral blood. The most active cells were glass wool-adherent, Fc receptor-positive cells, presumed to be macrophages.

In the glass wool-adherent population, they found heterogeneity with respect to the presence or absence of complement receptors. The complement receptor-positive, glass wool-adherent cell was also adherent to nylon wool but was not a B lymphocyte, and thus was assumed to be a monocyte. The complement receptor-negative, nylon wool-nonadherent, but Fc receptor-positive lymphoid cell was suggested to be a null lymphocyte and a true lymphoid K cell. As stated previously, the most active cell in ADCC was the adherent monocyte-macrophage, with very little activity associated with what these authors termed a lymphoid K cell. Rouse, Wardley and Babiuk (1976a), in order to ascertain the importance *in vivo* of ADCC as a recovery mechanism from herpes virus infection, studied the kinetics of ADCC and complement-dependent, antibody-mediated lysis in relation to the kinetics of development of viral infection in the target cell. Virus antigen was detected on the cell surface by immunofluorescence as early as 5 h after inoculation, while infectious virus could not be detected intracellularly before 9 h or extracellularly until 12 h. Using an infectious centre assay, virus transmission between contiguous cells was detectable from 9 h after inoculation. Virus-infected target cells were susceptible to lysis from 9 h onwards in an ADCC assay and from 11 h in antibody-complement lysis assays. When both effector cells and antibody were added to target cells soon after virus infection, there was a marked reduction in the level of dissemination of virus, as measured by the reduction in the number of plaques. In fact, at low levels of virus input, no plaques were recovered, indicating that the culture was cured of virus infection.

In further studies, Wardley, Rouse and Babiuk (1976) found that polymorphonuclear leucocytes are the most active effector cell type mediating ADCC in cattle. The high level of activity is reflected in the lower number of effector cells required for cytotoxicity and also in the fact that polymorphonuclear leucocytes produce ADCC at lower concentrations of antibody than macrophages or lymphocytes. The reasons for this high level of activity are not understood, but may be related to the type and/or density of Fc receptors or the potency of the innate killing mechanism present in these cells. The kinetics of ADCC in relation to virus dissemination have also been investigated using polymorphonuclear leucocytes. Cytotoxicity mediated by polymorphonuclear leucocytes is detectable earlier than cytotoxicity with macrophages and before virus dissemination, but again not before antigen expression on cell surfaces (Grewal, Rouse & Babiuk 1977).

Studies have also been undertaken to establish the physiological requirements for bovine polymorphonuclear leucocytes to mediate ADCC

(Wardley, Babiuk & Rouse 1976). These authors found that compounds which inhibit DNA, RNA or protein synthesis (mitomycin C, actinomycin D and puromycin) do not affect the level of ADCC, nor does a chelating agent (EDTA). Pretreatment of effector cells with silica or cytochalasin B suppresses the ADCC activity of polymorphonuclear leucocytes. Drugs that decrease cAMP and elevate cGMP levels (imidazole, acetyl methyl choline) also enhance ADCC, whereas drugs that increase cAMP (epinephrine, cholera toxin, dibutyryl cyclic AMP) all cause a significant decrease in ADCC. The alteration in the levels of ADCC by cyclic nucleotides may be due to their influence on microfilament formation and concentration in membranes. Cytochalasin B also affects microfilament formation, microvilli formation and membrane motility. Silica ingestion may exert its effect by alteration of membrane motility, depletion of cellular secretions or interference with the recognition and/ or lytic phase of ADCC (Wardley, Babiuk & Rouse 1976). A further finding of these studies was that ADCC by polymorphonuclear leucocytes can be enhanced by α-interferon.

Finally, ADCC in cattle can be facilitated by complement (C-ADCC) (Rouse et al. 1977). ADCC was obtained with complement under conditions of lower effector-to-target ratios, lower antibody concentrations or shorter duration of assay than required in the absence of complement. The same concentrations of antibody and complement produced no lysis in the absence of effector cells. Enhancement of ADCC by addition of complement was effective both with polymorphonuclear leucocytes as effector cells and with lymphoid K cells, which are found in peripheral blood mononuclear cells and are poor mediators of ADCC in the absence of complement. Rouse and colleagues (1977) suggested that binding of the effector cells via complement receptors may produce a more stable bond between the target and effector cells, resulting in more efficient lysis. Another finding was that immunoglobulin of the IgM isotype, normally incapable of mediating ADCC in cattle, could mediate ADCC in the presence of complement. The mechanism was not elucidated, but it was suggested that this could be an important effector mechanism *in vivo* during the early stages of viral infections in cattle.

In our laboratory, we are interested in the role of cellular immune mechanisms, both systemically and locally in the lung. Using lung lavage cells (predominantly alveolar macrophages) obtained from normal lungs in a 4 or 18 h ^{51}Cr-release assay, ADCC was demonstrated against target cells infected with both IBR virus and parainfluenza-3 (PI-3) virus. By cell fractionation on a percoll gradient, the ADCC activity was found to be associated with both the mononuclear, adherent, phagocytic cells (alveolar macrophages) and polymorphonuclear leucocytes present in the lung (Figures 1, 2A, 3A). Forman and Babiuk (1982) also observed ADCC mediated by alveolar macrophages but showed that ADCC was reduced dramatically 2 h after infection with IBR virus. They demonstrated that IBR virus infection of alveolar macrophages reduces the expression of both Fc and complement receptors 12 h after infection.

NK-cell activity in bovine immunity

Very little is known about the NK cell or natural cytotoxicity mechanisms in cattle. It is possible that the cytotoxic cells described as

Figure 1. Antibody-dependent cellular cytotoxicity (ADCC) by normal bovine lung lavage cells against bovine embryonic kidney cells infected with IBR virus (○) or PI-3 virus (■). Cytotoxicity on the equivalent uninfected targets is represented by the broken line. A constant number of target cells (10^4 per well), labeled with ^{51}Cr, was used in all assays. Lung lavage cells were obtained from the lungs of 1-year-old calves by flushing the airways with phosphate-buffered saline. The cells were washed three times in Hanks buffered salt solution and resuspended in RPMI 1640 culture medium supplemented with 10% foetal bovine serum. Convalescent sera were used as sources of antibody against IBR and PI-3 viruses. Antiserum to IBR virus was used at a dilution of 1:30, and antiserum to PI-3 virus at a dilution of 1:60.

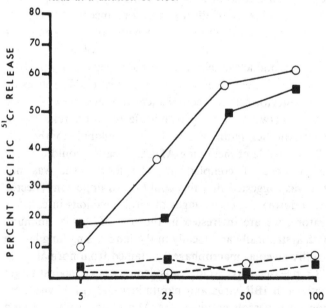

cytotoxic T lymphocytes by Rouse and Babiuk (1977) were in fact NK cells. They stated that cytotoxic T lymphocytes induced by IBR virus were not genetically restricted, although subsequently it has become clear that genetic restriction is an important prerequisite for defining these cells. While their findings on the specificity of the effector cells may be correct, NK-cell activity, a mechanism not fully characterized or described at the time of their study, may have been responsible for the observed

Figure 2. Antibody-dependent cellular cytotoxicity (A) and natural killer-cell activity (B) of different cell populations in normal bovine lung lavage cells against bovine embryonic kidney cells infected with IBR virus. Lung lavage cells were separated into polymorphonuclear and mononuclear leucocyte fractions by centrifugation on Hypaque-Ficoll. LLC: lung lavage cells; PMN: polymorphonuclear leucocytes; MNC: monouclear cells. Further details of the assay are as outlined in the legend to Figure 1.

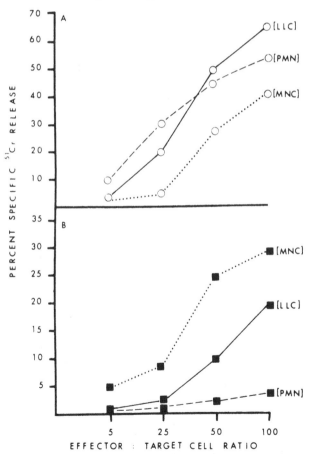

362   *M. P. Lawman and others*

Table 3. Effect of poly I:C treatment on natural killer cell-mediated cytotoxicity of bovine lung lavage cells against bovine embryonic kidney cells infected with IBR or PI-3 viruses. Effector cells pretreated with poly I:C were incubated overnight with 50µg of poly I:C (Sigma Chemicals) before the assay.

| Treatment of effector cell | Effector-to-target ratio | % specific ^{51}Cr release on different targets | | |
		IBR-infected	PI-3-infected	uninfected
No treatment	100	17.3	12.4	4.6
	50	9.6	14.0	1.3
	25	3.2	7.2	0
	5	0	3.1	0
Preincubated with poly I:C	100	29.3	38.4	13.2
	50	18.4	29.6	9.6
	25	9.6	12.4	3.6
	3	4.2	3.9	0

lack of genetic restriction (Fujimiya, Babiuk & Rouse 1978). A further argument against their interpretation has been the recent recording of genetically restricted, cell-mediated cytotoxicity in cattle undergoing immunization with *Theileria parva* (Emery et al. 1981). Natural cell-mediated cytotoxicity by bovine mononuclear cells against virus-infected cells has also recently been recorded (Campos, Rossi & Lawman 1982). These workers showed that bovine target cells infected with PI-3 virus were susceptible to natural cytotoxicity, while target cells infected with IBR virus were not as sensitive. Maximum cytotoxicity was recorded only when the effector cells were present for 20 h, suggesting that a period of activation was needed to stimulate effector cell function. Removal of adherent mononuclear cells on sephadex G-10 columns did not reduce the level of cytotoxicity against cells infected with PI-3 virus. The effector cell was partially characterized as nonphagocytic and nonadherent. These characteristics, together with the fact that target-cell lysis was not genetically restricted, indicate that the effector cell may be similar to NK cells described in other species. We are currently investigating the characteristics of these cells. Preliminary studies in cattle infected with IBR virus have shown that NK-like cells can be induced during experimental infection; their appearance correlates with induction of α-interferon (unpublished).

In studies conducted in our laboratory on immunity in the lung, antibody-independent lysis (natural cytotoxicity) was demonstrated against both IBR- and PI-3-infected target cells (Figure 4). By cell fractionation on percoll gradients, this natural cytotoxicity was only associated with the mononuclear cell population (Figure 2B, 3B). Selective depletion from the effector population of cells adherent to plastic and/ or to sephadex G-10 resulted in a significant reduction in natural cytotoxicity (Figure 5), indicating an adherent cell population as the possible effector cell. This is contrary to the finding that natural cytotoxicity in peripheral blood is mediated by a nonadherent cell (Campos, Rossi & Lawman 1982). An interesting observation was also

Figure 3. Antibody-dependent cellular cytotoxicity (A) and natural killer-cell activity (B) of different cell populations in normal bovine lung lavage cells against bovine embryonic kidney cells infected with PI-3 virus. LLC: lung lavage cells; PMN: polymorphonuclear leucocytes; MNC: mononuclear cells.

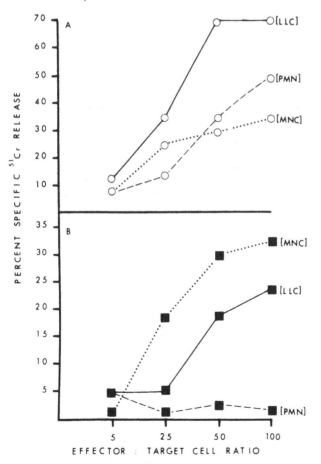

Figure 4. Natural killer cell-mediated cytotoxicity by normal bovine
lung lavage cells against bovine embryonic kidney cells infected with
IBR virus (○) or PI-3 virus (■). Cytotoxicity on the equivalent
uninfected targets is represented by the broken lines.

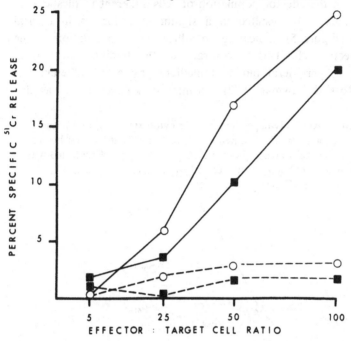

Figure 5. Natural killer cell-mediated cytotoxicity of normal bovine
lung lavage cells, unfractionated (open columns) or depleted of cells
adherent to sephadex G-10 (hatched columns), against bovine
embryonic kidney cells infected with IBR or PI-3 viruses.

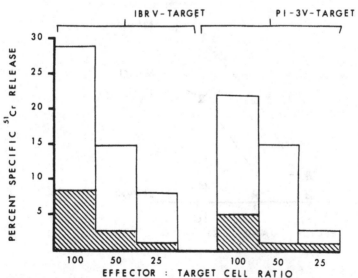

made concerning the possible involvement of interferon: preincubation of effector cells with poly I:C not only increased the level of cytotoxicity against virus-infected cells, but also resulted in significant killing of uninfected target cells (Table 3).

Acknowledgement

This work was supported by the United States Department of Agriculture grant SEA/82-CRSR-23-1067.

References

Abo, T., Cooper, M.D. & Balch, C.M. (1982). Characterization of human NK cells identified by the monoclonal HNK-1 (leu 7) antibody. In *NK cells and other natural effector cells* (ed. R.B. Herberman), pp. 31-38. New York: Academic Press.

Allison, A.C. (1974). Interactions of antibodies, complement components and various cell types in immunity against viruses and pyogenic bacteria. *Transplantation Reviews*, **19**, 3-55.

Austin, R. & Daniels, C.A. (1976). Inhibition by rheumatoid factor, anti-Fc and staphylococcal protein A of antibody-dependent cell-mediated cytolysis against herpes simplex virus-infected cells. *Journal of Immunology*, **117**, 602-7.

Babiuk, L.A., Wardley, R.C. & Rouse, B.T. (1975). Defence mechanisms against bovine herpes virus: relationship of virus-host cell events to susceptibility to antibody-complement lysis. *Infection & Immunity*, **12**, 958-63.

Blanden, R.V. (1974). T-cell response to viral and bacterial infection. *Transplantation Reviews*, **19**, 56-88.

Bloom, B.R. & Rager-Zisman, B. (1975). Cell-mediated immunity in viral infections. In *Viral immunity and immunopathology* (ed. A.L. Notkins), pp. 113-36. New York: Academic Press.

Brown, J. & Smalley, M.E. (1980). Specific antibody-dependent cellular cytotoxicity in human malaria. *Clinical & Experimental Immunology*, **41**, 423-29.

Butler, J.E. ed. (1982). *The ruminant immune system*. New York: Plenum Press.

Campos, M., Rossi, C.R. & Lawman, M.J.P. (1982). Natural cell-mediated cytotoxicity of bovine mononuclear cells against virus-infected cells. *Infection & Immunity*, **36**, 1054-59.

Cantor, H., Kasai, M., She, F.W., Leclerc, J.C. & Glimcher, L. (1979). Immunogenic analysis of natural-killer activity in the mouse. *Immunological Reviews*, **44**, 3-12.

Capron, M., Bazin, H., Joseph, M. & Capron, A. (1981). Evidence for IgE-dependent cytotoxicity by rat eosinophils. *Journal of Immunology*, **126**, 1764-67.

Cerottini, J. & Brunner, T. (1974). Cell-mediated cytotoxicity, allograft rejection and tumour immunity. *Advances in Immunology*, **18**, 67-132.

Chung, P.R., Asch, H.L. & Bruce, J.I. (1982). Antibody-dependent murine macrophage-mediated damage to *Schistosoma mansoni* schistosumula *in vitro*. *Cellular Immunology*, **74**, 243-59.

Delanduzuri, M.O., Kedar, E. & Fahey, J.L. (1974). Antibody-dependent cellular cytotoxicity to a syngeneic gross virus-induced lymphoma. *Journal of the National Cancer Institute*, **52**, 147-52.

Dickmeiss, E. (1974). Comparative study of antibody-dependent and direct lymphocyte-mediated cytotoxicity *in vitro* after alloimmunization in the human. 2. Chemical inhibitors. *Scandinavian Journal of Immunology*, **3**, 817-21.

Djeu, J.Y. (1982). Antibody-dependent cell-mediated cytotoxicity and natural killer-cell phenomenon. *Journal of the American Veterinary Medical Association*, **181**, 1043-48.

Djeu, J.Y., Heinbaugh, J.A. & Holden, H.T. (1979). Augmentation of mouse NK-cell activity by interferon inducers. *Journal of Immunology*, **122**, 175-81.

Durantez A. & Zighelboim, J. (1976). Studies of lymphocyte-dependent antibodies to leukaemia-associated antigens using frozen stored leukaemia target cells. *Transplantation*, **22**, 190-96.

Emery, D.L., Eugui, E.M., Nelson, R.T. & Tenywa, T. (1981). Cell-mediated immunity to *Theileria parva* (East Coast fever) during immunization and lethal infections in cattle. *Immunology*, **43**, 232-336.

Ferrarini, M. & Grossi, C.E. (1982). Could human large granular lymphocytes represent a new cell lineage? In *NK cells and other natural effector cells* (ed. R.B. Herberman), pp. 257-65. New York: Academic Press.

Forman, A.J. & Babiuk, L.A. (1982). Effect of infectious bovine rhinotracheitis virus infection on bovine alveolar macrophage function. *Infection & Immunity*, **35**, 1041-47.

Fujimiya, F., Babiuk, L.A. & Rouse, B.T. (1978). Direct lymphocyte cytotoxicity against herpes virus-infected cells. *Canadian Journal of Microbiology*, **24**, 1076-81.

Gale, R.P. & Zighelboim, J. (1974). Polymorphonuclear leucocytes in antibody-dependent cellular cytotoxicity. *Journal of Immunology*, **114**, 1047-51.

Garovoy, M.R., Strom, T.B., Gribik, M. & Carpenter, C.B. (1976). Antibody-dependent lymphocyte-mediated cytotoxicity (Ab-LMC), definition of the 'K cell' in the rat. *Transplantation*, **22**, 367-73.

Gelfand, E.W., Resch, K. & Prester, M. (1972). Antibody-mediated target-cell lysis by nonimmune cells: characterization of the antibody and effector-cell population. *European Journal of Immunology*, **2**, 419-24.

Ghebrehiwet, B., Medicus, R.G. & Müller-Eberhard, H.J. (1979). Potentiation of antibody-dependent cell-mediated cytotoxicity by target cell-bound C3b. *Journal of Immunology*, **123**, 1285-88.

Goldfarb, R.H., Timonen, T.T. & Herberman, R.B. (1982). The role of neutral serine proteases in the mechanism of tumour cell lysis by human natural killer cells. In *NK cells and other natural effector cells* (ed. R.B. Herberman), pp. 931-37. New York: Academic Press.

Greenberg, A.H., Shen L. & Roitt, I.M. (1973). Characterization of the antibody-dependent cytotoxic cell: a nonphagocytic monocyte? *Clinical & Experimental Immunology*, **15**, 251-59.

Greenberg, A.H., Hudson, L., Shen L. & Roitt, I.M. (1973). Antibody-dependent cell-mediated cytotoxicity due to a 'null' lymphoid cell. *Nature*, **242**, 111-13.

Greenberg, A.H., Shen L., Walker, L., Arnaiz-Villena, A. & Roitt, M. (1975). Characteristics of the effector cells mediating cytotoxicity against antibody-coated target cells. 2. The mouse nonadherent K cell. *European Journal of Immunology*, **5**, 474-80.

Grewal, A.S. & Rouse, B.T. (1979). Characterization of bovine leucocytes involved in antibody-dependent cell cytotoxicity. *International Archives of Allergy & Applied Immunology*, **60**, 169-77.

Grewal, A.S., Rouse, B.T. & Babiuk, L.A. (1977). Mechanism of resistance to herpes virus: comparison of effectiveness of different types in mediating antibody-dependent cell-mediated cytotoxicity. *Infection & Immunity*, **15**, 698-703.

Grossi, C.E. & Ferrarini, M. (1982). Morphology and cytochemistry of human large granular lymphocytes. In *NK cells and other natural effector cells* (ed. R.B. Herberman), pp. 1-9. New York: Academic Press.

Habu, S. & Okumura, K. (1982). Cell lineage of NK cell: evidence for T-cell lineage. In *NK cells and other natural effector cells* (ed. R.B. Herberman), pp. 209-15. New York: Academic Press.

Hagberg, M., Ahlstedt, S. & Hanson, L. (1982). Antibody-dependent cell-mediated cytotoxicity against *Escherichia coli* 0 antigens. *European Journal of Clinical Microbiology*, **1**, 59-65.

Hatcher, F.M. & Kuhn, R.E. (1982). Natural killer-cell activity against extracellular forms of *Trypanosoma cruzi*. In *NK cells and other natural effector cells* (ed. R.B. Herberman), pp. 1091-97. New York: Academic Press.

Hefland, S.L., Werkmeister, J. & Roder, J.C. (1982). The role of free oxygen radicals in the activation of the NK cytolytic pathway. In *NK cells and other natural effector cells* (ed. R.B. Herberman), pp. 1011-20. New York: Academic Press.

Herberman, R.B. ed. (1982). *NK cells and other natural effector cells*. New York: Academic Press.

Herberman, R.B., Ortaldo, J.R. & Bonnard, G.D. (1979). Augmentation by interferon of human natural and antibody-dependent cell-mediated cytotoxicity. *Nature*, **277**, 221-23.

Hoffman, T., Hirata, F., Bougnoux, P., Frazer, B.S., Goldfarb, R.H., Herberman, R.B. & Axelrod, J. (1981). Phospholipid methylation and phospholipase A$_2$ activation in cytotoxicity by human natural killer cells. *Proceedings of the National Academy of Sciences of the USA*, **78**, 3839-43.

Jewel, D.P. & Maclennan, I.C.M. (1973). Circulating immune complexes in inflammatory bowel disease. *Clinical & Experimental Immunology*, **14**, 219-26.

Joseph, M., Floriault, C., Capron, A., Vorng, H. & Viens, P. (1983). A new function for platelets: IgE-dependent killing of schistosomes. *Nature*, **303**, 810-12.

Kiessling, R. & Wigzell, H. (1979). An analysis of the murine NK cell. *Immunological Reviews*, **44**, 165-208.

Kiessling, R., Klein, E. & Wigzell, H. (1975). 'Natural' killer cells in the mouse. 1. Cytotoxic cells with specificity for mouse Moloney leukaemic cells: specificity and distribution according to genotype. *European Journal of Immunology*, **5**, 112-17.

Kipnis, T.L., James, S.C., Sher, A. & David, J.R. (1981). Cell-mediated cytotoxicity to *Trypanosoma cruzi*. 2. Antibody-dependent killing of bloodstream forms by mouse eosinophils and neutrophils. *American Journal of Tropical Medicine & Hygiene*, **30**, 47-53.

Klein, E., Masucci, M.C., Masucci, G. & Vanky, F. (1980). Natural and activated lymphocyte killers which affect tumour cells. In *Natural cell-mediated cytotoxicity against tumours* (ed. R.B. Herberman), pp. 909-20. New York: Academic Press.

Kohl, S., Lawman, M.J.P., Rouse, B.T. & Cahall, D.C. (1981). The effects of herpes simplex virus infection on murine antibody-dependent cellular cytotoxicity and natural-killer cytotoxicity. *Infection & Immunity*, **31**, 704-11.

Kohl, S., Starr, S.E., Oleske, J.M., Shore, S.L., Ashman, R.B. & Nahmias,

A.J. (1977). Human monocyte-macrophage-mediated antibody-dependent cytotoxicity to herpes simplex virus-infected cells. *Journal of Immunology*, **118**, 729-35.

Koo, G.C., Cayre, Y. & Mittl, L.R. (1982). Distinctive characteristics between splenic natural killer cells and prothymocytes. In *NK cells and other natural effector cells* (ed. R.B. Herberman), pp. 225-29. New York: Academic Press.

Kreth, H.W. & ter Meulen, V. (1977). Cell-mediated cytotoxicity against measles virus in SSPE. 1. Enhancement by antibody. *Journal of Immunology*, **118**, 291-95.

Kumar, S., Waghe, M. & Taylor, G. (1977). Tumour-specific antibodies reactive with cell-surface antigens in children with Wilm's tumour. *International Journal of Cancer*, **19**, 351-55.

Kuraibayashi, K., Gilks, S. & Kern, D.E. (1981). Murine NK cell cultures: effects of interleukin-2 and interferon on cell growth and cytotoxic reactivity. *Journal of Immunology*, **126**, 2321-27.

Lamon, E.W., Skurzak, H.M., Andersson, B., Whitten, H.D. & Klein, E. (1975). Antibody-dependent lymphocyte cytotoxicity in the murine sarcoma virus system: activity of IgM and IgG with specificity for MLV-determined antigen(s). *Journal of Immunology*, **114**, 1171-76.

Lamon, E.W., Shaw, M.W., Goodson, S., Lidin, B., Walia, A.S. & Fuson, E.W. (1977). Antibody-dependent cell-mediated cytotoxicity in the Moloney sarcoma virus system: differential activity of IgG and IgM with different subpopulations of lymphocytes. *Journal of Experimental Medicine*, **145**, 302-13.

Landolfo, S., Martinotti, G. & Martinetto, P. (1982). Mechanism of natural macrophage cytotoxicity against protozoa. In *NK cells and other natural effector cells* (ed. R.B. Herberman), pp. 1099-104. New York: Academic Press.

Lawman, M.J.P. (1982). Cell-mediated immunity: induction and expression of T-cell function. *Journal of the American Veterinary Medical Association*, **181**, 1022-29.

Maclennan, I.C.M. (1972). Antibody in the induction and inhibition of lymphocyte cytotoxicity. *Transplantation Reviews*, **13**, 67-90.

Merrill, J.E. (1983). Natural killer (NK) and antibody-dependent cellular cytotoxicity (ADCC) activities can be differentiated by their different sensitivities to interferon and prostaglandin El. *Journal of Clinical Immunology*, **3**, 42-50.

Mikulski, S. & Billing, R. (1977). Inhibition of lymphocyte-dependent antibody effector cell function by human anti-B lymphocyte alloantiserum. *Cellular Immunology*, **28**, 67-74.

Moller, E. (1965). Contact-induced cytotoxicity by lymphoid cells containing foreign isoantigens. *Science*, **147**, 873-79.

Moller, G. & Svehag, S.-E. (1972). Specificity for lymphoid-mediated cytotoxicity induced by *in vitro* antibody-coated target cells. *Cellular Immunology*, **4**, 1-19.

Murphy, J.W. & McDaniel, D.O. (1982). *In vitro* effects of natural killer (NK) cells on *Cryptococcus neoformans*. In *NK cells and other natural effector cells* (ed. R.B. Herberman), pp. 1105-12. New York: Academic Press.

Nencioni, L., Boraschi, D., Berti, B. & Taghabue, A. (1983). Natural and antibody-dependent cell activity against *Salmonella typhimurium* by peripheral and intestinal lymphoidal cells in mice. *Journal of Immunology*, **130**, 903-7.

Notkins, A.L. (1974). Commentary: immune mechanisms by which the spread of viral infections is stopped. *Cellular Immunology*, **11**, 478-83.

Ortaldo, J.R. (1982). Natural killer cells: a separate lineage? In *NK cells and other natural effector cells* (ed. R.B. Herberman), pp. 265-73. New York: Academic Press.

O'Toole, C.M., Tiptaft, R.C. & Stevens, A. (1982). HLA antigen expression on urothelial cells: detection by antibody-dependent cell-mediated cytotoxicity. *International Journal of Cancer*, **29**, 391-95.

Pearson, G.R. (1978). *In vitro* and *in vivo* investigations on antibody-dependent cellular cytotoxicity. *Current Topics in Microbiology & Immunology*, **80**, 65-96.

Pearson, G.R. & Orr, T.W. (1976). Antibody-dependent lymphocyte cytotoxicity against cells expressing Epstein-Barr virus antigen. *Journal of the National Cancer Institute*, **56**, 485-88.

Perlmann, P. & Holm, G. (1969). Cytotoxic effects of lymphoid cells. *Advances in Immunology*, **11**, 117-81.

Perlmann, P., Perlmann, H. & Wigzell, H. (1972). Lymphocyte-mediated cytotoxicity *in vitro* induction and inhibition by humoral antibody and nature of effector cells. *Transplantation Reviews*, **13**, 91-114.

Perlmann, H., Perlmann, P., Moretta, L. & Rönnholm, M. (1981). Regulation of IgG antibody-dependent cellular cytotoxicity *in vitro* by IgM antibodies. *Scandinavian Journal of Immunology*, **14**, 47-60.

Perussia, B. & Trinchieri, G. (1981). Inactivation of natural killer cell-cytotoxic activity after interaction with target cells. *Journal of Immunology*, **126**, 754-58.

Pollack, S. & Nelson, K. (1973). Effects of carrageenan and high serum dilutions on synergistic cytotoxicity to tumour cells. *Journal of Immunology*, **110**, 1440-43.

Quinnan, G.V. & Manischewitz, J.E. (1979). The role of natural killer cells and antibody-dependent cell-mediated cytotoxicity during murine cytomegalovirus infection. *Journal of Experimental Medicine*, **150**, 1549-54.

Resch, K., Gelfand, E.W. & Prester, M. (1974). Antibody-mediated target-cell lysis by nonimmune cells: the use of anti-immunoglobulin to distinguish effector cell populations. *Journal of Immunology*, **112**, 791-803.

Reynolds, C.W., Rees, R., Timonen, T. & Herberman, R.B. (1982). Identification and characterization of the natural killer (NK) cells in rats. In *NK cells and other natural effector cells* (ed. R.B. Herberman), pp. 17-23. New York: Academic Press.

Roder, J.D., Argov, S., Klein, M., Petersson, C., Kiessling, R., Anderson, K. & Hansson, M. (1980). Target-effector cell interaction in the natural killer cell system. 5. Energy requirements, membrane integrity and the possible involvement of lysosomal enzymes. *Immunology*, **40**, 107-16.

Rouse, B.T. & Babiuk, L.A. (1977). The direct anti-viral cytotoxicity of bovine lymphocytes is not restricted by genetic incompatibility of lymphocytes and target cells. *Journal of Immunology*, **188**, 618-24.

Rouse, B.T. & Babiuk, L.A. (1978). Mechanisms of recovery from herpes virus infections: a review. *Canadian Journal of Comparative Medicine*, **42**, 414-27.

Rouse, B.T., Wardley, R.C. & Babiuk, L.A. (1976a). The role of antibody-dependent cytotoxicity in recovery from herpes virus infections. *Cellular Immunology*, **22**, 182-86.

Rouse, B.T., Wardley, R.C. & Babiuk, L.A. (1976b). Antibody-dependent cell-mediated cytotoxicity in cows: comparison of effector cell activity against heterologous erythrocyte- and herpes virus-infected bovine target cells. *Infection & Immunity,* **13,** 1433-48.

Rouse, B.T., Grewal, A.S., Babiuk, L.A. & Fujimiya, Y. (1977). Enhancement of antibody-dependent cell cytotoxicity of herpes-infected cells by complement. *Infection & Immunity,* **18,** 660-65.

Saal, J.G., Rieser, E.P., Hadam, M. & Riethmuller, G. (1977). Lymphocytes with T-cell markers cooperate with IgG antibodies in the lysis of human tumour cells. *Nature,* **265,** 158-59.

Saksela, E., Imir, T. & Makela, O. (1975). Specific cytotoxic human and mouse lymphoid cells induced with antibody or antigen-antibody complex. *Journal of Immunology,* **115,** 1488-92.

Saksela, E., Carpen, O. & Virtanen, I. (1982). Cellular secretion associated with human natural killer-cell activity. In *NK cells and other natural effector cells* (ed. R.B. Herberman), pp. 983-88. New York: Academic Press.

Scornick, J.C. (1974). Antibody-dependent cell-mediated cytotoxicity. 2. Early interactions between effector and target cells. *Journal of Immunology,* **113,** 1519-26.

Scornick, J.C., Cosenza, H., Lee, W., Kohler, H. & Rowley, D.A. (1974). Antibody-dependent cell-mediated cytotoxicity. 1. Differentiation from antibody-independent cytotoxicity by normal IgG. *Journal of Immunology,* **113,** 1510-18.

Shen, L. & Fanger, M.W. (1981). Secretory IgA antibodies synergize with IgG in promoting ADCC by human polymorphonuclear cells, monocytes and lymphocytes. *Cellular Immunology,* **59,** 75-81.

Shore, S.L., Melewicz, F.M. & Gordon, D.S. (1977). The mononuclear cell in human blood which mediates antibody-dependent cellular cytotoxicity to virus-infected target cells. *Journal of Immunology,* **118,** 558-66.

Shore, S.L., Black, C.M., Melewicz, F.M., Wood, P.A. & Nahmias, A.J. (1976). Antibody-dependent cell-mediated cytotoxicity to target cells infected with type 1 and type 2 herpes simplex virus. *Journal of Immunology,* **116,** 194-201.

Sim, B.K., Kwa, B.H. & Mak, J.W. (1982). Immune responses in human *Brugia malayi* infections: serum-dependent cell-mediated destruction of infective larvae *in vitro*. *Transactions of the Royal Society of Tropical Medicine & Hygiene,* **76,** 362-70.

Skurzak, A.M., Klein, E., Yoshida, T.O. & Lamon, E.W. (1972). Synergistic or antagonistic effect of different antibody concentrations on *in vitro* lymphocyte cytotoxicity in the Moloney sarcoma virus system. *Journal of Experimental Medicine,* **135,** 997-1002.

Strom, T.B., Garovoy, M.R., Bear, R.A., Gribik, M. & Carpenter, C.B. (1975). A comparison of the effects of metabolic inhibitors upon direct and antibody-dependent lymphocyte-mediated cytotoxicity. *Cellular Immunology,* **20,** 247-56.

Sulica, A., Lakey, M., Gherman, M., Ghetie, V. & Sjoquist, J. (1976). Arming of lymphoid cells by IgG antibodies treated with protein A from *Staphylococcus aureus*. *Scandinavian Journal of Immunology,* **5,** 1191-97.

Sun, D. & Lohmann-Mathes, M.-L. (1982). Cells with natural-killer activity are eliminated by treatment with monoclonal specific anti-macrophage antibody plus complement. In *NK cells and other natural effector cells* (ed. R.B. Herberman), pp. 243-51. New York: Academic Press.

Sundsmo, J.S. & Müller-Eberhard, H.J. (1979). Neoantigen of the complement membrane-attack complex on cytotoxic human peripheral blood lymphocytes. *Journal of Immunology,* **122,** 2371-78.

Tagliabue, A., Baroschi, D., Alberti, S. & Luini, W. (1982). Large granular lymphocytes as effector cells of natural killer-cell activity in the mouse. In *NK cells and other natural effector cells* (ed. R.B. Herberman), pp. 23-31. New York: Academic Press.

Timonen, T., Ortaldo, J.R. & Herberman, R.B. (1981). Characteristics of human large granular lymphocytes and relationship to natural killer and K cells. *Journal of Experimental Medicine*, **153**, 569-82.

van Boxel, J.A., Paul, W.E., Frank, M.M. & Green, I. (1973). Antibody-dependent lymphoid cell-mediated cytotoxicity: role of lymphocytes bearing a receptor for complement. *Journal of Immunology*, **110**, 1027-36.

van Boxel, J.A., Paul, W.E., Green, I. & Frank, M.M. (1974). Antibody-dependent lymphoid cell-mediated cytotoxicity: role of complement. *Journal of Immunology*, **122**, 398-403.

Vose, B.M. & Moore, M. (1980). Natural cytotoxicity in humans: susceptibility of freshly isolated tumour cells to lysis. *Journal of the National Cancer Institute*, **65**, 257-63.

Wahlin, B., Perlmann, H. & Perlmann, P. (1976). Analysis by a plaque assay of IgG- and IgM-dependent cytolytic lymphocytes in human blood. *Journal of Experimental Medicine*, **144**, 1375-80.

Wardley, R.C., Babiuk, L.A. & Rouse, B.T. (1976). Polymorph-mediated antibody-dependent cytotoxicity: modulation of activity by drugs and immune interferon. *Canadian Journal of Microbiology*, **22**, 1222-28.

Wardley, R.C., Rouse, B.T. & Babiuk, L.A. (1976). Antibody-dependent cytotoxicity mediated by neutrophils: a possible mechanism of anti-viral defence. *Journal of the Reticuloendothelial Society*, **19**, 323-33.

Welsh, R.M., Zinkernagel, R.M. & Hallenbeck, L.A. (1979). Cytotoxic cells induced during LCMV infection of mice. 2. Specifications of the natural killer cells. *Journal of Immunology*, **122**, 475-81.

Wisloff, F. & Froland, S.S. (1973). Antibody-dependent lymphocyte-mediated cytotoxicity in man: no requirements for lymphocytes with membrane-bound immunoglobulin. *Scandinavian Journal of Immunology*, **2**, 151-57.

Wisloff, F., Froland, S.S. & Michaelsen, T.E. (1974). Antibody-dependent cytotoxicity mediated by human Fc receptor-bearing cells lacking markers for B and T lymphocytes. *International Archives of Allergy & Applied Immunology*, **47**, 139-54.

Wright, S.C. & Bonavida, B. (1981). Selective lysis of NK-sensitive target cells by a soluble mediator released from murine spleen cells and human peripheral blood lymphocytes. *Journal of Immunology*, **126**, 1516-21.

Wunderlich, J., Rosenburg, E., Connolly, J. & Park, J. (1975). Characteristics of a cytotoxic human lymphocyte-dependent antibody. *Journal of the National Cancer Institute*, **54**, 537-47.

Yamamoto, R.S., Weitzen, M.L., Miner, K.M., Devlin, J.J. & Granger, G.A. (1982). Role of lymphotoxins in natural cytotoxicity. In *NK cells and other natural effector cells* (ed. R.B. Herberman), pp. 969-76. New York: Academic Press.

Zagury, D., Maziere, J.C., Morgan, M., Fouchard, M. & Hosli, P. (1981). Polyenzymatic activation of cellular hydrolases induced by interferon: its role in cell lysis and other interferon-related phenomena. *Biomedicine*, **34**, 82-88.

Zarling, J.M., Eskra, L., Borden, E.C., Horozewicz, J. & Carter, W.A. (1979). Activation of human natural killer cells cytotoxic for human leukaemia cells by purified interferon. *Journal of Immunology*, **123**, 63-70.

Ziegler, H. & Henny, C. (1975). Antibody-dependent cytolytically active human leucocytes: an analysis of inactivation following *in vitro* interaction with antibody-coated target cells. *Journal of Immunology,* 115, 1500-4.

Zinkernagel, R.M. & Oldstone, M.B.A. (1976). Cells that express viral antigens but lack H-2 determinant are not lysed by other anti-viral immune attack mechanisms. *Proceedings of the National Academy of Sciences of the USA,* 73, 3666-70.

Humoral immune responses

20

Structure and function of immunoglobulins: the symbiotic interplay between parasites and antibodies

R. HAMERS, W. VAN DER LOO and

P. DE BAETSELIER

The immune system is manipulated extensively by pathogenic and parasitic
organisms of all kinds. This manipulation assumes a biological knowledge by
the parasite of the immune pathways of the host. Examples are emphasized in
which the interactive process between parasite and host immune system occur
primarily at the humoral antibody level by such diverse mechanisms as
proteolytic cleavage, immunomodulating antibodies or generation of antibody-
derived immunomodulating peptides. Among these, the existence of
immunomodulating antibodies raised against epitopes shared by the pathogen
and by cells of the immune system probably presents a widespread, but as yet
barely investigated, aspect of the host-parasite relation.

Introduction

In any discussion on the function and structure of
immunoglobulins it is tacitly assumed that these molecules have evolved
to help the vertebrate cope with external pathogens. Modern medicine
based on preventive vaccination has played a major part in emphasizing
this assumption. As it is experimentally possible to tilt the immune
mechanisms towards pathogen destruction, it has been concluded that
these mechanisms are necessarily derived to operate in that direction.
However, many parasites tilt the same mechanisms in a quite different
direction, such as immune avoidance, and, as Prof. A.J.S. Davies (this
volume) put it, this other direction might be the one of biological
relevance. Indeed, evolutionary strategies tend to accommodate the
different interacting parties: obviously both the host and the parasite will
gain from mechanisms aimed at avoiding lethal inflammatory reactions.
Thus, the immune system may have developed antibody secretion for

pathogen accommodation, rather than of pathogen destruction. We will essentially try to show that parasites can manipulate the immune system at different levels and that this leads to an evolutionary *modus vivendi* situation in which a quasi nonpathogen parasite survives in a quasi healthy host. Interactions between a parasite and the host's immune system can occur at all levels — from binding of antibodies to inactivation of antibodies, from binding of immunocompetent cells to immunomodulation, from immune evasion by hiding in cells to immune evasion by antigenic variation. Some of the mechanisms, such as antigenic variation in trypanosomes, are being thoroughly studied, others, such as immune modulation, are as yet barely touched. Many parasites have acquired an intimate knowledge both of their vertebrate and invertebrate hosts, and this knowledge of the host's immune system, acquired through evolution, allows the parasite to thrive.

We would like to illustrate with some examples how different pathogens appear to compete with the immune response of the host, and herein we shall focus on the interplay between antibodies and different parasites. Circulating antibodies are a recent evolutionary acquisition. Although specific and nonspecific immunity is well documented in invertebrates and even in the plant world, it is only in vertebrates that circulating antibodies appear as major mediators in immune mechanisms. These antibodies or immunoglobulins extended the capacity of the existing immune defence mechanisms and confirmed what seems to be a Promethean foresight into the structure of pathogens to be (Ohno et al. 1980).

Antibody molecules manifest two distinct biological activities, namely Fc-mediated functions and Fab-mediated functions, and parasites have

Figure 1. (On facing page.) Major sites of enzymatic and chemical cleavage of the IgG molecules. In the rabbit, all IgG belongs to allotype 11 (methionine 219) or allotype 12 (threonine 219). The methionine is located on the N-terminal side of the hinge and determines whether or not the molecule can be cleaved at that point with CNBr. In the allotype 12, a proportion of the IgG molecules has one of the hinge threonine residues glycosylated, conferring resistance to papain cleavage (Hamers-Casterman et al. 1979). Evidence for enzymatic obtention of the intact V_L region came very early when Baglioni showed excretion of V_L regions in myeloma patients (Baglioni et al. 1967). The acid cleavage of the immunoglobulin light chain was used as a major tool in sequence analysis of those light chains containing an acid-sensitive bond (Frazer, Poulsen & Haber 1972). The shaded domains indicate the portions of the molecule which have been studied following the different methods of cleavage.

to cope with these different molecular effectors. Hence we shall present evidence that parasite-mediated immunomodulation can occur both at the Fc and at the Fab level of the antibody molecule.

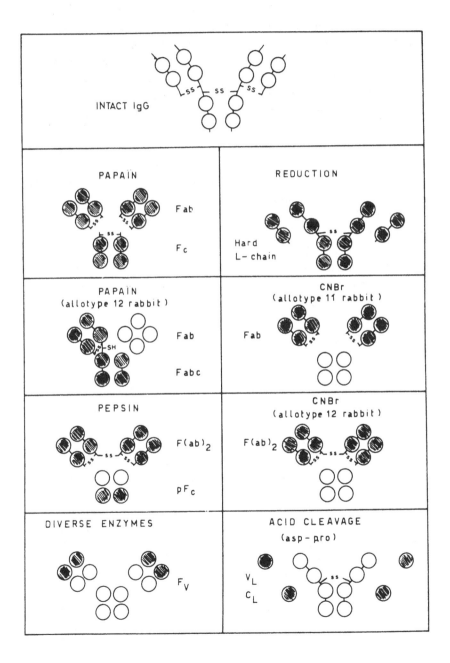

The structure of antibodies

Antibodies can be raised apparently against all possible molecules, material or man-made. As already mentioned, the antibody molecule has to fulfil two functions: it has to be specific for the antigen and it has to insert itself in numerous physiological mechanisms. Evolution has attributed these functions to two different regions of the molecule encoded by different genetic elements. The original work of Porter (1959), carried out by splitting the undenatured immunoglobulin G (IgG) molecule into functional domains, showed that the molecule consists of two antigen-binding fragments (Fab) and a third equally large crystallizable fragment (Fc). Two-thirds of the molecule is involved in antigen recognition, whereas the other functions seem to be confined in the Fc fragment which, being identical in all IgG molecules, can readily crystallize. Porter's other observation, that the IgG molecule is composed of heavy and light polypeptide chains, was subsequently shown to be valid for all immunoglobulins, and the crystallographic studies of the 1970s laid the basis for our present knowledge of the immunoglobulin structure (Figure 1).

The most striking feature of the structure of immunoglobulins is that they are all built up of homologous domains and that each domain is folded in approximately the same manner as a series of parallel and anti-parallel pleated sheets (the immunoglobulin fold) (Figure 2). It is worthwhile to examine how the antigen-combining site fragments and the Fc fragment perform their respective functions. The heavy chain of immunoglobulins is composed of four (IgG, IgA) or five (IgM, IgE) domains, of which one is subject to extensive variation correlated to antigen binding (the variable or v domain). The light chains of all classes are composed of a single v domain and a single constant domain. The antibody-combining site is determined by the folding and interaction of the v domains of the heavy and light chains. At first there seems to be a paradox between the capacity of diverse antibody molecules to bind totally different antigens and the 'immunoglobulin fold' structure which they so conservatively share. However, when sufficient data on amino-acid sequences became available, it became apparent that sequence differences between antibodies of different specificities were grouped in the so-called hypervariable regions of the v domains, and that the role of the 'immunoglobulin fold' was to bring those regions containing variant sequences together into a structurally well defined cavity, the antigen-binding site.

Most of the nonantigen-specific functions are determined by the domains of the Fc region (Table 1). A most important feature of the total molecule is its inherent flexibility. The domains are linked together by a single polypeptide chain so considerable interdomain movement is possible, which allows the antigen-binding sites to be presented to the antigen under various angles. Such a flexibility is clearly linked with antibody functions, such as cross-linking of cells, complement binding and cell arming. Even in crystals, this flexibility can be so extensive that, for instance, the whole Fc region of the molecule cannot be localized by X-ray diffractions. It is precisely this flexibility, due to rather loose polypeptide chains, which allows interdomain proteolytic cleavage (Figure 3). The Fc domain of the Ig molecule is responsible for the major nonimmunological functions, which can be as varied as binding the transport fragment in IgAs to allow secretion into the gut, genital and respiratory tracts, binding the joining piece on building up pentameric IgM, or allowing transplacental transport in rodents, lagomorphs (rabbits) and primates, but not in artiodactyls (pigs, camels, sheep and cattle) which have a multimembrane placenta. The most striking Fc function is probably complement fixation, in which a multiple-component system, 'complement', is activated following the binding of one of its components C_{1q} to immunoglobulins. The binding of C_{1q} initiates a cascade of enzymatic reactions which results in production of chemotactic and cytotoxic factors. The complement components are not related to antibodies, yet they are perfectly integrated in the immune function and

Figure 2. A schematic diagram of how the polypeptide chain of a v-region domain is folded. The folded (A) and partially unfolded (B) polypeptides are shown, with an arrow indicating the direction of folding. All domains have practically the same so-called 'immunoglobulin fold'. Folding and association of light and heavy chains bring the hypervariable residues into close proximity. The shaded areas of the chain show approximately where these hypervariable regions lie.

A B

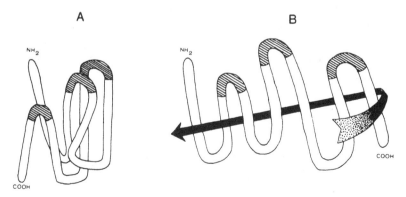

Table 1. Biological effector functions of the Fc region of antibodies.

Function	Antibody class involved	Fc domains involved
Complement fixation (C_{lq} binding)	IgG, IgM	$C\gamma2$, $C\mu$
FcR binding on:		
monocytes, B cells	IgG	$C\gamma2$, $C\gamma3$
macrophages, eosinophils	IgG (IgE)	$C\gamma2$, $C\gamma3$ ($C\epsilon$)
mast cells	IgE	$C\epsilon$
placental syncytiotrophoblast	IgG	$C\gamma2$, $C\gamma3$
Insertion of membrane Ig	IgM	additional C-terminal hydrophobic region
Binding of transport fragment for secretion	IgA	$C\alpha$
Protein A binding	IgG	$C\gamma2$, $C\gamma3$
Binding of joining piece for polymerization	IgM, IgA	$C\mu4$, $C\alpha3$

amplify its effects. A single antibody molecule binding to a red cell will generate 300 activated C_3 molecules which will in turn bind to the red cell or any other cell in the vicinity surrounding the widespread C_3 receptors. Such a system might appear to be dangerously nonspecific, and yet it is not. Activated complement components are extremely short lived and can only interact with structures to which they have time to diffuse within their useful chemical lifetime.

The genetic determination of antibody structure has been amply reviewed (Tonegawa 1983). Without going into details about how immunoglobulin structures are generated genetically, we would like to pay tribute to Todd's (1963) original paper, showing that the γ and μ heavy chains of rabbit immunoglobulin carry identical allotypes (later shown to be the v-region allotype) and that, consequently, the dogma '1 gene, 1 polypeptide chain' had to be abandoned. 'Several genes, one polypeptide chain', proved to be correct for the immunoglobulins. It is precisely this combination game in which several genes contribute to the final immunoglobulin gene, after a few rounds of *in vivo* DNA manipulation, that generates the diverse specificities of antibodies. Such a mechanism occurs within the cell nucleus and is largely protected from outside interference.

When did these mechanisms evolve and where do we expect to discover similar mechanisms which regulate symbiotic or parasitic relations between organisms? The analogy between immunoglobulin structure and the structure of the major histocompatibility (MHC) antigens pleads for the notion that the immunoglobulins share ancestral genes with the MHC. Histocompatibility and self-recognition have appeared in eukaryotes which emerged early in evolution. The gorgona of fan corals show an exquisite self-recognition, and no two distinct colonies will ever graft on each other.

Little or nothing is known of the molecular structures underlying self-recognition in these lower vertebrates, nor of the structures underlying acquired graft rejection as observed in annelids. However, similar cellular immunity of high specificity is present from the onset of vertebrate emergence, even in the *Agnatha* (lampreys and hagfish) which have not acquired humoral immunity. It is gratifying to find that the T-cell receptor of vertebrates responsible for antigen recognition, whilst not an immunoglobulin molecule as originally surmised, is a protein which once

Figure 3. Flexibility of the Ig molecule. The flexibility (indicated by arrows) around the switch and hinge region has now quite convincingly been demonstrated by X-ray crystallography (Huber 1979). Even when multiple interchain disulphides span the hinge region, as is the case for human IgG, this still allows a considerable scissor-type of flexing around the hinge which might give rise to coupled rotation of the Fc along the twofold axis of the molecule (Renneboog Squilbin 1974).

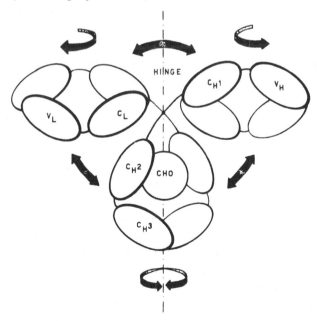

again shares an extensive structural homology with the antibody molecule (Yanagi et al. 1984; Hedrick et al. 1984). These observations suggest that, although humoral antibody production is a recent evolutionary step, the mechanisms for generating diversity, as measured by the related histocompatibility and graft-rejection mechanism, is potentially already available in the most primitive multicellular eukaryotes and is mediated through molecules sharing basic structural features with antibodies.

Immunomodulation at the Fc level

The simplest interaction between parasites and antibody molecules encountered in nature is the binding of antibody molecules on the parasite surface through Fc receptor-like structures. As a result, the pathogen may appear as a host component or cell by being completely covered with a layer of host immunoglobulin. Notorious examples of this kind of immune escape are the bacterium *Staphylococcus aureus*, in which Fc binding occurs via the protein A molecule in the cell wall, and *Schistosoma mansoni*, which expresses on shistosomula but not on cercaria a specific binding site for the Fc portion of IgG, possessing the specificity of Fcγ receptors described on eukaryotic cells (Torpier, Capron & Ouaissi 1979). Interestingly, it was subsequently demonstrated that the IgG molecules bound to the surface Fc receptor of *Shistosoma mansoni* shistosomulae undergo a proteolytic cleavage through excreted parasite proteases. The released peptides were found to decrease macrophage activation significantly and in particular IgE-dependent macrophage cytotoxicity, which is one of the main effector mechanisms involved in anti-shistosome immunity (Auriault et al. 1981). A tripeptide responsible for this inhibition has been characterized (Auriault et al. 1983) as Thr-Lys-Pro and differs in biological activity from the tetrapeptide Tuftsin (Thr-Lys-Pro-Arg), known to increase macrophage functions (Tzehoval et al. 1978). These observations imply that the parasites use proteases which cleave the adsorbed IgG in a specific way, leading to the release of inhibitory peptides. Thus, in this parasite system, Fc-mediated, adsorbed IgG is used by the parasite to modulate host-immune responses from the beginning of the infection. As the Fc receptor apparently plays a major role in macrophage functions, the host cannot afford to modify the Fc portion of the immunoglobulin to any great extent, and hence these three structures (Fc receptors of the host, Fc of the immunoglobulins and Fc receptors of the pathogen) are committed to evolve together. We have evidence that evolutionary pressure affects the Fc portion of the Ig molecules. In human IgG4, two subtypes exist, differing by the presence or deletion of a histidine in position 309 of the Fc portion of the heavy chain. In rabbit IgG, the same site shows genetic polymorphism, allotype e14 being characterized by a threonine, allotype e15 by an alanine in position 309. X-ray diffraction data (Huber 1979) and competition binding studies using the corresponding anti-allotype show that this residue is involved in protein A binding (C. Hamers-Casterman unpublished). Recent surveys of allotypes in wild rabbit populations in Australia, France, the Netherlands, England and Belgium show preferential associations between a light chain C_k allotype b4 and the e15 marker on the Fc part of the molecules. It appears that animals presenting the preferred combination of heavy and light chain genotypes have a significantly higher chance of survival (viability), at least in areas of myxomatosis outbreaks (coefficient of gametic association D = 0.015; $\chi_1^2 = 21$) (van der Loo et al. 1982). It must be realized that these two parts of the IgG molecules have apparently no physical contact from which constraints could ensue. The most plausible explanation is that environmental pressure affects the associations of genetic variants of the light chain (involved in the binding site) and of the Fc part of the molecule, playing an as yet unascribed essential function.

As already illustrated with *Shistosoma mansoni*, a second major way in which parasites interfere with immunoglobulins is by proteolytic cleavage. Mucosal bacteria develop IgA proteases which will cleave the mucosal immunoglobulins (Kilian et al. 1983) and some eukaryotic parasites can readily cleave bound IgG molecules. For instance, *Trypanosoma cruzi* will unarm the bound IgG by cleaving of the Fc fragment and hence presumably render the complement-fixation mechanism inoperative. It is difficult to say whether the hosts have responded to this challenge by protecting the antibody molecules from proteolytic cleavage. In man, two different IgA classes exist, one which can be cleaved by bacterial IgG proteases and the other which cannot. In the rabbit, genetic polymorphism of the IgG hinge region might well be a response to pathogen-linked proteolysis. In allotype d12, the methionine in position 219 in the hinge region is replaced by a threonine which can be glycosylated (Hamers-Casterman et al. 1979). The presence of a bulky carbohydrate residue blocks proteolytic cleavage of the exposed hinge peptides. As glycosylation occurs after synthesis and does not affect all molecules, IgG molecules with two hinge carbohydrates will be totally resistant to hinge proteolysis, whereas IgG with one or no carbohydrate will yield respectively an Fab-Fc fragment or the classical Fab and Fc fragments.

Finally, the most sophisticated form of immunomodulation at the Fc level is probably the selective induction of antibodies of a particular class or subclass. This notion is well documented for two other parasite-like relationships, namely foetal-maternal interactions and host-tumour interactions. In these two situations a selective modulation of the immune response occurs in a direction favourable to either the foetus or to the tumours by the production of so-called facilitating antibodies. Considering as an example the foetal-maternal situation, recent experiments in the murine system (Tartakovsky et al. 1981; 1983) have provided evidence regarding the differential recognition of MHC antigens, expressed on F1 foetal tissue, by female and male parental mice. Thus, striking sex-associated differences were found in the IgG isotype content of antisera raised against foetal bones: female mice produced more IgG1 than male mice, although both produced comparable amounts of IgG2. It was subsequently found that such IgG1 antibodies were directed mainly against the H-2K molecule, rather than the H-2D molecule of the MHC complex. Since H-2K determinants show a greater immunogenic strength than H-2D determinants and induce strong delayed-type hypersensitivity (DTH) reactions, the extra IgG1 molecules produced might protect the foetus from destructive recognition by effector cells of the mother, such as graft-*versus*-host inducing maternal immunocytes (Beer & Billingham 1973). Indeed, as described by Voisin (1980), IgG1 alloantibodies have enhancing properties, manifested not only by decreasing the intensity of immune reactions, but also by modifying their quality. Moreover, the IgG1 subclass of antibodies is devoid of complement-fixing activity and therefore does not mediate complement-dependent cytotoxicity. Induction of syngeneic-specific anti-tumour antibodies which mediate accelerated tumour growth has been described for various host-tumour murine model systems (reviewed in Witz 1977). The antibody responsible for this enhancing effect appears to belong preferentially to the IgG2 subclass. IgG2 anti-tumour antibodies can act facultatively either in a cytotoxic or a stimulatory manner, depending on variations in their concentration, the antigen density of the tumour-cell surface and physiological conditions (Witz 1977). Such pleiotropic activity of IgG2 anti-tumour antibodies might regulate the host-tumour interaction in a way favourable for controlled tumour proliferation.

To what extent are these observations relevant to parasitic infections? In fact, in a number of host/parasite systems, immunoglobulins of different isotypes are found to exert different functions. For instance, in the *S. mansoni*/mouse system, IgG1 antibodies are responsible for lethal effects on schistosomulae mediated by eosinophils plus antibody, whereas IgG2a is responsible for lethal effects mediated by complement killing (Ramalho-Pinto, de Rossi & Smithers 1979). Furthermore, in the *S.*

mansoni/rat system, two antibodies (IgG2a and IgE) and three cell populations (macrophages, eosinophils and platelets) appear as essential components of *in vitro* effector mechanisms (Capron et al. 1983). Conceivably the selective suppression of certain Ig isotypes may have interesting consequences for the survival of certain parasites within a host. A clear-cut example showing that induction of the blocking of anti-parasite antibodies modulates the efficiency of immune effectors is again provided by the *S. mansoni*/rat model (Capron et al. 1983). Indeed, the generation of monoclonal antibodies against shistosomulum surface antigens has led to the observation that a 38,000-dalton glycoprotein (GP 38) can be recognized by a protective IgG2a monoclonal antibody (which mediates IgG2a-dependent, eosinophil-mediated cytotoxicity) and an IgG2c monoclonal antibody devoid of any cytotoxic activity against shistosomula. *In vitro* and *in vivo* experiments have indicated that the IgG2c monoclonal was able to inhibit IgG2a-dependent, eosinophil-mediated cytotoxicity and to decrease strongly the protective role of IgG2a antibody. Obviously, during infections it might be of cardinal importance for the *Shistosoma* parasite to skew the humoral response towards selective production of IgG2c antibodies. Such selective suppression implies a modulation at the T-cell level, since it is well established that the induction of anti-parasite antibodies of various isotypes is T-cell dependent (Mitchell 1979). Extreme cases of T-cell immunodepression brought about by parasites can be found, for instance in lepromatous leprosy caused by *Mycobacterium leprae*, in which the T-cell response is quasi-annihilated (Harboe & Closs 1980). As another example, recent experiments in our laboratory have indicated that the antigen-specific T-cell proliferative response to *Trypanosoma brucei* is reduced during infection by almost 90%. Thus, the fine dissection of T-cell dependent events (including antigen *versus* mitogen effects) during parasitic infections will provide new information on activation of B cells with the capacity to synthesize Igs of various isotypes *in vivo*, the consequences of exaggerated responses on disease manifestations (hyper-γ-globulinaemias) and phenomena such as immunosuppression.

Immunomodulation through the Fab fragment: molecular mimicry

The concept of molecular mimicry is the sharing of a host determinant with a component found on a virus, bacterium or other microorganism. These common determinants are frequently detected by immunological means and may be responsible for the induction or initiation of autoimmunity. A striking example with regard to parasitology is the reported sharing of antigenic determinants between

Trypanosoma cruzi and neurons and cardiac muscle cells (Wood et al. 1982). Immune responses to this shared antigen may account for the pathology seen in South American trypanosomiasis.

An additional situation which has been largely neglected is the possible role of some anti-pathogen antibodies themselves in the modulation of the immune system. This hypothesis was put forward by Lyampert and colleagues (1976) and by ourselves some 8 years ago (De Baetselier et al. 1977). According to the Russian authors, autoimmune processes associated with streptococcal infection would be the end result of a two-stage process (Lyampert et al. 1976): (1) anti-streptococcal antibodies are induced, directed against a cross-reactive antigen between the group-A polysaccharide and an antigen present in epithelial cells of thymus and skin; (2) as a result of the interaction between anti-streptococcal antibodies and thymic epithelial cells, the immunosuppressive function of the thymus might be affected (i.e. maturation of suppressor T cells), leading to an outburst of autoreactive clones. These two events probably constitute the basic stage in the development of a generalized autoimmune reaction.

We proposed a similar hypothesis on account of the following observations. In the early 1970s, immunization of animals with different gram-positive cocci was a major breakthrough in obtaining monoclonal antibodies (Braun & Jaton 1974). Indeed, it was found that animals would respond to pathogenic *Streptococci*, *Pneumococci* or *Meningococci*, or even to nonpathogenic, gram-positive bacteria such as *Micrococcus luteus*, by synthesizing antibodies of very restricted electrophoretic mobility which could be used to determine the amino-acid sequences of variable regions. A major observation was that these restricted antibodies could be directed against various bacterial epitopes expressed either on the same bacteria or on different gram-positive cocci and that in a number of cases no corresponding antigen could be identified on the bacteria used for immunization. It was subsequently shown that the bacteria themselves were causing a restriction in the variety of antibodies which were concomitantly expressed. Other antigens injected simultaneously with the bacteria also gave rise to antibodies of restricted mobility. This was shown for diverse and unrelated antigens such as bovine serum albumin (De Baetselier et al. 1977), dinitro-phenyl (Montgomery & Pincus 1973) and lysozyme (Verloes & Kanarek 1981), all of which gave rise to antibodies of very restricted mobility and antigenicity (i.e. monoclonal) in exceedingly high yields. The effect produced by immunization with these bacteria did not limit itself to a clonal selection and amplification of the antibody repertoire. In rabbits, variable region genes, which are

Table 2. Hyperimmunizations or infections with gram-positive cocci induce aberrant humoral immune responses.

1. Induction of monoclonal or oligoclonal anti-bacterial antibodies.

2. Induction of antibodies of restricted heterogeneity with specificity for totally unrelated antigens (bovine serum albumin, dinitro-phenyl, tobacco mosaic virus, lysozyme).

3. Activation of quasi silent genes coding for rabbit allotypic variants.

4. Reactivation of antibody production against maternal allotypes (rabbit, human).

5. Induction of autoimmune antibodies (heart tissue, thymic epithelial tissue, skin epithelial tissue, rheumatoid factor).

usually silent, became expressed (van der Loo et al. 1977). Furthermore, in rabbits or in human patients acquiring streptococcal or related infections, other serological disorders were prevalent, namely anti-Ig autoantibody rheumatoid factors (Bokisch, Bernstein & Krause 1972). In some cases, other anti-Ig antibodies were produced which upon analysis turned out to be anti-allotype antibodies raised against nonshared genetic markers of maternal antibodies (anti-Gm in humans, anti-b4 allotypes in rabbits)(Speiser et al. 1969).

To explain such a pleiomorphic modulation phenomenon at the serological level (summarized in Table 2), we postulated that all these bacteria had developed antigenic determinants which mimic essential recognition structures of the immune system and that eliciting an immune response to these structures results in profound manipulation of the immune system itself.

Such a hypothesis implies several predictions which could be tested:
1. Anti-bacteria antibodies should modulate immune responses *in vitro* and *in vivo*.
2. Anti-bacteria antibodies should recognize antigens (receptors) on host cells, presumably of lymphoid origin.
3. Such antigens should play a major role in immune phenomena.
4. The antigen(s) should be common to different hosts.
Using the gram-positive bacterium *Micrococcus luteus* as a model system, we have tackled these different predictions experimentally.

Concerning prediction 1, we have been able to show that *in vitro* immunological systems can be modulated to a large extent by anti-bacteria antibodies. This has proven to be the case with polyclonal rabbit anti-*Micrococcus* antibodies in rabbits and in mice and by several monoclonal anti-*Micrococcus* antibodies in rabbits, mice and humans. Different T cell-dependent functions, such as T-cell proliferative responses to lectins, antigens or allogeneic cells, as well as *in vitro* antibody production, were found to be affected in an agonistic or antagonistic manner (De Baetselier, Hamers & Grooten 1981; De Baetselier et al. 1981). Moreover, the inoculation of newborn mice with rabbit anti-*Micrococcus* antibodies increased the basal level of IgM antibodies and depressed the basal level of IgG antibodies. An inhibitory effect of such antibodies on the induction of plaque-forming cells *in vivo* has also been reported (Verloes, Atassi & Kanarek 1979). Thus, there is accumulating experimental evidence that anti-*Micrococcus* antibodies manifest an immunomodulatory effect *in vivo* and *in vitro*.

Initial attempts to test prediction 2, namely reactivity against lymphoid membrane antigens, proved particularly frustrating and initially no clear-cut populations of lymphoid cells could be found to which anti-*Micrococcus* antibodies would bind. However, careful analysis revealed that malignant lymphoid cell lines will react with the anti-*Micrococcus* antibodies when they reach confluent growth (Grooten et al. 1980). This confluence antigen (Cag), reacting with the anti-*Micrococcus* antibodies, is clearly a cell cycle-dependent differentiation antigen and appears only on the cell surface when DNA synthesis is slowed down. Furthermore, this cryptic antigen can be unmasked on lymphoid tumour cells through interactions with macrophages and macrophage-derived factors (Hamers et al. 1983). Apparently the expression of the Cag marker in the cell membrane is a transient phenomenon dictated by physiological conditions. Such a situation is often encountered for functional membrane molecules of the immune system, such as T-cell antigen receptors (Puri, Shinitzky & Lonai 1980) and T-cell growth factor (TCGF) or interleukin-2 (IL-2) receptors (Palacios et al. 1982). In fact, binding of the anti-*Micrococcus* antibodies was subsequently observed on rabbit T lymphocytes, provided the cells were activated with antigen either *in vitro* or *in vivo*.

This last observation led us to probe prediction 3, namely that the membrane moiety recognized by anti-*Micrococcus* antibodies should be physiologically important. Indicative evidence for this prediction was the observation that anti-*Micrococcus* antibodies trigger rabbit T cells into mitogenesis, provided such cells had been activated either *in vivo* by antigen or *in vitro* by T-cell mitogens, such as PHA or concanavalin A (Vaeck et al. 1983; Hamers et al. 1983). A striking parallel thus appeared between the physiological effect of anti-*Micrococcus* antibodies and T-cell growth factors such as IL-2. To test this possibility we have analysed the effect of monoclonal anti-*Micrococcus* antibodies on the IL-2-dependent proliferation of continuous T-cell lines of human or murine origin. The results indicate that certain anti-*Micrococcus* antibodies exert agonistic or antagonistic effects on the IL-2-dependent proliferation of murine and human cytotoxic T cells. Some of the antibodies show significant binding to the membrane of these T-cell lines and, furthermore, interfere with the binding of a monoclonal anti-IL-2 receptor antibody (anti-TAC) (Papineau et al. 1984). Possibly the functional modulation of IL-2 receptors *in vivo* might lead to clonal selection and amplification of T-cell subsets leading to humoral abberations such as the activation of silent clones. Finally, since the above described effects were observed in at least three species, namely mice, rabbits and humans, implying that the anti-*Micrococcus*-recognized Cag marker is common to different hosts, prediction 4 is also fulfilled.

The presence of major antigens (Cag antigens) controlling important immunological signals leads to a final prediction that biologically unrelated parasites will develop cross-reacting decoying antigens and hence elicit similar immunomodulatory antibodies. A number of synergic interactions could be plausibly explained in this way, but remain to be documented. Our recent finding, however, that anti-lymphocyte binding anti-*Micrococcus* antibodies react with the free-living cercaria and the schistosomula of *S. mansoni*, a parasitic blood fluke, supports this idea. It is well known that animals infected with *Schistosoma* parasites display an immune resistance to reinfection and produce antibodies which react with the schistosomula, but not with the adult worms.

It is hence tempting to assume that some of these antibodies are immunoregulatory and that *Schistosoma* parasites have pushed immunomodulation to another level of sophistication in which the parasite, by differentiation to the adult stage, completely escapes immune reaction with the immunoregulatory antibodies and at the same time guarantees a certain level of health for its host by protecting it from reinfection.

References

Auriault, C., Joseph, M., Tartar, A. & Capron, A. (1983). Characterization and synthesis of a macrophage inhibitory peptide from the second constant domain of human immunoglobulin G. *FEBS Letters,* **153**, 11-15.

Auriault, C., Pestel, J., Joseph, M., Dessaint, J.P. & Capron, A. (1981). Interaction between macrophages and *Shistosoma mansoni* shistosomula: role of IgG peptides and aggregates on the modulation of β-glucuronidase release and the cytotoxicity against shistosomula. *Cell Immunology,* **62**, 15-27.

Baglioni, C., Cioli, D., Gorini, G., Ruffili, A. & Alescio-Zonta, L. (1967). Studies on fragments of light chain, chains of human immunoglobulins: genetic and biochemical implications. *Cold Spring Harbor Symposia on Qualitative Biology,* **32**, 147-59.

Beer, A.E. & Billingham R.E. (1973). Maternally acquired runt disease: immune lymphocytes from the maternal blood can traverse the placenta that causes runt disease in the progeny. *Science,* **179,** 240-43.

Bokisch, V.A., Bernstein, D. & Krause, R.M. (1972). Occurrence of 19 S and 7 S anti IgGs during hyperimmunization of rabbits with streptococci. *Journal of Experimental Medicine,* **136,** 799-815.

Braun, D.G. & Jaton, J.C. (1974). Homogeneous antibodies: induction and value as probe for the antibody problem. *Current Topics in Microbiology & Immunology,* **66,** 29-53.

Capron, A., Dessaint, J.P., Capron, M. & Joseph, M. (1983). Effector mechanisms against shistosomes. In *Progress in immunology V* (eds. Y. Yamamura & T. Tada), pp. 1305-16. New York: Academic Press.

De Baetselier, P., Hamers, R. & Grooten, J. (1981). Modulation of different immune functions by monoclonal anti-*Micrococcus* antibodies. In *Monoclonal antibodies and T-cell hybrids* (eds. G. Hammerling, U. Hammerling & J.F. Kearney), pp. 339-48. Amsterdam: Elsevier.

De Baetselier, P., Hamers-Casterman, C., van der Loo, W. & Hamers, R. (1977). Restriction of the anti-bovine serum albumin response in rabbits immunized with *Micrococcus lysodeikticus. Immunology,* **33,** 275-83.

De Baetselier, P., Vaeck, M., Hamers, R. & Grooten, J. (1981). Modulation of immune functions by monoclonal anti-carbohydrate antibodies. In *Mechanisms of lymphocyte activation* (eds. H. Resch & H. Kirchner), pp. 533-36. Amsterdam: Elsevier.

Frazer, K.J., Poulsen, K. & Haber, E. (1972). Specific cleavage between variable and constant domains of rabbit antibody light chains by dilute acid hydrolysis. *Biochemistry,* **11,** 4974-77.

Grooten, J., De Baetselier, P., Vercauteren, E. & Hamers, R. (1980). Anti-*Micrococcus* antibodies recognize an antigenic marker of confluent mouse lymphoid cell lines. *Nature,* **285,** 401-3.

Hamers, R., Grooten, J., Vaeck, M., Segal, S. & De Baetselier, P. (1983). Modulation of the immune response by anti-bacterial antibodies. *Clinical Respiratory Physiology,* **19,** 179-87.

Hamers-Casterman, C., Wittouck, E., van der Loo, W. & Hamers, R. (1979). Phylogeny of the rabbit γ-chain determinants: a d12-like antigenic determinant in *Pronolagus rupestris. Journal of Immunogenetics,* **6,** 373-81.

Harboe, M. & Closs, O. (1980). Immunological aspects of leprosy. In *Immunology 80: progress in immunology IV* (eds. M. Fougereau & J. Dausset), pp. 1231-43. London: Academic Press.

Hedrick, S.M., Nielsen, E.A., Kavaler, J., Cohen, D.I. & Davis, M.M. (1984). Sequence relationships between putative T-cell receptor polypeptides and immunoglobulins. *Nature,* **308,** 153-58.

Huber, R. (1979). Conformational flexibility and functional significance in some protein molecules. *Trends in Biochemical Science,* **5,** 271-75.

Kilian, M., Reinholdt, J., Mortensen, S.B. & Sorensen, C.H. (1983). Perturbation of the mucosal immune defence mechanisms by bacterial IgA proteases. *Bulletin of European Physiopathology & Respiration,* **19,** 99-104.

Lyampert, I.M., Beletskaya, L.V., Borodiyuk, N.A., Gnezditskaya, E.V. & Rassokhina, T.T. (1976). A cross-reactive antigen of thymus and skin epithelial cells common with the polysaccharide of group A streptococci. *Immunology,* **31,** 47-55.

Mitchell, G.F. (1979). Responses to infection with metazoan and protozoan parasites in mice. *Advances in Immunology,* **28,** 451-500.

Montgomery, P.C. & Pincus, J.H. (1973). Molecular restriction of anti-DNP antibodies induced by dinitrophenylated type III pneumococcus. *Journal of Immunology*, **111**, 42-51.

Ohno, S., Matsunaga, T., Epplen, J.T. & Hozumi, T. (1980). Interaction of viruses and lymphocytes in evolution, differentiation and oncogenesis. In *Immunology 80: progress in immunology IV* (eds. M. Fougereau & J. Dausset), pp. 577-98. London: Academic Press.

Palacios, R. (1982). Mechanism of T-cell activation: role and functional relationship of HLA-DR antigens and interleukins. *Immunological Reviews*, **63**, 73-110.

Papineau, M., De Baetselier, P., Cordier, G., Revillard, J.P. & Hamers, R. (1984). Modulation of interleukin-2-dependent human T-cell response by anti-*Micrococcus* antibodies. *Immunobiology*, **167**, 216-17.

Porter, R. (1959). The hydrolysis of rabbit γ-globulin and antibodies with crystalline papain. *Biochemical Journal*, **73**, 119-26.

Puri, J., Shinitzky, M. & Lonai, P. (1980). Concomitant increase in antigen binding and in T-cell membrane lipid viscosity induced by the lymphocyte-activating factor LAF. *Journal of Immunology*, **124**, 1937-42.

Ramalho-Pinto, F.J., de Rossi, R. & Smithers, S.R. (1979). Murine *Shistosomiasis mansoni*: anti-shistosomula antibodies and the IgG subclasses involved in the complement- and eosinophil-mediated killing of shistosomula *in vitro*. *Parasite Immunology*, **1**, 295-308.

Renneboog Squilbin, C. (1974). Further conformational study of the human IgG1 hinge peptide. *Macromolecules*, **6**, 838-44.

Speiser, P., Mickerts, D., Pausch, V. & Mayr, W.R. (1969). The transient nature of anti-Gm human anti-human γ-globulins: their specificity and function. In *Human anti-human γ-globulins* (eds. R. Grubb & G. Samuelson), pp. 151-60. Oxford: Pergamon.

Tartakovsky, B., De Baetselier, P., Feldman, M. & Segal, S. (1981). Sex-associated differences in the immune response against foetal major histocompatibility antigens. *Transplantation*, **32**, 395-97.

Tartakovsky, B., De Baetselier, P., Feldman, M. & Segal, S. (1983). Sex-associated differences in the immune response against foetal major histocompatibility antigens. 2. Association of the female IgG_1 with H-2K encoded alloantigens. *Transplantation*, **36**, 191-97.

Todd, C.W. (1963). Allotypy in rabbit 19 S protein. *Biochemical & Biophysical Research Communications*, **11**, 170-75.

Tonegawa, S. (1983). Somatic generation of antibody diversity. *Nature*, **302**, 575-81.

Torpier, G., Capron, A. & Ouaissi, M.A. (1979). Receptor for IgG(Fc) and human β_2-microglobulin on *S. mansoni* shistosomula. *Nature*, **278**, 447-49.

Tzehoval, E., Segal, S., Stabinsky, Y., Fridkin, M., Spirer, Z. & Feldman, M. (1978). Tuftsin (an Ig-associated tetrapeptide) triggers the immunogenic function of macrophages: implications for activation of programmed cells. *Proceedings of the National Academy of Sciences of the USA*, **75**, 3400-4.

Vaeck, M., Grooten, J., Hamers, R. & De Baetselier, P. (1983). The immunomodulatory effect of anti-*Micrococcus luteus* antibodies: effect on *in vitro* rabbit T-cell functions. *European Journal of Immunology*, **13**, 772-78.

van der Loo, W., De Baetselier, P., Hamers-Casterman, C. & Hamers, R. (1977). Evidence for quasi-silent germline genes coding for phylogenetically ancient determinants of the rabbit A-locus allotypes. *European Journal of Immunology*, **7**, 15-22.

van der Loo, W., Richardson, B., Ross, J., Garson, P., Wallage, M., James, A., Naessens, J., Kerremans, P., Arthur, C., Hamers, R. & Vanderveken,

M. (1982). Fitness of wild rabbits is influenced by the combined genotypes of the IgG heavy and light chain regions. In *Third international theriological congress* (eds. A. Myllmymaki & P. Pulliainen), p. 143. Helsinki: Helsingin Yliopiston Monistuspalvelu.

Verloes, R. & Kanarek, L. (1981). The obtention of high titre antibodies of restricted heterogeneity against lysozyme methyl ester in rabbits vaccinated with *Micrococcus lysodeikticus. Molecular Immunology*, **18**, 781-90.

Verloes, R., Atassi, G. & Kanarek, L. (1979). Influence of anti-*Micrococcus* immunoglobulins on the proliferation of normal immunocompetent cells in inbred mice. *Molecular Immunology*, **16**, 965-74.

Voisin, G.A. (1980). Role of antibody class in the regulatory facilitation reaction. *Immunological Reviews*, **49**, 1-57.

Witz, I.P. (1977). Tumour-bound immunoglobulins: *in situ* expressions of humoral immunity. *Advances in Cancer Research*, **25**, 95-141.

Wood, J.N., Hudson, L., Jessell, T.M. & Yamamoto, M. (1982). A monoclonal antibody defining antigenic determinants on subpopulations of mammalian neurons and *Trypanosoma cruzi* parasites. *Nature*, **296**, 34-38.

Yanagi, Y., Yasunobu, Y., Leggett, K., Clark, S.P., Aleksander, I. & Mak, T.W. (1984). A human T cell-specific cDNA clone encodes a protein having extensive homology to immunoglobulin chains. *Nature*, **308**, 145-53.

21

Biological properties of bovine immunoglobulins and systemic antibody responses

A.J. MUSOKE, F.R. RURANGIRWA and V.M. NANTULYA

The classification of bovine immunoglobulins is similar to that of nonruminant mammals, namely IgM, IgA, IgG_1, IgG_2 and IgE. There are minimal differences between IgG_1 and IgG_2 in complement fixation, ability to mediate passive cutaneous anaphylaxis, phagocytosis of coated erythrocytes by cultured monocytes and precipitation of antigens bearing unique determinants, such as ovalbumin. The functional properties of bovine IgM are identical to those of the IgM of other mammals. The antigens used in elucidating these properties, however, were of noninfective origin. These studies were extended to infectious agents such as African trypanosomes and *Theileria parva*. Cattle infected with *Trypanosoma congolense* and *T. brucei* produced high levels of specific IgM and IgG antibodies that could inhibit infectivity and mediate phagocytosis of homologous trypanosomes. The IgM produced in *T. brucei* infections differed from that produced against noninfectious agents in that it induced adherence of sensitized trypanosomes to, and phagocytosis by, cultured bovine monocytes in the absence of heterologous complement. The ability of trypanosome-infected cattle to respond to superinfection was depressed depending on the time interval between the first and the second inoculations. Cattle vaccinated with *Brucella abortus* vaccine during acute or chronic infection with *T. congolense* or *T. vivax* also showed reduced responses to the vaccine. The nature of these altered responses was reminiscent of antigenic competition. Cattle responded to *T. parva* infections by producing antibodies to the three stages of the parasite, namely sporozoites, macroschizonts and piroplasms. Only the anti-sporozoite antibodies could block entry of sporozoites into target cells.

Introduction

The major classes and subclasses of bovine immunoglobulins include IgG, IgM, IgA and IgE. The IgG class is subdivided into IgG_1 and IgG_2. Confirmation of a third subclass, first reported by Babel and Lang (1976), is still awaited.

Table 1. Physicochemical properties of bovine immunoglobulins.

Property	IgM	IgG₁	IgG₂	IgA
Heavy chain	μ	$\gamma 1$	$\psi 2$	α
Concentration (mg/ml)				
serum	0.6–5.0	6.0–15.0	5.0–13.0	0.05–1.0
colostrum	3.0–12.0	30.0–75.0	2.0–4.0	2.0–15.0
lacrimal gland	0.04	0.34	0.08	2.45
Molecular weight $\times 10^3$				
intact	900	146–163	146–150	385–340
heavy chain	75–78	55–60	55–60	60
light chain	22.5	22.5	22.5	22.5
Electrophoretic mobility	$\beta 2$	$\beta 2$	γ	$\beta 2$
Binding to protein A	–	–	+	–
Carbohydrate (total %)	10–12	2.8–3.1	2.6–3.0	6–10

The physicochemical properties of these immunoglobulins and associated proteins have been extensively reviewed by Butler (1969; 1981; 1983) and a summary is presented in Table 1. Information on the physicochemical properties of bovine IgE is not available, although its existence has been documented (Wells & Eyre 1971; Hammer, Kickhofen & Schmid 1971; Nielsen 1977). These authors showed that IgE is antigenically distinct from IgA, IgG and IgM; it possesses reagenic activity and activates mast cells. These properties are similar to those of human IgE which is well characterized.

The distribution of bovine IgG₁ and IgG₂ in external secretions is interesting as it differs from that in other species. Whereas in humans, rats and mice IgA predominates in external secretions such as colostrum and milk, in cattle and other ruminants IgG₁ forms the larger part of mammary and other secretions, except for saliva and lacrimal fluid which contain mostly IgA. The origin of IgG₁ in these secretions is still uncertain. Several investigators have suggested that it is passed mainly from serum to external secretions (Blakemore & Gamer 1956; Dixon, Weigle & Vazques 1961) with a small amount of local synthesis (Butler et al. 1972).

Anion-exchange resins can be used to purify the IgG subclasses because of the selective binding of these immunoglobulins. Under conditions of low ionic strength (about 1 to 5 mM), IgG₂ does not bind to the gel, whereas IgG₁ elutes under conditions of higher ionic strength (McGuire, Musoke & Kurtti 1979). However, it was found that, because of the total charge characteristics of the two immunoglobulins, IgG₂ was contaminated with with IgG₁ under low ionic strength and *vice versa*.

Table 2. Biological properties of bovine immunoglobulins.

Property	Antigen	Immunoglobulins		
		IgM	IgG$_1$	IgG$_2$
Complement fixation				
guinea pig complement	ovalbumin	n.d.	+	−
	DNP$_{19}$ ovalbumin	n.d.	+	−
	horse erythrocytes	+	n.d.	n.d.
bovine complement	ovalbumin	n.d.	+	+
	DNP$_{19}$ ovalbumin	n.d.	+	+
	horse erythrocytes	+	n.d.	n.d.
Homologous passive cutaneous anaphylaxis (bovine skin)	ovalbumin	n.d.	+	+
Heterologous passive cutaneous anaphylaxis (rat skin)	ovalbumin	n.d.	+	−
Adherence phagocytosis				
neutrophils	erythrocytes	+[a]	−	+
fresh monocytes	erythrocytes	+[a]	−	+
cultured monocytes	erythrocytes	+[a]	+	+

[a]With equine complement.

Staphylococcal protein A was used to achieve pure populations of these immunoglobulins since only IgG$_2$ binds to this protein. A combination of filtering euglobulin through Bio-gel 1.5 (Bio-Rad Laboratories, Richmond, California) and isoelectric focusing (LKB-Produkter AB, Bromma, Sweden) produced pure populations of IgM and also gave relatively pure IgA when used on lacrimal fluid (J.P.O Wamukoya & A.J. Musoke unpublished).

The biological properties of only three immunoglobulins (IgM, IgG$_1$ and IgG$_2$) have been studied in detail mainly because of the difficulty in isolating pure IgA or IgE in a state which allows interaction with antigen (Table 2). One striking property of both bovine IgG$_1$ and IgG$_2$ is that they are indistinguishable in their ability to fix bovine complement and to mediate passive cutaneous anaphylaxis in bovine skin (McGuire, Musoke & Kurtti 1979). They differ, however, in their ability to induce adherence to, and phagocytosis by, neutrophils and fresh monocytes: IgG$_2$ is the antibody class mediating this function (Table 2). Generally the biological properties of bovine immunoglobulins have received little attention and even those properties which are known were generated using immunoglobulins produced against inert antigens. This chapter attempts to compile information on the nature of bovine immune responses to infectious agents, such as the African trypanosome, an extracellular organism, and *Theileria parva*, a purely intracellular parasite.

African trypanosomiasis

African trypanosomiasis is an important vector-borne disease of humans and livestock in tropical Africa. In cattle the disease, commonly known as nagana, is caused by three trypanosome species: *Trypanosoma (Nannomonas) congolense*, *Trypanosoma (Duttonella) vivax*, and *Trypanosoma (Trypanozoon) brucei brucei*. The vector is the tsetse fly (*Glossina* spp.), although mechanical transmission by other blood-sucking Diptera, especially tabanid flies, can occur. The parasites live and multiply in the intravascular and extravascular spaces of the infected bovine host. Infection is characterized by a fluctuating parasitaemic course. Each parasitaemic wave in the same host consists predominantly of trypanosome populations whose protective surface antigens are distinctly different from those of populations in preceding parasitaemic waves. Clinically, signs of the disease are severe wasting and anaemia, and if left untreated death usually occurs after a period ranging from 4 weeks to 2 years. Trypanosomes are ingested by the vector while feeding on an infected animal. They undergo a developmental cycle that culminates in the emergence of the animal-infective endstage known as the metacyclic trypomastigote which is found in the mouth parts or salivary glands of the vector. The metacyclic trypomastigotes are then transmitted to another animal in the tsetse saliva when the fly next feeds. The metacyclic trypomastigotes are antigenically heterogeneous but their composition is characteristic for a given serodeme.

There is a marked increase in the concentration of serum IgM, IgG_1 and IgG_2 in infected animals (Houba & Allison 1966; Freeman et al. 1970; Kobayashi & Tizard 1976; Luckins & Mehlitz 1976; Nielsen et al. 1978), and there is also a rapid turnover of all the immunoglobulins. The halflife of IgG_1, IgG_2 and IgM in infected cattle has been estimated to fall from 17.4, 22.4 and 4.8 days to 1.9, 1.7 and 0.9 days, respectively (Nielsen et al. 1978). It has been suggested, however, that most of the immunoglobulins produced in this disease, are directed against antigens other than those of the infecting trypanosomes (Freeman et al. 1970; Houba, Brown & Allison 1969), such as red blood cells, trinitrophenyl conjugated to bovine serum albumin (TNP-BSA) and thymocytes (Kobayakawa et al. 1979; Sendashonga & Black 1982). The pronounced rise in serum immunoglobulins, which are supposedly polyspecific, thus led to the hypothesis that polyclonal B-cell activation is the stimulus for increased immunoglobulin synthesis in African trypanosomiasis (Hudson et al. 1976).

Since most of these studies were carried out in mice, while the few done in cattle consisted mainly of quantification of total immunoglobulins, we felt that there was a need to re-examine antibody specificity in trypanosome-infected cattle using absorption studies. A group of cattle was infected with *T. b. brucei* and sera collected at weekly intervals. Trypanosome stabilates from these animals were prepared daily. The stabilates were later inoculated into several irradiated rats and the trypanosomes arising in the rats were used to absorb antibodies from sera collected from the infected cattle (Musoke et al. 1981). The results of these studies revealed that the high immunoglobulin concentration in sera collected during the first 2 weeks of infection was reduced to preinfection levels after absorption. The concentration of immunoglobulins in the sera collected 3 weeks postinfection was reduced by 80%. Immunoglobulin levels were not affected when the same sera were absorbed using unrelated trypanosomes. It was, therefore, concluded that most, if not all, the immunoglobulins produced were directed at trypanosomal antigens. The failure to absorb all the immunoglobulins in sera collected 3 weeks postinfection was partly due to the absence of some antigenic variants in the trypanosome population used for absorption, and partly due to the fact that at this stage of infection the animals had responded to the internal trypanosomal antigens which would not be absorbed using live organisms. It was later suggested that, at least in the bovine host, the increase in serum immunoglobulins is due to specific responses to the antigens of the infecting trypanosomes rather than to nonspecific polyclonal stimulation of B cells (Musoke et al. 1981).

Antibodies to sheep erythrocytes (heterophile antibodies) were also detected in sera of infected animals. Interestingly, however, this activity was absorbed completely using trypanosomes (Musoke et al. 1981). These results further suggested that the multiplicity of antibodies to unrelated antigens in sera of infected individuals could reflect an extensive cross-reactivity of anti-trypanosomal antibody with various foreign antigens. Moreover, production of autoantibodies and rheumatoid-like factors in this disease (Houba, Brown & Allison 1969; Klein et al. 1970; Mansfield & Kreier 1972) could result from an autoimmune component of the immune response, rather than polyclonal B-cell activation. Unlike the situation in murine trypanosomiasis (Sendashonga & Black 1982), significant antibody activity against TNP-BSA was not detected in sera from *T. congolense*-infected cattle (Masake, Musoke & Nantulya 1983) providing further evidence that, at least in the bovine host, there is no significant evidence of polyclonal B-cell stimulation.

Biological properties of antibodies produced in infection
Neutralization of infectivity of trypanosomes
In most infected animals there were recurrent peaks of IgM and IgG antibody activity against the variable surface glycoproteins (VSG) of the infecting trypanosomes as well as of those trypanosomes arising during the course of infection (Nantulya et al. 1979; Masake, Musoke &

Nantulya 1983), as detected by radioimmunoassay and indirect immunofluorescence. The ability of the antibodies from the recurrent peaks to neutralize the infectivity of homologous trypanosomes was investigated (Nantulya et al. 1979; Musoke et al. 1981; Masake, Musoke & Nantulya 1983). Neutralization assays were carried out using infection serum and purified VSG-specific IgM and IgG antibodies isolated from the serum by affinity chromatography (Musoke et al. 1981). The results of these assays showed that antibodies from the second antibody peak neutralized infectivity better that those from the first peak. IgM antibodies from both peaks were equally potent and about 20 µg were required to neutralize 1×10^4 trypanosomes. In contrast, 200 µg of IgG_1 antibody from the first peak were required to neutralize an equivalent number of trypanosomes compared with 4 to 8 µg from the second peak.

Three possible explanations were offered for the recurrent peaks of antibody activity against the infecting trypanosomes whose avidity resembled that of antibodies produced in a secondary immune response (Nantulya et al. 1979). Firstly, it was suggested that the variable antigen type of the infecting clones could have been re-expressed in the same host during the same infection. This has also been suggested by Feinnes (1947) who thought that trypanosomes may 'run out' of the ability to synthesize new variable antigens and so have to revert to types which have already been formed. Secondly, there could have been an influx of the original trypanosomes into the bloodsteam from tissue foci where they could have been multiplying. Thirdly, a secondary immune response can sometimes be elicited with an antigen that resembles the primary immunogen. Most of the antibodies made then react more strongly with the first rather than the second immunogen, a phenomenon known as 'original antigenic sin' (Fazekas de St. Groth & Webster 1966).

Our observations suggested that the second and subsequent rises in antibody could be due to the emergence in the same host of trypanosomes whose variable antigen type closely resembled, but was not identical to, that of the infecting trypanosome population (Nantulya et al. 1979). Subsequent investigations by Barbet, Davis and McGuire (1982) provided evidence to support this hypothesis. These workers isolated from a deer mouse a trypanosome clone that was similar with regard to the protective surface epitope of its VSG but was not identical to the infecting clone.

Interaction of trypanosomes, anti-trypanosome antibodies and phagocytic cells

Several reports have suggested that the mononuclear phagocytic system might play an important role in the elimination of trypanosomes from the blood circulation. For example, in mice infected with *T. brucei* there is histological evidence for the expansion of the mononuclear phagocytic system with macrophages from the liver, lymph nodes and bonemarrow displaying the morphological features of increased activity (Murray 1974). Similar observations have been recorded for cattle infected with *T. congolense* or *T. vivax* (Feinnes 1947; Murray 1974). While the role played by anti-trypanosome antibodies and mononuclear phagocytic cells in clearance of parasitaemia has received some attention in laboratory animal models (Takayanagi, Nakatake & Enriquez 1974; Takayanagi & Nakatake 1977; Stevens & Moulton 1978; MacAskill et al. 1980; Cook 1981; Lumsden & Herbert 1967), there is little or no information on their role in cattle. For this reason, we investigated the effect of bovine anti-trypanosome antibodies on the interaction between *T. brucei* organisms and bovine peripheral blood monocytes and the possible role of this interaction in controlling the course of trypanosomal infections (Ngaira et al. 1983).

When freshly isolated bovine peripheral blood monocytes from uninfected animals were incubated with trypanosomes for 5 min in the presence of homologous heat-inactivated ($56°C$ for 30 min) immune sera or VSG-specific IgM or IgG_1 antibodies from infected cattle, they did not bind the parasites. After *in vitro* cultivation for at least 3 h, however, adherence of trypanosomes sensitized with immune sera or purified IgM antibodies was observed. Binding of IgG_1-sensitized trypanosomes to cultured monocytes could be demonstrated until after 7 days of culture. Preinfection sera did not induce this adherence. The adherence of IgM-coated trypanosomes to cultured monocytes was completely inhibited by 10 mM ethylenediamine tetracetic acid, trisodium salt (Na_3H EDTA). Investigation by electron microscopy revealed that trypanosomes sensitized with immune whole serum or VSG-specific IgM or IgG_1 antibodies were rapidly engulfed and digested by the cultured monocytes.

The finding that freshly isolated bovine peripheral blood monocytes from noninfected animals could neither bind nor phagocytose trypanosomes coated with unfractionated heat-inactivated bovine immune serum or purified IgM or IgG_1 antibodies confirmed the absence of receptors for IgM and IgG_1 on freshly isolated monocytes (McGuire, Musoke & Kurtti 1979; Rossi & Kiesel 1977). We also found that IgG_1 antibody-coated equine or sheep erythrocytes could not adhere to freshly isolated bovine peripheral blood monocytes from a normal animal until the cells had been in culture for at least 5 days. Bovine IgG_2, which has been shown to possess a receptor for peripheral blood monocytes (McGuire, Musoke & Kurtti 1979), could not be tested in the present study since VSG-specific antibodies of this IgG subclass were not detected in sera from cattle infected with the *T. brucei* used in these investigations (Musoke et al. 1981).

However, the results we obtained with cultured monocytes differed from those of Rossi and Kiesel (1977), McGuire, Musoke and Kurtti (1979) and Birmingham and Jeska (1980). Whereas these authors found that IgM-sensitized erythrocytes could not adhere to bovine cultured monocytes even in the presence of bovine complement, in our studies IgM-coated trypanosomes were consistently bound and phagocytosed by bovine monocytes without supplementation of the assay system with bovine complement.

Two possible explanations for this discrepancy have been suggested (Ngaira et al. 1983). Firstly, it has been observed that uncoated (procyclic forms) *T. brucei* and *T. congolense* readily adhere to and are phagocytosed by bovine monocytes *in vitro* without prior sensitization with antibodies (H. Hirumi personal communication). This would imply that African trypanosomes possess ligands for bovine mononuclear phagocytic cells in their plasma membranes. Thus, it is possible that disturbances in the trypanosome surface coat caused by the action of VSG-specific IgM antibodies could lead to the exposure of cell membrane determinants which mediate adherence of parasites to the mononuclear phagocytic cells. Secondly, conformational changes in the Fc region occur when IgG antibody molecules combine with antigen (Henney, Stanworth & Gell 1965; Henney & Stanworth 1966; Feinstein & Rowe 1965; Valentine & Green 1967), and it is conceivable that similar changes might occur in other immunoglobulins. It is equally possible that structurally different antigens might give rise to different conformations of the antibody molecules. Different configurations of the antibody molecule could lead to different exposures of binding sites in the Fc region, one of which could be a specific binding site for bovine monocytes.

The finding that adherence of IgM-coated trypanosomes to bovine monocytes was completely inhibited by Na_3H EDTA showed that in our assay system the IgM-induced adherence was Ca^{++}-dependent. Involvement of Ca^{++} in this interaction suggests that adherence of IgM-coated parasites to bovine monocytes was mediated through a specific combining site for monocytes in the Fc region of the IgM molecules rather than through a C3b receptor. Lay and Nussenzweig (1969) have similarly reported a Ca^{++}-dependent adherence of 19S antibody-coated sheep erythrocytes to mouse macrophages, which indicates that the mouse IgM antibody molecules could have bound to the macrophages directly through a specific combining site different from the C3b receptor.

On examining the infected cattle, we found that peripheral blood monocytes from *T. brucei-* or *T. congolense-*infected animals required a much shorter *in vitro* cultivation period before they bound and phagocytosed antibody-coated trypanosomes than monocytes from uninfected animals, indicating that the mononuclear phagocytic system of the bovine host is activated during infection with African trypanosomes. Thus, it was suggested that the bovine mononuclear phagocytic system, acting in concert with VSG-specific antibodies, could be a major mechanism of parasite clearance in the infected host (Ngaira et al. 1983).

Antibodies to metacyclic trypanosomes in sera from chronic infections

When cattle were infected with bloodstream *T. congolense* or *T. brucei* and allowed to maintain a chronic infection, the sera collected 2 to 3 months postinfection and assessed by immunofluorescence and neutralization assays revealed the presence of neutralizing antibodies against the entire repertoire of metacyclic variable antigen types (VATs) of the corresponding serodemes (Nantulya et al. 1984). In addition, the potency of the antibodies from these animals increased during the course of infection. The animals eventually eliminated the trypanosomes from the peripheral blood circulation and demonstrated a serodeme-specific resistance against subsequent challenge by infected tsetse.

Although these animals may have continued to harbour trypanosomes in some privileged tissue sites, the apparently complete restoration of their physical well-being, including the return of the packed red cell volume (PCV) to preinfection levels, following the elimination of trypanosomes from the peripheral blood circulation suggested that spontaneous self-cure had occurred, at least in some animals. Such a self-cure phenomenon could be related to the process of antigenic variation. Cattle infected with *T. brucei* or *T. congolense* produce high levels of neutralizing and phagocytosis-promoting antibodies against the infecting and subsequent VATs that arise during infection (Musoke et al. 1981; Masake, Musoke & Nantulya 1983; Ngaira et al. 1983). If immunity builds up against the entire repertoire of the pathogenic bloodstream VATs expressed in the bovine host by the infecting trypanosomes, spontaneous self-cure might occur. The animals which did not recover from infection might have had less effective immune responses and/or parasite clearance mechanisms, giving rise to prolonged parasitaemia and eventually death.

The occurrence of anti-metacyclic antibodies in cattle infected with bloodstream clones, which has also been reported in rabbits (Barry et al. 1978; Le Ray, Barry & Vickerman 1978), suggests that some of the bloodsteam VATs that arise during the course of a chronic infection possess surface epitopes which are identical or similar to those of metacyclic trypanosomes. Recent studies using monoclonal antibodies to metacyclic forms of *T. brucei* and *T. rhodesiense* have shown that such bloodstream VATs do indeed exist (Esser et al. 1981; Nantulya et al. 1983). The rising titre of neutralizing anti-metacylic antibodies in sera obtained 3 to 4 months postinfection is an indication that the bloodstream VATs responsible for the production of anti-metacylic antibodies reappear or that other closely related VATs emerge during the course of infection. The immunity exhibited by these animals against homologous challenge could, therefore, be directed against the metacyclic and/or the bloodstream VATs.

Under field conditions, *T. brucei* and *T. congolense* are transmitted principally by tsetse, although mechanical transmission may also be possible. The observation that serodeme-specific immunity to tsetse-transmitted challenge readily develops in livestock has significant practical implications. It indicates that under natural conditions of challenge livestock could, over a course of time, acquire resistance against both bloodstream and metacyclic VATs of the local trypanosome serodemes. Several workers have suggested that cattle maintained using trypanocidal drugs in areas of trypanosomiasis challenge acquire immunity to trypanosomiasis, as shown by the progressive lengthening of the interval between drug treatment and reinfection (summarized by Murray, Morrison & Whitelaw 1982).

Altered immune responses

Experimental African trypanosomiasis in rodents is characterized by functional lymphoid disorders resulting in concurrent immunodepression and hypergammaglobulinaemia (Goodwin et al. 1972; Greenwood, Whittle & Molyneux 1973). This apparent anomaly has also been demonstrated in bovine trypanosomiasis (Holmes et al. 1974; Scott et al. 1977; Rurangirwa et al. 1978; 1979; Whitelaw et al. 1979). In an attempt to explain this paradox, three hypotheses have been proposed for the mouse system.

The first presupposes that B cells are polyclonally stimulated, causing excess production of immunoglobulin, including autoantibodies. This leads to clonal exhaustion and immunodepression (Greenwood 1974; Hudson et al. 1976; Assoku, Tizard & Nielsen 1977; Askonas et al. 1979). The implication is that B-cell clones are polyclonally stimulated to produce IgM for whatever antigenic determinants the clones are genetically committed to recognize. However, this does not seem to be the case in bovine trypanosomiasis. Rurangirwa and colleagues (1979; 1983) could not detect antibodies to *Leptospira biflexa* or *Brucella abortus* in sera of animals infected with *T. congolense* and/or *T. vivax*. Following immunization with *L. biflexa* and *B. abortus*, however, trypanosome-infected animals were able to mount a humoral immune response against both antigens. This indicated that the infected animals had cells capable of making antibodies to *L. biflexa* and *B. abortus* but the relevant B cells had not been polyclonally stimulated during infection to produce such antibodies.

These findings have been corroborated by studies by Tabel and colleagues (1981) who could not detect any increase in heterophile antibodies to chicken and sheep red blood cells in infected cattle. Instead, there was a decrease in the level of these antibodies compared to preinfection levels at the decline of the first parasitaemic wave. These authors concluded that there was no polyclonal activation of the B cells in bovine trypanosomiasis. Our additional interpretation of their results is that the decline in the level of heterophile antibody could have been due to absorption by cross-reacting antigens released by disintegrating trypanosomes, as suggested by the absorption studies of Musoke and colleagues (1981).

Suppressor cells, both T cells and macrophages, have been demonstrated *in vitro* in murine trypanosomiasis (Jayawardena & Waksman 1977; Mansfield 1978) and the second hypothesis suggests that they may be responsible for the immunodepression observed in murine trypanosomiasis. Whereas this may be true in murine models, it has not yet been demonstrated in bovine trypanosomiasis. The observation that cattle infected with *T. congolense* or *T. vivax* respond as well as uninfected animals when simultaneously treated with diminazene aceturate and immunized with *L. biflexa* (Rurangirwa et al. 1979) indicates that suppressor cells are not responsible for immunodepression in bovine trypanosomiasis. However, this finding suggests that only live trypanosomes may mediate immunodepression, perhaps through direct action or by an immunodepressive factor.

Recently, a third hypothesis has suggested that hypergammaglobulinaemia is due to the production of specific antibodies against the numerous VATs which arise during trypanosome infection in cattle (Musoke et al. 1981), while immunodepression may be due to antigenic competition between trypanosomal and the heterologous test antigens (Nantulya et al. 1982). This is supported by the findings of Rurangirwa and colleagues (1979) who demonstrated that when cattle infected with *T. congolense* were immunized against *L. biflexa* and *B. abortus* and simultaneously treated with diminazene aceturate, they mounted a normal immune response. This suggests that the trypanocidal treatment eliminated the trypanosomal antigens while the other antigens were introduced. This is reminiscent of sequential antigenic competition in which T cells or macrophages stimulated by trypanosomes of one VAT may release nonspecific inhibitors which depress immune responses to unrelated antigens (Taussig 1975). When the trypanosomes are killed, the nonspecific inhibitors are no longer produced so the animal is capable of mounting a normal immune response against test antigens.

In our studies on the capacity of trypanosome-infected cattle to mount an immune response to a simultaneous or subsequent challenge with other trypanosomes, using various clones of *T. congolense* and *T. brucei*, we demonstrated that antigenic competition was a major contributor to immunodepression in bovine trypanosomiasis (Nantulya et al. 1982). On the basis of these findings we proposed that the immunodepression seen in bovine trypanosomiasis may be a result of antigenic competition (Nantulya et al. 1982).

Theileriosis

East Coast fever (ECF) caused by *Theileria parva* is one of the most important diseases of cattle in eastern Africa in economic terms, causing the death of at least 4 million cattle annually. The disease is characterized by two intracellular stages of the parasite in the vertebrate host: a schizont stage within the lymphoid cells and a piroplasm stage within the erythrocytes. The stage infective to the mammalian host, the sporozoite, is found in the salivary glands of the tick vector *Rhipicephalus appendiculatus*.

Cattle that recover from ECF either spontaneously or by treatment acquire life-long resistance to reinfection with the same *T. parva* strain. Serological tests, including indirect immunofluorescence, complement fixation, indirect haemagglutination and radioimmunoassay, show that sera from recovered animals contain high antibody titres against the schizont, piroplasm and sporozoite stages of the parasite (Burridge 1971; Duffus & Wagner 1974; Wagner, Duffus & Burridge 1974; Musoke et al. 1982).

Biological activity of antibodies to Theileria parva

Until recently the biological activity of anti-theilerial antibodies has received little attention. This may have been a result of an earlier report in which Muhammed, Lauerman and Johnson (1975) failed to transfer resistance against *Theileria parva* passively using serum rich in antibodies against piroplasms. This observation led to the belief that immunity against ECF is solely cell mediated. Later studies by Emery (1981) in which immunity was adoptively transferred between twins using thoracic duct lymphocytes and studies by Eugui and Emery (1981) which demonstrated genetically restricted cytotoxic responses in recovered cattle reinforced this belief.

Recently however, Musoke and colleagues (1982; 1984) presented evidence which suggested a role for antibodies in immunity to ECF. It was observed that immune sera from cattle exposed to ECF experimentally or under field challenge contained neutralizing antibodies against sporozoites of *T. parva*. The authors postulated from these results that cell-mediated responses could be acting with antibodies against the sporozoites. It was suggested that when sporozoites are deposited by ticks into the skin of the immune animal, a proportion are killed by sporozoite-specific antibody while those escaping this antibody-mediated attack proceed to invade target cells and develop into schizonts which may be eliminated by cytotoxic cells.

Antibodies may kill sporozoites through the phagocytic system. Recent findings have shown that IgG_2 antibodies induce phagocytosis by homologous neutrophils and monocytes much more efficiently than IgG_1 antibodies (McGuire, Musoke & Kurtti 1979). Alternatively, IgG_2 antibodies could operate through the complement system. Recently, complement has been demonstrated in the skin of guinea pigs at sites where salivary gland antigens of the tick *Dermacentor andersoni* were deposited (Allen, Khalil & Graham 1979). If these findings hold true for bovine skin at sites where sporozoites of *T. parva* are deposited by *R. appendiculatus*, then a mechanism of killing involving IgG_2 antibodies and complement can readily be envisaged since IgG_2 is also known to fix bovine complement (McGuire, Musoke & Kurtti 1979).

The observation that animals which recover from infection with one stock of *T. parva* are not always immune to challenge with other stocks of the parasite has led to the prevailing belief that there are several strains of *T. parva*. Monoclonal antibodies have detected antigenic differences between schizont forms of the parasite (Irvin et al. 1983), but these differences are not evident among sporozoites (Musoke et al. 1982; 1984; Dobbelaere et al. 1984). These workers have shown that both sera collected from recovered animals and monoclonal antibodies raised against sporozoites can neutralize the infectivity of sporozoites of several strains of *T. parva*, indicating that these antibodies are responding to a common antigenic determinant present on the sporozoites of different strains. This has raised the possibility of a vaccine against ECF based on this common determinant.

References

Allen, J.R., Khalil, H.M. & Graham J.E. (1979). The location of tick salivary antigens, complement and immunoglobulin in the skin of guinea pigs infected with *Dermacentor andersoni* larvae. *Immunology*, **38**, 467-72.

Askonas, B.A., Corsini, A.C., Clayton, C.E. & Ogilvie, B.M. (1979). Functional depletion of T- and B-memory cells and other lymphoid cell subpopulations during trypanosomiasis. *Immunology*, **36**, 313-21.

Assoku, R.K.G., Tizard, I.R. & Nielsen, K.H. (1977). Free fatty acids, complement activation and polyclonal B-cell stimulation as factors in the immunopathogenesis of African trypanosomiasis. *Lancet*, **2**, 956-58.

Babel, C.L. & Lang, R.W. (1976). Identification of a new immunoglobulin subclass in three ruminant species. *Federation Proceedings*, **35**, 372.

Barbet, A.F., Davis, W.C. & McGuire, T.C. (1982). Cross-neutralization of two different trypanosome populations derived from a single organism. *Nature*, **300**, 453-56.

Barry, J.D., Jadjuk, S.J., Vickerman, K. & Le Ray, D. (1978). Detection of multiple variable antigen types in metacyclic populations of *Trypanosoma brucei*. *Transactions of the Royal Society of Tropical Medicine & Hygiene*, **73**, 205-8.

Birmingham, J.R. & Jeska, E.L. (1980). The isolation, long-term cultivation and characterization of bovine peripheral blood monocytes. *Immunology*, **41**, 807-14.

Blakemore, F. & Gamer, R.J. (1956). The maternal transference of antibodies in the bovine. *Journal of Comparative Pathology*, **66**, 287.

Burridge, M.J. (1971). Application of the indirect fluorescent antibody test in experimental East Coast fever (*Theileria parva* infection in cattle). *Research in Veterinary Science*, **12**, 338-41.

Butler, J.E. (1969). Bovine immunoglobulins: a review. *Journal of Dairy Science*, **52**, 1895-1909.

Butler, J.E. (1981). A concept of humoral immunity among ruminants and an approach to its investigation. In *The ruminant immune system* (eds. J.E. Butler, K. Nielsen & J.R. Duncan), pp. 3-55. New York: Plenum Press.

Butler, J.E. (1983). Bovine immunoglobulins: an augmented review. *Veterinary Immunology & Immunopathology*, **4**, 43-152.

Butler, J.E., Maxwell, C.F., Pierce, C.S., Hylton, M.B., Asoksky, R. & Kiddy, C.A. (1972). Studies on the relative synthesis and distribution of IgA and IgG$_1$ in various tissues and body fluids of the cow. *Journal of Immunology*, **109**, 38-48.

Cook, R.M. (1981). Attachment of *Trypanosoma brucei* to rabbit peritoneal exudate cells. *International Journal for Parasitology*, **11**, 149-56.

Dixon, F.J., Weigle, W.O. & Vazques, J.J. (1961). Metabolism and mammary secretion of serum proteins in the cow. *Laboratory Investigation*, **10**, 216-37.

Dobbelaere, D.A.E., Spooner, P.R., Barry, W.C. & Irvin, A.D. (1984). Monoclonal antibody neutralizes the sporozoite stage of different *Theileria parva* stocks. *Parasite Immunology*, **6**, 361-70.

Duffus, W.P.H. & Wagner, G.G. (1974). The specific immunoglobulin response in cattle immunized with *Theileria parva* (Muguga) stabilate. *Parasitology*, **69**, 31-42.

Emery, D.L. (1981). Adoptive transfer of immunity to infection with *Theileria parva* (East Coast fever) between cattle twins. *Research in Veterinary Science,* 30, 364-67.

Esser, K.M., Schoenbechler, M.J., Gingrich, J.B. & Diggs, C.L. (1981). Monoclonal antibody analysis of *Trypanosoma rhodesiense* metacyclic antigen types. *Federation Proceedings,* 40, 1011.

Eugui, E.M. & Emery, D.L. (1981). Genetically restricted cell-mediated cytotoxicity in cattle immune to *Theileria parva. Nature,* 290, 251-54.

Fazekas de St. Groth, S. & Webster, R.G. (1966). Disquisitions on original antigenic sin. 2. Proof in lower creatures. *Journal of Experimental Medicine,* 124, 347-61.

Feinnes, R.N.T.W. (1947). Immunity and preimmunity in cattle trypanosomiasis. *Veterinary Record,* 59, 291-92.

Feinstein, A. & Rowe, A.J. (1965). Molecular mechanism of formation of an antigen-antibody complex. *Nature,* 205, 147-49.

Freeman, T., Smithers, S.R., Targett, G.A.T. & Walker, P.J. (1970). Specificity of immunoglobulin G in rhesus monkeys infected with *Schistosoma mansoni, Plasmodium knowlesi* and *Trypanosoma brucei. Journal of Infectious Diseases,* 121, 401-6.

Goodwin, L.G., Green D.G., Guy, M.W. & Voller, A. (1972). Immunosuppression during trypanosomiasis. *British Journal of Experimental Pathology,* 53, 40-43.

Greenwood, B.M.(1974). Possible role of a B-cell mitogen in hypergammaglobulinaemia in malaria and trypanosomiasis. *Lancet,* 1, 435-36.

Greenwood, B.M., Whittle, H.C. & Molyneux, D.H. (1973). Immunosuppression in Gambian trypanosomiasis. *Transactions of the Royal Society of Tropical Medicine & Hygiene,* 67, 846-50.

Hammer, D.K., Kickhofen, B. & Schmid, T. (1971). Detection of homocytotropic antibody associated with a unique immunoglobulin class in the bovine species. *European Journal of Immunology,* 1, 249-58.

Henney, C.S. & Stanworth, D.R. (1966). Effect of antigen on the structural configuration of homologous antibody following antigen-antibody combination. *Nature,* 210, 1071-72.

Henney, C.S., Stanworth, D.R. & Gell, P.G.H. (1965). Demonstration of the exposure of new antigenic determinants following antigen-antibody combination. *Nature,* 205, 1079-81.

Holmes, P.H., Mammo, E., Thomson, A., Knight, P.A., Lucken, R., Murray, P.K., Murray, M., Jennings, F.W. & Urquhart, G.M. (1974). Immunosuppression in bovine trypanosomiasis. *Veterinary Record,* 95, 86-7.

Houba, V. & Allison, A.C. (1966). M-antiglobulins (rheumatoid factor-like globulins) and other γ-globulins in relation to tropical parasitic infections. *Lancet,* 1, 848-52.

Houba, V., Brown, K.N. & Allison, A.C. (1969). Heterophile antibodies, M-antiglobulins and immunoglobulins in experimental trypanosomiasis. *Clinical & Experimental Immunology,* 4, 113-23.

Hudson, K.M., Byner, C., Freeman, J. & Terry, R.J. (1976). Immuno-depression, high IgM levels and evasion of the immune response in murine trypanosomiasis. *Nature,* 256-58.

Irvin, A.D., Dobbelaere, D.A.E., Mwamachi, D.M., Minami, T., Spooner, P.R. & Ocama, J.G.R. (1983). Immunization against East Coast fever: correlation between monoclonal antibody profiles of *Theileria parva* stocks and cross immunity *in vitro. Research in Veterinary Science.* 35, 341-46.

Jayawardena, A.N. & Waksman B.H. (1977). Suppressor cells in experimental trypanosomiasis. *Nature,* 265, 539-41.

Klein, F., Mattern, P., Kornman, H.J. & Bosch, V.D. (1970). Experimental induction of rheumatoid factor-like substances in animal trypanosomiasis. *Clinical & Experimental Immunology*, 7, 851-63.

Kobayakawa, T., Louis, J., Izui, S. & Lambert, P-H. (1979). Autoimmune response to DNA, red blood cells and thymocyte antigens in association with polyclonal antibody synthesis during experimental African trypanosomiasis. *Journal of Immunology*, 122, 296-301.

Kobayashi, A. & Tizard, I.R. (1976). The response to *Trypanosoma congolense* infection in calves. Determination of immunoglobulins IgG_1, IgG_2, IgM and C3 levels and the complement-fixing antibody titres during the course of infection. *Tropenmedizin & Parasitologie*, 27, 411-17.

Lay, W.H. & Nussenzweig, V. (1969). Ca^{++}-dependent binding of antigen-19s antibody complexes to macrophages. *Journal of Immunology*, 102, 1172-78.

Le Ray, D., Barry, J.D. & Vickerman, K. (1978). Antigenic heterogeneity of metacyclic forms of *Trypanosoma brucei*. *Nature*, 273, 300-2.

Luckins, A.G. & Mehlitz, D. (1976). Observations on serum immunoglobulin levels in cattle infected with *Trypanosoma brucei*, *T. vivax* and *T. congolense*. *Annals of Tropical Medicine & Parasitology*, 70, 479-80.

Lumsden, W.H.R. & Herbert, W.J. (1967). Phagocytosis of trypanosomes by mouse peritoneal macrophages. *Transactions of the Royal Society of Tropical Medicine & Hygiene*, 61, 142.

MacAskill, J.A., Holmes, P.H., Whitelaw, D.D., McConnell, I., Jennings, F.W. & Urquhart, G.M. (1980). Immunological clearance of ^{75}Se-labeled *Trypanosoma brucei* in mice. 2. Mechanisms in immune animals. *Immunology*, 40, 629-35.

Mansfield, J.M. (1978). Immunobiology of African trypanosomiasis. *Cellular Immunology*, 39, 204-10.

Mansfield, J.M. & Kreier, J.P. (1972). Autoimmunity in experimental *Trypanosoma congolense* infections of rabbits. *Infection & Immunity*, 5, 648-56.

Masake, R.A., Musoke, A.J. & Nantulya, V.M. (1983). Specific antibody responses to the variable surface glycoproteins of *Trypanosoma congolense* in infected cattle. *Parasite Immunology*, 5, 345-55.

McGuire, T.C., Musoke, A.J. & Kurtti T. (1979). Functional properties of bovine IgG_1 and IgG_2: interaction with complement, macrophages, neutrophils and skin. *Immunology*, 38, 249-56.

Muhammed, S.I., Lauerman, L.H. & Johnson, L.W. (1975). Effect of humoral antibodies on the course of *Theileria parva* (East Coast fever of cattle). *American Journal of Veterinary Research*, 36, 399-402.

Murray, M. (1974). The pathology of African trypanosomiases. In *Progress in immunology II* (eds. L. Brent & E.J. Holborow), vol. 4, pp. 181-92. Amsterdam: North-Holland.

Murray, M., Morrison, W.I. & Whitelaw, D. (1982). Host susceptibility to African trypanosomiasis: trypanotolerance. *Advances in Parasitology*, 21, 1-68.

Musoke, A.J., Nantulya, V.M., Rurangirwa, F.R. & Buscher, G. (1984). Evidence for a common protective antigenic determinant on sporozoites of several *Theileria parva* strains. *Immunology*, 52, 231-38.

Musoke, A.J., Nantulya, V.M., Barbet, A.F., Kironde, F. & McGuire, T.C. (1981). Bovine immune response to African trypanosomes: specific antibodies to variable surface glycoproteins of *Trypanosoma brucei*. *Parasite Immunology*, 3, 97-106.

Musoke, A.J., Nantulya, V.M., Buscher, G., Masake, R.A. & Otim, B. (1982). Bovine immune response to *Theileria parva*: neutralizing antibodies to sporozoites. *Immunology*, **45**, 663-68.

Nantulya, V.M., Musoke, A.J., Barbet, A.F. & Roelants, G.E. (1979). Evidence for reappearance of *Trypanosoma brucei* variable antigen types in relapse populations. *Journal of Parasitology*, **65**, 673-79.

Nantulya, V.M., Musoke, A.J., Moloo, S.K. & Ngaira, J.M. (1983). Analysis of the variable antigen composition of *Trypanosoma brucei brucei* metacyclic trypanosomes using monoclonal antibodies. *Acta Tropica*, **40**, 19-24.

Nantulya, V.M., Musoke, A.J., Rurangirwa, F.R. & Moloo, S.K. (1984). Resistance of cattle to tsetse-transmitted challenge with *Trypanosoma brucei* or *Trypanosoma congolense* after spontaneous recovery from syringe-passaged infections. *Infection & Immunity*, **43**, 735-38.

Nantulya, V.M., Musoke, A.J., Rurangirwa, F.R., Barbet, A.F., Ngaira, J.M. & Katende, J.M. (1982). Immune depression in African trypanosomiasis: the role of antigenic competition. *Clinical & Experimental Immunology*, **47**, 234-42.

Ngaira, J.M., Nantulya, V.M., Musoke, A.J. & Hirumi, K. (1983). Phagocytosis of antibody-sensitized *Trypanosoma brucei in vitro* by bovine peripheral blood monocytes. *Immunology*, **49**, 393-400.

Nielsen, K. (1977). Bovine reaginic antibody. 3. Cross-reaction of anti-human IgE and anti-bovine reaginic immunoglobulin antisera with sera from several species of mammals. *Canadian Journal of Comparative Medicine*, **41**, 345-48.

Nielsen, K., Sheppard, J., Holmes, W. & Tizard, I. (1978). Changes in serum immunoglobulins, complement and complement components in infected animals. *Immunology*, **35**, 817-26.

Rossi, C.R. & Kiesel, G.K. (1977). Bovine immunoglobulin G subclass receptor sites on bovine macrophages. *American Journal of Veterinary Research*, **38**, 1023-25.

Rurangirwa, F.R., Tabel, H., Losos, G.J. & Tizard, I.R. (1979). Suppression of antibody response to *Leptospira biflexa* and *Brucella abortus* and recovery from immunosuppression after Berenil treatment. *Infection & Immunity*, **26**, 822-26.

Rurangirwa, F.R., Musoke, A.J., Nantulya, V.M. & Tabel, H. (1983). Immune depression in bovine trypanosomiasis: effects of acute and chronic *Trypanosoma congolense* and chronic *Trypanosoma vivax* infections on antibody response to *Brucella abortus* vaccine. *Parasite Immunology*, **5**, 267-76.

Rurangirwa, F.R., Tabel, H., Losos, G., Masiga, W.N. & Mwambu, P. (1978). Immunosuppressive effect of *Trypanosoma congolense* and *Trypanosoma vivax* on the secondary immune response of cattle to *Mycoplasma mycoides* subsp. *mycoides*. *Research in Veterinary Science*, **25**, 395-97.

Scott, J.M., Pegram, R.G., Holmes, P.H., Pay T.W.F., Knight, P.A., Jennings, F.W. & Urquhart, G.M. (1977). Immunosuppression in bovine trypanosomiasis: field studies using foot-and-mouth disease vaccine and clostridial vaccine. *Tropical Animal Health & Production*, **9**, 159-65.

Sendashonga, C.N. & Black, S.J. (1982). Humoral responses against *Trypanosoma brucei* variable surface antigen are induced by degenerating parasites. *Parasite Immunology*, **4**, 245-57.

Stevens, D.R. & Moulton, J.E. (1978). Ultrastructural and immunological aspects of the phagocytosis in *Trypanosoma brucei* by mouse peritoneal macrophages. *Infection & Immunity,* **19**, 972-82.

Tabel, H., Losos, G.J., Maxie, M.G. & Minder, C.E. (1981). Experimental bovine trypanosomiasis (*Trypanosoma vivax* and *T. congolense*). 3. Serum levels of immunoglobulins, heterophile antibodies and antibodies to *T. vivax. Tropenmedizin & Parasitologie,* **32**, 149-53.

Takayanagi, T. & Nakatake, Y. (1977). *Trypanosoma gambiense*: the binding activity of antiserum to macrophages. *Experimental Parasitology,* **42**, 21-26.

Takayanagi, T., Nakatake, Y. & Enriquez, G.L. (1974). Attachment and ingestion of *Trypanosoma gambiense* to the rat macrophage by specific antiserum. *Journal of Parasitology,* **60**, 336-39.

Taussig, M.J. (1975). Antigenic competition. In *The immune system: a course on the molecular and cellular basis of immunity* (eds. M.J. Hobart & I. McConnell), pp. 164-78. Oxford: Blackwell Scientific Publications.

Valentine, R.C. & Green, N.M. (1967). Electron microscopy of an antibody-hapten complex. *Journal of Molecular Biology,* **27**, 615-17.

Wagner, G.G., Duffus, W.P.H. & Burridge, M.J. (1974). The specific immunoglobulin responsible in cattle immunized with isolated *Theileria parva* antigens. *Parasitology,* **69**, 43-53.

Wells, P.W. & Eyre, P. (1971). Preliminary characterization of bovine homocytotrophic antibody. *Immunochemistry,* **9**, 88-91.

Whitelaw, D.D., Scott, J.M., Reid, H.W., Holmes, P.H., Jennings, F.W. & Urquhart, G.M. (1979). Immunosuppression in bovine trypanosomiasis: studies with louping-ill vaccine. *Research in Veterinary Science,* **26**, 102-7.

22

The physiology of secretory immunity

J.G. HALL

Although it is often convenient to consider secretory immunity separately from the more familiar systemic immunity, the distinction has little physiological validity. The immune responses that provide the blood and the external secretions with antibodies are expressions of the same general phenomena. This is true even for man and the common laboratory rodents, where the principal immunoglobulin in the secretions, IgA, is demonstrably different from the conventional IgG and IgM immunoglobulins that predominate in the blood. In cattle, sheep and goats, where the external secretions contain abundant IgG_1 apparently indistinguishable from that which occurs in the blood, there is even less reason for making a rigid distinction between systemic and secretory immunity. Nonetheless, secretory immunity does have an extra dimension that gives it particular interest to those who are concerned with the study of the partitioning of macromolecules between different compartments of the body. This extra dimension is represented by the active transcellular transport of secretory immunoglobulins from blood or tissue fluids to the external surfaces. This process is carried out by the epithelial cells of the alimentary, respiratory and genital tracts; it is initiated by receptor-mediated endocytosis and can be studied in a number of model systems. Some of these are described and related to the cellular basis of antibody production.

Introduction

The term 'secretory immunity' is taken here to cover the processes leading to the appearance of immunoglobulin (Ig) antibodies in the external secretions which bathe the mucosae of the alimentary, respiratory, genito-urinary and naso-lacrimal tracts and which populate colostrum and milk. An essential consideration is the means by which Ig macromolecules are exported across epithelia which are generally, and necessarily, impermeable to molecules of this size. In many mammals, including man, the bulk of secreted immunoglobulin is of the IgA class. The process of its secretion begins when dimers (or higher polymers) of

IgA unite with a glycoprotein receptor, called the secretory component (SC), which is arrayed on the baso-lateral portions of the cytoplasmic membrane of appropriate epithelial cells. The IgA-SC adducts become associated with klathrin-coated pits and are internalized in endocytic vesicles, transported across the cytoplasm and discharged *en bloc* on to the external surface of the epithelium (Figure 1) where they constitute secretory IgA. Cattle and sheep possess a secretory immune system that works in this way but it is a subsidiary one: IgA is not always the

Figure 1. Diagramatic representation of the production of dimeric IgA (two monomeric units joined by a 'J', or joining chain) by a submucosal plasma cell, the binding of the IgA dimer to secretory component (SC) displayed on the baso-lateral aspect of the enterocyte, the endocytosis of the IgA-Sc adduct and its transport to the lumen of the gut where the secretory IgA becomes associated with the boundary layer of mucus that overlies the surface of the epithelium. Reproduced by permission of the *African Journal of Immunology*.

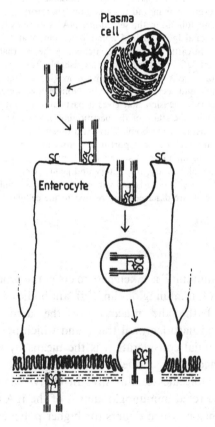

predominant Ig in their secretions. IgG_1 is often present in similar or higher concentrations and, although the mechanism of its transport is presumed to be analogous to that of IgA, a detailed description of the process in subcellular and molecular terms has proved elusive.

Another difficulty is the use of the terms 'local' or 'regional' to describe immune responses at mucous surfaces. These terms encapsulate the observation that secretory antibodies are often formed by plasma cells which lie immediately beneath the mucosae. Although the actual production of the antibody may appear to be a local event, the idea that a significant immune response can take place in an isolated anatomical site, without the participation of several populations of migrant cells and without the systemic dissemination of cells and antibodies, is surely wrong. In considering, for example, a localized group of plasma cells beneath the intestinal epithelium, it is obvious that some of the Igs which are produced may be transported directly to the lumen of the gut by the overlying enterocytes. However, there is no active mechanism which can convey an Ig molecule from the plasma cell to the enterocyte. The Igs reach the enterocytes by simple diffusion through the tissue fluid, and, particularly in the postabsorptive state when the net fluid flow is inwards rather than outwards, some of the Ig is inevitably swept into the regional lymphatics and conveyed to the systemic blood circulation. Also, antigens which penetrate the mucosae in sufficient quantity to cause any sort of detectable response usually cause the production of some antibodies of the IgM and IgG classes, and most of these will remain in the blood serum until removed by the normal processes of catabolism. In spite of these considerations, there are several claims in the literature for the detection of specific antibodies in secretions which are absent in the blood. In my opinion, most such claims result from the use of insensitive and inappropriate techniques, combined sometimes with an oversimplified view of physiological reality. Nonetheless, there may be cases where, from a purely pragmatic standpoint, the production of secretory antibodies does appear to be a local event. If the mucosal surface involved and the volume of secretions produced are sufficiently small, a high local concentration of antibody may be detected. A good example would be a subconjuctival immune response in the eye. Inevitably, some of the antibody formed would be carried to the blood by the local lymphatics, but would undergo such dilution in the intravascular and extravascular protein pool that it would be difficult to detect.

I have laboured these difficulties in order to emphasize that there are a number of factors which complicate and make difficult a comprehensive review of secretory immunity, particularly in relation to sheep and cattle.

In the brief review set out below there are inevitably many arguments that would fail if transferred from the general to the particular or *vice versa*. This difficulty can be partially overcome by digesting a large mass of recent work (McGhee & Mestecky 1983; Bourne 1981; Ciba Foundation 1977; Heremans 1974), but there is no denying the fact that many basic questions are, at the moment, unanswerable.

The traffic of lymphocytes through the gut: the cellular basis of humoral immunity in the intestine

In spite of the differences mentioned above, there seems to be a surprising degree of similarity between different species in the way lymphocytes, particularly the large lymphoid immunoblasts, migrate to and from the gut. In considering these cellular migrations, which are ultimately responsible for populating the submucosa with plasma cells, it is usual to consider the gut-associated lymphoid tissue (GALT) as a whole. The GALT is comprised of first, the diffuse submucosal lymphoid tissue in the *lamina propria* of the intestine (principally the small intestine); second, the aggregates of organized lymphoid tissue in the intestinal wall, the Peyer's patches; and third, the mesenteric lymph nodes. This arrangement is typical of monogastric omnivores such as rats and man, but other species have special features. In sheep, and probably in cattle and goats, the Peyer's patches are well developed in early life and include a long continuous patch in the ileum which tends to undergo considerable involution as sexual maturity approaches. In addition, there are many lymph nodes associated with fore-stomachs; the status of these as functional parts of the GALT has not yet been thoroughly assessed. The GALT of rabbits also differs in that there are large accumulations of lymphoid tissue in the appendix and *sacculus rotundus*. Such differences should be borne in mind when interpreting experimental results which come, for the most part, from work on rats and sheep.

The large lymphoid immunoblasts

Gowans and Knight (1964) originally showed that many of the large lymphoid immunoblasts in the thoracic duct lymph of rats (which is composed principally of intestinal lymph from the mesenteric nodes) had the capacity to 'home' back to the *lamina propria* of the gut after being labeled with ^{3}H-thymidine *in vitro* and injected intravenously into syngeneic recipients. This observation has been amply confirmed and reviewed (Bienenstock & Befus 1980; Hall 1980). Almost identical results were obtained from experiments on sheep (Hall, Hopkins & Orlans 1977) in which 30% of an intravenous dose of syngeneic immunoblasts from intestinal lymph localized in the small intestine, while immunoblasts from somatic lymph nodes localized in the spleen and lungs but not the gut. The close agreement between the results of experiments on animals as different as rats and sheep is comforting for its own sake and interesting because it suggests that the isotype of the immunoglobulin being made may not be an important factor in guiding the immunoblasts to their destination. Whereas most Ig-synthesizing immunoblasts in the thoracic duct lymph of mice (Mandel & Asofsky 1968) and rats (Williams & Gowans 1975) make IgA, the corresponding cells in the intestinal lymph of sheep are just as likely to make IgG or IgM (Hall, Hopkins & Orlans 1977). It is hard to believe that these cells would behave so uniformly in terms of their extravasation in the gut, if the production of a particular class of Ig were a crucial factor in the process. In any case, formal experiments in rodents have shown that the presence of IgA in or on immunoblasts does not influence their homing to the gut (McWilliams, Phillips-Quagliata & Lamm 1975). Furthermore, T immunoblasts from intestinal lymph, which neither display nor contain immunoglobulin, are also capable of homing to the gut (Sprent 1976; Rose, Parrott & Bruce 1976), though they may be more likely to go to the Peyer's patches than the *lamina propria.*

Another factor which does not seem to be involved in mediating the extravasation of immunoblasts is the presence of antigenic material, specific or otherwise, in the gut. Immunoblasts from the thoracic duct lymph of rats home just as readily to the antigen-free gut of unsuckled, newborn rats (Halstead & Hall 1972) or to subcutaneous implants of sterile foetal gut (Moore & Hall 1972) as they do to the antigen-laden gut of adults. Similar results have been obtained in the foetuses of sheep (Hall, Hopkins & Orlans 1977). However, local antigen may play a role in promoting the further proliferation of immunoblasts of appropriate specificity once they have extravasated in the gut (Husband, Monie & Gowans 1977).

Unfortunately, in spite of all this experimental work, no one understands what really does mediate the extravasation in the gut of this class of lymphborne immunoblasts. Shortly after they have extravasated, these immunoblasts turn into classical plasma cells (Gowans & Knight 1964) with concentric, lamellar endoplasmic reticulum (Hall, Parry & Smith 1972) where, during a life span of a few days (Mattioli & Tomasi 1973), they produce immunoglobulin antibodies before dying off and making room for the newcomers which arrive continually.

The immunoblasts in the intestinal lymph arise, like immunoblasts elsewhere, by the transformation of small lymphocytes in response to antigenic stimulation (Gowans et al. 1962). This antigenic stimulation can take place either in the wall of the gut itself or in the regional mesenteric node (Hall, Hopkins & Orlans 1977). In both situations it is likely that dendritic macrophage-like cells are important in presenting the antigen to the lymphocytes (Hall 1979), but it is less clear how antigen, other than actively invasive helminths, protozoa, bacteria or viruses, actually penetrates the gut. Specialized epithelial M cells which overlie the Peyer's patches (Owen & Nemanic 1978) may be important in sampling the gut contents by a phagocytic mechanism and may pass the antigenic message to the lymphocytes and true macrophages below them. In certain types of transient damage to the epithelium, antigen may leak between epithelial cells (Draper et al. 1983) and so contact the lymphoid tissue in the *lamina propria*. These two routes hardly seem sufficient to account for all the observed phenomena, for it is certainly a fact that the GALT of all mammals living in a normal environment is subject to continual stimulation by dietary and microbial antigens in the gut. As a result, the mesenteric nodes always display the histological stigmata of antigen-driven proliferation, and significant numbers of immunoblasts are always present in the intestinal lymph. However, most of the lymphocytes in intestinal lymph are the normal, small recirculating lymphocytes, and these too show some special features *vis a vis* their counterparts in lymph efferent from peripheral somatic nodes.

The recirculation of small lymphocytes through the gut of sheep

Until the mid-1970s, it was believed generally that the pool of circulating lymphocytes recirculated more or less randomly through the spleen, lymph nodes, Peyer's patches and, indeed, through all organized lymphoid tissue except the thymus. However, in 1976 it was discovered that small lymphocytes collected from the intestinal lymph of sheep and labeled *in vitro* with ^{51}Cr recirculated preferentially through the GALT after they had been injected intravenously. The converse was also true: lymphocytes collected from lymph efferent from peripheral somatic lymph nodes (PSLN) recirculated readily through other PSLNs and tended to avoid recirculation via the GALT (Scollay, Hopkins & Hall 1976). These studies were performed using unfractionated lymph cells, i.e. a mixture of B and T cells. Later, similar results were obtained using lymph cells from which most of the B lymphocytes had been removed (Cahill et al. 1977). This recirculatory bias is a postnatal phenomenon: although lymphocyte recirculation is in vigorous progress in the sheep foetus *in utero* (Pearson, Simpson-Morgan & Morris 1976), the recirculating cells do not distinguish between the GALT and other lymphoid tissue (Cahill et al. 1979). Recirculating lymphocytes may acquire the ability to make this distinction through postnatal contact with antigen or, more likely, because an additional cohort of recirculating cells with gut-seeking properties is discharged into the circulation during the explosive postnatal development of GALT in lambs (Cole & Morris 1971): these are matters for speculation. However, it seems relatively certain that in sheep, and I suspect in other ruminants, there are subsets of small lymphocytes and large lympoid immunoblasts with a special propensity for migrating to or through the submucosa of the gut. Nothing conclusive has been discovered yet about the nature of the receptors on these cells or the complementary groups on the endothelial cells of the blood vessels that mediate these phenomena. Similarly, the general molecular basis of lymphocyte recirculation is still completely obscure.

Migration of lymphocytes to other mucosae

In a number of publications, Bienenstock and his colleagues have advanced the idea of a generalized, mucosal-associated lymphoid tissue (MALT) (Bienenstock & Befus 1980). The key element is that the mucosae are provided with a common pool of recirculating lymphocytes which are the precursors of the immunoblasts and plasma cells which produce secretory Igs (usually IgA) in the alimentary, respiratory, genito-urinary and naso-lacrimal tracts. It is envisaged that sensitization to a given antigen in, say, the gut will generate a cohort of immunoblasts which, besides populating the *lamina propria* of the intestine, will be carried by the lymphatics to the blood and thus disseminated to all mucosal sites. There is some evidence for this, particularly in the case of the gut-mammary gland axis (e.g. Weisz-Carrington et al. 1979), but the evidence is generally circumstantial and occasionally intuitive. There is also a good deal of evidence against it. For example, in studies of immunoblasts in the intestinal lymph of sheep (Hall, Hopkins & Orlans 1977) no evidence could be found that these cells extravasated to any significant extent in the respiratory tract, the salivary glands, the lacrimal glands or the conjuctivae. Similarly, in our current studies on lung lymph from sheep, the immunoblasts in efferent lymph from the caudal mediastinal node show little tendency to home to the gut, even though some of them can be shown to be making IgA. Moreover, the small lymphocytes that recirculate through the lung-associated lymphoid tissue have the same recirculatory pattern as those efferent from PSLNs, i.e. they recirculate well through PSLNs but tend to avoid the gut. Thus, in sheep, the lymphocytes that migrate through the lung- and bronchial-associated lymphoid tissue seem to be distinct from those that recirculate through the GALT, even though both sets of cells are concerned directly with mediating the production of secretory Igs. It is unlikely that these results are merely the reflection of species differences between sheep and the usual small laboratory animals. Experiments on rats (Spencer, Gyure & Hall 1983) have confirmed that IgA-producing immunoblasts from the intrathoracic lymph nodes, which drain the lungs and bronchi, show no tendency to home to the gut.

Overall then, the evidence for and against a common mucosal immune system (as originally conceived) is fairly conflicting. If there really is a special functional link between the lymphoid tissues associated with various mucosae, it is likely to be more subtle and complicated than its more vociferous proponents are accustomed to allow.

IgA switching, T-cell dependence and immunological memory in secretory immunity

The question arises as to why, given the great mobility of lymphocytes, the production of a particular immunoglobulin isotype, IgA, is limited to certain sites, such as the gut, and excluded from others. Why does a popliteal node, with all the resources of the recirculating pool at its disposal, never (a risky word) generate any cells that make IgA? The recirculatory bias that tends to divert gut-associated small lymphocytes away from PSLNs, such as the popliteal, is by no means so extreme that it can deny all potential IgA-producing cells to the node. Clearly, in such nodes, either something is lacking so that IgA cannot be switched on or something is present which suppresses IgA production. The latter possibility is enjoying a considerable vogue, and the regulation of IgA production is nearly always discussed in terms of the influence of T cells (Elson 1983). I cannot accept that T cells play a primary role in switching on or suppressing IgA production. Where thymus-independent antigens are concerned, athymic nude rats mount IgA responses which are indistinguishable from those of normal rats (Andrew & Hall 1982a), and recently we have found additional evidence against the primary role of T cells in the regulation of IgA production. By adoptively transferring syngeneic lymphoid cells to recipient rats whose lymphoid systems have been destroyed by γ-irradiation, we have been able to dissect some of the influences that regulate the production of IgA. We found that lymphocytes from any source, e.g. PSLNs or spleen, were perfectly able to produce IgA provided they were placed in the environment of the GALT (Denham et al. 1984). We concluded that the crucial element in switching uncommitted lymphoid cells to IgA production was a radioresistant (i.e. nonlymphoid) influence in the microenvironment of the GALT. What this influence actually is, is anybody's guess. Our guess, and it is no more than that, is that the specialized dendritic macrophages which are known to be abundant in the tissue fluid of the intestine (Hall, Hopkins & Orlans 1977; Hall 1979) and in the GALT (Mayrhofer, Pugh & Barclay 1983) of rats may play an important role.

For many years it has been debated whether secretory immune responses exhibited immunological memory. The evidence was conflicting and bedevilled by the difficulties of accurately measuring antibodies in small volumes of secretions that included viscous mucus, surfactants and proteolytic enzymes. These difficulties were largely overcome when it was discovered that after an antigen was injected into the Peyer's patches of rats specific antibodies of the IgA class appeared in the bile in impressive titres, while other isotypes of Ig were present only in trace amounts (Hall et al. 1979). This system, described below, made the natural history of IgA responses much more susceptible to experimental investigation than was possible previously. It also became fairly easy to establish that secretory immune responses do indeed have an immunological memory which is very similar to that shown by the more familiar humoral responses in the blood (Andrew & Hall 1982b). It seems, then, that secretory immune responses are no more and no less dependent on thymic influences than conventional serological responses, and that their memory is every bit as good.

Hepatobiliary transport of secretory immunoglobulins

I have pointed out that a substantial proportion of the immunoglobulins produced by submucosal plasma cells in the gut must inevitably be absorbed into the regional lymphatics. Direct measurement bears this out: in rats (Orlans et al. 1978), dogs (Vaerman & Heremans 1970) and sheep (Hall, Hopkins & Orlans 1977; Quin, Husband & Lascelles 1975) the concentration of IgA in the intestinal lymph is higher than in the blood, even though the concentration of total proteins is lower. These results are a clear demonstration of the 'refluxing' of locally synthesized (and potentially secretable) Ig from mucosae into lymph, but it poses a problem. Large volumes of this IgA-rich lymph flow continually into the blood, yet the IgA content of blood remains low. Where does all the IgA go? The continual extraction from the blood of such large quantities of an intact macromolecule could be accomplished only by an organ with a rich blood supply that flowed through vessels with an endothelium sufficiently fenestrated to permit the rapid egress of large molecules. This specification applies only to the liver; if the liver extracts

IgA from the blood, it is likely that it would secrete IgA in the bile. This proved to be the case: kinetic studies in rats showed that up to 30% of an intravenous dose of ^{125}I-labeled IgA dimer was secreted into the bile within 1 h of injection (Orlans et al. 1978). An electron microscopic study (Birbeck et al. 1979) showed that this was carried out by hepatocytes, and further biochemical and immunochemical studies showed that hepatocytes displayed SC on their surfaces (Orlans et al. 1979) and that the IgA-SC adducts were transported across the hepatocytes' cytoplasm in endocytic vesicles (Mullock et al. 1979). It is important to stress that only dimers or higher polymers of IgA are capable of being transported in this way. Monomeric IgA, which is scanty in most animals but abundant in man, cannot unite with SC and so cannot be transported. Also, once the dimeric IgA has combined with SC and been secreted, it cannot reundergo the secretory process; secretory IgA isolated from bile and reinjected remains in the circulation because its receptor for SC is already occupied (Reynolds et al. 1980). In most of these studies it was convenient to use myeloma material as a source of IgA, but the system works just as effectively with genuine polyclonal antibodies (Reynolds et al. 1980) or the products of IgA-secreting hybridomas (Dean et al. 1982).

Although for practical purposes IgA is the only intact immunoglobulin in rat bile, there are traces of IgM, and this suggests that IgM may compete, albeit feebly, for SC. After artificially elevating the serum concentration of IgM several fold by implanting an IgM-secreting myeloma, we were able to demonstrate that substantial amounts of IgM appeared in the bile; moreover, the IgA was displaced from the bile and accumulated to unnaturally high levels in the blood (Peppard, Jackson & Hall 1983). Thus when the preponderance of IgM over IgA is sufficiently great, it can compete successfully *in vivo* for SC. This is a happy circumstance for individuals with congenital or acquired deficiency of IgA; their mucosae are not left unprotected, because IgM is secreted in its stead and seems to represent a moderately effective defence. Indeed, IgM is probably the only secretory immunoglobulin in lower vertebrates (Wang & Fudenberg 1974).

The biological role of secretory antibody is primarily to prevent by steric hindrance the engagement of pathogenic bacteria, viruses and toxins with the appropriate receptors on epithelial cells. As long as enteric pathogens are prevented from getting a foothold, they are effectively condemned to excretion. In this context it makes sense to have freshly ingested food doused with biliary antibody as soon as it enters the duodenum. However, the biological role of IgA does not end here. If antigens or microorganisms from the diet do succeed in penetrating the epithelium, IgA still has a role to play. Immune complexes formed by the combination of IgA with such antigens are secreted in the bile just as effectively as uncomplexed IgA, and in this way immune complexes formed from dietary antigens can be cleared from the circulation and excreted rapidly in the bile. Further, because IgA does not bind complement, there are no damaging allergic consequences (Hall & Andrew 1980; Peppard et al. 1981).

Comparative aspects of the hepatobiliary transport of IgA

A considerable amount of data has been published on comparative aspects of the hepatobiliary transport of IgA (Hall, Gyure & Payne 1980; Orlans et al. 1983; Hall et al. 1981), but the picture is far from complete. An interesting feature is that some animals, such as rats and rabbits, not only transport their own IgA into bile very efficiently, but also transport heterologous material just as well. Other animals, such as guinea pigs, dogs, cats and hamsters, transport their own IgA, but cannot transport heterologous IgA. Ruminants (or at any rate sheep) only transport relatively small amounts of their own IgA, albeit quite actively, and cannot transport heterologous material. The presence or absence of this functional homology does not follow any clear evolutionary pattern; chickens, for example (Rose et al. 1981), are able to transport human IgA from blood to bile, though many mammals cannot do so. These findings are summarized in Table 1.

The status of man in the IgA transport league is uncertain. One complication is the fact that 75% or more of the IgA in human serum is monomeric. The biological function of the monomer is unknown, and only a minority of IgA plasma cells in man make such material. However, because the monomer cannot be secreted, it accumulates in the blood and predominates over dimer. Thus, from the point of view of a transport physiologist, clinical laboratory measurements of total serum IgA are uninterpretable. The investigator wants to know the concentration of dimeric IgA in the blood compared with that in the secretions, but such information is difficult to obtain, particularly in the case of bile. Bile collected from inflamed or otherwise diseased biliary tracts is usually offered to the investigator, and this is quite untypical of the real situation. Obviously, pure bile from normal, healthy people is by definition almost unobtainable, but such specimens as we have obtained (Orlans et al. 1983) suggest that active transport of IgA goes on in man as it does in many other mammals. However, we can say little about the extent or kinetics of the hepatobiliary transport of IgA in man. Also, no one has succeeded in demonstrating the presence of SC on human hepatocytes;

Table 1. Classification of domestic and experimental animals in terms of hepatobiliary transport of IgA. The position of man in this ranking is conjectural.

Rapid and substantial active transport of homologous and heterologous IgA dimer from blood to bile	Substantial active transport of homologous, but not heterologous, IgA dimer from blood to bile	Some active transport of homologous, but not heterologous, IgA dimer from blood to bile
Rat	cat	sheep
Mouse	dog	(?goat)
Rabbit (?man)	pig	(?cattle)
Chicken	guinea pig	
	hamster	

there is ultrastructural evidence that the transport of IgA from blood to bile in man is carried out by specialized cells rather than hepatocytes (Smith et al. 1981).

Hepatobiliary transport of secretory immunoglobulins in sheep

Sheep bile is rather intractable material and care must be taken in its collection. We have always collected bile from the common bile duct by inserting a cannula as high up as possible to avoid contamination with pancreatic enzymes. Also it has been our practice to ligate the cystic duct and excise the gall bladder. In these ways we have sought to collect pure hepatic bile from unanaesthetized sheep, and such material has been the main subject of our investigations. Even this material is quite unstable. Although when freshly collected it is a clear, emerald-coloured fluid, it soon becomes turbid and brownish, so samples that cannot be analysed immediately should be snap frozen and stored at -20°C. Because it contains powerful chromogens the usual colorimetric assays for protein tend to be unsuitable for untreated bile, giving wildly unreproducible results. In order to overcome this, we precipitated the proteins with 10% TCA, washed the precipitate with acid alcohol to remove most of the chromogens and bleached the remainder with dilute hydrogen peroxide. The precipitate was then redissoved in 1.0 NaOH and assayed for protein by the Biuret or Lowry method against a standard of serum proteins. Individual proteins were measured in untreated bile by radial immunodiffusion, except for IgA which was measured by 'rocket' immunoelectrophoresis. The results of these studies are shown in Table 2. The figures are average values and their accuracy in absolute terms is questionable, but they show a number of things clearly. First, bile like other secretions of sheep, contains t least as much IgG and IgM

Table 2. The content of total proteins, albumin and immunoglobulins in the blood serum, intestinal lymph plasma and hepatic bile of sheep. Material was collected from eight sheep and mean values are shown.

Fluid	Total protein	Individual proteins as % of total			
		Albumin	IgG	IgM	IgA
Blood serum	78.4 (61−88)	69.8	30.2	6.6	0.1
Intestinal lymph plasma	53.5 (39−56)	72.8	25.1	3.7	1.0
Hepatic bile	4.8 (1.7−8.9)	21.8	8.0	12.6	5.1

as IgA. Nonetheless, relative to albumin, the concentration of IgA is some 163 times higher in bile than in blood, whereas that of IgG is essentially the same in both fluids. This can only mean that IgA is actively transported from blood to bile or that it is produced locally. Local production is easily discounted: the liver of healthy mammals contains no lymphoid cells, and, although the biliary tract has some lymphoid cells in a rudimentary lamina, plasma cells are very sparse compared with the gut. A litre of sheep bile may be produced every 24 h, with a content of at least 250 µg per ml. Thus at least 250 mg of IgA leaves the liver in the bile every day: this represents a substantial proportion of the IgA carried by intestinal lymph in the same period. Those who seek to explain the presence of IgA in bile on the basis of local production must ask themselves whether the biliary tract, with a surface area of about 100 cm^2 and scanty submucosal plasma cells, could begin to produce the amounts of IgA made by the gut, which has the surface area of a tennis court and infinitely more plasma cells per unit area. No doubt the plasma cells in the biliary tract do contribute their widow's mite of IgA to the bile, but this is a drop in the ocean. On the basis of currently available data, the only tenable conclusion is that the IgA in sheep bile is actively transplanted from the blood. It is true, however, that sheep are much less efficient transporters of IgA than rats, as shown by the following experiments.

Dimeric (or polymeric) IgA was isolated from pooled, intestinal lymph plasma of sheep by repeated column chromatography. This was a laborious process but ultimately successful. The isolated sheep IgA was labeled with ^{125}I and injected intravenously into a rat (Figure 2). Twenty percent of the injected material appeared in the rat's bile within 3 h of injection, i.e. the sheep IgA was transported about half as well as isologous material. However, when the ^{125}I-sheep IgA was injected into a sheep, only 5% was recovered in the bile during the experimental period, while human and rat IgA were not recovered in sheep bile at all. Nonetheless, at the peak of biliary excretion of labeled IgA, the specific radioactivity of the bile was much greater than that of blood (Figure 3), indicating that the transport of IgA into bile is an active process. SDS-PAGE techniques showed that both the heavy and light chains of the labeled IgA were of normal size and undamaged.

It was possible to show by means of an antiserum to sheep SC, that the labeled IgA that appeared in the bile had acquired SC as it traversed the liver, but as yet we have not been able to demonstrate the presence of SC on sheep hepatocytes. It is possible that SC is only sparsely expressed on the sinusoidal facets of sheep hepatocytes and this could account for the somewhat meagre transport of IgA. Also, it is possible that sheep IgM competes more successfully with IgA for SC than is the case in rats. The amounts of IgM in sheep bile certainly suggest that this could be so.

Antibody activity in the bile of sheep

The studies described above showed that there is abundant immunoglobulin in the bile of sheep, some of it of the IgA class. We were interested to find out whether immunization with the GALT of sheep would produce the high titres of specific biliary IgA antibodies that it does in rats (Hall et al. 1979; Andrew & Hall 1982a; 1982b).

Figure 2. The specific radioactivity (cps/ml) of blood (●) and bile(▲) measured for 4 h after the i.v. injection of ^{125}I-dimeric sheep IgA into a rat. During this experiment 20% of the injected dose of radioactivity was recovered in the bile.

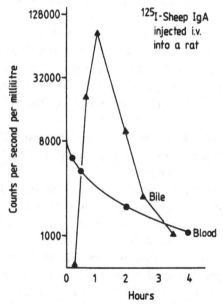

Accordingly, we undertook a series of experiments on sheep in which we infiltrated the wall of the intestine with antigen (usually a suspension of killed bacteria). At the same time the PSLNs were stimulated with another antigen and thereafter daily samples of blood and hepatic bile were collected and frozen, pending the assay of the antibodies. Ten such experiments were performed successfully. They all gave similar results: and a typical set of findings is shown in Figure 4. Although substantial titres of antibodies appeared in the bile after day 7, they were lower than those in the blood. Also, antibodies directed against the antigen used to stimulate the PSLNs appeared in the bile. The isotype of the antibody in the bile and serum was determined by radioimmunoassay using appropriate antiglobulin reagents. The results are shown in Table 3.

These results show that the biliary antibodies which can be induced in sheep are composed of all the major immunoglobulin isotypes. It is likely that the IgA antibodies were actively transported from blood to bile, but this cannot be deduced directly from the experiments. The biliary titres merely reflect the serum titres and do not suggest *a priori* that antibody activity in general is transported actively into the bile. In some

Figure 3. The specific radioactivities (cps/ml) of blood (●) and bile (▲) measured for 11 h after the i.v. injection of ^{125}I-dimeric sheep IgA into a sheep. During the experiment 5% of the injected dose of radioactivity was recovered in the bile.

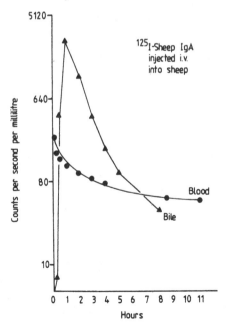

424   *J. G. Hall*

Table 3. The distribution of antibody between the various isotypes of Ig in bile and blood serum 7 days after the GALT of a sheep had been immunized with a suspension of killed *Brucella abortus*. The organisms were exposed to immune bile or serum, washed and then treated with affinity-purified antiglobulin reagents labeled with [125]I.

	Blood serum			Hepatic bile		
	IgG	IgM	IgA	IgG	IgM	IgA
cpm bound	51×10^4	38×10^4	5×10^4	38×10^4	20×10^4	23×10^4
Ratios relative to IgA	2.0	1.3	1.0	1.5	0.8	1.0

experiments, the antibodies to the antigen used to immunize the nonintestinal lymphoid tissue were very much lower in bile than in the blood. Although this suggested that the injection of antigen into the GALT is the best way to populate the bile with antibodies, it was not possible to demonstrate this in every case. Another feature which distinguished the responses in sheep from those in rats was the relatively high levels of IgA antibodies that apparently occurred in the blood of sheep. In rats, the clearance of IgA antibodies from the blood is very rapid (Reynolds et al. 1980), and in practice, it is often difficult to detect antibodies of this class in the blood, even though large quantities pass through the blood-vascular compartment every day. The fact that significant amounts of specific antibody of the IgA class were detected in the blood of immunized sheep is thus, perhaps, another demonstration of the relative inefficiency with which IgA is transported from blood to bile in this species.

Concluding remarks

It is impossible to come to a succinct conclusion about the mechanisms of secretory immunity in ruminants in general, or sheep in particular. There are too few accurate quantitative data. This in part reflects the lack of freely available, well-characterized antiglobulin reagents. Myelomata do not seem to occur in ruminants, and without them it is difficult to produce a comprehensive range of antisera. Monoclonal antibodies may help to close this gap but, at the time of writing, commercially oriented laboratories have judged the potential profits of anti-ruminant Ig reagents to be too low to justify their production. Thus, investigators who are interested in ruminants have to make their own reagents and this diverts time and energy from original experiments, which is a pity when there is so much that could be done.

It is, perhaps, premature to spend time arguing about the details of lymphocyte migration through various mucosae when, in large animals such as ruminants, the questions can be answered directly by relatively straightforward experiments. However, the results of such studies lose some of their value unless they are accompanied by accurate measurements of the Ig classes in the various tissue compartments or cells involved. This is not to say that substantial progress has not been made: a good deal is known, for example, about aspects of secretory immunity that relate to the mammary gland, lactation and suckling. I have excluded this important topic from my review for two reasons: first, I have made no contributions to it, and, second, the subject will be dealt with by Professor A.K. Lascelles (this volume). His earlier review of this subject (Lascelles & MacDowell 1974) must still be the starting point for those who are interested in secretory immunity in ruminants.

Acknowledgements

My own work has been and is supported by project and program grants from the Medical Research Council and the Cancer Research

Figure 4. Graphic representation of the results of an experiment in which a sheep was immunized in the GALT by a direct injection of a suspension of killed *Brucella abortus* at time 0. At the same time, the peripheral somatic lymph nodes (PSLN) were stimulated antigenically by multiple subcutaneous injections of killed *Salmonella typhi* organisms, displaying the 'O' antigen. The titres of agglutinins against *B. abortus* (aBAB) and *S. typhi* O (aSTO) in blood (●○) and bile (▲△) were measured daily for the next fortnight.

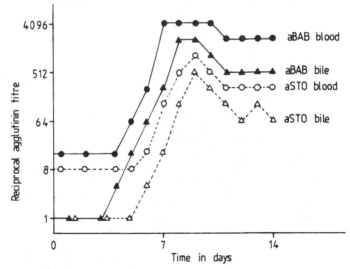

Campaign. None of it would have been possible without the help of many other people. I have been lucky in having Roland Scollay, John Reynolds and Sylvia Denham as post-doctoral colleagues, and my erstwhile research students, Marilyn Smith, Tessa Halstead, Tony Moore, John Hopkins, Liz Andrew and Jo Spencer, have all contributed substantially. Dr Jane Peppard and the late Dr Eva Orlans, assisted by Andrew Payne, provided the immunochemical expertise, and Christine Wynstanley performed many of the biochemical analyses on sheep bile.

References

Andrew, E.M. & Hall, J.G. (1982a). IgA antibodies in the bile of rats. 1. Some characteristics of the primary response. *Immunology*, **45**, 169-74.

Andrew, E.M. & Hall, J.G. (1982b). IgA antibodies in the bile of rats. 2. Evidence for immunological memory in secretory immunity. *Immunology*, **45**, 177-82.

Bienenstock, J. & Befus, A.D. (1980). Mucosal immunity. *Immunology*, **41**, 249-70.

Birbeck, M.S.C., Cartwright, P., Hall, J.G., Orlans, E. & Peppard, J. (1979). The transport by hepatocytes on immunoglobulin A from blood to bile visualized by autoradiography and electron microscopy. *Immunology*, **137**, 477-84.

Bourne, F.J. ed. (1981). *The mucosal immune system*, Current Topics in Veterinary Medicine & Animal Science 12. London: Martinus Nijhoff.

Cahill, R.N.P., Poskitt, D.C., Frost, H. & Trnka, Z. (1977). Two distinct pools of recirculating T lymphocytes: migratory characteristics of nodal and intestinal T lymphocytes. *Journal of Experimental Medicine*, **145**, 420-28.

Cahill, R.N.P., Poskitt, D.C., Hay, J.B., Heron, I. & Trnka, Z. (1979). The migration of lymphocytes in the foetal lamb. *European Journal of Immunology*, **9**, 251-54.

Ciba Foundation (1977). Immunology of the gut. *Ciba Foundation Symposium*, **46**.

Cole, G.J. & Morris, B. (1971). The growth and development of lambs thymectomized *in utero*. *Australian Journal of Experimental Biology & Medical Science*, **49**, 33-53.

Dean, C.J., Gyure, L.A., Styles, J.M., Hobbs, S.M., North, S.M. & Hall, J.G. (1982). Production of IgA-secreting hybridomas: a monoclonal rat antibody of the IgA class with specificity for RTI^c. *Journal of Immunological Methods*, **53**, 307-12.

Denham, S., Spencer, J., Barfoot, R. & Hall, J.G. (1984). IgA antibodies in the bile of rats. 4. Synthesis of secretory antibodies in irradiated rats reconstituted with spleen cells: influence of microenvironment on the regulation of IgA production. *Immunology*, **51**, 45-50.

Draper, L.R., Gyure, L.A., Hall, J.G. & Robertson, D. (1983). Effect of alcohol on the integrity of the intestinal epithelium. *Gut*, **24**, 399-404.

Elson, C.O. (1983). T cells specific for IgA switching and for IgA B-cell differentiation. *Immunology Today*, **4**, 189-93.

Gowans, J.L. & Knight, E.J. (1964). The route of recirculation of lymphocytes in the rat. *Proceedings of the Royal Society of London*, **159**, 257-62.

Gowans, J.L., McGregor, D.D., Cowen, D.M. & Ford, C.E. (1962). Initiation of immune responses by small lymphocytes. *Nature*, **196**, 651-57.

Hall, J.G. (1979). Lymphocyte recirculation and the gut: the cellular basis of humoral immunity in the intestine. *Blood Cells*, **5**, 479-92.

Hall, J.G. (1980). An essay on lymphocyte recirculation and the gut. *Monographs in Allergy*, **16**, 100-11.

Hall, J.G. & Andrew, E.M. (1980). Biliglobulin: a new look at IgA. *Immunology Today*, **1**, 100-4.

Hall, J.G., Parry, D.M. & Smith, M.E. (1972). The distribution and differentiation of lymphborne immunoblasts after intravenous injection into syngeneic recipients. *Cell & Tissue Kinetics*, **5**, 269-81.

Hall, J.G., Hopkins, J. & Orlans, E. (1977). Studies on the lymphocytes of sheep. 3. Destination of lymphborne immunoblasts in relation to their tissue of origin. *European Journal of Immunology*, **7**, 30-37.

Hall, J.G., Gyure, L.A. & Payne, A.W.R. (1980). Comparative aspects of the transport of immunoglobulin A from blood to bile. *Immunology*, **41**, 899-902.

Hall, J.G., Gyure, L.A., Payne, A.W.R. & Andrew, E. (1981). Comparative aspects of secretory immunity. In *The mucosal immune system* (ed. F.J. Bourne), pp. 31-34. The Hague: Martinus Nijhoff.

Hall, J.G., Orlans, E., Reynolds, J., Dean, C.J., Peppard, J.V., Gyure, L.A. & Hobbs, S. (1979). The occurrence of specific antibodies of the IgA class in the bile of rats. *International Archives of Allergy & Applied Immunology*, **59**, 75-84.

Halstead, T.E. & Hall, J.G. (1972). The homing of lymphborne immunoblasts to the gut of neonatal rats. *Transplantation*, **14**, 339-46.

Heremans, J.F. (1974). Immunoglobulin A. In *The antigens* (ed. M. Sela), vol. 2, pp. 365-522. New York: Academic Press.

Husband, A.J., Monie, H.J. & Gowans, J.L. (1977). The natural history of cells producing IgA in the gut. *Ciba Foundation Symposium*, **46**, 29-54.

Lascelles, A.K. & MacDowell, G.H. (1974). Localized humoral immunity with particular reference to ruminants. *Transplantation Reviews*, **19**, 170-208.

Mandel, M.A. & Asofsky, R. (1968). Studies on thoracic duct lymphocytes in mice. *Journal of Immunology*, **100**, 363-69.

Mattioli, C.A. & Tomasi, T.B. (1973). The life span of IgA plasma cells from mouse intestine. *Journal of Experimental Medicine*, **138**, 452-60.

Mayrhofer, G., Pugh, C.W. & Barclay, A.N. (1983). The distribution, ontogeny and origin of rat Ia-positive cells with dendritic morphology and of Ia antigen in epithelia with special reference to the intestine. *European Journal of Immunology*, **13**, 112-16.

McGhee, J.R. & Mestecky, J. eds. (1983). The secretory immune system. *Annals of the New York Academy of Sciences*, **409**.

McWilliams, M., Phillips-Quagliata, J.M. & Lamm, M.E. (1975). Characterization of mesenteric lymph node cells homing to gut-associated lymphoid tissue in syngeneic mice. *Journal of Immunology*, **115**, 54-58.

Moore, A.R. & Hall, J.G. (1972). Evidence for a primary association between immunoblasts and the small gut. *Nature*, **239**, 161-62.

Mullock, B.M., Hinton, R.H., Dobrota, M., Peppard, J.V. & Orlans, E. (1979). Endocytic vesicles in liver carry polymeric IgM from serum to bile. *Biochimica et Biophysica Acta*, **587**, 381-91.

Orlans, E., Peppard, J.V., Reynolds, J. & Hall, J.G. (1978). Rapid active transport of immunoglobulin A from blood to bile. *Journal of Experimental Medicine*, **147**, 588-92.

Orlans, E., Peppard, J.V., Fry, F., Hinton, R.H. & Mullock, B.M. (1979). Secretory component as the receptor for polymeric IgA on rat hepatocytes. *Journal of Experimental Medicine*, **150**, 1577-81.

428   *J. G. Hall*

Orlans, E., Peppard, J.V., Payne, A.W.R., Fitzharris, B.M., Mullock, B.M., Hinton, R.H. & Hall, J.G. (1983). Comparative aspects of the hepatobiliary transport of IgA. *Annals of the New York Academy of Sciences,* **409**, 411-26.

Owen, R.L. & Nemanic, P. (1978). Antigen-processing structures in the mammalian intestinal tract: an SEM study of lymphoepithelial organs. *Scanning Electron Microscopy,* **2**, 367-74.

Pearson, L.D., Simpson-Morgan, M.W. & Morris, B. (1976). Lymphopoiesis and lymphocyte recirculation in the sheep foetus. *Journal of Experimental Medicine,* **143**, 167-75.

Peppard, J.V., Jackson, L.E. & Hall, J.G. (1983). The occurrence of secretory IgM in the bile of rats. *Clinical & Experimental Immunology,* **53**, 623-26.

Peppard, J.V., Orlans, E., Payne, A.W.R. & Andrew, E. (1981). The elimination of circulating complexes containing polymeric IgA by excretion in the bile. *Immunology,* **42**, 83-89.

Quin, J.W., Husband, A.J. & Lascelles, A.K. (1975). The origin of the immunoglobulins in intestinal lymph of sheep. *Australian Journal of Experimental Biology & Medical Science,* **53**, 205-14.

Reynolds, J., Gyure, L.A., Andrew, E. & Hall, J.G. (1980). Studies on the transport of polyclonal IgA antibody from blood to bile in rats. *Immunology,* **39**, 463-67.

Rose, M.L., Parrott, D.M.V. & Bruce, R.G. (1976). Migration of lymphoctyes to the small intestine. 1. Effect of *Trichinella spiralis* infection on the migration of mesenteric T lymphoblasts in syngeneic mice. *Immunology,* **31**, 723-30.

Rose, M.E., Orlans, E., Payne, A.W.R. & Hesketh, P. (1981). The origin of IgA in chicken bile: its rapid active transport from blood. *European Journal of Immunology,* **11**, 561-64.

Scollay, R.G., Hopkins, J. & Hall, J.G. (1976). Possible role of surface Ig in the nonrandom recirculation of small lymphocytes. *Nature,* **260**, 528-29.

Smith, P.D., Nagura, H., Nakane, P.K. & Brown, W.R. (1981). IgA in human hepatic bile and liver. *Journal of Immunology,* **80**, 1476-80.

Spencer, J., Gyure, L.A. & Hall, J.G. (1983). IgA antibodies in the bile of rats. 3. The role of intrathoracic lymph nodes and the migration patterns of their blast cells. *Immunology,* **48**, 687-92.

Sprent, J. (1976). Fate of H-2-activated T lymphocytes in syngeneic hosts. *Cellular Immunology,* **21**, 278-302.

Vaerman, J.-P. & Heremans, J.F. (1970). Origin and molecular size of immunoglobulin A in the mesenteric lymph of the dog. *Immunology,* **18**, 27-38.

Wang, A.C. & Fudenberg, H. (1974). IgA and evolution. *Journal of Immunogenetics,* **1**, 3-13.

Weisz-Carrington, P., Roux, M.E., McWilliams, M., Phillips-Quagliata, J.M. & Lamm, M.E. (1979). Organ and isotype distribution of plasma cells producing specific antibody after oral immunization: evidence for a generalized secretory immune system. *Journal of Immunology,* **123**, 1705-8.

Williams, A.F. & Gowans, J.L. (1975). The presence of IgA on the surface of rat thoracic duct lymphocytes which contain internal IgA. *Journal of Experimental Medicine,* **141**, 335-45.

23

The mucosal immune system with particular reference to ruminant animals

A.K. LASCELLES, K. J. BEH, T.K. MUKKUR and
D.L. WATSON

Extensive work conducted over the past 20 years has established that IgA is
overwhelmingly the predominant immunoglobulin in external secretions in
nonruminant species. In cattle and sheep, however, IgG — especially IgG$_1$ —
is quantitatively much more strongly represented in external secretions. The
partition of immunoglobulin in interstitial fluid, whether synthesized locally or
derived from the blood, between lymph and external secretion is discussed
and differences between secretory organs highlighted. Data are presented on
the effects of route of immunization and type of adjuvant used on the class
specificity of the immune response. Results have shown that intraperitoneal
injection of irritant adjuvant elicits a substantial antibody-containing cell
(ACC) response in intestinal and hepatic lymph, with approximately 30% of
the ACC IgA specific, whereas injection with bland adjuvant elicits only
trivial ACC response in intestinal lymph. The ACC response in hepatic lymph
is in part due to stimulation by antigen after absorption from the peritoneal
cavity. The importance of antibody of different immunoglobulin classes in
external secretions in protection against diseases is considered, with the
interim conclusion that IgI$_1$ and IgG$_2$ are more important than IgA in
mucosal defence of ruminant animals. It is submitted that more work is
required on optimization of the antibody response to ensure that adequate
quantities of antibodies of appropriate immunoglobulin class specificity reach
external secretion. A major effort is also needed to explore the significance of
intraepithelial lymphocytes in immune surveillance and their possible
relationship with other nonepithelial cells in the mucosa.

Introduction

Intense interest in the importance of mucosal immunity as a
distinct part of the body's immune system has only developed in the last
two decades. Immediately before this period, the general concept of a
separate local immune system was viewed with considerable suspicion,
even disbelief, by many scientists. This view prevailed even though

isolated reports in the older literature had established fairly clearly the concept of local as distinct from humoral immunity. Work conducted in rabbits early in this century revealed that oral administration of killed or live cultures of *Shigella* gives rise to specific intestinal resistance to infection which is entirely unrelated to serum antibody level (Davies 1922; Besredka 1927). Approximately 20 years later, Burrows and colleagues working with *Vibrio cholerae* in guinea pigs emphasized the separation of the local and humoral immune system and demonstrated that resistance to oral challenge with living *Vibrio cholerae* organisms was correlated with levels of faecal antibody (Burrows, Elliott & Havens 1947; Burrows & Havens 1948). Thus for the first time the association between protection and the actual level of local antibody was demonstrated. Other studies on the genital tract and mammary gland also suggested that a separate local immune system existed in these organs although the evidence, especially with regard to the mammary gland, was equivocal.

Extensive work conducted over the last 20 years has established that IgA is the universal mucosal immunoglobulin, at least in nonruminant species. In cattle and sheep however, IgG, and especially IgG_1, is much more important, and accordingly in a review of the mucosal immune system of ruminant animals it is appropriate to consider not only secretory IgA but also local antibody belonging to other immunoglobulin classes. Some attention is also given to emerging concepts of T-cell regulation of antibody responses to mucosally applied antigen and the possible function of the nonepithelial cells within the intestinal epithelium, even though most of the work to date has been conducted in laboratory animals.

Immunoglobulins in external secretions

The immunoglobulin in external secretions is either made locally by cells of the lymphocyte-plasma cell series or is derived from the blood plasma and transferred across the epithelium from the interstitial fluid compartment by selective or passive processes. Table 1 presents the range of concentrations of immunoglobulins in blood serum and in various external secretions of cattle and sheep.

A feature of the mucosal immune system of ruminants, which sets it apart from that of other species, is the prominence of IgG_1 relative to IgA. This is particularly striking in the ruminant mammary gland where IgG_1 is overwhelmingly predominant in colostrum on account of its selective transfer from blood serum shortly before parturition (Brandon, Watson & Lascelles 1971). Selective transfer continues into lactation in a much less intense fashion (Watson & Lascelles 1973a). Of all the

Table 1. Concentrations of immunoglobulins in blood serum and various secretions of sheep and cattle. Values in mg/ml represent the average from data presented by various authors, with ranges where appropriate in brackets. From Lascelles & McDowell 1974; Cripps & Lascelles 1976; Morgan et al. 1981; Butler 1983.

Source	Species	IgG_1	IgG_2	IgM	IgA
Serum	cattle	11.2 (6.0–15.1)	9.20 (5.0–13.5)	3.05 (10.6–4.3)	0.37 (0.06–1.0)
Serum	sheep	17.9 (22.6–12.1)	6.30 (4.1–8.8)	2.40 (1.7–3.6)	0.16 (0.05-0.32)
Colostral whey	cattle	48.2 (39.0–61.1)	3.98 (2.1–4.3)	7.10 (5.0–9.9)	4.70 (3.8–6.3)
Milk whey	cattle	0.40 (0.28–0.51)	0.06 (0.03–0.08)	0.15 (0.10–0.19)	0.11 (0.05-0.32)
Parotid saliva	sheep	0.0017	0.0003	0.0001	0.0012
Submaxillary saliva	sheep	0.0572	0.0100	0.0049	0.5050
Thiry Vella loop fluid	adult sheep	2.97	1.48	0.30	4.87
Normal intestinal fluid	cattle	0.25	0.06	trace	0.24
Bile	cattle	0.10	0.09	0.05	0.08
Tears	cattle	0.32	0.01	0.18	2.72
Nasal secretion	cattle	1.56		0.40 (0.23–0.56)	2.81 (1.8–3.9)

secretory organs, the ruminant mammary gland has the least well developed IgA system, although it can be stimulated somewhat by infusion of local antigen during the dry period. If antigen is infused some weeks before parturition, locally produced antibody, a substantial proportion of which is IgA, will be secreted into milk in the ensuing lactation (Lascelles, Outteridge & MacKenzie 1966).

The data in Table 1 show that the intestine of ruminants has a much more prominent IgA system, but IgG is also strongly represented, compared with nonruminant species where immunoglobulin in intestinal secretion is almost exclusively IgA. There is considerable variation in reported relative proportions of IgG and IgA in the intestinal secretion of ruminant animals, probably due to differences in anatomical sites of sampling and methods of collection leading to varying degrees of proteolytic digestion of immunoglobulins on the one hand or increases in permeability of the mucosa on the other (Lascelles, Beh & Mukkur 1982). In sheep, IgA has been shown to be a relatively minor immunoglobulin of parotid saliva but is the major immunoglobulin in saliva collected from the submaxillary gland (Cripps & Lascelles 1976). In accord with these findings the parotid gland contains virtually no plasma cells of any immunoglobulin class specificity whereas the submaxillary gland contains numerous IgA staining cells (Watson & Lascelles 1973b).

Transport of IgG from blood into parotid and submaxillary saliva is clearly selective relative to IgG_2 (Cripps & Lascelles 1976), the magnitude of selective transfer being similar to that reported for the lactating mammary gland (Watson & Lascelles 1973a). In contrast, transfer of IgG_1 (relative to IgG_2) into the secretion of gut loops of sheep is nonselective (Cripps, Husband & Lascelles 1974) although Curtain, Clark and Dufty (1971) reported some degree of selective transfer of IgG_1 into the intestinal contents of cattle. The high concentrations of IgG_2 in the secretion from gut loops of sheep, all of which is derived from blood, suggests the existence of a common γ-receptor on the intestinal epithelium. However, preliminary data (T.K.S. Mukkur & A.K. Lascelles unpublished) indicate that IgG_2 is not transferred into intestinal secretion more readily than ovine serum albumin.

A great deal of our knowledge on the local immune system of the respiratory tract has been derived from studies by clinical immunologists working with human subjects. In general, it has been established that a prominent local IgA system is located mainly in the upper reaches of the respiratory tract, with a virtual absence around the alveolar region where antibody would appear to be almost entirely of humoral origin. In cattle and sheep, the relative abundance of IgA in nasal secretion (Mach & Pahud 1971; Morgan et al. 1981) and IgG in broncho-alveolar washings suggests that these species conform to this pattern.

Origin of immunoglobulin at mucous surfaces

Considerable difficulty has been experienced in the past in establishing unequivocally the local origin of antibody in external secretions. Immunoglobulin may reasonably be judged to be of local origin if three criteria are satisfied. Firstly, the secretion:blood serum concentration ratio for the particular immunoglobulin should be higher than the ratio for other immunoglobulins or serum albumin. Secondly, cells of the lymphocyte-plasma cell series should be present in tissue near

Table 2. Intestinal secretion/blood plasma ratios for IgG_1, IgG_2 and IgA determined following intravenous inoculation of sheep with radioisotope-labeled immunoglobulins. The proportion of each immunoglobulin in intestinal secretion derived from plasma is computed by expressing the specific activity for secretion as a percentage of the specific activity for plasma. Values presented are means $\pm$ SE. From Cripps, Husband & Lascelles (1974).

	IgG_1	IgG_2	IgA
Number of animals	4	4	3
Secretion/plasma ratio	0.175 ± 0.016	0.212 ± 0.035	0.572±0.072
Percentage plasma derived	93.96 ± 2.04	102.65 ± 8.87	2.21 ± 0.54

the glandular epithelium and a significant proportion of these cells should be shown to contain the immunoglobulin in question, by either immunofluorescence or immunoperoxidase techniques. Finally, in order to make the critical distinction between local synthesis on the one hand and selective transport of serum-derived immunoglobulin across the glandular epithelium on the other, it is necessary to inject intravenously purified immunoglobulin labeled with an isotopic marker and compare specific radioactivities in serum and secretion. Estimates of the proportion of locally derived immunoglobulin in secretions into Thiry Vella intestinal loops, or the putatively more physiological double re-entrant intestinal loops, have been made by expressing the specific radioactivity (cpm/mg immunoglobulin) of immunoglobulin in secretion as a percentage of specific activity of the immunoglobulin in blood plasma (Table 2). It can be seen that virtually all the IgA (the IgA labeled was dimeric IgA derived from intestinal lymph of sheep) was made locally in the intestine whereas all the IgG_2 and the great majority of IgG_1 were derived from the blood. This is perhaps surprising in view of the large numbers of IgG_1-containing plasma cells in the *lamina propria* of the small intestine of sheep and cattle. In this regard studies of intestinal lymph are of interest. It is generally accepted that the concentrations of various proteins in interstitial fluid are similar to those in lymph and that in most cases virtually all the protein in lymph is derived from the capillary filtrate (Yoffey & Courtice 1980). In studies on sheep (Quin, Husband & Lascelles 1975), comparison of the intestinal lymph:blood plasma concentration ratios for various immunoglobulins with that for albumin suggest that IgG_1, IgG_2 and especially IgA are added to the capillary filtrate forming in the intestine or are added to the lymph passing through the mesenteric lymph node. More precise estimates of the regional contribution of immunoglobulin were obtained by use of labeled proteins. Three labels

— ^{125}I and ^{131}I for immunoglobulins and T1824 (Evans blue dye) for albumin — were used routinely in each animal, allowing precise comparisons between the concentrations of any two immunoglobulin classes or subclasses and albumin and thus providing an estimate of the amount of protein derived from serum by capillary filtration. Estimates of the percentage of each protein derived from blood plasma using these methods are given in Table 3. While Beh's (1977) studies showed that virtually all the IgA in intestinal lymph is derived from the *lamina propria*, the extent of local contribution of IgG$_1$ and IgG$_2$ by immunoglobulin-secreting lymphoid cells in the *lamina propria* relative to those in the mesenteric lymph node is not yet known. However, the presence of large numbers of IgG$_1$-containing plasma cells in the *lamina propria* in the absence of substantial intestinal secretion of IgG$_1$, together with evidence for a significant local contribution of IgG$_1$ from the region drained by the intestinal lymph duct, suggests that a great deal of the IgG$_1$ made locally in the *lamina propria* contributes to the circulating pool. Figure 1 provides an interpretative illustration, without passing judgement on the question of selective transport of IgG into intestinal secretion.

Secretory IgA

The so-called secretory IgA (S.IgA) is present in external secretions. It is composed of an IgA dimer held together by a covalently bonded J chain (molecular mass of approximately 16,500 daltons) (Halpern & Koshland 1970; Koshland 1975; Butler 1983) which is now known to be a component of all polymeric immunoglobulins. Bovine IgM possesses covalently and noncovalently bonded J chains (Komar & Mukkur 1975). The IgA dimer and J chain are synthesized by the plasma cells and secreted into interstitial fluid (Hanson & Brandtzaeg 1980). The dimeric IgA-J-chain complex has a strong affinity for another polypeptide chain (molecular mass of approximately 80,000 daltons), now called the secretory component (SC), which is made by the glandular epithelial cells. SC is synthesized as a large transmembrane protein. A portion of the SC remains in the membrane after proteolytic cleavage liberates the free form, which in the bovine has a molecular mass of 74,000 daltons (Butler 1983). The portion of SC remaining in the epithelial membrane serves as a receptor for the dimeric IgA-J-chain complex; this IgA-J-chain-SC complex is ferried across the glandular epithelial cell to be released at the apical surface into the lumen of the secretory organ (Figure 1). Thus IgA in external secretions occurs almost exclusively as S.IgA with a sedimentation rate of 11S and a molecular mass of 385,000 to 450,000 daltons, whereas in the serum of ruminant animals it is mainly IgA dimer with J chain. In nasal secretions of cattle, IgA has also been found to occur in trimeric form with a molecular mass of 642,000 daltons (Komar, Abson & Mukkur 1975). It is of interest that IgM with J chain also has an affinity for SC and, as with IgA, the attachment is by a covalent bond (Brandtzaeg 1975).

The affinity of dimeric IgA for transmembrane SC provides a selective mechanism of transport for dimeric IgA (Fisher et al. 1979; Socken & Underdown 1978). In addition, studies in other species have shown that S.IgA is more resistant to proteolysis than dimeric IgA (Brown, Newcomb & Ishizaka 1970; Shuster 1971). While it has been formally demonstrated,

Figure 1. This diagram is an interpretation of work carried out on transport of blood-derived and locally synthesized immunoglobulins of various isotypes into intestinal secretion and the draining lymph. Dimeric IgA and J chain synthesized by plasma cells in the *lamina propria* bind selectively with secretory component displayed on the basal or intercellular membrane of the epithelial cell and is then ferried, probably in vesicles, across the cell. Some dimeric IgA is removed from the interstitial fluid compartment in the lymph. Most of the IgG which is locally synthesized is thought to be removed in the lymph, with a smaller amount transported into intestinal secretion. Some of the IgG_1, IgG_2 and IgM in the capillary filtrate is transported into external secretion while the remainder will return to the circulation by way of the lymphatic system.

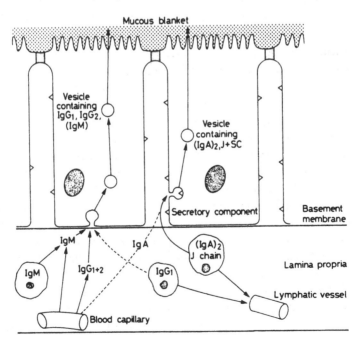

Table 3. The concentration of immunoglobulins and albumin in sheep plasma and lymph and their lymph-to-plasma label ratios, determined following intravenous injection of radioisotope-labeled immunoglobulins and Evans blue-labeled albumin. Values are presented as the mean ± SE. Significance is determined by analysis of variance of the difference between the amount derived from plasma and the lymph concentration. From Quin, Husband & Lascelles (1975).

	IgG$_1$	IgG$_2$	IgA	Albumin
Number of sheep tested	4	3	3	3
A. plasma concentration mg/ml	21.56±0.59	5.47±1.06	0.16±0.02	22.70±2.47
B. lymph concentration mg/ml	11.20±1.26	2.14±0.64	0.48±0.10	11.18±2.06
C. lymph : plasma label ratio	0.38±0.4	0.28±0.04	0.30±0.04	0.49±0.05
Amount locally synthesized mg/ml (B-AxC)	3.23±0.89	0.6±0.21	0.44±0.10	1.00±0.42
% derived from plasma	74.8±8.2	73.4±4.4	8.3±3.5	99.2±1.5
Significance of local synthesis	0.001	0.001	0.001	not significant

for the intestine at least, that most of the IgA in secretion from isolated loops is made locally by cells of the lymphocyte-plasma cell series (Table 2), it is clear that dimeric IgA, derived from the bloodstream and appearing in the capillary filtrate of an external secretory organ synthesizing transmembrane SC receptor, would also be largely transported into external secretions (Bienenstock & Befus 1983). The relative importance of blood-derived dimeric IgA in any particular secretion (with abundant SC receptor) would clearly be a function of its plasma concentration and capillary permeability on the one hand and the relative abundance of mucosal IgA plasma cells on the other. The IgA content of parotid saliva in sheep is of interest in this regard (Watson & Lascelles 1973b; Cripps & Lascelles 1976). There are essentially no plasma cells in parotid salivary tissue and virtually all the IgA in the saliva would be expected to be derived from blood. The concentration of IgA in saliva is considerably higher than can be accounted for by passive diffusion from blood, indicating that IgA is selectively transferred presumably by virtue of the SC receptor on the glandular epithelium.

The high concentration of IgA in intestinal lymph compared with blood serum, virtually all of which is dimeric IgA in ruminant animals (Beh, Watson & Lascelles 1974), indicates that most of the IgA in serum is derived from the small intestine. Beh, Watson and Lascelles reported that only lymph from the involuting mammary gland contains higher levels of IgG than blood, but in view of the low flow rate of lymph it was concluded that this source does not contribute substantially to the circulating pool. The situation in other species, such as the mouse, human and pig, differs somewhat, in that 7S.IgA derived from lymph nodes and spleen is much more abundant in serum (Butler 1983) and would be expected to have transport dynamics more closely comparable to IgG than dimeric IgA (Lascelles & McDowell 1974).

In recent years it has been established that in rodents IgA reaches the intestinal lumen by way of the bile. It is evident that IgA oligomer is selectively transported from blood, largely by virtue of its affinity for secretory component located on the surface of hepatocytes (Bienenstock & Befus 1983) and IgA immune complexes are effectively removed from blood into the bile of rodents in this way. Thus in rodents the bile seems to serve as an IgA 'pump', contributing to the S.IgA in the intestinal lumen and also to the clearance of circulating antigens. In the ruminant the concentration of IgA in bile is relatively low and the IgA 'pump' would appear to be poorly developed (Hall, Gyure & Payne 1980). Nevertheless the bile:serum concentration ratio for IgA in ruminants is higher than the ratio for other immunoglobulins, suggesting either a degree of selective transport and/or local synthesis of IgA. In this connection K.J. Beh (unpublished) has detected IgA-containing plasma cells in the submucosa of the gall bladder and biliary ducts, suggesting that at least some of the IgA in bile is of local origin.

Cellular basis of mucosal immunity

Lymphatic recirculation and homing

The thoracic duct collects lymph from the small and large intestine, with smaller but significant contributions from the liver and hindquarters (Yoffey & Courtice 1980). Most of the lymphoid cells in the thoracic duct are derived from the *lamina propria* of the small intestine and the mesenteric lymph node chain, as well as the hepatic lymph nodes. In contrast to lymph collected from other regions, lymph from the intestine contains a substantial number of large lymphocytes (up to 10%) which arise by cell division in the region drained by the intestinal lymph, as shown by studies on ^{3}H thymidine incorporation. Unlike small lymphocytes in thoracic duct lymph and lymph from other regions, the greater part of the large dividing cell population does not recirculate continuously between blood and lymph (Gowans & Knight 1964). Gowans and Knight reported that ^{3}H thymidine-labeled large lymphocytes from the thoracic duct lymph of rats preferentially seed the lymphoid tissue of the gut following intravenous injection of syngeneic recipients. Some years later essentially similar findings were reported by Griscelli, Vassalli and McCluskey (1969) who in addition showed that lymphocytes from peripheral lymph nodes home to similar nodes following injection into syngeneic recipients. A substantial proportion of the large dividing cell population in sheep contains IgA (Beh & Lascelles 1974). In laboratory rodents at least another population of cells has been recognized which lacks internal IgA but bears IgA on the cell surface, and it is suggested that these cells are probably precursors of IgA-containing cells.

There are three possible sites of origin of IgA-containing cells in thoracic duct lymph: (1) Peyer's patches, (2) intestinal *lamina propria* and (3) mesenteric lymph node. Craig and Cebra (1971) have demonstrated that Peyer's patches in the rabbit are an enriched source of potential IgA-producing cells and that cells from Peyer's patches seed the gut more readily than lymphoid cells from the peripheral blood or popliteal lymph nodes. In suitably primed rats, the removal of Peyer's patches from an isolated loop of intestine effectively prevents the generation of antibody-containing cells of the IgA class in thoracic duct lymph following mucosal antigenic stimulation (Husband, Moni & Gowans 1977). Ablation studies have demonstrated that IgA-containing cells in thoracic duct lymph do not originate in mesenteric lymph nodes, but it has been suggested that the mesenteric lymph node may play a role in expanding the population (Husband & Gowans 1978). Studies in sheep on efferent and afferent lymph from the distal end of the small intestine have shown that IgA-containing cells originate from Peyer's patches and/or *lamina propria*, and not from the mesenteric lymph node (Beh 1977).

In the rodent, it has been shown that a proportion of the large cells are T blasts which home to the *lamina propria* of the adult intestine or into foetal intestinal isografts (Guy-Grand, Griscelli & Vassalli 1974; Guy-Grand, Griscelli & Vassalli 1978; Sprent 1976), whereas T blasts from peripheral lymph nodes only localize in the intestine under special conditions (Rose, Parrott & Bruce 1976).

Studies by Cahill and colleagues (1977) showed that small lymphocytes also exhibit preferential migration. These workers, employing regional lymph-collection methods in unanaesthetized sheep, demonstrated that ^{51}Cr-labeled, nylon-wool-fractionated small lymphocytes (putative T lymphocytes) from the thoracic duct or peripheral lymph return to the gut and peripheral lymph node respectively and thence to the lymph, following their injection intravenously into the same sheep. These results extend those of Scollay, Hall and Orlans (1976) and are essentially confirmed by Chin and Hay (1980). Other studies on recirculation of labeled lymphocytes in mesenteric lymphadenectomized sheep indicate that the cells responsible for preferential intestinal migration are present in efferent and afferent lymph (Hall 1980).

Analysis of the mucosal T-cell population

The functional significance of the mucosal-associated T-cell population has not been studied in ruminant animals for obvious reasons. It seems important in a review of the mucosal immune system in ruminant animals to be aware of the general thrust of studies in the laboratory rodent, illustrating the importance and complexity of T-cell influences on the mucosal immune response. Recent *in vitro* data (Kawanishi, Saltzman & Strober 1983) have shown that murine Peyer's patches possess a special T-cell subset which causes switching of surface (s) IgM-bearing cells to sIgA-bearing cells, whereas T cells from the spleen fail to stimulate expansion of sIgA-bearing cells in a B-cell population derived from Peyer's patches. Thus it is evident that these 'switching' T cells are in fact responsible for IgA B-cell enrichment in Peyer's patches.

It is also evident that T-cell subsets are responsible for the regulation of terminal maturation (after switching) of B cells bearing IgA or other immunoglobulin classes. These regulatory T-cell subsets are responsible for the generation of the IgA response and suppression of systemic IgG and IgM responses following exposure to orally administered antigen (Clough, Mims & Strober 1971; Mattingly & Waksman 1978; Ebersole, Taubman & Smith 1979; Bienenstock & Befus 1980; Challacombe & Tomasi 1980; Richman et al. 1981). It is apparent that T-suppressor cells are induced in Peyer's patches after oral administration of T cell-dependent antigens and that subsequent migration of these cells leads to systemic suppression of the IgG and IgM responses while concurrent local secretory IgA immunity is induced.

More recently a contrasuppressor T-cell subset in Peyer's patches has been demonstrated in *in vitro* studies (Green et al. 1982). Contrasuppressor cells can protect IgA helper cells from high levels of suppressor activity. In this way it is postulated that an IgA response is permitted to occur in the mucosal microenvironment when the systemic immune response is highly suppressed. Notwithstanding, parenteral administration of cholera toxin or toxid to rats and rabbits often results in suppression of the mucosal response in the intestine (Yardley et al.

1978; Pierce & Koster 1980). The suppression is reported to be largely due to the suppressive effects of cells derived from the spleen on the primary response in the Peyer's patches (Koster & Pierce 1983).

Thus the situation is extremely complex, but it is already clear that an understanding of T cell-regulating influences involved in mucosal immune responses will assist in determining the basis for the variability in response to exposure to oral antigen. It is appropriate at this stage to use the experimentally amenable laboratory animal.

The common mucosal immune system

It is now accepted that deposition of antigen in the gut leads to the production of IgA antibody in secretion at sites distant from the intestine. Bohl and colleagues appear to be the first to have reported this effect (Bohl et al. 1972; Saif, Bohl & Gupta 1972). In their studies on pigs, they described the appearance of IgA antibody in the colostrum and milk following infection or oral vaccination with transmissible gastroenteritis virus. The existence of this phenomenon was subsequently confirmed by other workers in humans and other monogastric species and was extended to include lacrimal and salivary secretions (Bienenstock & Befus 1980). It is of interest that cells originally thought to be IgA plasma cells in human colostrum (Goldblum et al. 1975; Ahlstedt et al. 1975) have now been shown to be colostral macrophages replete with secretory IgA (Pittard, Polmar & Fanaroff 1977; Crago et al. 1979).

There are significant numbers of T and B lymphocytes in colostrum and milk of various species, including ruminants, which are responsive to mitogens, antigens and in mixed leucocyte culture (Head & Beer 1978; Parmely, Beer & Billingham 1976; Smith & Schultz 1977; Outteridge & Lee 1981). The relative proportion of T and B lymphocytes and lymphocytes with C3b receptors in blood and secretion indicate a selective recruitment of lymphocytes into the secretions of humans (Head & Beer 1979). The transfer of IgA-specific B cells probably accounts for much of this. However, while selective recruitment of cells occurs into the ruminant mammary gland, jejunal stimulation with antigen in cows gives rise to a predominantly IgG_1 antibody response in mammary secretion (Chang, Winter & Norcross 1981). As only a few IgG_1-producing cells are found in this tissue, it would appear that the IgG_1 antibody is largely of humoral origin.

Bienenstock and colleagues first drew attention to the similarities between bronchus-associated lymphoid tissue (BALT) and gut-associated lymphoid tissue (GALT) and went on to develop the concept of the common mucosal immune system (Bienenstock, Johnston & Perey 1973a;

1973b). This now well-known concept prescribes that GALT and BALT are the sources of IgA precursor cells sensitized to ingested or inhaled antigens. These cells enter the circulation via the thoracic duct and seed various submucosal sites. However, it is evident that GALT is much more important than BALT in quantitative terms as the central source of IgA precursor cells.

Stimulation of mucosal immunity

Feeding T cell-dependent antigens to rodents (Thomas & Parrott 1974; Andr et al. 1975; Swarbrick, Stokes & Soothill 1979) and pigs (Newby et al. 1979) tends to induce oral tolerance — a state in which the systemic immune response is suppressed. On the other hand, feeding many T cell-independent antigens, such as endotoxins of gram-negative bacteria and bacterial polysaccharides, does not induce oral tolerance, but actually increases the immune response to the same antigen (Chidlow & Porter 1977; Stokes et al. 1979). However, studies in sheep with the T cell-dependent antigens ovalbumin and ferritin have demonstrated that repeated local infusions of antigen into Thiry loops result in the appearance of IgG antibody in blood serum, although it was not determined whether some degree of suppression of the response occurs. Prolonged infusion of these antigens eventually induces a detectable local antibody response which is mainly IgA specific (Husband & Lascelles 1974).

It has been reported in rats and dogs that the capacity of different forms of cholera toxin to bind to cell surfaces correlates with their ability to induce an IgA immune response (Pierce 1978; Pierce & Cray 1982). The most effective form for inducing an anti-toxin response is the holotoxin which activates adenyl cyclase as well as binding to the cell membrane. Recent results of Lycke, Lindholm and Holmgren (1983), based on studies with the cholera B toxin subunit which lacks the adenyl cyclase-stimulating activity, suggest that this property of cholera toxin may be responsible for stimulating the end stage of differentiation of IgA-committed B cells in the *lamina propria*.

Replicating microorganisms are generally believed to be more effective in eliciting a local immune response than their nonreplicating counterparts. This has been demonstrated recently with *Escherichia coli* immunization in pigs where smaller doses of live organisms produce higher levels of mucosal antibody than large numbers of killed organisms (Evans et al. 1980). It is not clear whether the difference is due to replicating organisms achieving a greater antigenic mass over time or whether during the course of colonization the replicating organism

Table 4. Output and immunoglobulin class distribution of ACC in the efferent lymph from different lymph nodes in sheep at the peak of response, 7 to 8 days after intraperitoneal injection of ovalbumin in Freund's complete adjuvant.

Regional lymph	Total ACC output per hour ($\times 10^{-5}$)	Distribution of ACC among Ig classes (%)		
		IgG	IgA	IgM
Popliteal	0	–	–	–
Intestinal	120±57	27	25	46
Coeliac	55±15	38	28	34
Hepatic	40±22	37	38	25

becomes more intimately associated with the epithelium, thus facilitating transfer of antigen across the epithelium of the Peyer's patches or *lamina propria*. It is also possible that the replicating organisms may induce regional permeability changes with an enhanced transfer of larger molecules from the lumen by virtue of a mild irritant effect.

Studies in rats have shown that the most efficient experimental regime for inducing large numbers of antibody-containing cells (ACC) of the IgA class in the *lamina propria* is to inject antigen (cholera toxid) in Freund's complete adjuvant by the intraperitoneal route followed by an oral boost of antigen 2 weeks later (Pierce & Gowans 1975; Husband & Gowans 1978). Large numbers of ACC appeared in the thoracic duct lymph shortly after the oral boost, and subsequently these cells extravasated into the *lamina propria* and proliferated under the influence of locally instilled antigen (see also Husband 1982).

Stimulation of IgA responses in ruminant animals

In sheep, intraperitoneal injection of antigen in Freund's complete adjuvant, without an oral boost, stimulates a vigorous ACC response in intestinal lymph, with a smaller proportion of the ACC of the IgA isotype than reported in rats (Beh, Husband & Lascelles 1979). This work on specific 'central stimulation' of the IgA system raises questions about the influence of the route of administration and the nature of adjuvant on the isotype specificity of the ACC response in lymph from different regions. These questions were addressed by comparing the output of ACC of different isotypes in lymph from various regions following regional injection of ovalbumin in Freund's complete and incomplete adjuvant and Adjuvant 65, a nonirritant, vegetable oil-based adjuvant.

The ACC output in hepatic, coeliac and intestinal lymph is presented in Table 4, together with results for popliteal lymph following

Table 5. Output and immunoglobulin class distribution of ACC in the popliteal and intestinal lymph of sheep at the peak of response, 6 days after subcutaneous injection of ovalbumin in Freund's complete adjuvant in the hind leg.

Regional lymph	Total ACC output per hour $(\times10^{-5})$	Distribution of ACC among Ig classes (%)			
		IgG_1	IgG_2	IgM	IgA
Popliteal	122±48	86	10	4	3
Intestinal	0	–	–	–	–

intraperitoneal injection of ovalbumin in Freund's complete adjuvant. The coeliac lymphatic duct drains lymph formed in the abomasum and possibly the first few centimetres of the duodenum. The duct enters the *cisterna chyli* directly or joins with the hepatic efferent duct.

As expected, intraperitoneal immunization failed to elicit a response in the popliteal lymph, but substantial responses occurred in lymph from the abdominal viscera (Beh & Lascelles 1981). ACC comprised 2 to 4% of total lymphocytes at the peak of the response, and cells of IgA specificity constituted a substantial proportion of ACC in lymph from the three drainage regions. Essentially similar results were obtained when Freund's incomplete adjuvant was used instead of Freund's complete adjuvant. In contrast, when antigen was injected subcutaneously in the hind limb with Freund's complete adjuvant, only a very small proportion of the substantial ACC output in the popliteal lymph was of the IgA isotype (Table 5).

In an attempt to account for the substantial IgA ACC response in intestinal and hepatic lymph following the injection of antigen in Freund's adjuvants, it was suggested that the irritant adjuvant creates a permeability change in the serosa permitting the penetration of antigen into Peyer's patches or *lamina propria* where IgA precursor cells reside (Beh, Husband & Lascelles 1979). This hypothesis was partially tested by comparing the ACC responses in intestinal and hepatic lymph following the intraperitoneal injection of antigen in either irritant Freund's adjuvant or nonirritant Adjuvant 65. In contrast to the vigorous ACC response in intestinal lymph following injection of antigen in Freund's adjuvant, the response in the animals treated with Adjuvant 65 was small. Indeed there were insufficient ACC to determine immunoglobulin class distribution accurately (Beh & Lascelles 1984).

Autopsy at intervals after intraperitoneal immunization with antigen in Freund's adjuvant revealed peritoneal thickening of extensive areas of intestinal and hepatic serosa together with small granulomata containing

Table 6. Density and immunoglobulin class specificity of ACC in Thiry Vella loops of four sheep. Sheep were immunized with ovalbumin in Freund's incomplete adjuvant by intraperitoneal injection. Two loops were prepared in each sheep 11 days later, and after 2 more days ovalbumin in saline was infused into one loop. The response was measured after 6 more days. Values are means ± SE. Class specifications are percentage of ACC staining with respective antisera. From Husband, Beh & Lascelles (1979).

	ACC/cm	% IgG	% IgA
Immunized loop	1831 ± 419	38.0 ± 3.9	53.5 ± 3.6
Nonimmunized loop	45 ± 9	55.4 ± 6.9	46.1 ± 0.1

adjuvant. In contrast, Adjuvant 65 was rapidly removed without any evidence of irritation. The basis for the ACC response in hepatic lymph has been examined in recent experiments and the results show that a significant component of the response can be attributed to the return of some of the antigen/oil emulsion to the circulation by way of the diaphragmatic lymphatic vessels and stimulation of the hepatic lymphoid apparatus following uptake from the blood (Beh & Lascelles 1984).

Maximization of the local immune response

The IgA system in the ruminant mammary gland is poorly developed and can only be readily discerned following local infusion of antigen during the dry period, but even then IgA in colostrum and milk secreted during the ensuing lactation represents only a small proportion of total immunoglobulin. Although the IgA system of the intestine of sheep and cattle is not as highly developed as in other species, it is much more prominent than the system in the ruminant mammary gland.

Local immunization of isolated gut loops in sheep which have received a prior intraperitoneal injection of antigen in Freund's adjuvant gives rise to a massive ACC response in the *lamina propria* (Table 6), with approximately 50% of the ACC of the IgA class (Husband, Beh & Lascelles 1979). On the other hand, preliminary findings indicate that intraperitoneal priming only results in a relatively small enhancement of the antibody response in mammary glands to subsequent local infusion of antigen. The relatively small enhancement of the response in the mammary gland may reflect a deficiency of appropriate T-helper cells in this tissue in the ruminant, thereby limiting the expansion of extravasated IgA precursor cells.

DEAE dextran has been shown to be an effective intestinal adjuvant when infused with ovalbumin or killed *Brucella* (Beh 1979). This adjuvant markedly stimulates the ACC response in regional lymph from the distal end of the ileum. On the other hand, the same adjuvant administered locally into sheep mammary glands during the dry period does not produce a measurably enhanced local immune response during the following lactation (K.J. Beh & A.K. Lascelles unpublished).

Functional role of antibody at mucous surfaces

It has long been accepted that antibody at mucous surfaces often plays a key role in preventing initial establishment and penetration of pathogens. However, the relative importance of blood serum-derived and locally produced antibody (or simply antibody in ingested milk not absorbed from the gut) and the class specificity of antibody have recently preoccupied immunologists working with domestic animals.

In general, secretory IgA antibody is believed to have a role in the exclusion of food allergens (Walker & Isselbacher 1977), in viral neutralization in the absence of complement — preventing colonization by viruses which remain superficially on the mucosa (Tomasi & Grey 1972) —, and in immunity to *E. coli* infections in suckling and weaned pigs (Porter et al. 1974). Its role in ruminants is less clear than in monogastric species. However, neutralizing activity in the nasal secretions of cattle, following natural or intranasal exposure to live parainfluenza (PI-3) virus or a mixed live vaccine comprising PI-3 and infectious bovine rhinotracheitis (IBR) viruses, is almost exclusively associated with IgA (Morein 1972; Mukkur, Komar & Savina 1975). Furthermore, it was recently reported that live IBR vaccines administered intranasally are clearly superior to an inactivated polyvalent calf pneumonia vaccine (which includes IBR virus) administered systemically (Frerichs et al. 1982). On the other hand, resistance in calves to *Mycoplasma* infection of the respiratory tract, which appears to depend on the presence of antibody of the IgG class in the lung at the time of challenge, can be induced by parenteral administration of killed organisms incorporated in adjuvant (Howard, Gourlay & Taylor 1981).

The significance of enhancement of the IgA response by initial intraperitoneal injection of antigen in Freund's adjuvant has not been evaluated adequately in relation to infectious diseases of the intestine and mammary gland, although there is evidence that such a vaccination regime may be of significance against challenge with *Salmonella* in sheep (Husband 1978). Another important aspect of the functional significance of IgA is the relationship of this immunoglobulin to phagocytic cells in the context of microbial infections at epithelial surfaces. One report suggested that IgA could act as an opsonin against *E. coli* (Wernet et al. 1971) but this result could not be repeated in other laboratories (Porter, Parry & Allen 1977). Most evidence in the literature suggests that IgA plays no active role in phagocytosis of microbes (Klebanoff & Clark 1978), whereas some workers have shown that polymeric IgA activity actually inhibits neutrophil function (Kemp, Cripps & Brown 1980; Van Epps & Williams 1976). On the other hand, there is some recent evidence that subpopulations of neutrophils may bear IgA membrane receptors (Lawrence, Weigle & Spiegelberg 1975; Shen & Fanger 1981). Clearly, this feature of IgA needs to be determined for each relevant disease and species and also using populations of cells from the particular tissues of interest.

The role of IgA responses in immunity to helminth parasites has yet to be resolved. While a substantial IgA ACC response in gastric (coeliac) lymph of sheep has been found associated with reduced numbers of worms and arrested development of survivors following a large challenge of animals immune to *Osteragia*, a direct role of IgA in parasite immunity has not yet been established (Smith et al. 1983). It is of interest that previously immunized sheep receiving only a small challenge dose of infective *Osteragia* larvae did not exhibit an ACC response in gastric lymph, and this was associated with evidence of a much lower level of resistance to the parasite (Smith et al. in press).

Unquestionably the most important immunoglobulin isotype in young ruminants is IgG_1 derived from colostrum, which is crucial for protection of the newborn against various pathogens. Young animals are adequately protected against colibacillosis and rotavirus infections by maternally derived IgG_1. Snodgrass and colleagues demonstrated conclusively the effectiveness of passive IgG_1 antibody acting locally in the gut lumen of lambs and calves challenged with rotaviruses (Snodgrass & Wells 1978; Snodgrass et al. 1980; Fahey et al. 1981). Recent data indicate that anti-K99 pilus antibody of the IgG class derived from the colostrum of immunized ewes is more efficient on a molar basis than IgM derived from the same source (Altmann & Mukkur 1983).

It has now been established in sheep (Watson 1975), cattle (McGuire, Musoke & Kurtti 1979; Howard, Taylor & Brownlie 1980) and goats (Micusan & Borduas 1977) that IgG_2 is the only immunoglobulin isotype which is cytophilic for neutrophils. Furthermore, this property of IgG_2 is of crucial importance in resistance to those microbial infections in which phagocytosis is an important effector mechanism, e.g. staphylococcal mastitis (Watson 1976) and theileriosis (Musoke et al. 1982). Clearly immunization procedures need to be devised to accentuate IgG_2 synthesis in these situations. In this connection, it has been shown that subcutaneous injection of sheep with live *Staphylococcus aureus* vaccine promotes IgG_2 synthesis rather than IgG_1, whereas killed vaccines favour IgG_1 synthesis. The live immunization regime gives rise to substantial protection against intramammary staphylococcal challenge (Watson & Lee 1978; Watson & Kennedy 1981), contrasting with the relatively poor protection provided by killed vaccines. In conformity with the above findings, it has been reported that selective immunodeficiency of IgG_2 in the Red Danish breed of cattle is associated with a striking increase in susceptibility to infection with pyogenic bacteria (Nansen 1972).

Nonepithelial cells in the gut epithelium

The humoral aspects of mucosal immunity, including its cellular basis, has preoccupied research workers: studies have been concerned mainly with analysing the functions of cells from Peyer's patches and the *lamina propria* in relation to IgA production. The large population of lymphoid cells situated in the epithelium itself between the basement membrane and the lumenal surface has been recognized for some time, but serious attention has been devoted only recently to considering its natural history and functional significance. The presence in a substantial proportion of intraepithelial lymphocytes of basophilic granules resembling those in mast cells has prompted the suggestion that granular intraepithelial lymphocytes give rise to mast cells (Guy-Grand, Griscelli & Vassalli 1978). These researchers suggested that the cells are of T-lymphocyte lineage, a contention supported by recent studies drawing attention to similarities of surface antigens (Janossy et al. 1980; Lyscom & Brueton 1982).

There is mounting evidence that intraepithelial lymphocytes are related to natural killer (NK) cells. Thus intraepithelial lymphocytes from guinea pigs and mice have been reported to exhibit NK-cell activity against tumour targets (Arnaud-Battandier et al. 1978; Tagliabue et al. 1981). Studies in mice further show that intraepithelial lymphocytes are virtually identical with NK cells isolated from the spleen with regard to their nonadherence, partial susceptibility to high concentrations of anti-Thy 1.2 antibodies plus complement, target-cell specificity and the effect of interferon B. Similar cells were not found in Peyer's patches. Essentially similar properties were reported for isolated rat intraepithelial cells by Flexman, Shellam and Mayrhofer (1983), except that the cells were not as responsive to interferon as were NK cells of splenic origin. NK-cell activity has been shown in the lungs of mice (Puccetti et al. 1980), and the possible role of intraepithelial lymphocytes as a mucosal cellular surveillance mechanism has been postulated (see also Herberman & Ortaldo 1981).

Studies by Mayrhofer and colleagues indicate that granular and nongranular forms of intraepithelial cells are not thymus derived but that their development is under thymic influence (Mayrhofer 1980; Mayrhofer & Bazin 1981) as indeed are the proliferative responses of mucosal mast cells (Mayrhofer 1979; Miller & Nawa 1979a; 1979b; Nawa & Miller 1979; Haig et al. 1982). Nawa and Miller have also demonstrated the striking expansion of the mucus-secreting goblet-cell population in the intestinal epithelium following injection of immune (immunoglobulin-negative) putative T lymphocytes from the thoracic duct lymph, and to a lesser extent immune serum, into rats infected with *Nippostrongylus brasiliensis*. The changes were associated with parasite expulsion.

Studies in the rat have shown that globule leucocytes and mast cells in the *lamina propria*, in contrast to those in connective tissue, contain intracellular IgE (Mayrhofer, Bazin & Gowans 1976). Mayrhofer and colleagues also demonstrated that plasma IgE, which increases enormously in concentration in animals infected with *N. brasiliensis*, is produced mainly in the mesenteric lymph node, thus arguing against the notion that IgE is a secretory immunoglobulin. It would thus appear that IgE on and in mast cells is derived from the capillary filtrate.

There are suggestions that intraepithelial lymphocytes bear IgE receptors (Befus & Bienenstock 1982) which may be associated with their suggested immune surveillance role. Parasitic infections of the gut are associated with increases in concentration of intraepithelial lymphocytes especially the granular forms (Ferguson 1977; Flexman, Shellam & Mayrhofer 1983). In contrast, recent studies in sheep have shown much fewer intraepithelial lymphocytes (and plasma cells in the *lamina propria*) in ileal segments isolated from the intestinal tract before birth. When isolated segments were reconnected into the intestinal tract the number of intraepithelial lymphocytes returned to normal (Reynolds & Morris 1983).

Eosinophils are also normal inhabitants of the gut epithelium and *lamina propria* and their numbers also increase in parasitic diseases, but their detailed role in expulsion of a parasitic burden has still to be resolved. On the other hand, the neutrophil is not a common inhabitant of the intestinal epithelium, but chemotactic influences in the lumen will stimulate a massive migration of these cells through the epithelial layer (Bellamy & Nielsen 1974). Neutrophils may well execute a protective function in the intestinal mucosa as they do in other external secretory organs such as the mammary gland, but apart from an isolated report

(Takeuchi 1971) of bacterial phagocytosis by neutrophils on the intestinal mucosal surface there is really no basis for a definitive conclusion at this time.

Concluding remarks

When attempting to design a vaccination strategy for particular diseases involving mucous surfaces, it is clearly helpful to understand the pathogenesis of the diseases and the nature of the effector mechanisms involved in their resolution. It is evident that mucosal defence, usually mediated by antibody belonging to immunoglobulin classes other than IgA, plays a crucial role in maintaining the health of ruminant animals. More work needs to be done on the optimization of antibody responses to ensure that adequate quantities of antibody of appropriate immunoglobulin class specificity reach external secretion. In particular, the influence of oral presentation of antigen on systemic immune responses and the delivery of IgG antibody to the mucosae, especially the intestinal mucosa, is worthy of detailed attention.

Continuing study is required to elucidate the functional significance and modulation of various cells in the intestinal mucosa — mast cells/globule leucocytes, phagocytic cells and goblet cells — in the resolution of microbial and parasitic diseases. A serious effort must be made to assess the significance of the large population of intraepithelial lymphocytes in immune surveillance and to determine their relationship to the other cells in the intestinal mucosa.

References

Ahlstedt, S., Carlsson, B., Hanson, L.A. & Goldblum, R.M. (1975). Antibody production by human colostral cells. 1. Immunoglobulin class specificity and quantity. *Scandinavian Journal of Immunology*, **4**, 535-39.

Altmann, K. & Mukkur, T.K.S. (1983). Passive immunization of neonatal lambs against infection with enteropathogenic *Escherichia coli* via colostrum of ewes immunized with crude and purified K99 pili. *Research in Veterinary Science*, **35**, 234-39.

Andr, C., Heremans, J.F., Vaerman, J.P. & Cambiaso, C.L. (1975). A mechanism for the induction of immunological tolerance by antigen feeding: antigen-antibody complexes. *Journal of Experimental Medicine*, **142**, 1509-19.

Arnaud-Battandier, F., Bundy, B.M., O'Neill, M., Bienenstock, J. & Nelson, D.L. (1978). Cytotoxic activities of gut mucosal lymphoid cells in guinea pigs. *Journal of Immunology*, **121**, 1059-65.

Befus, D. & Bienenstock, J. (1982). Factors involved in symbiosis and host resistance at the mucosa-parasite interface. *Progress in Allergy*, **31**, 76-77.

Beh, K.J. (1977). The origin of IgA-containing cells in intestinal lymph of sheep. *Australian Journal of Experimental Biology & Medical Science*, **55**, 263-74.

Beh, K.J. (1979). Antibody-containing cell response in lymph of sheep after intraintestinal infusion of ovalbumin with and without DEAE-dextran. *Immunology*, **37**, 279-86.

Beh, K.J. & Lascelles, A.K. (1974). Class specificity of intracellular and surface immunoglobulin of cells in popliteal and intestinal lymph from sheep. *Australian Journal of Experimental Biology & Medical Science*, **52**, 505-14.

Beh, K.J. & Lascelles, A.K. (1981). The effect of route of administration of antigen on the antibody-containing cell response in lymph of sheep. *Immunology*, **42**, 577-82.

Beh, K.J. & Lascelles, A.K. (1984). The antibody-containing cell response in hepatic and intestinal lymph following intraperitoneal and intravenous administration of antigen in different adjuvants. *Veterinary Immunology & Immunopathology*, **5**, 15-26.

Beh, K.J., Watson, D.L. & Lascelles, A.K. (1974). Concentration of immunoglobulins and albumin in lymph collected from various regions of the body of the sheep. *Australian Journal of Experimental Biology & Medical Science*, **52**, 81-86.

Beh, K.J., Husband, A.J. & Lascelles, A.K. (1979). Intestinal response of sheep to intraperitoneal immunization. *Immunology*, **37**, 385-88.

Bellamy, J.E.C. & Nielsen, N.O. (1974). Immune-mediated emigration of neutrophils into the lumen of the small intestine. *Infection & Immunity*, **9**, 615-19.

Besredka, A. (1927). *Local immunization: specific dressings* (trans. H. Plotz). London: Bailliere, Tindall & Cox.

Bienenstock, J. & Befus, A.D. (1980). Review: mucosal immunology. *Immunology*, **41**, 249-70.

Bienenstock, J. & Befus, A.D. (1983). Some thoughts on the biologic role of immunoglobulin A. *Gastroenterology*, **84**, 78-185.

Bienenstock, J., Johnston, N. & Perey, D.Y.E. (1973a). Bronchial lymphoid tissue. 1. Morphologic characteristics. *Laboratory Investigation*, **28**, 686-92.

Bienenstock, J., Johnston, N. & Perey, D.Y.E. (1973b). Bronchial lymphoid tissue. 2. Functional characteristics. *Laboratory Investigation*, **28**, 693-98.

Bohl, E.H., Gupta, R.K.P., Olquin, M.V.F. & Saif, L.J. (1972). Antibody responses in serum, colostrum and milk of swine after infection or vaccination with transmissible gastroenteritis virus. *Infection & Immunity*, **6**, 289-301.

Brandon, M.R., Watson, D.L. & Lascelles, A.K. (1971). The mechanism of transfer of immunoglobulin into mammary secretion of cows. *Australian Journal of Experimental Biology & Medical Science*, **49**, 613-23.

Brandtzaeg, P. (1975). Human secretory immunoglobulin M: an immunochemical and immunohistochemical study. *Immunology*, **29**, 559-70.

Brown, W.R., Newcomb, R.W. & Ishizaka, K. (1970). Proteolytic degradation of exocrine and serum immunoglobulin. *Journal of Clinical Investigation*, **49**, 1374-80.

Burrows, W. & Havens, I. (1948). Studies on immunity to Asiatic cholera. 5. The absorption of immune globulin from the bowel and its excretion in the urine and faeces of experimental animals and human volunteers. *Journal of Infectious Diseases*, **82**, 231-50.

Burrows, W., Elliott, M.E. & Havens, I. (1947). Studies on immunity to Asiatic cholera. 4. The excretion of coproantibody in experimental enteric cholera in the guinea pig. *Journal of Infectious Diseases*, **81**, 261-81.

Butler, J.E. (1983). Bovine immunoglobulins: an augmented review. *Veterinary Immunology & Immunopathology*, **4**, 43-152.

Cahill, R.N.P., Poskitt, D.C., Frost, H. & Trnka, Z. (1977). Two distinct pools of recirculating T lymphocytes: migratory characteristics of nodal and intestinal T lymphocytes. *Journal of Experimental Medicine*, **145**, 420-28.

Challacombe, S.J. & Tomasi, T.B. (1980). Systemic tolerance and secretory immunity after oral immunization. *Journal of Experimental Medicine*, **152**, 1459-72.

Chang, C.C., Winter, A.J. & Norcross, N.L. (1981). Immune response in the bovine mammary gland after intestinal, local and systemic immunization. *Infection & Immunity*, **31**, 650-59.

Chidlow, J.W. & Porter, P. (1977). Uptake of maternal antibody by the neonatal pig following intramuscular and intramammary vaccination of the preparturient sow. *Research in Veterinary Science*, **23**, 185-90.

Chin, W. & Hay, J.B. (1980). A comparison of lymphocyte migration through intestinal lymph nodes, subcutaneous lymph nodes and chronic inflammatory sites of sheep. *Gastroenterology*, **79**, 1231-42.

Clough, J.D., Mims, L.H. & Strober, W. (1971). Deficient IgA antibody responses to arsanilic acid bovine serum albumin (BSA) in neonatally thymectomized rabbits. *Journal of Immunology*, **106**, 1624-29.

Crago, S.S., Prince, S.J., Pretlow, T.G., McGhee, J.R. & Mestecky, J. (1979). Human colostral cells. 1. Separation and characterization. *Clinical & Experimental Immunology*, **38**, 585-97.

Craig, S.W. & Cebra, J.J. (1971). Peyer's patches: an enriched source of precursors for IgA-producing immunocytes in the rabbit. *Journal of Experimental Medicine*, **134**, 188-200.

Cripps, A.W. & Lascelles, A.K. (1976). The origin of immunoglobulins in salivary secretion of sheep. *Australian Journal of Experimental Biology & Medical Science*, **54**, 191-95.

Cripps, A.W., Husband, A.J. & Lascelles, A.K. (1974). The origin of immunoglobulins in intestinal secretion of sheep. *Australian Journal of Experimental Biology & Medical Science*, **52**, 711-16.

Curtain, C.C., Clark, B.L. & Dufty, J.H. (1971). The origins of the immunoglobulins in the mucous secretions of cattle. *Clinical & Experimental Immunology*, **8**, 335-44.

Davies, A. (1922). An investigation into the serological properties of dysentery stools. *Lancet*, **2**, 1009-12.

Ebersole, J.L., Taubman, M.A. & Smith, D.J. (1979). Thymic control of secretory antibody responses in the rat. *Journal of Immunology*, **123**, 19-24.

Evans, P.A., Newby, T.J., Stokes, C.R., Patel, D. & Bourne, F.J. (1980). Antibody response of the lactating sow to oral immunization with *Escherichia coli*. *Scandinavian Journal of Immunology*, **11**, 419-29.

Fahey, K.J., Snodgrass, D.R., Campbell, I., Dawson, A. M. & Burrells, C. (1981). IgG₁ antibody in milk protects lambs against rotavirus diarrhoea. *Veterinary Immunology & Immunopathology*, **2**, 27-33.

Ferguson, A. (1977). Intraepithelial lymphocytes of the small intestine. *Gut*, **18**, 921-37.

Fisher, M.M., Nagy, B., Bazin, H. & Underdown, B.J. (1979). Biliary transport of IgA: role of secretory component. *Proceedings of the National Academy of Sciences of the USA*, **76**, 2008-12.

Flexman, J.P., Shellam, G.R. & Mayrhofer, G. (1983). Natural cytotoxicity, responsiveness to interferon and morphology of intraepithelial lymphocytes from the small intestine of the rat. *Immunology*, **48**, 733-41.

Frerichs, G.N., Woods, S.B., Lucas, M.H. & Sands, J.J. (1982). Safety and efficacy of live and inactivated infectious bovine rhinotracheitis vaccines.

Veterinary Record, 111, 116-22.

Goldblum, R.M., Ahlstedt, S., Carlsson, B., Hanson, L.A., Jodah, U., Lidin-Janson, G. & Sohl-Akerlund, A. (1975). Antibody-forming cells in human colostrum after oral immunization. *Nature*, 257, 797-99.

Gowans, J.L. & Knight, E.J. (1964). The route of recirculation of lymphocytes in the rat. *Proceedings of the Royal Society of London B*, 159, 257-82.

Green, D.R., Gold, J., St. Martin, S., Gershon, R. & Gershon, R.K. (1982). Microenvironmental immunoregulation: possible role of contrasuppressor cells in maintaining immune responses in gut-associated lymphoid tissues. *Proceedings of the National Academy of Sciences of the USA*, 79, 889-92.

Griscelli, C., Vassalli, P. & McCluskey, R.T. (1969). The distribution of large dividing lymph node cells in syngeneic recipient rats after intravenous injection. *Journal of Experimental Medicine*, 130, 1427-51.

Guy-Grand, D., Griscelli, C. & Vassalli, P. (1974). The gut-associated lymphoid system: nature and properties of the large dividing cells. *European Journal of Immunology*, 4, 435-43.

Guy-Grand, D., Griscelli, C. & Vassalli, P. (1978). The mouse gut T lymphocyte, a novel type of T cell: nature, origin and traffic in mice in normal and graft-*versus*-host conditions. *Journal of Experimental Medicine*, 148, 1661-77.

Haig, D.M., McKee, T.A., Jarrett, E.E.E., Woodbury, R. & Miller, H.R.P. (1982). Generation of mucosal mast cells is stimulated *in vitro* by factors derived from T cells of helminth-infected rats. *Nature*, 300, 188-90.

Hall, J.G. (1980). Effect of skin painting with oxazalone on the local extravasation of mononuclear cells in sheep. *Ciba Foundation Symposium*, 71, 197-209.

Hall, J.G., Gyure, L.A. & Payne, A.W.R. (1980). Comparative aspects of the transport of immunoglobulin A from blood to bile. *Immunology*, 41, 899-902.

Halpern, M.S. & Koshland, M.E. (1970). Novel subunit in secretory IgA. *Nature*, 228, 1276-78.

Hanson, L.A. & Brandtzaeg, P. (1980). The mucosal defence system. In *Immunological disorders in infants and children* (eds. E.R. Stiehm & V.A. Fulginiti), 2nd ed., pp. 137-64. Philadelphia: Saunders.

Head, J.R. & Beer, A.E. (1978). The immunologic role of viable leucocytic cells in mammary exosecretions. In *Lactation* (ed. B.L. Larson), pp. 337-64. New York: Academic Press.

Head, J.R. & Beer, A.E. (1979). *In vivo* and *in vitro* assessment of the immunologic role of leucocytic cells in milk. In *Immunology of breast milk* (eds. P.L. Ogra & D. Dayton), pp. 207-25. New York: Raven Press.

Herberman, R.B. & Ortaldo, J.R. (1981). Natural killer cells: their role in defences against disease. *Science*, 214, 24-30.

Howard, C.J., Taylor, G. & Brownlie, J. (1980). Surface receptors for immunoglobulin on bovine polymorphonuclear neutrophils and macrophages. *Research in Veterinary Science*, 29, 128-30.

Howard, C.J., Gourlay, R.N. & Taylor, G. (1981). Immunity to mycoplasma infections of the calf respiratory tract. In *The ruminant immune system* (ed. J.E. Butler), pp. 711-26. New York: Plenum Press.

Husband, A.J. (1978). An immunization model for the control of infectious enteritis. *Research in Veterinary Science*, 25, 173-77.

Husband, A.J. (1982). Kinetics of extravasation and redistribution of IgA specific antibody-containing cells in the intestine. *Journal of Immunology*, 128, 1355-59.

Husband, A.J. & Gowans, J.L. (1978). The origin and antigen-dependent distribution of IgA-containing cells in the intestine. *Journal of Experimental Medicine*, **148**, 1146-60.

Husband, A.J. & Lascelles, A.K. (1974). The origin of antibody in intestinal secretion of sheep. *Australian Journal of Experimental Biology & Medical Science*, **52**, 791-99.

Husband, A.J., Moni, H.J. & Gowans, J.L. (1977). The natural history of the cells producing IgA in the gut. *Ciba Foundation Symposium*, **46**, 29-54.

Husband, A.J., Beh, K.J. & Lascelles, A.K. (1979). IgA-containing cells in the ruminant intestine following intraperitoneal and local immunization. *Immunology*, **37**, 597-601.

Janossy, G., Tidman, N., Selby, W.S., Thomas, J.A., Granger, S., Kung, P.S. & Goldstein, G. (1980). Human T lymphocytes of inducer and suppressor type occupy different microenvironments. *Nature*, **288**, 81-84.

Kawanishi, H., Saltzman, L.E. & Strober, W. (1983). Mechanisms regulating IgA class-specific immunoglobulin production in murine gut-associated lymphoid tissues. 1. T cells derived from Peyer's patches that switch sIgM B cells to sIgA B cells *in vitro*. *Journal of Experimental Medicine*, **157**, 433-50.

Kemp, A.S., Cripps, A.W. & Brown, S. (1980). Suppression of leucocyte chemokinesis and chemotaxis by human IgA. *Clinical & Experimental Immunology*, **40**, 388-95.

Klebanoff, S.J. & Clark, R.A. (1978). *The neutrophil: function and clinical disorders*. Amsterdam: North-Holland.

Komar, R. & Mukkur, T.K.S. (1975). Isolation and characterization of J chain from bovine colostral immunoglobulin M. *Canadian Journal of Biochemistry*, **53**, 943-49.

Komar, R., Abson, E.C. & Mukkur, T.K.S. (1975). Isolation and characterization of immunoglobulin A (IgA) from bovine nasal secretions. *Immunochemistry*, **12**, 323-27.

Koshland, M.E. (1975). Structure and functions of the J chain. *Advances in Immunology*, **20**, 41-69.

Koster, F.T. & Pierce, N.F. (1983). Parenteral immunization causes antigen-specific cell-mediated suppression of an intestinal IgA response. *Journal of Immunology*, **131**, 115-19.

Lascelles, A.K. & McDowell, G.H. (1974). Localized humoral immunity with particular reference to ruminants. *Transplantation Reviews*, **19**, 170-208.

Lascelles, A.K., Beh, K.J. & Mukkur, T.K.S. (1982). Techniques for immunological studies of the gastrointestinal tract with particular reference to sheep. In *Techniques in the life sciences: digestive physiology*. Shannon, Ireland: Elsevier.

Lascelles, A.K., Outteridge, P.M. & MacKenzie, D.D.S. (1966). Local production of antibody by the lactating mammary gland following antigenic stimulation. *Australian Journal of Experimental Biology & Medical Science*, **44**, 169-80.

Lawrence, D.A., Weigle, W.O. & Spiegelberg, H.L. (1975). Immunoglobulins cytophilic for human lymphocytes, monocytes and neutrophils. *Journal of Clinical Investigation*, **55**, 368-376.

Lycke, N., Lindholm, L. & Holmgren, J. (1983). IgA isotype restriction in the mucosal but not in the extramucosal immune response after oral immunizations with cholera toxin or cholera B subunit. *International Archives of Allergy & Applied Immunology*, **72**, 119-27.

Lyscom, N. & Brueton, M.J. (1982). Intraepithelial, *lamina propria* and Peyer's patch lymphocytes of the rat small intestine: isolation and characterization in terms of immunoglobulin markers and receptors for monoclonal antibodies. *Immunology*, **45**, 775-83.

Mach, J.P. & Pahud, J.J. (1971). Secretory IgA, a major immunoglobulin in most bovine external secretions. *Journal of Immunology*, **106**, 552-63.

Mattingly, J.A. & Waksman, B.H. (1978). Immunologic suppression after oral administration of antigen. 1. Specific suppressor cells formed in rat Peyer's patches after oral administration of sheep erythrocytes and their systemic migration. *Journal of Immunology*, **121**, 1878-83.

Mayrhofer, G. (1979). The nature of the thymus dependency of mucosal mast cells. 2. The effect of thymectomy and of depleting recirculating lymphocytes on the response to *Nippostrongylus brasiliensis*. *Cellular Immunology*, **47**, 312-22.

Mayrhofer, G. (1980). Thymus-dependent and thymus-independent subpopulations of intestinal intraepithelial lymphocytes: a granular subpopulation of probable bonemarrow origin and relationship to mucosal mast cells. *Blood*, **55**, 532-35.

Mayrhofer, G. & Bazin, H. (1981). Nature of the thymus dependency of mucosal mast cells. 3. Mucosal mast cells in nude mice and nude rats, in B rats and in a child with the Di George syndrome. *International Archives of Allergy & Applied Immunology*, **64**, 320-31.

Mayrhofer, G., Bazin, H. & Gowans, J. L. (1976). Nature of cells binding anti-IgE in rats immunized with *Nippostrongylus brasiliensis*: IgE synthesis in regional nodes and concentration in mucosal mast cells. *European Journal of Immunology*, **6**, 537-45.

McGuire, T.C., Musoke, A.J. & Kurtti, T. (1979). Functional properties of bovine IgG[1] and IgG[2]: interaction with complement, macrophages, neutrophils and skin. *Immunology*, **38**, 249-56.

Micusan, V.V. & Borduas, A.G. (1977). Biological properties of goat immunoglobulins G. *Immunology*, **32**, 373-81.

Miller, H.R.P. & Nawa, Y. (1979a). *Nippostrongylus brasiliensis*: intestinal goblet-cell response in adoptively immunized rats. *Experimental Parasitology*, **47**, 81-90.

Miller, H.R.P. & Nawa, Y. (1979b). Immune regulation of intestinal goblet-cell differentiation. Specific induction of nonspecific protection against helminths. *Nouvelle Revue Française d'Hématologie*, **21**, 31-45.

Morein, B. (1972). Immunity against parainfluenza-3 virus in cattle: immunoglobulins in serum and nasal secretions after subcutaneous and nasal vaccination. *Zeitschrift fr Immunittsforschung*, **144**, 63-74.

Morgan, K.L., Bourne, F.J., Newby, T.J. & Bradley, P.A. (1981). Humoral factors in the secretory immune system of ruminants. In *The ruminant immune system* (ed. J.E. Butler), pp. 391-411. New York: Plenum Press.

Mukkur, T.K.S., Komar, R. & Savina, L.R. (1975). Immunoglobulins and their relative neutralizing efficiency in cattle immunized with infectious bovine rhinotracheitis-parainfluenza-3 (IBR-PI-3) virus vaccine. *Archives of Virology*, **48**, 195-201.

Musoke, A.J., Nantulya, V.M., Büscher, G., Masake, R.A. & Otim, B. (1982). Bovine immune response to *Theileria parva*: neutralizing antibodies to sporozoites. *Immunology*, **45**, 663-68.

Nansen, P. (1972). Selective immunoglobulin deficiency in cattle and susceptibility to infection. *Acta Pathologica et Microbiologica Scandinavica B*, **80**, 49-54.

Nawa, Y. & Miller, H.R.P. (1979). Adoptive transfer of the intestinal mast-cell response in rats infected with *Nippostrongylus brasiliensis*. *Cellular Immunology*, **42**, 225-39.

Newby, T.J., Stokes, C.R., Huntley, J., Evans, P. & Bourne, F.J. (1979). The immune response of the pig following oral immunization with soluble protein. *Veterinary Immunology & Immunopathology*, **1**, 37-47.

Outteridge, P.M. & Lee, C.S. (1981). Cellular immunity in the mammary gland with particular reference to T and B lymphocytes and macrophages. In *The ruminant immune system* (ed. J.E. Butler), pp. 513-34. New York: Plenum Press.

Parmely, M.J., Beer, A.E. & Billingham, R.E. (1976). *In vitro* studies on the T-lymphocyte population of human milk. *Journal of Experimental Medicine*, **144**, 358-70.

Pierce, N.F. (1978). The role of antigen form and function in the primary and secondary intestinal immune responses to cholera toxin and toxoid in rats. *Journal of Experimental Medicine*, **148**, 195-206.

Pierce, N.F. & Gowans, J.L. (1975). Cellular kinetics of the intestinal immune response to cholera toxoid in rats. *Journal of Experimental Medicine*, **142**, 1550-63.

Pierce, N.F. & Koster, F.T. (1980). Priming and suppression of the intestinal immune response to cholera toxoid/toxin by parenteral toxoid in rats. *Journal of Immunology*, **124**, 307-11.

Pierce, N.F. & Cray, W.C. (1982). Determinants of the localization, magnitude, duration of a specific mucosal IgA plasma cell response in enterically immunized rats. *Journal of Immunology*, **128**, 1311-15.

Pittard, W.B., Polmar, S.H. & Fanaroff, A.A. (1977). The breast milk macrophage: a potential vehicle for immunoglobulin transport. *Journal of the Reticuloendothelial Society*, **22**, 597-603.

Porter, P., Parry, S.H. & Allen, W.D. (1977). Significance of immune mechanisms in relation to enteric infections of the gastrointestinal tract in animals. *Ciba Foundation Symposium*, **46**, 55-75.

Porter, P., Kenworthy, R., Noakes, D.E. & Allen, W.D. (1974). Intestinal antibody secretion in the young pig in response to oral immunization with *Escherichia coli*. *Immunology*, **27**, 841-53.

Puccetti, P., Santoni, A., Riccardi, C. & Herberman, R.B. (1980). Cytotoxic effector cells with the characteristics of natural killer cells in the lungs of mice. *International Journal of Cancer*, **25**, 153-58.

Quin, J.W., Husband, A.J. & Lascelles, A.K. (1975). The origin of the immunoglobulins in intestinal lymph of sheep. *Australian Journal of Experimental Biology & Medical Science*, **53**, 205-14.

Reynolds, J.D. & Morris, B. (1983). The influence of gut function on lymphoid cell populations in the intestinal mucosa of lambs. *Immunology*, **49**, 501-9.

Richman, L.K., Graeff, A.S., Yarchoan, R. & Strober, W. (1981). Simultaneous induction of antigen-specific IgA helper T cells and IgG suppressor T cells in the murine Peyer's patch after protein feeding. *Journal of Immunology*, **128**, 2079-83.

Rose, M.L., Parrott, D.M.V. & Bruce, R.G. (1976). Migration of lymphoblasts to the small intestine. 2. Divergent migration of mesenteric and peripheral immunoblasts to sites of inflammation in the mouse. *Cellular Immunology*, **27**, 36-46.

Saif, L.J., Bohl, E.H. & Gupta, R.K.P. (1972). Isolation of porcine immunoglobulins and determination of the immunoglobulin classes of transmissible gastroenteritis viral antibodies. *Infection & Immunity*, **6**, 600-9.

Scollay, R., Hall, J. & Orlans, E. (1976). Studies on the lymphocytes of sheep. 2. Some properties of cells in various compartments of the recirculating lymphocyte pool. *European Journal of Immunology*, **6**, 121-25.

Shen, L. & Fanger, M.W. (1981). Secretory IgA antibodies synergize with IgG in promoting ADCC by human polymorphonuclear cells, monocytes and lymphocytes. *Cellular Immunology*, **59**, 75-81.

Shuster, J. (1971). Pepsin hydrolysis of IgA-delineation of two populations of molecules. *Immunochemistry*, **8**, 405-11.

Smith, J.W. & Schultz, R.D. (1977). Mitogen- and antigen-responsive milk lymphocytes. *Cellular Immunology*, **29**, 165-73.

Smith, W.D., Jackson, F., Jackson, E., Williams, J. & Miller, H.R.P. (in press). Manifestations of resistance to ovine ostertagiasis associated immunological responses in the gastric lymph. *Journal of Comparative Pathology*.

Smith, W.D., Jackson, F., Jackson, E. & Williams, J. (1983). Local immunity and *Ostertagia circumcincta*: changes in the gastric lymph of immune sheep after a challenge infection. *Journal of Comparative Pathology*, **93**, 479-88.

Snodgrass, D.R. & Wells, P.W. (1978). Passive immunity in rotaviral infections. *Journal of the American Veterinary Association*, **173**, 565-68.

Snodgrass, D.R., Fahey, K.J., Wells, P.W., Campbell, I. & Whitelaw, A. (1980). Passive immunity in calf rotavirus infections: maternal vaccination increases and prolongs immunoglobulin G₁ secretion in milk. *Infection & Immunity*, **28**, 344-49.

Socken, D.J. & Underdown, B.J. (1978). Comparison of human, bovine and rabbit secretory component-immunoglobulin interactions. *Immunochemistry*, **15**, 499-506.

Sprent, J. (1976). Fate of H-2-activated T lymphocytes in syngeneic hosts. 1. Fate in lymphoid tissues and intestines traced with ³H-thymidine, ¹²⁵I-deoxyuridine and ⁵¹chromium. *Cellular Immunology*, **21**, 278-302.

Stokes, C.R., Newby, T.J., Huntley, J.H., Patel, D. & Bourne, F.J. (1979). The immune response of mice to bacterial antigens given by mouth. *Immunology*, **38**, 497-502.

Swarbrick, E.T., Stokes, C.R. & Soothill, J.F. (1979). The absorption of antigens after oral immunization and the simultaneous induction of specific systemic tolerance. *Gut*, **20**, 121-25.

Tagliabue, A., Luini, W., Soldateschi, D. & Boraschi, D. (1981). Natural killer activity of gut mucosal lymphoid cells in mice. *European Journal of Immunology*, **11**, 919-22.

Takeuchi, A. (1971). Penetration of the intestinal epithelium by various microorganisms. *Current Topics in Pathology*, **54**, 1-27.

Thomas, H.C. & Parrott, D.M.V. (1974). The induction of tolerance to a soluble protein antigen by oral administration. *Immunology*, **27**, 631.

Tomasi, T.B. & Grey, H.M. (1972). Structure and function of immunoglobulin A. *Progress in Allergy*, **16**, 81-213.

Van Epps, D.E. & Williams, R.C. (1976). Suppression of leucocyte chemotaxis by human IgA myeloma components. *Journal of Experimental Medicine*, **144**, 1227-42.

Walker, W.A. & Isselbacher, K.J. (1977). Intestinal antibodies. *New England Journal of Medicine*, **297**, 767-73.

Watson, D.L. (1975). Cytophilic attachment of ovine IgG₂ to autologous polymorphonuclear leucocytes. *Australian Journal of Experimental Biology & Medical Science*, **53**, 527-29.

Watson, D.L. (1976). The effect of cytophilic IgG₂ on phagocytosis of ovine polymorphonuclear leucocytes. *Immunology*, **31**, 159-65.

Watson, D.L. & Kennedy, J.W. (1981). Immunization against experimental staphylococcal mastitis in sheep: effect of challenge with a heterologous strain of *Staphylococcus aureus*. *Australian Veterinary Journal*, **57**, 309-13.

Watson, D.L. & Lascelles, A.K. (1973a). Mechanisms of transfer of immunoglobulins into mammary secretion of ewes. *Australian Journal of Experimental Biology & Medical Science,* 51, 247-54.

Watson, D.L. & Lascelles, A.K. (1973b). Comparisons of immunoglobulin secretion in the salivary and mammary glands of sheep. *Australian Journal of Experimental Biology & Medical Science,* 51, 255-58.

Watson, D.L. & Lee, C.G. (1978). Immunity to experimental staphylococcal mastitis: comparison of live and killed vaccines. *Australian Veterinary Journal,* 54, 374-78.

Wernet, P., Breu, H., Knop, J. & Rowley, D. (1971). Anti-bacterial action of specific IgA and transport of IgM, IgA and IgG from serum into the small intestine. *Journal of Infectious Diseases,* 124, 223-26.

Yardley, J.H., Keren, D.F., Hamilton, S.R. & Brown, G.D. (1978). Local (immunoglobulin A) immune response by the intestine to cholera toxin and its partial suppression with combined systemic and intraintestinal immunization. *Infection & Immunity,* 19, 589-97.

Yoffey, J.M. & Courtice, F.C. (1980). *Lymphatics, lymph and the lymphomyeloid complex.* London: Academic Press.

Role of immune responses in protection against infectious diseases

24

Effector mechanisms against viral infections

L.D. PEARSON and K.A. KNISLEY

Differences in structure and mode of replication in at least 20 different groups
of viruses present the animal host with a complexity of immunogenic stimuli.
The biological importance of a particular viral antigen is dependent on its
function during replication and its accessibility to the immune system. Host
resistance is mediated by humoral factors and lymphoid cells of the immune
system. The earliest effector mechanisms for expression of resistance to viral
infection include the alternate pathway of complement activation, interferon
production and attack of infected cells by natural killer cells. Recovery from
several different primary viral infections involves neutralization and
elimination of viruses by antibodies, as well as specific elimination of cells
infected with virus by antibody-dependent cellular cytotoxicity or by activated
T cells. Antibodies and rapid recall responses by T cells help to prevent
reinfection by viruses.

Introduction

A virus infection can produce a diversity of immunological
responses involving humoral and cellular effector mechanisms. Recent
review articles concerning effector mechanisms for anti-viral immunity
have dealt with antibody-mediated immunity (Cooper 1979; Sissons &
Oldstone 1980a; 1980b), virus-specific cytotoxic lymphocytes (Burakoff
et al. 1980; Zinkernagel & Doherty 1979) or a combination of these
(Greenspan, Schwartz & Doherty 1983; Onions 1983; Sissons & Oldstone
1980c). This chapter will review the general nature of viral antigens, the
opportunities that the host has to respond to biologically important
antigens, and the humoral and cellular anti-viral effector mechanisms
that have been determined, with particular emphasis on ruminants.

Viral structure and antigenic components

Structural components

Viruses consist of nucleic acid cores of either RNA or DNA encapsulated by a proteinaceous coat called the capsid. The structural units of the capsid are called capsomeres and these are composed of proteins. The capsids of different virus groups have different arrangements, including cubic symmetry with a icosahedral shape (e.g. adenovirus), winding of the capsomeres around the nucleic acid to form an elongated helical structure (e.g. influenza virus), or a complex structure (e.g. vaccinia). Viruses with helical symmetry become surrounded by an envelope derived from host cell membranes. The phospholipids, glycolipids and cholesterol in the envelope are derived from the host cell, but most of the proteins are virus-specific, i.e. their production is governed by viral nucleic acid. Some viruses with cubic symmetry (e.g. herpes viruses and togaviruses) become enveloped, but several other with this structure exist as naked virions (e.g. adenovirus). A matrix protein supports the virus envelope, except in the togaviruses. The most complex structures are found in retroviruses, in which a helical nucleocapsid is surrounded by an outer icosahedral capsid, and pox viruses, which have an elaborate substructure of membranes and enzymes, required for replication (Onions 1983).

Antigenic components

The surface structural components are the most accessible to the immune system. Nonenveloped or naked virions may have surface projections, known as fibres, that are antigenic but not glycosylated. Surface projections made up of glycoprotein spikes have been identified in enveloped viruses. Histocompatibility antigens, presumably coded for by the host cell, have been identified in the envelopes of rhabdoviruses and retroviruses (Griffin & Compans 1979; Bubbers & Lilly 1977). Matrix proteins and ribonucleoprotein antigens become exposed to the host during the course of a viral infection.

Virus and host cell interactions

Infections with nonenveloped virus

Viruses are obligate intracellular parasites which require a living cell for their replication. For an infection to occur, the surface components of the virus must first attach to complementary receptors on the host cell membrane. Virions enter the cell either by penetrating the membrane or by phagocytosis. Once inside the cell, the capsid is subjected to the destructive activities of lysosomal enzymes and the nucleic acid is released.

Once freed, the viral genome directs the synthesis of proteins which are required for assembling new virion particles. The rapid release of large numbers of nonenveloped viruses after assembly generally results in lysis and death of the host cell. This type of virus does not code for the insertion of proteins in the host cell membranes. Virion components that are not completely assembled at the time of lysis of the host cell (i.e. internal virion enzymes such as nucleoproteins) become exposed to the immune system.

Infections with enveloped virus

Enveloped viruses enter the host cell following the interaction of the glycoprotein spikes on the virus envelope with receptors on the host cell membrane. Enveloped viruses with fusion proteins merge into the plasma membrane of the cell. Lysosomal enzymes release the viral nucleic acid from the capsids once they are inside the cell. These viruses gain their envelope by budding through the nuclear (herpes viruses) or plasma membrane. The budding process may cause the host cell to lyse. However, some enveloped viruses may be continuously released from the cell without causing destruction or a significant alteration in cellular function (e.g. paramyxoviruses and retroviruses). Enveloped viruses usually code for proteins that are inserted into the host cell membrane. Matrix protein for influenza virus may also be expressed at the cell surface (Hackett et al. 1980). Novel, virus-specific antigens that are not found in virions or in any intermediates involved in processing the gag gene have been detected in the plasma membranes and as soluble antigens released from cells infected with feline leukaemia virus (Neil et al. 1980).

Latency

Some viruses do not complete their replication cycle immediately. These viruses insert their genomes into cells and complete replication only in response to certain stimuli. Herpes viruses are known to remain latent in neurons for years (Yates 1982). Exposure to ultraviolet light, temperature changes or immunosuppressive agents may stimulate some of the viruses suddenly to complete their replication cycle, lyse host cells and, as a result, cause clinical signs. Feline leukaemia virus remains latent in the bonemarrow. This latency may be the result of humoral and cellular immune responses to viral antigens at the plasma membrane of infected cells (Rojko et al. 1981).

Oncogenicity

Several DNA viruses and retroviruses of the RNA group are known to be associated with the occurrence of neoplastic disease. Cells infected with these viruses may become transformed (they are capable of growing into neoplasms following inoculation into syngeneic recipients). Cells transformed by DNA viruses do not replicate new virions, whereas cells infected wih retroviruses may continually release virus. Cells transformed by DNA or RNA viruses may express virus-coded antigens in the plasma membranes in the absence of virus production; the feline oncornavirus cell membrane antigen (FOCMA) expressed in cells transformed by feline leukaemia virus is an example (Essex et al. 1971). At least 16 cellular genes, sometimes called 'proto-oncogenes', can become activated or retrieved by retrovirus infections so that expression of the oncogenes can transform the host cells (Weinberg 1984).

Table 1. MHC restriction of cell-mediated cytotoxicity in species other than mice.

Species	Virus	Restricted	Reference
Man	measles (acute)	yes (?)	Kreth, ter Meulen & Eckert 1979
	influenza (IVG)	yes	McMichael et al. 1977 Biddison, Shaw & Nelson 1979
Dog	vacinia	no	Ho, Babiuk & Rouse 1978
	canine distemper	yes[a]	Shek, Shultz & Apel 1980
Rabbit	vaccinia	no	Woan, Yp & Tompkins 1978
Syrian hamster	vaccinia	no	Zinkernagel, Althage & Jensen 1977 Nelles & Streilein 1980
Cattle	vaccinia	no	Rouse & Babaiuk 1977
Sheep	fibrosarcoma	no	Theilen, Pederson & Higgins 1979
	vaccinia	yes	Schmaljohn 1981
Chicken	Rous sarcoma	yes[a]	Wainberg et al. 1974
Rat	vaccinia	yes[a]	Zinkernagel, Althage & Jensen 1977
	LCM	yes	Zinkernagel, Althage & Jensen 1977

[a]*In vitro* generation.

Soluble effectors of resistance to viral infections

Neutralization

As a result of antibody binding to certain surface antigens of virions, critical steps in the cycle required for virus replication may not be completed. The loss of infectivity resulting from the interaction of virus and antibody is termed neutralization and the process has been reviewed in detail by Mandel (1978; 1979) and Cooper (1979). Steps in the replication cycle where antibody may interfere include adsorption, penetration and uncoating (Mandel 1978).

Determinants on single proteins at the surface of the virion which combine with neutralizing antibodies are known as 'critical' sites. An example of how the fine structure of critical sites can be mapped is provided by Bruck and colleagues (1982). They used 15 different monoclonal antibodies to identify eight independent antigenic regions on the gp51 of bovine leucosis virus. Antigenic sites designated A, B, C and D on a 35,000-dalton fragment of gp51 were not biologically active. Three antigenic sites (F, G and H) on the remaining 15,000-dalton fragment were involved in neutralization of infectivity or syncytia formation; furthermore, the G site was exposed as a target for antibody and complement-dependent cytotoxicity. Another example of such critical sites involves the hemagglutinin spikes found on influenza virus. Nonneutralizing antibodies may combine with other determinants (noncritical sites) on the virion surface, but not cause neutralization. An example of noncritical sites is the neuraminidase spikes of influenza virus. Mandel (1978; 1979) has reviewed attempts to determine the number of antibody molecules required to neutralize effectively the infectivity of a virion. This was done by checking for infectivity at various time points after mixing virus and antibody. A linear curve for a plot of log (remaining infectivity/original infectivity) against time, beginning at time zero, indicates 'one hit' kinetics. A single-hit mechanism does not rule out the possibility that more than one molecule combines with a particle before activation, but it does rule out the participation of these molecules in the inactivating event. A lag in the onset of neutralization can be achieved by lowering the temperature, and the rate of interactions can be affected by decreasing the antibody concentration, indicating that several antibody molecules are required per virion to effect neutralization.

Other lines of evidence support this requirement for several antibody molecules per virion to cause neutralization. Infectious antibody-virus complexes that form with low concentrations of antibody can be neutralized by adding antiglobulin or early components of complement (Oldstone 1975). Synergism may occur between antibodies that react with different determinants on the same virion.

The fraction of the virus that remains nonneutralized even after a prolonged time period is termed the persistent fraction. The attachment of antibody to a virion is reversible and the virus can regain infectivity following dissociation (Lafferty 1963). Antibodies obtained from serum taken early after the onset of infection or shortly after immunization are often of low avidity and the persistent fraction is largest when serum neutralization tests are conducted with such antibodies. Mandel (1978) reviewed nonneutralizability and concluded that it is a relative condition that must consider the virus, antibody and host cell used for the assay. For example, foot-and-mouth disease virus is neutralized 100-fold easier on mouse cells compared with pig kidney, and rabbit pox virus is neutralized more easily on rabbit skin cells than on mouse brain or the chorioallantoic membrane (Mandel 1978). First neutralization results from polio virus were the same for HeLa and monkey kidney cells, but antiglobulin neutralization is only seen with HeLa cells. The ease with which virus grows in the cell line is not the only significant factor.

Complement can produce antibody-initiated neutralization by three mechanisms (Cooper 1979). First, the physical bulk of complement protein on the viral surface may impair adsorption or penetration of a susceptible cell. Second, complement may facilitate agglutination of virions and reduce the number of available infectious particles. Third, complement may lyse viruses with lipid-containing envelopes or membranes and cause irreversible structural damage. Neutralization of equine arteritis virus-antibody complexes is enhanced when purified components of complement are added in sequence (Radwan & Crawford 1974). Similarly, significant neutralization was noted when C1 and C4 were added to herpes virus and antibody complexes (Daniels et al. 1970). Further addition of C2 and C3 increased the degree of neutralization. The early complement components may act by masking cell receptors on the virion surface (Onions 1983) or by aggregation of virions by C3 (Oldstone, Cooper & Larson 1974).

Lysis
The envelope of viruses may be damaged after specific antibody combines with surface antigens and activates the classical pathway of the complement system. Lytic damage has been demonstrated for avian

infectious bronchitis virus, equine arteritis virus, retroviruses, influenza virus, lymphocytic choriomeningitis virus, rubella virus, Sendai virus and Sindbis virus, as reviewed by Cooper (1979).

The alternate pathway of the complement system can be activated by certain viruses in the absence of immunoglobulins. Oncornaviruses have been inactivated and lysed by human, but not by guinea pig or by rabbit, complement in the absence of detectable antibody (Welsh et al. 1975). Welsh (1977) reported that, depending on the cell-passage history, serum from nonimmunized humans may or may not inactivate lymphocytic choriomeningitis or Newcastle disease virus. Viruses may assimilate host-cell components on to the virion surface during replication and thus become susceptible to neutralization by 'natural' antibodies and complement. Welsh (1977) concluded that the likely mechanism for inactivation of lymphocytic choriomeningitis virus was by lysis of the virion, but that Newcastle disease virus could be inactivated by the classical or alternate pathways independently in the absence of any detectable antibodies in human serum. Each pathway was virucidal in the absence of the other, but inactivation did not occur when both factors B and C4 were depleted from the serum.

Cells infected with viruses that cause expression of virus-specified antigens at their surface are subject to lysis by antibody and complement. For example, our laboratory showed that sheep fibroblasts infected with the parapox virus, orf, are capable of being lysed within 4 h of the viruses and cells making initial contact. New virions are not released in appreciable numbers until 18 h after infection (DeMartini, Pearson & Fiscus 1978). Cells in the efferent lymph of sheep stimulated with Kunjin virus or Ross River virus secreted lytic antibody and were able to lyse, in the presence of complement, goose erythrocytes coated with either virus (Pearson et al. 1976).

Interference with replication

A heterologous group of proteins and glycoproteins known as interferons (IFN) are associated with viral infections. Virus-infected leucocytes, including macrophages, produce a family of at least ten different proteins known as the α-IFN. Infected fibroblasts release up to five different proteins known as β-IFN. Immune IFN or γ-IFN is produced largely by T cells (Lyt 2,3-positive in mice, Fc-positive in man) but may also be produced by B cells, macrophages or natural killer (NK) cells. Besides inducing inhibition of viral replication, γ-IFN immunoregulates by enhancing the following factors: Fc-positive receptors to promote antibody-dependent cellular cytotoxicity (ADCC), expression of class I and class II major histocompatibility (MHC) antigens, T-cell cytotoxicity, release of interleukin-2 (IL-2), macrophage activation, NK-cell activity and production of new NK cells. Immune responses inhibited by γ-IFN include activation of B cells and antibody synthesis and release, mixed lymphocyte reactions, graft-*versus*-host disease, allograft rejection and delayed-type hypersensitivity reactions (Bloom 1980; Tizard 1984).

The exact nature of the interaction between the virus or viral antigens and the host cells that leads to the induction of IFN is not known; a dozen or more genes may be derepressed to give rise to antigenically related IFNs that are thought to vary in their function (Stanton & Barron 1984). Following the binding with cell-membrane receptors, secondary cell 'signal molecules' initiate the synthesis of anti-viral effector molecules by a derepressional event. One mechanism that helps to achieve anti-viral activity is the inhibition of viral mRNA translation. Type I or α-IFN may reach significant levels soon after infection, but inhibitory levels of α-IFN are transient unless additional stimulation for production occurs. For example, high serum IFN levels were noted on postinoculation day 3 with bovine respiratory syncytial virus in calves and decreased to undetectable amounts by day 6 (El Azhary, Silim & Roy 1981). Interferon was detected in the nasal secretions of calves as early as 24 h after they were inoculated with infectious bovine rhinotracheitis (IBR) virus and continued to be detectable until day 7 in five out of six calves (Cummins & Rosenquist 1980). Several factors may affect the titration of bovine interferon and these include pH of the fluids containing IFN, cell type used to assay the activity, age of culture and the suppressive effect of contaminating noncytopathogenic bovine viral diarrhoea virus (Rossi, Kiesel & Hoff 1980). The action of α-IFN is nonspecific in that it inhibits infection with viruses other than the one that incited its production (Cummins & Rosenquist 1983). Type II or γ-IFN is a lymphokine that is released from previously sensitized lymphocytes. Both types of IFN may interfere with viral replication and also enhance NK-cell activity similarly. The IFNs are structurally similar polypeptides but the significance of little or no glycosylation of α-IFN compared to γ-IFN is not known (Stanton & Barron 1984).

Interaction of antibodies with viral antigens at the plasma membrane of infected cells can modify the replication cycle of certain viruses. Anti-measles virus antibody added to infected cell cultures facilitates rapid removal of measles virus antigens from the surface of infected cells (Joseph & Oldstone 1975). The modulated cells continue to express measles virus antigens internally but they are refractory to immune lysis. Removal of anti-viral antibody allows the re-expression of measles virus antigens at the cell surface, the formation of syncytia and cell death (Fujinami & Oldstone 1979).

Cellular effectors of resistance to viral infection

Infections with enveloped viruses, in which virus-specified antigens are presented in the host cell membrane, permit recognition of the infected cells by activated effector cells, including cytotoxic T cells and NK cells. Such virus-coded cell-surface antigens may not always induce an antibody response.

Cytotoxic T cells

When thymic function is inhibited by thymectomy, animal hosts become more susceptible to infections by enveloped viruses, but not to nonenveloped viruses (Blanden 1970; Zisman, Hirsch & Allison 1970; Rager-Zisman & Allison 1973), which suggests an important role in cell-mediated immune responses for enveloped viruses. The induction of T cell-mediated immunity may be accompanied by the release of soluble mediators, including γ-IFN, proliferation of cells and the development of delayed hypersensitivity as well as cytotoxic T cells.

In mice, cytotoxic responses result from interactions between distinct T-cell subsets. For herpes virus the response is induced by Ia-positive, virus-infected macrophages (Zisman, Hirsch & Allison 1970). Schmid, Larsen and Rouse (1982) noted that Ia-positive cells must also produce interleukin-1 (IL-1) to induce an anti-herpes simplex virus cytotoxic lymphocyte response in mice. Induction of cytotoxic lymphocytes for reovirus also requires Ia expression on the infected cell (Letvin, Kauffman & Finberg 1981). The presence of Ia on the accessory cell may not be a universal requirement, however, since Hapel, Bablanian and Cole (1980) reported that Ia antigen was not required for induction of responses to a poxvirus system. In mice T-cell subsets can be distinguished on the basis of different cell-surface antigens (McKenzie & Potter 1979; Swain & Dutton 1980). Lyt 1-positive 2-negative 3-negative T cells of mice, are stimulated by presentation of antigen in association with I-region MHC products and provide T-cell help. Lyt 1-negative 2-positive 3-positive T cells can become cytotoxic effector cells after recognizing antigen in the context of MHC products coded for by the K and D region of the H-2 complex (Zinkernagel & Doherty 1974; Doherty & Zinkernagel 1975; Zinkernagel & Doherty 1979). Interaction of precytotoxic cells with antigen stimulates the expression of a receptor for the lymphokine IL-2 which is released from helper T cells (Watson & Mochizuki 1980). IL-2 appears to be required for the activation and maintenance of cytotoxic T cells (Wagner et al. 1980).

Three basic models for T-cell recognition of antigen have been proposed. The dual-recognition model (Cohn & Epstein 1978; Zinkernagel et al. 1978; Williamson 1980) proposes that separate T-cell receptors recognize MHC product and antigen. The associative recognition model proposes that while the MHC product and antigen are distinctly separate entities, they are still physically close enough to be recognized by a single receptor. The third basic model has been called 'altered self' and states that the MHC product becomes modified in some way, such as a conformational change caused by expression of the virus-specified antigen, and the altered MHC product then becomes recognizable by a single cytotoxic T-cell receptor (Matzinger 1981).

The requirement for recognition of MHC product and antigen by the cytotoxic T cell has been called genetic restriction or MHC restriction. Genetically restricted cytotoxicity has been observed in several species (Table 1). There have been reports that genetic restriction does not apply to IBR or vaccinia virus infections in cattle (Rouse & Babiuk 1977). The same laboratory also concluded that genetic restriction did not apply to anti-viral cytotoxicity in the dog (Ho, Babiuk & Rouse 1978). However, these reports should be examined and interpreted with caution until defined effector-cell populations and MHC-typed cells are available (Onions 1983). Our laboratory has evidence that genetic restriction does apply to direct lymphocyte killing of vaccinia virus-infected fibroblasts in sheep (Schmaljohn 1981; Pearson, Schmaljohn & DeMartini 1981). Autologous vaccinia-infected skin fibroblasts were preferentially lysed by peripheral blood leucocytes 7 to 9 days postinfection over allogeneic or zenogeneic infected target cells in all ten sheep tested. Unlike the experiments reported for cattle (Rouse & Babiuk 1977) and dogs (Ho, Babiuk & Rouse 1978), we did not trypsinize the target cells several hours after infection. Vaccinia virus interrupts protein synthesis and trypsinization alters the MHC products on the plasma membrane (Ertl & Koszinkowski 1976). We have found that the *in vitro* growth and function of sheep killer lymphocytes are enhanced by supplementing the media with sheep IL-2 (T-cell growth factor) and that the cytotoxic effector cells generated *in vitro* are genetically restricted when killing vaccinia-infected fibroblasts (Knisley & Pearson 1983).

Cytotoxic T cells appear early after infection and are often able to kill target cells infected with different serotypes of the virus (Hackett et al. 1980). They are thought to have an important role in effecting recovery from primary infections with enveloped viruses.

Natural killer (NK) cells

A subpopulation of nonadherent mononuclear cells taken from animals that have not been immunized are capable of killing certain normal cells, neoplastic cells or virus-infected cells. Cells capable of such killing have been called NK cells and the topic has been reviewed recently by Welsh (1981). Although the NK cells appear to represent a morphologically homogeneous cell subset described as large granular lymphocytes, their specific haemopoietic lineage has not been unambiguously identified. The NK cells may represent a stage of differentiation of the myelomonocytic lineage or a distinct lineage; however, several surface features and functions suggest a kinship to the T-cell lineage (Casali & Trinchieri 1984). The NK cells have features that distinguish them from other lymphocytes. Human NK cells are nonadherent, surface Ig negative, have a Fc receptor of low avidity, often rosette with sheep red blood cells, and display the spotted pattern typical of lymphocytes stained for acid alpha-naphthyl acetate esterase rather than staining diffusely like macrophages (Saksela et al. 1979). The human NK cell may belong to the T-γ subset that displays suppressor activity (Moretta et al. 1977). Distinctive features of mouse NK cells are that they have Lyt 5 antigen which is also seen on thymocytes but not on B cells or macrophages (Cantor et al. 1979) and they have either low or no expression of Thy 1 antigen (Herberman, Nun & Holden 1978). Very little is known about the differentiation and ontogenesis of NK cells because of the lack of specific markers and the difficulty of identifying immature cells on the basis of their cytotoxic function (Casali & Trinchieri 1984).

Activity of NK cells has been found in nonadherent, mononuclear cell populations from chickens (Lam & Linna 1977), cats (McCarty & Grant 1980) and pigs (Huh, Kim & Amos 1981). Several viruses (Coxsackie, C-type retroviruses, Kunjin, lactic dehydrogenase, lymphocytic choriomeningitis, Minute virus of mice, mouse adenovirus, mouse hepatitis, Newcastle, Pichinde, polyoma, Semliki Forest and Sendai) have been noted to induce NK-cell activity in mice (Welsh 1978). Virus-infected cells are more susceptible to NK activity than uninfected targets, and cold-competition or target-absorption studies indicate that the same population of NK cells is capable of killing a variety of target cells (Santoli, Trinchieri & Lief 1978; Weston et al. 1981).

Most studies designed to detect *in vivo* NK-cell activity have been performed in mice. Various viral infections in mice, including those by lymphocytic choriomeningitis, lactic dehydrogenase, mumps, Moloney sarcoma, mouse hepatitis, Newcastle disease, Pichinde, Semliki Forest, vaccinia and vesicular stomatitis virus, greatly enhance NK-cell activity (Casali & Trinchieri 1984). The induction of NK-cell activity during lymphocytic choriomeningitis infection of mice occurs concomitantly with increasing levels of interferon and elimination of viral particles. Acute infections with Epstein-Barr, measles, mumps or cytomegaloviruses activate NK cells in humans (Casali & Trinchieri 1984). Although the association between height of NK-cell activity and control of the virus infection *in vivo* is interesting, definitive proof that NK cells are responsible for the control of the infection awaits clearer definition.

Macrophages

Macrophages are important in antigen presentation in the generation of both humoral and cellular immunity. In the mouse IL-1 is released from macrophages to stimulate Lyt 1-positive 2-negative 3-negative helper T cells to release IL-2 which results in the proliferation of activated T-effector cells (Smith et al. 1980; Smith, Gilbride & Favata 1980; Larsson, Iscove & Covtinho 1980; Watson et al. 1982). Macrophages remove antibody-neutralized virus from the circulation and degrade the complexes to low molecular weight substances. As effector cells, macrophages may become activated and exhibit nonspecific direct cell-mediated cytotoxicity by secreting monokines, as yet uncharacterized, that are cytotoxic for surrounding cells. Macrophages may act as effector cells in ADCC reactions against virus-infected cells through their Fc receptor for antibody (Evans & Alexander 1972; Bloom & Rager-Zisman 1975).

Neutrophils

Neutrophils were found to inhibit plaque formation by the herpes virus IBR by a mechanism that was not dependent on antibody and which did not require direct contact between the neutrophil and the IBR virus-infected target cell (Rouse et al. 1977). Infection of polymorphonuclear leucocytes with bovine viral diarrhoea virus impairs the myeloperoxidase, hydrogen peroxide and halide anti-bacterial systems, but does not alter oxidative metabolism (Roth, Kaeberle & Griffith 1981). Impairment of neutrophil function may partially explain the increased susceptibility of cattle to secondary bacterial infection during infection with bovine viral diarrhoea virus.

Combined soluble and cellular effectors of resistance to viral infections

Antibody-dependent cellular cytotoxicity (ADCC)

Specificity for the reaction of ADCC is dependent on the Ig bound to the target cell that is expressing virus-specified antigen. Unlike T-cell cytotoxicity, ADCC has no constraints of genetic restriction. Effector cells bearing Fc receptors bind to the Fc portion of the bound IgG. Cells that function as effectors include monocytes and macrophages (Kohl et al. 1977; Shore, Melewicz & Gordon 1977), polymorphonuclear leucocytes (Gale & Zighelboim 1975) and T cells (Santoli & Koprowski 1979). The most efficient effector cell in most species is a mononuclear cell in the peripheral blood that is characterized by being Fc receptor-positive and belonging to the null-cell subset, i.e. in man it is nonadherent, E-rosette negative and surface-Ig negative (Greenberg et al. 1973). However, exceptions to this generality have been noted in several species. Neutrophils were found to be the most efficient effectors for ADCC to IBR virus-infected cells in cattle (Grewal, Rouse & Babiuk 1977; 1980; Rouse 1981). Fewer antibody molecules per target cell are required for ADCC when compared to antibody- and complement-mediated lysis. For example, measles virus infected HeLA cells required about 5×10^5 antibody molecules per cell for ADCC as compared to at least 5×10^6 molecules per cell for complement-mediated lysis (Perrin, Tishon & Oldstone 1977). Thus ADCC can be effective at an earlier stage of infection than antibody plus complement.

The way in which effector cells bring about lysis is not completely understood for ADCC. Antibody must first bind to the target cell in most systems (Greenberg et al. 1977). The effector cells can cause lysis within a few minutes after interacting with the Fc portions of bound immunoglobulin molecules (Ziegler & Henney 1975). The kinetics of target-cell killing as measured by ^{51}Cr release indicate that one effector cell interacts with one target cell. Unlike T-cell killing, in which the effector cell may kill several successive target cells, the ADCC effector cell remains attached to the target cell and is thereby inhibited from participating in further cytolytic functions (Ziegler & Henney 1975).

The significance of ADCC in the living animal is not clear. Teleologically it is appealing to envisage circulating effector cells which attach to target cells marked for death with relatively few antibody molecules and rapidly lyse them in order to stop the viral infection quickly.

Interferon and NK cells

Short-term incubation of lymphocytes in the presence of added interferon markedly increases NK-cell activity without augmenting ADCC (Trinchieri & Santoli 1978; Trinchieri, Santoli & Koprowski 1978). Herberman and colleagues (1979) reported that prolonged exposure to increased levels of interferon increased NK and ADCC activity. Interferon acts to enhance NK-cell activity in at least three different ways: firstly, by increasing the number of large granular lymphocytes able to bind to the target cells; secondly, by accelerating the kinetics of lysis; and thirdly, by increasing the recycling ability of active NK cells (Timonen, Ortaldo & Herberman 1982).

One should interpret reports of ADCC or T-cell cytotoxicity against virus-infected cells with caution because interferon can be produced by the infected target cells and T cells during the course of the assay and possibly augment NK cytotoxicity during the same time frame (Onions 1983).

Conclusions

The most rapid expression of host resistance to viral infections is mediated by innate effectors such as the alternate pathway of complement activation and NK cells. Production of the various forms of interferon within the first few days of infection can inhibit viral replication in neighbouring cells as well as augment NK-cell killing of infected cells. Low levels of antibody bound to virus-infected cells are sufficient to allow circulating leucocytes with Fc receptors to bind and selectively lyse the infected cell. T cell-mediated cytotoxicity kills virus-infected cells and is a major factor in a host animal's recovery from an enveloped virus infection. Through the mechanisms of ADCC, antibody-complement mediated lysis and modulation of virus replication, specific antibody plays an important role in recovery from viral infections. Neutralizing antibody functions as a major barrier to reinfection by the same virus.

Acknowledgement

This work was supported in part by the Science and Education Administration, Collaborative Research Animal Health and Disease Research Funds, Regional Research Project W-112 from the Colorado State University Experiment Station and US Public Health Service Grant No. 5R23 AI 14893-02.

References

Biddison, W.E., Shaw, S. & Nelson, D.L. (1979). Virus specificity of human influenza virus-immune cytotoxic T cells. *Journal of Immunology*, **122**, 660-64.

Blanden, R.V. (1970). Mechanisms of recovery from a generalized viral infection: mousepox. *Journal of Experimental Medicine*, **132**, 1035-54.

Bloom, B.R. (1980). Interferons and the immune system. *Nature*, **284**, 593-95.

Bloom, B.R. & Rager-Zisman, B. (1975). Cell-mediated immunity in viral infections. In *Viral immunology and immunopathology* (ed. A.L. Notkins), pp.116-18. New York: Academic Press.

Bruck, C., Portetelle, D., Burny, A. & Zavada, J. (1982). Topographical analysis by monoclonal antibodies of BLV-gp51 epitopes involved in viral functions. *Virology*, **122**, 353-62.

Bubbers, J.E. & Lilly, F. (1977). Selective incorporation of H-2-antigenic determinants into Friend virus particles. *Nature*, **266**, 458-59.

Burakoff, S.J., Reiss, C.S., Finberg, R. & Mescher, M.F. (1980). Cell-mediated immunity to viral glycoproteins. *Review of Infectious Diseases*, **2**, 62-77.

Cantor, H., Kasai, M., Shen, F.W., Leclerc, J.C. & Glimcher, L. (1979). Immunogenic analysis of 'natural killer' activity in the mouse. *Immunological Reviews*, **44**, 3-12.

Casali, P. & Trinchieri, G. (1984). Natural killer cells in viral infection. In *Concepts in viral pathogenesis* (eds. A.L. Notkins & M.B.A. Oldstone), pp. 11-19. New York: Springer Verlag.

Cohn, M. & Epstein, R. (1978). T-cell inhibition of humoral responsiveness. 2. Theory on the role of restrictive recognition in immune regulation. *Cellular Immunology*, **39**, 125-153.

Cooper, N.R. (1979). Humoral immunity to viruses. In *Virus-host interactions: immunity to viruses* (eds. H. Fraenkel-Conrat & R.R. Wagner), Comprehensive Virology 15, pp. 123-70. New York: Plenum.

Cummins, J.M. & Rosenquist, B.D. (1980). Protection of calves against rhinovirus infection by nasal secretion induced by infectious bovine rhinotracheitis virus. *American Journal of Veterinary Research*, **41**, 161-65.

Cummins, J.M. & Rosenquist, B.D. (1983). Partial protection of calves against parainfluenza-3 virus infection by nasal secretion interferon induced by infectious bovine rhinotracheitis virus. *American Journal of Veterinary Research*, **43**, 1334-38.

Daniels, C.A., Boros, T., Rapp, H.J., Synderman, R. & Notkins, A.L. (1970). Neutralization of sensitized virus by purified components of complement. *Proceedings of the National Academy of Sciences of the USA*, **56**, 528-35.

DeMartini, J.C., Pearson, L.D. & Fiscus, S.A. (1978). Chromium-51-release assay of antibody and complement-mediated cytotoxicity for contagious ecthyma virus-infected cells. *American Journal of Veterinary Research*, **39**, 1922-26.

Doherty, P.C. & Zinkernagel, R.M. (1975). H-2 compatibility is required for T cell-mediated lysis of target cells infected with lymphocytic choriomeningitis virus. *Journal of Experimental Medicine*, **141**, 502-7.

El Azhary, M.A.S.Y., Silim, A. & Roy, R.S. (1981). Interferon, fluorescent antibody, and neutralizing antibody responses in sera of calves inoculated with bovine respiratory syncytial virus. *American Journal of Veterinary Research*, **42**, 1378-82.

Ertl, H.C. & Koszinkowski, U.H. (1976). Modification of H-2-antigenic sites by enzymatic treatment influences virus-specific target-cell lysis. *Journal of Immunology*, **117**, 2112-18.

Essex, M., Klein, G., Snyder, P. & Harrold, J.B. (1971). Correlation between humoral antibody and regression of tumours induced by feline sarcoma virus. *Nature*, **233**, 195-96.

Evans, R. & Alexander, P. (1972). Mechanisms of immunologically specific killing of tumour cells by macrophages. *Nature*, **236**, 168-70.

Fujinami, R. & Oldstone, M.B.A. (1979). Anti-viral antibody reacting on the plasma membrane alters measles virus expression inside the cell. *Nature*, **279**, 529-30.

Gale, R.P. & Zighelboim, J. (1975). Polymorphonuclear leucocytes in antibody-dependent cellular cytotoxicity. *Journal of Immunology*, **114**, 1047-51.

Greenberg, A.H., Haidson, L., Shen, L. & Roitt, I.M. (1973). Antibody-dependent cell-mediated cytotoxicity due to a null lymphoid cell. *Nature*, **242**, 111-13.

Greenberg, S.B., Criswell, B.S., Six, H.R. & Couch, R.B. (1977). Lymphocyte cytotoxicity to influenza virus-infected cells. *Journal of Immunology*, **119**, 2100-6.

Greenspan, N.S., Schwartz, D.H. & Doherty, P.C. (1983). Role of lymphoid cells in immune surveillance against viral infection. In *Advances in host defence mechanisms* (eds. J.L. Gallin & A.S. Fauci), vol. 2, pp. 101-41. New York: Raven Press.

Grewal, A.S., Rouse, B.T. & Babiuk, L. (1977). Mechanisms of resistance to herpes viruses: comparison of the effectiveness of different cell types in mediating antibody-dependent cell-mediated cytotoxicity. *Infection & Immunity*, **15**, 698-703.

Grewal, A.S., Rouse, B.T. & Babiuk, L. (1980). Mechanisms of recovery from viral infections: destruction of infected cells by neutrophils and complement. *Journal of Immunology*, **124**, 312-19.

Griffin, J.A. & Compans, R.W. (1979). Effect of cytochalasin B on the maturation of enveloped viruses. *Journal of Experimental Medicine*, **150**, 379-91.

Hackett, C.J., Askonas, B.A., Webster, R.G. & van Wyke, K. (1980). Quantitation of influenza virus antigens on infected target cells and their recognition by cross-reactive cytotoxic T cells. *Journal of Experimental Medicine*, **151**, 1014-25.

Hapel, A.J., Bablanian, R. & Cole, G. (1980). Induction requirements for the generation of virus-specific T lymphocytes. 2. Poxvirus and H-2 antigens associate without cellular or virus-directed protein synthesis, and remain immunogenic in cell membrane fragments. *Journal of Immunology*, **124**, 1990-96.

Herberman, R.B., Nun, M.E. & Holden, J. (1978). Low density of Thy 1 antigen on mouse effector cells mediating natural cytotoxicity against tumour cells. *Journal of Immunology*, **121**, 304-9.

Herberman, R.B., Djeu, J.Y., Kay, H.D., Ortaldo, J.R., Riccardi, C., Bonnard, G.D., Holden, H.T., Fagnani, R., Santoni, A. & Puccetti, P. (1979). Natural killer cells: characteristics and regulation of activity. *Immunological Reviews*, **44**, 43-70.

Ho, C.K., Babiuk, L.A. & Rouse, B.T. (1978). Immune effector cell activity in canines: failure to demonstrate genetic restriction in direct anti-viral cytotoxicity. *Infection & Immunity*, **19**, 18-26.

Huh, N.D., Kim, Y.B. & Amos, D.B. (1981). Natural killing (NK) and antibody-dependent cellular cytotoxicity (ADCC) in specific pathogen-free miniature swine and germ-free piglets: two distinct effector cells for NK and ADCC. *Journal of Immunology*, **127**, 2190-93.

Joseph, B.S. & Oldstone, M.B.A. (1975). Immunologic injury in measles virus infection. 2. Suppression of immune injury through antigenic modulation. *Journal of Experimental Medicine*, **142**, 864-76.

Knisley, K.A. & Pearson, L.D. (1983). Ovine T-cell growth factor: *in vitro* generation of cytotoxic cells. *Federation Proceedings*, **42**, 1239.

Kohl, S., Starr, S.E., Oleske, J.M., Shore, S.L., Ashman, R.B. & Nahmias, A.J. (1977). Human monocyte-macrophage-mediated antibody-dependent cytotoxicity to herpes simplex virus-infected cells. *Journal of Immunology*, **118**, 729-35.

Kreth, H.W., ter Meulen, V. & Eckert, G. (1979). Demonstration of HLA-restricted killer cells in patients with acute measles. *Medical Microbiology & Immunology*, **165**, 203-24.

Lafferty, K.J. (1963). The interaction between virus and antibody. *Virology*, **21**, 76-90.

Lam, K.M. & Linna, T.J. (1977). Protection of newly hatched chickens from Marek's disease (JMV) by normal spleen cells from older animals. In *Advances in comparative leukaemia research* (eds. P. Bentvelzen, J. Hilgers & D.S. Yohn). Amsterdam: Elsevier.

Larsson, E., Iscove, N.N. & Coutinho, A. (1980). Two distinct factors are required for induction of T-cell growth. *Nature*, **283**, 664-66.

Letvin, N.L., Kauffman, R.S. & Finberg, R. (1981). T-lymphocyte immunity to viruses: cellular requirements for generation and role in clearance of primary infections. *Journal of Immunology*, **127**, 2334-39.

Mandel, B. (1978). Neutralization of animal viruses. *Advances in Virus Research*, **23**, 205-68.

Mandel, B. (1979). Interaction of viruses with neutralizing antibodies. In *Virus-host interactions: immunity to viruses* (eds. H. Fraenkel-Conrat & R.R. Wagner), Comprehensive Virology 15, pp. 37-121. New York: Plenum.

Matzinger, P. (1981). A one receptor view of T-cell behaviour. *Nature*, **292**, 497-501.

McCarty, J.M. & Grant, C.K. (1980). Cytotoxic effector cell populations in cats. In *Feline leukaemia virus* (eds. W.D. Hardy Jr., M. Essex & A.J. McClelland), pp. 203-10. Amsterdam: Elsevier.

McKenzie, I.F.C. & Potter, T. (1979). Murine lymphocyte surface antigens. *Advances in Immunology*, **27**, 281-338.

McMichael, A.J., Ting, A., Szeerink, H.J. & Askonas, B.A. (1977). HLA restriction of cell-mediated lysis of influenza virus-infected human cells. *Nature*, **270**, 524-26.

Moretta, L., Webb, S.R., Grossi, C.E., Lydyard, P.M. & Cooper, M.D. (1977). Functional analysis of two human T-cell subpopulations: help and suppression of B-cell responses by T cells bearing receptors for IgM and IgG. *Journal of Experimental Medicine*, **146**, 184-200.

Neil, J.C., Smart, J.E., Hayman, M.J. & Jarrett, O. (1980). Polypeptides of feline leukaemia virus: a glycosylated gag-related protein is released into culture fluids. *Virology*, **105**, 250-53.

Nelles, M.J. & Streilein, J.W. (1980). Immune responses to acute virus infection in the Syrian hamster. 1. Studies on genetic resistance of cell-mediated cytotoxicity. *Immunogenetics*, **10**, 185-99.

Oldstone, M.B.A. (1975). Virus neutralization and virus-induced immune complex disease. *Progress in Medical Virology,* **19**, 84-119.

Oldstone, M.B.A., Cooper, N.R. & Larson, D.L. (1974). Formation and biologic role of polyoma virus-antibody complexes. *Journal of Experimental Medicine,* **140**, 549-95.

Onions, D.E. (1983). The immune response to virus infections. *Veterinary Immunology & Immunopathology,* **4**, 237-77.

Pearson, L.D., Schmaljohn, A.L. & DeMartini, J.C. (1981). Genetic restriction of cell-mediated cytotoxicity for vaccinia-infected fibroblasts by sheep leucocytes. In *Advances in experimental medicine and biology* (eds. J.E. Butler, J.R. Duncan & K. Nielson), vol. 137, p. 759. New York: Plenum.

Pearson, L.D., Doherty, P.C., Hapel, A. & Marshall, I.D. (1976). The responses of the popliteal lymph node of the sheep to Ross River and Kunjin viruses. *Australian Journal of Experimental Biology & Medical Science,* **54**, 371-79.

Perrin, L.H., Tishon, A. & Oldstone, M.B.A. (1977). Immunologic injury in measles virus infection. *Journal of Immunology,* **118**, 282-90.

Radwan, A.L. & Crawford, T.B. (1974). The mechanisms of neutralization of sensitized equine arteritis virus by complement components. *Journal of General Virology,* **25**, 229-37.

Rager-Zisman, B. & Allison, A.C. (1973). Effects of immunosuppression on Coxsackie B-3 virus infection in mice, and passive protection by circulating antibody. *Journal of General Virology,* **19**, 339-51.

Rojko, J.L., Hoover, E.A., Quackenbush, S.L. & Olsen, R.G. (1981). Latency and reactivation of feline leukaemia virus infection. In *Advances in comparative leukaemia research* (eds. D.S. Yohn & J.R. Blakeslee), pp. 225-26. Amsterdam: Elsevier.

Rossi, C.R., Kiesel, G.K. & Hoff, E.J. (1980). Factors affecting the assay of bovine type I interferon on bovine embryonic lung cells. *American Journal of Veterinary Research,* **41**, 552-56.

Roth, J.A., Kaeberle, M.L. & Griffith, R.W. (1981). Effects of bovine viral diarrhoea virus infection on bovine polymorphonuclear leucocyte function. *American Journal of Veterinary Research,* **42**, 244-50.

Rouse, B.T. (1981). Role of neutrophils in anti-viral immunity. In *Advances in experimental medicine and biology* (eds. J.E. Butler, J.R. Duncan & K. Nielson), vol. 137, pp. 263-78. New York: Plenum.

Rouse, B.T. & Babiuk, L.A. (1977). The direct anti-viral cytotoxicity by bovine lymphocytes is not restricted by genetic incompatibility of lymphocytes and target cells. *Journal of Immunology,* **118**, 618-24.

Rouse, B.T., Wardley, R.C., Babiuk, L.A. & Mukkur, T.K.S. (1977). The role of neutrophils in anti-viral defence: *in vitro* studies on the mechanism of anti-viral inhibition. *Journal of Immunology,* **118**, 1957-61.

Saksela, E., Timonen, T., Ranki, A. & Hary, P. (1979). Morphological and functional characterization of isolated effector cells responsible for human natural killer activity to foetal fibroblasts and to cultured cell line targets. *Immunological Reviews,* **44**, 1-123.

Santoli, D. & Koprowski, H. (1979). Mechanisms of activation of human natural killer cells. *Immunological Reviews,* **44**, 125-63.

Santoli, D., Trinchieri, G. & Lief, F.S. (1978). Cell-mediated cytotoxicity against virus-infected target cells in humans. *Journal of Immunology,* **121**, 526-31.

Schmaljohn, A.L. (1981). Vaccinia virus-induced cell-mediated cytotoxicity in sheep blood and lymph. Ph.D. thesis. Colorado State University.

Schmid, D.S., Larsen, H.S. & Rouse, B.T. (1982). Role of Ia-antigen expression and secretory function of accessory cells in the induction of cytotoxic T-lymphocyte responses against herpes simplex virus. *Infection & Immunity*, **37**, 1138-47.

Shek, W.R., Shultz, R.D. & Appel, M.J.G. (1980). Natural and immune cytolysis of canine distemper virus-infected target cells. *Infection & Immunity*, **28**, 724-34.

Shore, S.L., Melewicz, F.M. & Gordon, D.S. (1977). The mononuclear cell in human blood which mediates antibody-dependent cellular cytotoxicity to virus-infected target cells. *Journal of Immunology*, **118**, 558-66.

Sissons, J.G.P. & Oldstone, M.B.A. (1980a). Antibody-mediated destruction of virus-infected cells. *Advances in Immunology*, **29**, 209-60.

Sissons, J.G.P. & Oldstone, M.B.A. (1980b). Killing of virus-infected cells: the role of anti-viral antibody and complement in limiting virus infection. *Journal of Infectious Diseases*, **142**, 442-48.

Sissons, J.G.P. & Oldstone, M.B.A. (1980c). Killing of virus-infected cells by cytotoxic lymphocytes. *Journal of Infectious Diseases*, **142**, 114-19.

Smith, K.A., Gilbride, K.J. & Favata, M.F. (1980). Lymphocyte-activating factor promotes T-cell growth factor production by cloned murine lymphoma cells. *Nature*, **287**, 853-55.

Smith, K.A., Lachman, L.B., Oppenheim, J.I. & Favata, M.F. (1980). The functional relationship of the interleukins. *Journal of Experimental Medicine*, **151**, 1151-56.

Stanton, G.D. & Baron, S. (1984). Interferon and viral pathogenesis. In *Concepts in viral pathogenesis* (eds. A.L. Notkins & M.B.A. Oldstone), pp. 4-10. New York: Springer Verlag.

Swain, S.L. & Dutton, R.W. (1980). Mouse T-lymphocyte subpopulations: relationships between function and Lyt antigen phenotype. *Immunology Today*, **1**, 61-65.

Theilen, G.H., Pederson, N.C. & Higgins, J. (1979). Role of regional and distant lymph nodes in rejection of feline sarcoma virus-induced tumours in sheep. *Journal of the National Cancer Institute*, **63**, 389-97.

Timonen, T., Ortaldo, J.R. & Herberman, R.B. (1982). Analysis by a single cell cytotoxicity assay of natural killer (NK) cell frequencies among human large granular lymphocytes and of the effects of IFN on their activation. *Journal of Immunology*, **128**, 2514-21.

Tizard, I.R. (1984). *Immunology: an introduction*, p. 213-14. Philadelphia: Saunders.

Trinchieri, G. & Santoli, D. (1978). Anti-viral activity induced by culturing lymphocytes with tumour-derived or virus-transformed cells. *Journal of Experimental Medicine*, **147**, 1314-33.

Trinchieri, G., Santoli, D. & Koprowski, H. (1978). Spontaneous cell-mediated cytotoxicity in humans: role of interferon and immunoglobulins. *Journal of Immunology*, **120**, 1849-55.

Wagner, H., Hardt, C., Heegk, K., Pfizenmaier, K., Solbach, W., Barlett, R., Stockinger, H. & Rollinghoff, M. (1980). T-T-cell interactions during cytotoxic T-lymphocyte responses: T cell-derived helper factor (interleukin-2) as a probe to analyse CTL responsiveness and thymic maturation of CTL progenitors. *Immunological Reviews*, **51**, 215-55.

Wainberg, M.A., Markson, Y., Weiss, D.N. & Doljanski, F. (1974). Cellular immunity against Rous sarcomas of chickens: preferential reactivity against autochthonous target cells as determined by lymphocyte adherence and cytotoxicity tests *in vitro*. *Proceedings of the National Academy of Sciences of the USA*, **71**, 3565-69.

Watson, J. & Mochizuki, D. (1980). Interleukin-2: a class of T-cell growth factors. *Immunological Reviews*, **51**, 257-78.

Watson, J., Frank, M.B., Mochizuki, D. & Gillis, J. (1982). The biochemistry and biology of interleukin-2. In *Lymphokines* (ed. S.B. Mizel), pp. 113-14. New York: Academic Press.

Weinberg, R.A. (1984). Cellular oncogenes and the pathogenesis of cancer. In *Concepts in viral pathogenesis* (eds. A.L. Notkins & M.B.A. Oldstone), pp. 178-86. New York: Springer Verlag.

Welsh, R.M. (1977). Host-cell modification of lymphocytic choriomeningitis virus and Newcastle disease virus altering viral inactivation by human complement. *Journal of Immunology*, **118**, 348-54.

Welsh, R.M. (1978). Mouse natural killer cells: induction, specificity, and function. *Journal of Immunology*, **121**, 1631-35.

Welsh, R.M. (1981). Natural cell-mediated immunity during viral infections. In *Current topics in microbiology and immunology* (ed. O. Haller), vol. 92, pp. 83-106. New York: Springer Verlag.

Welsh, R.M., Cooper, N.R., Jensen, F.C. & Oldstone, M.B.A. (1975). Human serum lyses RNA tumour viruses. *Nature*, **257**, 612-14.

Weston, P.A., Jensen, P.J., Levy, N.L. & Koren, H.S. (1981). Spontaneous cytotoxicity against virus-infected cells: relationship to NK against uninfected cell lines and to ADCC. *Journal of Immunology*, **126**, 1220-24.

Williamson, A.R. (1980). Three-receptor, clonal expansion model for selection of self-recognition in the thymus. *Nature*, **283**, 527-32.

Woan, M.C., Yip, D.M. & Tompkins, W.A.F. (1978). Autochthonous, allogeneic, and xenogeneic cells as targets for vaccinia-immune lymphocyte cytotoxicity. *Journal of Immunology*, **120**, 312-16.

Yates, W.D.G. (1982). A review of infectious bovine rhinotracheitis, shipping fever pneumonia and viral-bacterial synergism in respiratory disease of cattle. *Canadian Journal of Comparative Medicine*, **46**, 225-63.

Ziegler, H.K. & Henney, C. (1975). Antibody-dependent cytolytically active human leucocytes: an analysis of inactivation following *in vitro* interaction with antibody-coated target cells. *Journal of Immunology*, **115**, 1500-4.

Zinkernagel, R.M. & Doherty, P.C. (1974). Immunological surveillance against altered self-components by sensitized T lymphocytes in lymphocytic choriomeningitis. *Nature*, **251**, 547-48.

Zinkernagel, R.M. & Doherty, P.C. (1979). MHC-restricted cytotoxic T cells: studies on the biological role of polymorphic major transplantation antigens determining T-cell restriction-specificity, function and responsiveness. *Advances in Immunology*, **27**, 51-177.

Zinkernagel, R.M., Althage, A. & Jensen, F.C. (1977). Cell-mediated immune responses to lymphocytic choriomeningitis and vaccinia virus in rats. *Journal of Immunology*, **119**, 1242-47.

Zinkernagel, R.M., Callahan, G.N., Althage, A., Cooper, S., Klein, P.A. & Klein, J. (1978). On the thymus in the differentiation of 'H-2 self-recognition' by T cells: evidence for dual recognition? *Journal of Experimental Medicine*, **147**, 882-96.

Zisman, B., Hirsch, M.J. & Allison, A.C. (1970). Selective effect of anti-macrophage serum, silica and anti-lymphocyte serum on pathogenesis of herpes virus infection of young adult mice. *Journal of Immunology*, **104**, 1155-59.

25

Immunogenicity of different physical forms of viral antigenic subunits

B. MOREIN

Whole virus particles are often strongly immunogenic, but when they are
dissociated the immune response is low to surface antigens in the viral
envelope which are normally important in protective immunity. On the other
hand, whole killed virus particles from the Paramyxoviridae family are not
suitable in vaccines because they do not stimulate an immune response
against the fusion protein in the viral envelope. However, these problems can
be circumvented by arranging the surface antigens in different physical forms.
At present there are four defined forms in which we can arrange viral
envelope proteins: (1) the monomer form, where one envelope protein is in
complex with several detergent molecules, (2) the micelle, which is a complex
of envelope proteins aggregated by their hydrophobic parts towards the
centrum of the complex, (3) the virosome, which is a lipid vesicle (liposome)
with the envelope protein integrated in the lipid wall by the hydrophobic
moiety, and (4) the iscom, a new type of complex which has been constructed
with the envelope proteins attached to a glycoside matrix. Results are
summarized of studies carried out mainly in mice to compare the
immunogenicity of these viral antigen preparations. The monomers are poorly
immunogenic, inducing little or no antibody and a very low degree of
protective immunity. When inoculated together with micelles they suppress
the immune response to the micelles. The micelles are generally strongly
immunogenic, inducing protective immunity and high antibody responses. In
lambs, however, adjuvant had to be included in the micelle preparation
containing the envelope proteins of parainfluenza-3 (PI-3) virus to induce
protective immunity. The virosomes have similar immunogenicity to the
micelles, but are more difficult to make and are unstable. Using bovine PI-3
virus envelope proteins, iscoms were found to induce about 10 times higher
antibody responses in mice than micelles and correspondingly high levels of
protection against challenge. Iscoms prepared with measles viral envelope
proteins were also found to elicit a significant antibody response against the
fusion protein.

Introduction

Vaccines have been a major contributory factor in the control of infectious diseases for 200 years. There has generally been little change in the type of vaccine used against viral diseases over the years. Basically, two kinds of virus vaccines have been used: attenuated live vaccines and killed vaccines. In the case of live vaccines, virulent strains of virus have been attenuated by growing them in an unnatural host or host cell until they no longer produce disease in the natural host, but grow well enough to invoke an immune response. There are also other means of attenuating virus strains. For killed vaccines, virus is grown in large quantities and then killed (inactivated), usually by treatment with a chemical, such as formaldehyde, B-propiolactone or an imine.

The live and killed vaccines both have advantages and drawbacks. Live vaccines are comparatively inexpensive to produce, and may also be easier to administer. For example, in flocks of chickens, they can be given in the drinking water, thus reducing the costs of handling. However, they need careful control for the presence of other infectious agents. Moreover, live vaccines are often fragile and sensitive to extreme environmental conditions, so that care is required in their transport and storage in order to maintain infectivity. A typical example of this kind of problem is the transport of vaccines to distant places in tropical countries where facilities for keeping the vaccine cold are lacking. Another potential problem is inadequate attenuation: some sensitive individuals in the population that lack resistance may develop pathogenic infections. This has happened with smallpox vaccine which caused severe disease in T cell-defective children. Similarly, live polio vaccine has caused paralytic poliomyelitis (Mortimer 1978). Live vaccines may also be 'overattenuated' and have no or low immunogenicity. Under ideal conditions, however, live vaccines are more effective than killed vaccines, inducing a more complete immune response. Good examples are the vaccines against Marek disease, rinderpest and canine distemper, while attempts to make live foot-and-mouth disease vaccine have not been successful.

Killed vaccines are also used extensively. Their main advantages are that they are safe, if the inactivation process is correctly applied, and are more stable than live vaccines. A drawback is that they generally are more expensive than live vaccines. Usually at least two inoculations are needed to induce protective immunity, and the duration of immunity is shorter than that obtained with live microorganisms. In several cases, it has not been possible to make effective killed vaccines, for example against canine distemper virus. In other cases, the amount of antigen required in a killed vaccine is so high that the vaccine is toxic. This is particularly the case with bacterial vaccines, where endotoxins are a limiting factor (e.g. *Escherichia coli* vaccines). It is also well known that vaccines containing *Mycobacteria* often have side effects. Even killed whole virus vaccines may be toxic if high doses are used. As an example, whole influenza virus vaccines, if given to children below school age, can cause a toxic reaction (Center for Disease Control 1984). However, good results have been obtained with a number of killed vaccines, including foot-and-mouth disease virus and parvo viruses. Killed polio vaccine has successfully been used to eradicate polio in Finland, the Netherlands and Sweden and is an alternative to the attenuated poliovirus vaccine. The attenuated vaccine was found to be ineffective when administered to some individuals in tropical countries, possibly due to the presence of inhibitors to the attenuated virus in the alimentary tract.

During the last 20 years, studies have been conducted on mechanisms of immunity and pathogenesis in many infectious diseases, so that much more is now known about the antigenic components of the microorganisms responsible for induction of protective immunity. With regard to enveloped viruses, the peplomers generally seem to be the important components in a vaccine. These are proteins, mostly glycoproteins, which are inserted in the envelope or membrane of the virus and extend outwards.

More intensive work to produce a vaccine containing the relevant antigenic subunits started with influenza virus. In early vaccines, the virus particles were disrupted with detergent (mostly cholates or deoxycholates) and organic solvents such as ether. Since most virus components were left in these vaccines, they are more correctly referred to as 'split vaccines' rather than subunit vaccines. These split vaccines did not give side effects, but unfortunately they were poorly immunogenic. Subsequently, many attempts have been made to produce subunit vaccines from different kinds of viruses with limited success. In the first attempts, little consideration was given to the physical form in which the relevant antigens were presented, although studies had indicated that physical form greatly influences immunogenicity. Brand and Skehel (1972), in an attempt to make a subunit vaccine from the influenza virus, cleaved the peplomers near the lipid bilayer of the virion with the proteolytic enzyme bromelain. The resulting peplomer product, in a monomeric form, was poorly immunogenic. Similar results were obtained with Gp 71 monomeric envelope protein of Friend leukemia virus (Hunsman, Moenning & Schafer 1976). More recently, in studies with rabies virus, it was found that the monomeric form of the soluble glycoprotein was poorly immunogenic (Dietzschold et al. 1983). Thus, to date, subunits isolated from viruses have been found to be less immunogenic than the intact virus particle. Some results indicate that the arrangement of antigens influences their ability to elicit an immune response. Also, adjuvants have been used to overcome the problem of poor immunogenicity of inactivated or subunit vaccines. The most widely used adjuvants are aluminium hydroxide, aluminium phosphate, saponins or an active substance derived from a saponin, named Quil A (Dalsgaard 1978), and mineral oils. Generally these adjuvants increase immunogenicity, but the really effective adjuvants — saponin, Quil A and the mineral oils — are toxic or give rise to severe local reactions when they are used in effective doses. This has limited their use in vaccines both for humans and for animals. The low efficacy of subunit and sometimes whole virus vaccines indicates that the presentation of antigens to the immune system is not optimal. This, together with the problems which occur when adjuvants are used, suggests that new strategies are required for the preparation of vaccines against viral diseases.

This chapter will outline the different physical forms in which viral peplomers can be arranged and their relative efficacy in inducing the appropriate immune responses. Finally, the problem will be considered of obtaining optimal immunity with the new generation of vaccines based on peptides obtained by recombinant DNA techniques or produced synthetically.

Immunogenicity of monomers, micelles and virosomes

The influence on the immune response of the physical form in which antigens are presented has been tested in mice using virus peplomer proteins. Initially, three different physical forms were prepared of the peplomer of Semliki Forest virus and these were tested for their ability to induce protective immunity and antibody in mice (Morein et al. 1978; Balcarova, Helenius & Simons 1981).

Preparation of micelles

The three different forms tested were the monomer, the micelle and the virosome. In all three cases, the peplomers are solubilized from the virus membrane by use of a detergent (Morein et al. 1978).

The peplomer is an amphiphilic protein, i.e. it has a lipid-soluble region which is the hydrophobic part inserted in the lipid bilayer of the virion, while the hydrophilic part is extended outwards from the virus particle. Appropriate detergents are generally nonionic, since they are mild and do not denature protein, in contrast to ionic detergents (Helenius & Simons 1975). The virus is solubilized with an excess of detergent (e.g.

2% Triton X-100), and then the detergent is removed by sucrose density gradient centrifugation (Helenius & Simons 1975; Tanford & Reynolds 1976). The centrifugation procedure for preparation of micelles according to the method developed by Helenius and Bonsdorff (1976) is illustrated in Figure 1. The solubilized virus is layered on a sucrose gradient containing a zone of 15% sucrose with 1% Triton X-100 above a sucrose gradient of 20% to 50% devoid of detergent. Under optimal conditions, the viral nucleocapsids with the attached internal proteins sediment to the bottom of the gradient. During sedimentation through the zone of Triton X-100, the peplomers are delipidated and the detergent is removed when the peplomers enter the detergent-free sucrose zone. The Triton X-100 does not enter this zone due to its buoyant density. Concomitant with the removal of detergent, the peplomers associate by their hydrophobic regions in the centre of a complex with the hydrophilic parts extending outwards. Thus, a micelle complex is formed which is structurally similar to detergent micelles. The micelles of the Semliki Forest virus proteins have an octameric structure with a molecular mass of 9.5 million daltons and a sedimentation value of 29S. They are practically free from lipid or detergent.

Preparation of monomers or peplomers

To prepare monomers or peplomers, the same centrifugation method is used as for the preparation of protein micelles, except that the 20 to 50% sucrose gradient contains 0.05% Triton X-100 which prevents the peplomers from self-associating. Complexes are formed consisting of one peplomer with 75 Triton X-100 molecules around the hydrophobic region. The molecular mass is approximately 100,000 daltons with a sedimentation value of 4S.

Reconstruction of viral peplomers with phospholipids into virosomes

Solubilized virus peplomers can be reconstituted with phospholipids to form liposomes in which peplomers are exposed on the external surface. Such liposomes with virus peplomers have been called virosomes (Almeida et al. 1975). To prepare Semliki Forest virus virosomes, the virus was solubilized with the detergent Triton X-100, which is effective and mild, but difficult to remove. Triton X-100 was exchanged for β-octylglucoside which has a 100-fold higher critical micellar concentration than Triton X-100 and is, therefore, well suited for reconstitution although it is not an efficient solubilizing agent. The detergent was exchanged by centrifuging Triton X-100-solubilized Semliki Forest virus into a 10 to 40% sucrose gradient containing octylglucoside.

Table 1. Vaccination against Semliki Forest virus encephalomyelitis in BALB/c mice: the protective effect induced by different forms of purified spike protein vaccine. The vaccination experiments were performed on 5- to 6-week-old female BALB/c mice. The mice were vaccinated with a single dose of 10 μg spike protein in 100 μl Eagle's minimal essential medium. One-third of the dose was injected subcutaneously and two-thirds intraperitoneally. Two weeks later groups of 10 mice were challenged intraperitoneally with virulent strain L10.H6.Cl of Semliki Forest virus at doses ranging from 50 to 10,000 PFU (Experiment 1) 50 to 100,000 PFU (Experiment 2) and 50 to 10 million PFU (Experiment 3). The infectivity was determined by plaque titration in BHK-21 cells. The amount of protein in the SDS in the vaccine preparations was measured by the method of Lowry and colleagues (1951) using 0.1% SDS in the reaction mixture. From Morein et al. (1978). Reprinted by permission of *Nature*.

Physical form of vaccine	Dose (protein μg)	50% lethal dose of challenge virus ($\log_{10}$ PFU)		
		Expt 1	Expt 2	Expt 3
(control)		2.3	2.1	2.6
4.5S complex (monomer)	10	n.d.	2.7	3.4
29S complex (octamer)	10	>3.5	>5.3	>6.3
Virosome (multimer)	10	>4.0	>5.1	>6.6

The mixture of peplomers, octylglucoside, sucrose and phospholipid was dialysed against a buffer resulting in the removal of sucrose and detergent (Helenius, Fries & Kartenbeck 1977). The protein-lipid ratio at the reconstitution of the Semliki Forest virus peplomers was 1:1 (w/w). Two classes of lipid vesicles were formed preferentially — small protein-rich and large lipid-rich vesicles. In the small vesicles, all peplomers were accessible to protease treatment, while in the large vesicles, 30% of the peplomers were inaccessible. The protease-accessible proteins are considered to extend outward from the vesicles, while the inaccessible proteins are directed inward. For further details on the mechanism by which the two types are formed see Helenius, Fries and Kartenbeck (1977).

Parainfluenza-3 (PI-3) virus peplomers were not reconstituted into the lipid vesicles using a ratio of 1:1 of protein to lipid. At this ratio, the peplomers self-associated, but at a protein-lipid ratio of 1:3 or less, the PI-3 virus peplomers were incorporated into liposomes (Morein et al. 1982). A modification of the preparation of virosomes using preformed liposomes is described by Oxford and colleagues (1981).

Vaccination experiments
Semliki Forest virus

The three different preparations of Semliki Forest virus antigen were tested in vaccination experiments in mice. Ten BALB/c mice were each immunized with 10 μg of Semliki Forest virus peplomers, either in the form of protein monomers, protein micelles or virosomes (Morein et al. 1978). The mice were inoculated with one-third of the dose intraperitoneally and two-thirds subcutaneously. Two weeks later they were challenged with a virulent strain of Semliki Forest virus which causes a fatal encephalomyelitis in susceptible mice.

The challenge dose ranged from 50 to 10 million times the 50% lethal dose (LD_{50}), using 10 mice per group. Nonvaccinated BALB/c mice in

Figure 1. Sucrose density gradient centrifugation procedure used in preparation of micelles containing PI-3 viral proteins. To the right, SDS-polyacrylamide gel electrophoresis of the resulting preparation is shown. Two proteins (H and F) are present in the protein micelles.

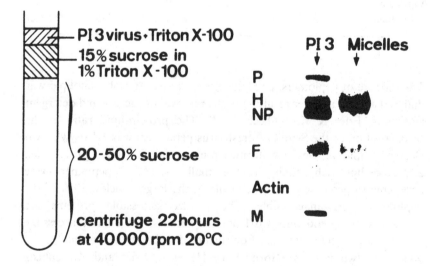

groups of 10 were used to determine the LD_{50}. The results are presented in Table 1. Mice vaccinated with the monomers only resisted challenge doses up to about 6 LD_{50} units of virus. In contrast, the mice vaccinated with multimer forms, i.e the micelles or the virosomes, resisted challenge with more than 10,000 LD_{50} units of virus. The antibody response of the immunized mice was also measured in a solid phase radioimmunoassay (Morein et al. 1978; Balcarova, Helenius & Simons 1981). The results were similar to those obtained in the challenge experiments: the multimer forms, i.e. micelles and virosomes, induced high antibody titres, while no detectable antibody response was found in mice vaccinated with monomers. Nude BALB/c mice vaccinated with micelles did not show a detectable response, indicating the role of T cells in antibody production (Morein et al. 1978).

Parainfluenza-3(PI-3) virus

Monomer and micelle forms prepared from the peplomers of PI-3 virus have also been tested for immunogenicity in mice. The antibody response following immunization was measured in the serum with an enzyme-linked immunosorbent assay (ELISA) and with a haemagglutination inhibition (HI) test (Morein et al. 1983). The results were similar to those

Figure 2. Antibody response in serum from BALB/c mice immunized with micelles prepared from peplomers of PI-3 virus. Antibody of the IgG class was measured with an ELISA technique. The results are expressed in $\log_2$ µg/µl as the mean + the standard devition for each group of mice. $\triangle$ is 10 µg micelles; $\bigcirc$ is 1 µg micelles; $\bullet$ is 0.1 µg micelles; $\blacktriangle$ is nonvaccinated controls. From Morein et al. (1983). Reprinted by permission of the *Journal of General Virology*.

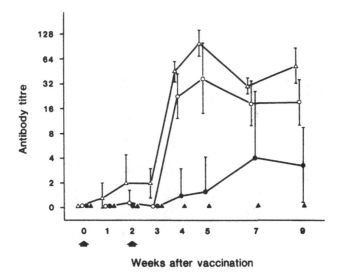

obtained with envelope proteins from Semliki Forest virus. Comparatively high titres were found in the mouse sera following two immunizations with 1 or 10 µg PI-3 envelope proteins in micelles (50 µg of IgG anti-PI-3 per ml mouse serum, as shown in Figure 2), while only a low serum antibody response was seen after two immunizations of mice with 10 µg protein in the monomer form (about 5 µg IgG anti-PI-3 antibody per ml serum). Mice immunized twice with 5 µg of PI-3 virus peplomers, comprising 2.5 µg in the form of monomers and 2.5 µg in the form of protein micelles, showed an antibody response of the same magnitude as mice immunized with monomers alone. This response should be compared with the 10-fold higher antibody response induced by 1 µg peplomers given as micelles only. These results suggest that the monomer form is not only poorly immunogenic, but also exerts an immunosuppressive effect (Morein et al. 1983).

The micelles containing PI-3 peplomers were also tested in lambs (Morein et al. 1983). Two groups of lambs were immunized twice intramuscularly at 3-week intervals with 15 µg PI-3 virus micelles. Seven lambs received the first dose of micelles emulsified in an oil adjuvant and the second dose without adjuvant. Another six lambs were given both doses without adjuvant. Only the lambs inoculated with micelles in

Figure 3. Recovery of PI-3 virus from nasal swabs after challenge of lambs which had been vaccinated with micelles containing PI-3 virus peplomers. Group A is nonvaccinated lambs; Group B is lambs vaccinated twice with protein micelles without adjuvant; Group C is lambs vaccinated twice with protein micelles, with adjuvant included in the first vaccination. From Morein et al. (1983). Reprinted by permission of the *Journal of General Virology*.

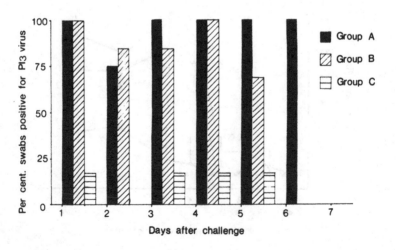

adjuvant exhibited a significant antibody response, as indicated by the HI test, and they were protected against challenge with a PI-3 virus strain pathogenic for sheep. The protection was measured by virus excretion and by evaluation of lung lesions (Figures 3 and 4). The efficacy of the micelles seems to be similar to that of formalin-killed virus (Wells et al. 1976).

Immunization with viral antigens in iscoms
Preparation of iscoms
A new type of complex has been formed using a matrix or ligand molecule, a glycoside extracted from the bark of the South African tree *Molina saponaria quillaria*. This new type of complex was found to be highly immunogenic and was designated iscom, which is derived from the expression 'immunostimulating complex'. It can be prepared in one step by gradient centrifugation. Virus solubilized with a detergent (A layer in Figure 5) such as Triton X-100 is centrifuged through a layer of sucrose containing Triton X-100 (B layer in Figure 5), where lipid and detergent is trapped while the peplomers migrate into the part of the sucrose gradient containing the glycoside (C layer in Figure 5) in a concentration above the critical micellar concentration. We suggest that the iscom is formed when the peplomers in the form of monomers are

Figure 4. Lung lesions found at necropsy following challenge with PI-3 virus of control lambs and lambs vaccinated with micelles of PI-3 virus, with or without adjuvant. Black areas represent areas with lesions. From Morein et al. (1983). Reprinted by permission of the *Journal of General Virology*.

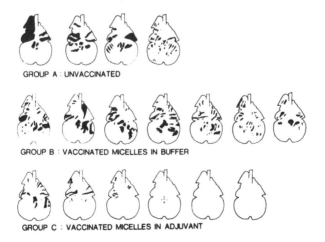

GROUP A : UNVACCINATED

GROUP B : VACCINATED MICELLES IN BUFFER

GROUP C : VACCINATED MICELLES IN ADJUVANT

488   *B. Morein*

migrating out of the sucrose layer containing detergent into the sucrose containing the glycoside micelles. The micelles then capture the membrane proteins by hydrophobic interaction. The iscom was first prepared with the peplomers from PI-3 virus and was characterized by its sedimentation coefficient, its protein profile and by its morphology as shown by electron microscopy. The sedimentation coefficient of the iscom was lower (19S) than the micelle containing the same proteins (30S), but higher than the monomer form of the peplomer, which has a sedimentation coefficient of 4S in sucrose gradient. In sodium dodecylsulphate polyacrylamide gel electrophoresis (Laemmli 1970) of the PI-3 virus, only the peplomers, i.e. the haemagglutinin-neuraminidase and fusion envelope proteins, were detected. Electron microscopy showed a globular structure with a diameter of about 35 nm built up by 12 nm circular subunits (Figure 6).

Immunogenicity of iscoms

Iscoms containing the peplomers of PI-3 were tested in immunization experiments in mice (for details see Figure 7). When mice were immunized twice at 3-week intervals with 5 µg of the PI-3 peplomers, 10 times higher antibody titres were obtained when the peplomers were included in iscoms, rather than in micelles. Ten times more peplomers were required in the micelle preparation than in the iscom preparation to induce a similar antibody response, and one immunization with

Figure 5. Preparation of PI-3 virus iscoms. Two hundred µg of solubilized virus in 100 µl 0.05 M Tris and 0.1 M NaCl (TN) was applied to a layer of 200 µl 8% sucrose in TN and 1% Triton X-100, which was layered over a 5 ml sucrose gradient in TN containing 0.2% Quil A. The centrifugation was performed in a SW50 rotor at 150,000 g for 4 h at 20°C. Fractions of 250 µg were collected. The fractions containing peplomer proteins were traced by ELISA and pooled. The sucrose was removed by dialysis.

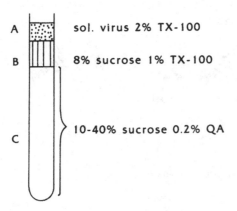

A   sol. virus 2% TX-100

B   8% sucrose 1% TX-100

C   10-40% sucrose 0.2% QA

peplomers in iscoms induced the same magnitude of antibody response as two immunizations with micelles.

PI-3 and other viruses of the Paramyxoviridae family, including rinderpest, canine distemper, measles, mumps and Sendai viruses, have two different envelope proteins, one responsible for the attachment of the virion to the cell surface. This activity may be visualized by agglutination of red blood cells (e.g. for PI-3, measles and rinderpest viruses). The second protein has the ability to open the plasma membrane of the cell after attachment of the virion through a process of fusion between the virion and the cell. In order to induce protective immunity, it is claimed that immune responses are required against both envelope proteins of the virion. Vaccines prepared from killed viruses of the Paramyxoviridae family have generally induced high antibody to the protein responsible for attachment, but not to the protein causing fusion. Such killed vaccines have failed to give complete protection (see Merz, Scheid & Choppin 1980; Choppin et al. 1981).

For example, the killed paramyxovirus vaccine for measles induces high haemagglutination inhibition titres and high virus-neutralizing antibody. People immunized with this vaccine are protected against

Figure 6. Electron micrograph of iscoms containing the envelope proteins of PI-3 virus.

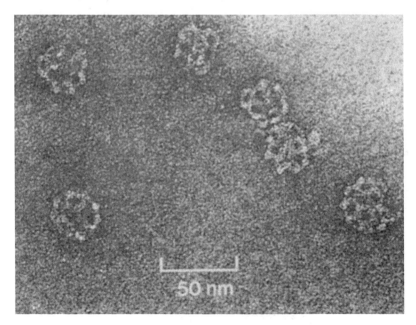

490   *B. Morein*

natural challenge with measles virus, but occasionally develop an atypical
measles, characterized by rash and giant cell pneumonia (Norrby, Enders-
Ruckle & ter Meulen 1975; Merz, Scheid & Choppin 1980). This is said
to be related to the failure of killed virions to induce an immune response
to the fusion protein (Norrby, Enders-Ruckle & ter Meulen 1975). The
authors concluded that the fusion protein is nonimmunogenic due to
denaturation caused by the inactivation process.

In a recent experiment, rats were immunized with a measles vaccine
utilizing betapropiolactone-killed virus or with the envelope proteins
included in iscoms which were produced from the betapropiolactone-
killed virus preparation. Five µg of envelope protein was used in both
preparations, in a dose applied subcutaneously at one time. Antibody
responses were monitored by the neutralization, haemagglutination
inhibition and haemolysis inhibition tests (Morein et al. 1984). Both
preparations induced high neutralizing antibody of similar magnitude,
but the iscoms induced about 10 times higher haemagglutination

Figure 7. The antibody response of BALB/c mice to vaccination with
PI-3 virus protein micelles or iscoms. Antibody of the IgG class was
measured with an ELISA technique. The results are expressed as the
mean + standard deviation for each group of 5 mice. ○ received 5
µg of protein micelles; ● received 0.5 µg of protein micelles; △
received 5 µg of iscoms; ▲ received 0.5 µg of iscoms; and □ were
nonvaccinated controls. At the end of the experiment the HI titre
was determined. From Morein et al. (1984). Reprinted by permission
of *Nature*.

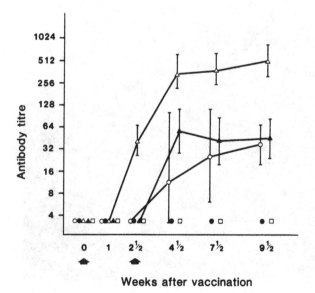

Weeks after vaccination

inhibition titres in the serum of the rats. The most striking result, however, was that the iscom preparation induced high serum haemolysis inhibition titres, i.e. it evoked antibody to the fusion protein, while the killed virus preparation did not (Figure 8). In this case, the discrepancy in antibody response between the two different preparations cannot be due to a denaturing process, since the iscoms were derived from the killed virus preparation. The iscoms probably exposed the two envelope proteins equally well, while the fusion protein was partially hidden in the whole virion (Armstrong et al. 1982).

Iscoms have also been prepared containing the envelope proteins haemagglutinin (H) and neuraminidase (N) of influenza viruses. Their immunogenicity has been tested in mice in doses of 0.1 to 5 µg, using an equine strain HEq2 NEq2 or the human strain PR8 (H1N1). In guinea pigs and horses from 3 to 30 µg of the HEq1 NEq1 and HEq2 NEq2 strains were used. The iscoms were compared with an experimental micelle vaccine and with commercial horse influenza vaccines, based on killed whole virions. In all animals tested, the iscoms induced about 10 times

Figure 8. Antibody response of groups of three Wistar rats to a single intramuscular vaccination with B-propiolactone (BPL)-killed measles virus or with measles virus iscoms. Titres were measured in a virus neutralization (VN) test, an HI test and an HLI test after absorption with Tween-ether treated measles virus antigen, as described by Norrby, Enders-Ruckle and ter Meulen (1975). The symbols indicate values of individual animals vaccinated: ● received BPL-killed virus containing 5 µg of glycoprotein per dose; ○ received measles virus iscoms containing 5 µg of glycoprotein per dose.

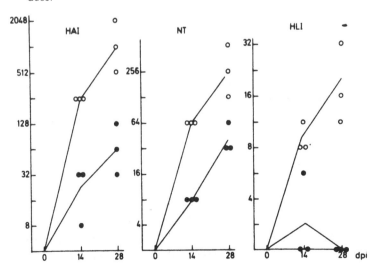

higher antibody titres as measured with ELISA or haemagglutination inhibition tests. It also seems that one immunization with iscoms is as efficient as two immunizations with micelles or killed whole virions. Interestingly, doses as low as 0.1 µg in mice and 1.5 µg in horses induced antibody to the peplomers in the iscoms. In challenge experiments with the PR8 strain, mice were protected by two 1 µg doses of iscoms and were partially protected by two 0.1 µg doses.

Why are the iscoms more effective than the micelles or than the whole virion? One probable reason is that all the surface proteins are equally well exposed on the iscom particle. Whether a particle contains 10 or 200 peplomers probably makes no difference as regards its immunogenicity. What may be important is that several peplomers are arranged in the form of a particle. If this is so, it would be more economical from the point of view of antigenicity to have a particle carrying few proteins. Furthermore, the iscoms expose the peplomer more openly than virus particles do, as judged by electron microscopy. It would be of interest to see if a peplomer presented in an iscom reveals more antigenic sites than a peplomer in a virion or a micelle. In the iscom the antigens are tied to a glycoside which has been shown to be an effective adjuvant, for instance in foot-and-mouth virus vaccines (Dalsgaard 1978). However, the amount of glycoside in the iscom preparation is far below the concentration needed to exert its adjuvant activity in a conventional vaccine, in some cases more than several hundred times lower. In spite of this, the glycoside may exert an adjuvant activity by virtue of the fact that it is bound to the antigens. In experiments reported by Arnon and colleagues (1980), adjuvant covalently linked to a synthetic antigen showed activity at much lower doses than when administered unlinked to the antigen. The iscom complexes are not only interesting from the point of view of their application to the production of vaccines: they are also of value for studies of immunological responses to membrane proteins in general, for example to proteins from viruses, bacteria, parasites or mammalian cells. The iscoms have also been found to be more effective than complete virions in *in vitro* stimulation of lymphocytes as measured by ^{3}H-thymidine uptake (B. Morein unpublished).

The next generation of vaccines

It is anticipated that in the coming years a new generation of vaccines will be produced by the use of recombinant DNA techniques and/or *in vitro* synthesis of polypeptides. Recombinant DNA products or synthetic peptides which react with neutralizing antibodies have been identified from a number of viruses. Such peptides may induce protective immunity if they are inoculated together with an oil adjuvant. However, as already stated, adjuvants may cause undesirable side effects. Further, the amount of antigen required to induce the appropriate immune response is often several hundredfold higher than when the intact organism is used. Synthetic peptides also need to be linked to carriers such as keyhole limpet haemocyanin or bovine serum albumin, which may not be acceptable in a widely applied vaccine.

Examples of recombinant DNA products expressed and produced by *E. coli* cells are the haemagglutinin of fowl plague (influenza) virus (Emtage et al. 1980), the g-protein of rabies (Yelverton et al. 1983) and the surface glycoprotein of hepatitis B virus (Fujisawa et al. 1983). Kleid and colleagues (1981) achieved the most successful results with the protein VP1 of foot-and-mouth disease virus which induced protection in pigs and cattle against challenge with an homologous strain. Perhaps the best experimental vaccination results with synthetic peptides were also obtained in protection experiments against foot-and-mouth disease. Bittle and colleagues (1982) induced protection against this disease with a peptide predicted from the viral nucleotide sequence.

There are two examples of recombinant DNA products which are highly immunogenic because of the particular way in which they have been produced. Both involve hepatitis B virus. First, the gene for the hepatitis B surface antigen has been cloned in yeast cells, and these express the gene and secrete a product with the same morphology as the 22-nm particle found in the blood of infected patients (Valenzuela et al. 1982). The particle produced by the yeast has the same immunogenicity as the 22-nm particle derived from human blood, which is used for vaccine production. The second example is the insertion of the cloned hepatitis B surface-antigen gene into vaccinia virus. The vaccinia virus is then applied on the skin by scarification, and infected epidermal cells produce both hepatitis B surface antigen and vaccinia viral antigens, resulting in the induction of immune responses to both viruses (Mackett, Smith & Moss 1982).

These two examples illustrate the importance of the mode of presentation of viral subunit antigens in order to achieve maximum immunogenicity. As further immunologically important gene products of microorganisms are identified, more information will be required regarding the physical forms in which such antigens should be presented to the immune system and how adjuvants can best be used to obtain optimal immune responses.

References

Almeida, J.D., Brand, C.M., Edwards, C.D. & Heath, T.D. (1975). Formation of virosomes from influenza subunits and liposomes. *Lancet*, **11**, 899-901.

Armstrong, M.A., Frazer, K.B., Dermott, E. & Shirodaria, P.V. (1982). Immunoelectronmicroscopic studies on haemagglutinin and haemolysin of measles virus in infected HE2p cells. *Journal of General Virology*, **59**, 187-92.

Arnon, R., Sela, M., Parant, M. & Chedid, L. (1980). Anti-viral response elicited by a completely synthetic antigen with built-in adjuvanticity. *Proceedings of the National Academy of Sciences of the USA*, **77**, 6769-72.

Balcarova, J., Helenius, A. & Simons, K. (1981). Antibody response to spike protein vaccines prepared from Semliki Forest virus. *Journal of General Virology*, **53**, 85-92.

Bittle, J.L., Houghten, R.A., Alexander, H., Shinnick, T.M., Sutcliffe, J.G., Lerner, R.A., Rowlands, D.J. & Brown, F. (1982). Protection against foot-and-mouth disease by immunization with a chemically synthesized peptide predicted from the viral nucleotide sequence. *Nature*, **298**, 30-33.

Brand, C.M. & Skehel, J.J. (1972). Crystalline antigen from the influenza virus envelope. *Nature*, **238**, 145-47.

Center for Disease Control. (1984). Prevention and control of influenza. *ACJP Recommendation*, **33**, 253-66.

Choppin, P.W., Richardson, C.D., Merz, D.C. & Scheid, A. (1981). An adhesion and microorganism pathogenicity. *Ciba Foundation Symposium*, **80**, 252-69.

Dalsgaard, K. (1978). A study of the isolation and characterization of the saponin Quil A. Evaluation of its adjuvant activity with special reference to the application in the vaccination of cattle against foot-and-mouth disease. *Acta Veterinaria Scandinavica*, **Suppl.69**, 1-40.

Dietzschold, B., Wiktor, T.J., Wunner, W.H. & Varrichio, A. (1983). Chemical and immunological analysis of the rabies soluble glycoprotein. *Virology*, **124**, 330-37.

Emtage, J.S., Tacon, W.C.A., Catlin, G.H., Jenkins, B., Porter, A.G. & Carey, N.H. (1980). Influenza antigenic determinants are expressed from haemagglutinin genes cloned in *Escherichia coli*. *Nature*, **283**, 171-74.

Fujisawa, Y., Ito, Y., Sasada, R., Ono, Y., Igarashi, K., Marumoto, K., Kikuchi, M. & Sugino, Y. (1983). Direct expression of hepatitis B surface-antigen gene in *E. coli*. *Nucleic Acids Research*, **11**, 3581-91.

Helenius, A. & Simons, K. (1975). Solubilization of membranes by detergents. *Biochimica et Biophysica Acta*, **415**, 29-79.

Helenius, A. & von Bonsdorff, C.H. (1976). Semliki Forest virus membrane proteins: preparation and characterization of spike complexes soluble in detergent-free medium. *Biochimica et Biophysica Acta*, **436**, 895-99.

Helenius, A., Fries, E. & Kartenbeck, H. (1977). Reconstitution of Semliki Forest virus membrane. *Journal of Cell Biology*, **75**, 866-80.

Hunsmann, G., Moenning, V. & Schafer, W. (1976). Properties of mouse leukemia viruses. 9. Active and passive immunization of mice against Friend leukemia with isolated viral gp71 glycoprotein and its corresponding antiserum. *Virology*, **66**, 327-29.

Kleid, D.G., Yansura, D., Small, B., Dowbenko, D., Moore, D.M., Grubman, M.J., McKercher, P.D., Morgan, D.O., Robertson, B.H. &

Bachrach, H.L. (1981). Cloned viral protein vaccine for foot-and-mouth disease: responses in cattle and swine. *Science,* 214, 1125-29.

Laemmli, U.K. (1970). Cleavage of structural proteins during the assembly of the head of bacteriophage T4. *Nature,* 227, 680-85.

Lowry, O.H., Rosebrough, N.J., Farr, A.L. & Randall, R.J. (1951). Protein measurement with the Folin phenol reagent. *Journal of Biological Chemistry,* 193, 265-75.

Mackett, M., Smith, G.L. & Moss, B. (1982). Vaccinia virus: a selectable eucaryotic cloning and expression vector. *Proceedings of the National Academy of Sciences of the USA,* 79, 7415-19.

Merz, D.C., Scheid, A. & Choppin, P.W. (1980). Importance of antibodies to the fusion glycoprotein of paramyxoviruses in the prevention of spread of infection. *Journal of Experimental Medicine,* 151, 275-88.

Morein, B., Sharp, M., Sundquist, B. & Simons, K. (1983). Protein subunit vaccines of parainfluenza-type 3 virus: immunogenic effect in lambs and mice. *Journal of General Virology,* 64, 1557-69.

Morein, B., Helenius, A., Simons, K., Pettersson, R., Kaariainen, L. & Schirrmacher, V. (1978). Effective subunit vaccines against an enveloped virus. *Nature,* 276, 715-18.

Morein, B., Sundquist, B., Hoglund, S., Helenius, A. & Simons, K. (1982). Protein micelles and virosomes from the surface glycoproteins of parainfluenza-3 virus. In *Protides of the biological fluids* (ed. H. Peeters), pp.1010-4. Oxford: Pergamon Press.

Morein, B., Sundquist, B., Hoglund, S., Dalsgaard, K. & Osterhaus, A. (1984). Iscom, a novel structure for antigenic presentation of membrane proteins from enveloped viruses. *Nature,* 308, 457-60.

Mortimer, E.A. (1978). Immunization against infectious disease. *Science,* 200, 902-7.

Norrby, E., Enders-Ruckle, G. & ter Meulen, V. (1975). Differences in the appearance of antibodies to structural components of measles virus after immunization with inactivated and live virus. *Journal of Infectious Diseases,* 132, 262-69.

Oxford, J.S., Hockley, D.J., Heath, T.D. & Pattersson, S. (1981). The interaction of influenza virus haemagglutinin with phospholipid vesicles: morphological and immunogical studies. *Journal of General Virology,* 52, 329-43.

Tanford, C. & Reynolds, J.A. (1976). Characterization of membrane proteins in detergent solutions. *Biochimica et Biophysica Acta,* 457, 133-70.

Valenzuela, P., Medina, A., Rutter, W.J., Ammerer, G. & Hall, B.D. (1982). Synthesis and assembly of hepatitis B virus surface-antigen particles in yeast. *Nature,* 298, 347-50.

Wells, P.W., Sharp, J.M., Burrels, L., Rushton, B. & Smith, W.D. (1976). The assessment in sheep of an inactivated vaccine of parainfluenza-3 virus incorporating double-stranded RNA as adjuvant. *Journal of Hygiene,* 77, 255-61.

Yelverton, E., Norton, S., Obijeski, J.F. & Goeddel, D.V. (1983). Rabies virus glycoprotein analogs: biosynthesis in *Escherichia coli. Science,* 219, 614-20.

26

Mucosal mast cells, basophils, immediate hypersensitivity reactions and protection against gastrointestinal nematodes

H.R.P. MILLER

Three different examples of worm expulsion are considered in order to analyse the protective functions of local gastrointestinal anaphylactic reactions. The example of self-cure in sheep has, unfortunately, received relatively little attention since Stewart's (1953; 1955) original studies. The second example, of spontaneous expulsion of primary nematode infections, has been intensively studied, although the data remain controversial. For certain host-parasite relationships there is evidence to suggest that basophil-mast cell-derived mediators are detrimental to worm survival. In the third example, of rapid worm expulsion from a primed host, there is again a suggestion that local hypersensitivity reactions are involved in the rejection process. It is now clear that neither enumeration of mast cells and basophils nor measurement of amine levels in the parasitized mucosa provide sufficient information on the functional activity of these cells or on the turnover of inflammatory mediators at the site of infection. New approaches will be necessary to determine the roles of so-called 'preformed' and 'secondarily formed' mediators since it is possible that both are synthesized *de novo* during infection. Until such methodology has been developed, the role of local hypersensitivity mechanisms in protection against gastrointestinal nematodes is unlikely to be fully understood.

Introduction

Gastrointestinal nematode infections are a major cause of disease in ruminants. Similarly, in areas where standards of hygiene and medical attention are inadequate, there is a high incidence of intestinal nematodiasis in man. Whilst such infections are only rarely catastrophic, they can cause serious debilitation and are a major source of economic loss in domestic animals. Thus there are sound medical and economic reasons for attempting to reduce the extent of parasitic gastroenteritis.

Because there have been several recent reviews describing mechanisms of resistance against gastrointestinal nematodes (Befus & Bienenstock 1982; Miller 1984), this chapter will be concerned with a relatively poorly defined component of the immune response to these parasites — the potential significance of immediate-type hypersensitivity reactions.

The association between helminthiasis and immediate hypersensitivity reactions and the possibility that such reactions might be involved in protecting the host against helminth infections have been recognized for a number of years (Stewart 1953; reviewed in Jarrett & Miller 1982). Stewart's original observations on the self-cure reactions against *Haemonchus contortus* in sheep have generated considerable interest and research during the last 30 years, and yet the question as to whether immediate hypersensitivity reactions play a beneficial role remains unanswered.

The principal effector cells in immediate hypersensitivity reactions are mast cells and basophils, and the inflammatory mediators which are released from these cells include histamine, 5-hydroxytryptamine (5HT or serotonin), slow reacting substance of anaphylaxis (SRS-A), prostaglandins and several granule-associated enzymes (Metcalfe, Kaliner & Donlon 1981). A variety of stimuli, not all of which are necessarily associated with the classically described IgE-mediated type I immediate hypersensitivity, may cause mediator release from basophils and mast cells (Askenase 1980). Moreover, T cell-mediated reactions with a delayed time course, as well as antibodies, are involved in the recruitment of mast cells and basophils to the site of infection (reviewed in Askenase 1980; Miller 1980).

Thus, the course of development of hypersensitivity reactions is complex, involving cell-mediated as well as immediate hypersensitivity reactions. Given the large number of mediators which may be released from mast cells and basophils and the effect that these may have on tissue integrity and recruitment of bloodborne effector cells, it is not surprising that the protective role of type I hypersensitivity reactions remains in dispute.

Before beginning this review, it is important to define the type of response observed when nematode parasites are immunologically rejected. For example, worm expulsion following primary infection may occur within weeks or months of initial challenge depending on the species of parasite and host. In this review the immune expulsion of nematodes during primary infection will be referred to either as spontaneous cure (Wakelin 1978) or primary expulsion.

The kinetics of expulsion are altered in animals which have been primed by an earlier infection: immune elimination of the worms occurs over a shorter period of time and begins sooner after challenge in the immune than in the naive host (Jarrett, Jarrett & Urquhart 1968). In the field, animals are commonly exposed to repeated challenge and some become highly refractory to further infection. Examples in the sheep include both gastric and intestinal nematode larvae which apparently fail to establish in the mucosa and are rapidly expelled. This phenomenon has been described in much more detail in laboratory rodents and is generally referred to as 'rapid expulsion', the worms being expelled within 24 h and often within 3 to 4 h of challenge (reviewed in Miller 1984).

The final example of worm expulsion is that of 'self-cure'. Originally introduced by Stoll (1929), the term self-cure refers to the drop in faecal egg counts in sheep infected with adult *H. contortus* and grazing on pastures heavily contaminated with the parasite. Subsequent studies have revealed that self-cure involves the rapid elimination of the adult worm burden by a sudden intake of infective larvae.

Historical perspective

Although the occurrence of immediate hypersensitivity reactions in response to nematodes was recognized early in this century (reviewed in Andrews 1962) and infiltration of the intestinal mucosa with 'connective-tissue basophils' and eosinophils was reported in *Nippostrongylus brasiliensis*-infected rats (Taliaferro & Sarles 1939), Stewart (1953) was the first to suggest that allergic reactions *per se* might render tissues inhospitable to nematode parasites.

The phenomenon of self-cure, in which expulsion of already established adult *H. contortus* worms is precipitated by an intake of infective larvae, was examined by Stewart (1953; 1955). He found that blood histamine levels were raised in sheep that manifested the self-cure reaction 2 to 5 days after larval challenge (Stewart 1953). However, the titre of *Haemonchus*-specific antibodies in serum did not increase until after a drop in faecal egg counts had occurred (Stewart 1950). Treatment of sheep with the anti-histaminic Anthisan for periods of 4 to 11 days after the administration of larvae prevented self-cure on 5 out of 15 occasions, although the subsequent rise in antibody titres was unaffected (Stewart 1953). The intake of *H. contortus* larvae was associated with the self-cure of unrelated species of nematodes, both in the stomach (*Ostertagia circumcincta* and *Trichostrongylus axei*) and in the intestine (*Trichostrongylus colubriformis*) (Stewart 1953; 1955). Direct observation of the abomasum in parasitized sheep, following injection of massive doses of exsheathed larvae into the abomasal lumen, revealed increased peristalsis and segmentation within 10 min of injection, and the wall of the abomasum became pale and oedemateous within 1 h (Stewart 1955). Nevertheless, and somewhat surprisingly, there was no increased blood histamine within 4 h after the injection of larvae. Nor was there any deviation from the normal values of histamine in extracts of the abomasal wall, although the presence of oedema was confirmed histologically (Stewart 1955). Histological examination of *H. contortus-* and *T. colubriformis*-infected tissues following the completion of self-cure revealed local lesions of oedema and cellular infiltration (Stewart 1953).

The responses described by Stewart (1950; 1953; 1955), although variable and not always repeatable, tended to implicate local allergic reactions in the self-cure phenomenon. He proposed that there was a change in local environmental conditions at the time of the self-cure reaction (Stewart 1955) and that serum antibodies were irrelevant to this response, as implied by the timing of their appearance in the blood.

The role of local allergic reactions in the gut was given further prominence by studies in *N. brasiliensis*-infected rats, where it was noted that mucosal permeability, although evident during primary infection, was substantially augmented by intravenous injection of worm antigen (Urquhart et al. 1965). The other effects noted included the secretion of mucus and hyperaemia of the intestinal wall. These changes, although they could be detected by day 7 after inoculation, increased in severity with time after infection (Urquhart et al. 1965). The effect was passively transferable with hyperimmune serum and could be blocked with the anti-histaminic promethazine hydrochloride and with the corticosteroid betamethasone (Urquhart et al. 1965). It was hypothesized that either the physical changes associated with anaphylaxis were alone responsible for worm expulsion or that they promoted the translocation of parasite-specific antibody into the intestinal lumen (Barth, Jarrett & Urquhart 1966). The latter hypothesis was tested in naive rats infected with adult worms and subjected to a heterologous anaphylactic shock. No effect was detected unless the rats had been inoculated with immune serum before the induction of anaphylaxis (Barth, Jarrett & Urquhart 1966). The results indicated that local anaphylaxis promoted pathotopic transfer of specific anti-worm antibodies into the gut lumen.

A contrasting result was obtained by Panter (1969) who showed that a heterologous anaphylactic shock induced at the time of challenge was alone capable of preventing the establishment of a large proportion of *Nematospiroides dubius* larvae in naive mice; she also observed that intravenous injection of *N. dubius* antigen into primed mice resulted in changes very similar to those described for *N. brasiliensis* (Urquhart et al. 1965) and included accumulation of Evan's blue and excessive secretion of mucus in the intestine.

Whatever the significance of antibody translocation as a component of resistance, the results of three separate investigations indicated that local hypersensitivity reactions are associated with the immune expulsion of nematodes from the gastrointestinal tract. The stage was thus set for a much more detailed analysis of the response, the principal components being anaphylactic antibodies, mast cells and basophils, as well as the cellular mechanisms which recruit these effector cells to the gastrointestinal tract.

Anaphylactic antibodies

While IgE is the immunoglobulin isotype most commonly recognized in skin-sensitizing and anaphylactic reactions, other subclasses of antibody, such as IgG_1 in guinea pigs and mice and IgG_{2a} in rats (reviewed in Askenase 1977), also participate in immediate hypersensitivity reactions. Early reports suggested that an IgG_{1a} isotype was responsible for anaphylactic activity in sheep (Curtain 1969; Hogarth-Scott 1969), but this has not yet been confirmed. There are preliminary studies of an IgE isotype in cattle (Gershwin & Dygert 1983), although it has yet to be properly characterized.

Several different approaches have been used to assess the importance of anaphylactic antibodies. These include: (1) the temporal relationship between raised titres of anaphylactic antibodies and the onset of nematode expulsion, (2) the protective capacities of passively transferred anaphylactic antibodies, (3) specific immunosuppression of IgE responses and its effect on worm expulsion, (4) the role of anaphylactic antibodies in promoting cell-mediated helminthotoxicity and (5) studies of anaphylactic antibodies in resistant animals.

There seems to be little or no correlation between the time course of the protective response and titres of parasite-specific IgE (Ogilvie 1967; Jarrett, Haig & Bazin 1976; Befus et al. 1982). However, anaphylactic sensitivity often develops before the onset of worm expulsion (Urquhart et al. 1965) and it may be that anaphylactic antibodies of the IgG isotype develop early in the primary response, whereas IgE is more significant following secondary challenge, when worm expulsion can occur very rapidly (Castro et al. 1976; Love, Ogilvie & McLaren 1976). An alternative explanation, yet to be confirmed, is that circulating antigens secreted by the parasites interfere with, or block, the passive cutaneous anaphylactic reactivity of test sera obtained during primary infection (Hogarth-Scott 1972).

Attempts to assess the protective role of anaphylactic antibodies by the passive transfer of parasite-specific reagin-rich sera have met with variable success. In some instances, partial protection has been achieved with IgG fractions alone (Jones, Edwards & Ogilvie 1970; Pritchard et al. 1983), suggesting that if anaphylactic antibodies are functionally important in these experiments they are of the IgG isotype. However, where passive immunization conferred more than 95% protection against *N. brasiliensis* in the rat, heat inactivation of the serum before transfer caused a small but significant reduction in its protective capacity (Miller 1979). This experiment, together with a study in which the passive transfer of sera rich in IgE and IgG was more effective in eliminating *Trichinella spiralis* from mice than were sera which had low titres of parasite-specific anaphylactic antibodies (Gabriel & Justus 1979), indicates a protective role for such antibodies. Similarly, the passive protection conferred against the metacestodes of *Trichostrongylus taeniaformis* in the rat is apparently mediated by the anaphylactic isotypes IgE and IgG_{2a} (Musoke & Williams 1975; Musoke, Williams & Leid 1978).

One of the more convincing experiments demonstrating the protective capacities of anaphylactic antibodies in the rat showed significantly reduced protection against the tissue-dwelling helminth *Shistosoma mansoni* when either IgE or IgG_{2a} were selectively depleted from the passively transferred sera (Capron, Capron & Dessaint 1980). Comparable experiments have not yet been carried out for gastrointestinal nematodes.

Suppression of antibody synthesis by pretreating rats or mice with anti-μ or anti-ε antisera has also been used to assess the protective functions of anaphylactic antibodies. Rats immunosuppressed with anti-ε antibodies were unable to mount parasite-specific IgE responses; although there was decreased protection against the establishment of newborn *T. spiralis* larvae in the muscles, no such effect was demonstrated against adult worms on the intestine (Dessein et al. 1981). Suppression of immunoglobulin responses in mice with anti-μ antiserum had no effect on spontaneous cure of *N. brasiliensis* (Jacobson, Reed & Manning 1977). No parasite-specific IgE responses were detected in mice treated in this way, and yet the worms were expelled at the same rate as in infected control mice.

The results of both of these experiments suggest that IgE does not play an essential role in the primary expulsion process. Nevertheless, *N. brasiliensis* is not a parasite of the mouse, and the validity of this model in the analysis of enteric resistance mechanisms against nematodes has been questioned (Ha, Reed & Crowle 1983).

Anaphylactic antibodies are known to promote antibody-dependent, cell-mediated killing of helminths *in vitro* where both granulocytes and macrophages have been shown to kill schistosomulae and nematode larvae (reviewed in Capron, Capron & Dessaint 1980). There is, at present, no direct evidence to suggest that such mechanisms operate against adult gut-dwelling nematodes, although tissue-invading larval stages of parasites such as *Oesophagostomum* spp. and *N. dubius*, which enter the submucosa, may be targets for antibody-dependent killing since massive accumulations of eosinophils surround dead and dying worms in immune animals (Liu 1965; Elek & Durie 1966).

Titres of IgE are high in rodents which are immune to enteric nematodes (Jarrett & Miller 1982) and it could be argued that the protective response against reinfection is likely to involve anaphylactic antibodies because elimination of infective larvae can occur within 30 to 240 min after challenge (Russell & Castro 1979; Miller, Huntley & Wallace 1981). There is, at present, no direct evidence to support this hypothesis, although there is indirect evidence to suggest that mast-cell activation occurs during rapid expulsion of infective larvae.

Thus, the precise role of IgE and other anaphylactic antibodies in resistance to gastrointestinal nematodes remains to be defined, even in laboratory rodents where there is a considerable body of knowledge on the structure and function of IgE. In ruminants, a role of anaphylactic antibodies can only be inferred from indirect observations on the nature of the hypersensitivity responses to infective larvae.

Mast cells and basophils
Mast cells

Mast cells in the gastrointestinal tract of various species, including humans, differ from connective-tissue mast cells morphologically, histochemically and in their fixation properties (Enerback 1981; Strobel, Miller & Ferguson 1981; Crowle & Phillips 1983). Such mast cells are commonly called mucosal mast cells. They have been most fully defined in the rat. Functional and biochemical studies have shown that rat mucosal mast cells contain less histamine and serotonin than connective-tissue mast cells (Wingren et al. 1983), have a distinctive granule-specific serine protease (rat mast cell protease II) (Woodbury, Gruzenski & Lagunoff 1978; Woodbury et al. 1978), and are generally less responsive to histamine-releasing agents than tissue mast cells (Befus et al. 1982; Pearce et al. 1982). Mucosal mast cells in the mouse are also distinct from tissue mast cells, both morphologically and in their fixation properties (Crowle & Phillips 1983). There is, however, some controversy as to the relationship in the mouse between mucosal mast cells, granulated intraepithelial lymphocytes and a cell of unknown lineage — the globule leucocyte (reviewed in Miller 1980;

Table 1. The properties of mucosal mast cells, transitional cells and globule leucocytes isolated from parasitized ovine abomasum, as indicated by various staining techniques. From Huntley et al. (1984).

	MMC	TC	GL
Basic Dyes	+ +[a]	+ - + +	± - + +
Dopamine	+ +	+ +	± - + +
Serine esterase	+ - + +	+ +	± - + +
Intracellular Ig	+	+	± - +
Surface Ig	+ +	+ +	+ +

[a] Intensity of reaction: ± = little or none;
+ + = intense staining.

Askenase 1980), all of which contain basophilic granules. In particular, Ruitenberg and Elgersma (1979) believe that globule leucocytes are unrelated to mucosal mast cells and arise *sui generis* from an unknown precursor cell.

Globule leucocytes are particularly prominent in parasitized ruminants and are characterized by large acidophilic granules or globules (Gregory 1979). Histochemical studies suggest that globule leucocytes could be related to plasma cells because they contain globulins (Dobson 1966). However, ultrastructural observations do not support this view (Miller, Murray & Jarrett 1967; Murray, Miller & Jarrett 1968); evidence has been obtained that globule leucocytes are modified mucosal mast cells in which granules are depleted of proteoglycan and monoamines. This view has received further support from more recent studies of mucosal mast cells and globule leucocytes isolated from parasitized ovine gastric mucosa (Huntley, Wallace & Miller 1982; Huntley, Newlands & Miller 1984) where mucosal mast cells, globule leucocytes and a cell-type intermediate between the two — the transitional cell (Murray, Miller & Jarrett 1968)–were found to have very similar properties (Table 1). These results further emphasize the probable origin of globule leucocytes from mucosal mast cells, and the relationship has been confirmed in the rat where rat mast-cell protease II, the predominant granule-associated enzyme in mucosal mast cells, has also been located within globule leucocytes (Woodbury & Miller 1982; Huntley et al. 1984). Similarly, ovine globule leucocytes, mucosal mast cells and transitional cells contain a serine esterase (Table 1).

In vitro studies have now shown that mucosal mast cells in the rat arise from a bonemarrow-precursor cell (Haig et al. 1982; 1983). Differentiation and growth of mucosal mast cells is promoted by a factor derived from *Nippostrongylus*-primed T cells cultured in the presence of worm antigen or from concanavalin A-stimulated T cells. The mast cells derived in this fashion were identified as mucosal mast cells by their content of rat mast-cell protease II. No such marker is yet available for murine mast cells but *in vitro* growth of mast-cell subsets was originally described by Ginsburg and colleagues (1981) and others have recently suggested that bonemarrow-derived mast cells exhibit many properties typical of mucosal mast cells (Crapper & Schrader 1983; Sredni et al. 1983).

Mucosal mast cells and worm expulsion

Nematode parasite infections in laboratory animals are invariably associated with the development of mucosal mastocytosis (Askenase 1980; Miller 1980). Mastocytosis in both rats and mice is thymus-dependent (Ruitenberg & Elgersma 1976; Mayrhofer 1979) and can, in *N. brasiliensis*-infected rats, be adoptively transferred with immune mesenteric lymph node cells (Befus & Bienenstock 1979; Nawa & Miller 1979), thoracic duct lymphocytes (Nawa & Miller 1979) or T cell-enriched thoracic duct lymphocytes (Nawa & Miller 1979). Thus, these studies are in agreement with *in vitro* results which indicate that T cells influence the recruitment and differentiation of mucosal mast cells. Indirect support for this view comes from studies by Alizadeh and Wakelin (1982b) who showed that the time of onset of mastocytosis in *T. spiralis*-infected mice is determined by cells of bonemarrow origin. However, it should also be noted that mucosal mastocytosis can be passively transferred with serum from rats immune to *N. brasiliensis* (Miller 1979; Befus & Bienenstock 1979). The mechanisms involved in this process are not understood.

Attempts to assess the role of mucosal mast cells in the expulsion of nematodes have, until recently, concentrated on the timing of the onset of mastocytosis and its relationship to primary worm expulsion, but in many instances there is little or no correlation between these two events (reviewed in Ogilvie & Love 1974; Askenase 1980). Further doubt on the role of mucosal mast cells has arisen from studies of congenitally mast cell-deficient W/W^v mice where primary expulsion of *N. brasiliensis* occurs either at a normal rate (Uber, Roth & Levy 1980) or at a slower rate with delayed onset (Crowle & Reed 1981). In neither of these experiments were mast cells detected in the mucosa during worm expulsion.

The requirement for mast cells is more apparent in *T. spiralis* infections. Primary expulsion of this nematode is substantially delayed in W/W^v

mice and can be speeded up by restoration of the deficiency with bonemarrow from normal littermates (Ha, Reed & Crowle 1983). Studies indicate that the immune system of these mice functions normally; their major defect is in the mast-cell precursor which is derived from bonemarrow (Reed, Crowle & Ha 1982).

The assumption implicit in all of these studies is that an increase in the number of mucosal mast cells indicates their increased functional activity. Yet there is no evidence to support such an assumption, and, as might be expected from the conflicting data on mast cell numbers, the results of quantifying monoamines such as histamine (Keller 1971; Jones at al. 1974) are also conflicting. More recently it has been shown that mucosal mastocytosis in rats precedes the accumulation of rat mast-cell protease II in cells within the mucosa (Woodbury & Miller 1982), and comparative histochemical studies suggest that maturing mast cells contained little or none of this protease. Consequently, it has been hypothesized that immature cells could be secreting, rather than storing rat mast-cell protease II (Woodbury & Miller 1982). Experiments to test this hypothesis have shown that the protease is released into the systemic circulation as early as 7 days after primary infection of rats with either *T. spiralis* or *N. brasiliensis* (Woodbury et al. 1984) and reaches a peak on day 10 just before the onset of worm expulsion. Since this protease predominates in mucosal mast cells (Woodbury, Gruzenski & Lagunoff 1978; Woodbury & Miller 1982), it is reasonable to conclude that the systemically released enzyme arises from these cells.

During the *in vitro* growth of mucosal mast cells from rat bonemarrow, rat mast-cell protease II is released into the medium at a stage when very few cells with classical mast-cell morphology can be detected (Haig et al. 1983). Similarly, among the mast cells recovered from the rat peritoneal cavity, the immature cells secrete histamine most actively (Beavan et al. 1983). Ultrastructurally, mucosal mast cells have depleted granules even as they develop (Miller 1971a), and, histochemically, the immature, poorly granulated cells have a reduced content of glycosaminoglycan and monoamine (Miller & Walshaw 1972). Thus, there is sufficient evidence to suggest that immature mast cells are functionally active, especially since they have abundant endoplasmic reticulum and prominent golgi complexes (Miller 1971b), and thus have a potential for protein synthesis and secretion. In conclusion, mucosal mast-cell precursors may be as active as the fully mature cells, at least in the rat; the mere quantification of granulated mast cells may have little relevance to their functional activity.

This argument is essentially in agreement with the proposition by Askenase (1980) that mucosal mast cells crucial to parasite expulsion may arise and discharge immediately and may not, therefore, be recognizable. Such an argument is less readily acceptable in the case of W/W^v mice where expulsion occurs in the complete absence of mast cells. Nevertheless, it has recently been shown that stimulation of the skin of such mice with a tumour promoter is followed 12 h later by the accumulation of bonemarrow-derived cells synthesizing histidine decarboxylase (Taguchi et al. 1982). The nature of these cells has not been discovered, but it is clear that even mast cell-deficient mice can mobilize cells with histamine-synthesizing capacity within a very short time after an appropriate stimulus.

Basophils

In contrast to mast cells which normally reside within the tissues, basophils differentiate in bonemarrow and, after an appropriate stimulus, are recruited via the blood to tissue sites (Askenase 1977). Very little is known about basophils in rodents and ruminants, but their biology has been studied intensively in the human and guinea pig (Askenase 1977). Guinea pig basophils can be recruited to tissue sites by both T cells and IgG$_1$ antibody (Askenase 1980). Like mast cells, basophils have Fc receptors for IgG$_1$ and IgE anaphylactic antibodies.

The most extensive studies on the role of basophils during nematode parasitic infections have been those of Rothwell and colleagues, who examined the responses of guinea pigs to *T. colubriformis*, a nematode parasite of sheep. Increasing numbers of basophils infiltrated the small intestinal *lamina propria* before and during the expulsion of primary infections (Rothwell & Dineen 1972; Rothwell 1975) and the numbers remained high 2 months later when previously infected guinea pigs were rechallenged (Rothwell 1975). Unlike the response in rats infected with *N. brasiliensis* where an anamnestic increase in mucosal mast cells occurred after rechallenge (Mayrhofer 1979), intestinal infiltration with basophils was not significantly increased after rechallenge with *T. colubriformis* (Rothwell 1975).

The infiltrating basophils were observed intraepithelially (Rothwell 1975), and basophils in the *lamina propria* had reduced numbers of granules during both primary and secondary challenge. Ultrastructural analysis showed that many of the basophils within the tissues had electron-lucent halos around the granule matrices which had lost their orderly periodicity, changing to a fibrillar or amorphous appearance

(Huxtable & Rothwell 1975). The evidence for basophil degranulation is thus rather similar to observations of granular depletion in mucosal mast cells (Miller 1971b).

Subsequent studies of two lines of guinea pigs bred for resistance and susceptibility to *T. colubriformis* showed that the resistant strain developed an intestinal mastocytosis at an earlier stage than the susceptible strain, but few basophils were noted at the relatively early times of examination (days 7 and 10 after infection) (Handlinger & Rothwell 1981). It was suggested that the mast-cell response which preceded worm expulsion in the resistant line could be responsible for the recruitment of effector eosinophils to the site of infection (Handlinger & Rothwell 1981).

Basophil infiltration of the jejunum in *N. brasiliensis*-infected rats preceded mucosal mastocytosis by several days (Miller 1971a). Rat intestinal basophils could not be identified by conventional techniques nor in paraffin sections by the special fixation technique used to identify mucosal mast cells (Miller 1969). These cells were only identifiable in 1 μm plastic embedded sections and at the ultrastructural level (Miller 1971a), and in both instances they were morphologically distinguishable from mast cells (Miller 1971a; Ferguson & Miller 1979). They were observed intraepithelially (Miller 1969) and were abundant in the mucosa 10 to 12 days after infection (Miller 1969), at the time of onset of worm expulsion. The kinetics of blood basophilia (Ogilvie, Askenase & Rose 1980; Roth & Levy 1980) were consistent with the timing of appearance of basophils in the intestinal mucosa. Since different strains of rats were used in two experiments (Miller 1971a; Ogilvie, Askenase & Rose 1980), it is likely that basophil infiltration is not strain-dependent in parasitized rats.

Preliminary histological studies of parasitized ruminant mucosae have so far revealed only a limited infiltration with basophils (H.R.P. Miller unpublished). These cells can apparently be detected in sheep by conventional techniques. Basophils are thought to be extremely rare in mice (Askenase 1977), but careful analysis of peripheral blood leucocytes in normal and repeatedly immunized mice have shown that basophilia can develop in several mouse strains (Urbina, Ortiz & Hurtado 1981). These cells have an atypical appearance and contain relatively few granules so they may not be readily identified within the tissues (Urbina, Ortiz & Hurtado 1981). It would be premature to suggest on the basis of studies in the rat that basophils do not play a role in immunity to nematodes in the mouse. Much further work is required to determine whether mast cell-deficient W/W^v mice are also deficient in basophils or if the bonemarrow-derived cells which synthesize histidine decarboxylase (Taguchi et al. 1982) are in any way related to basophils.

Inflammatory mediators and the intestinal response to nematodes

Several different approaches have been followed in attempts to assess the importance of inflammatory mediators in the intestinal response to nematodes. These include measuring the tissue concentrations of different mediators, testing the effects of these mediators on the parasites themselves either *in vivo* or *in vitro*, and using drugs to block or inhibit the mediator or to deplete it from the tissues.

Mediators derived from mast cells and basophils may be preformed and stored within the granules, or they may be formed secondarily as a consequence of membrane perturbation (Lewis & Austen 1981). Preformed mediators include: monoamines (histamine, 5HT and, in ruminants, dopamine), proteoglycans (heparin or chondroitin sulphate), chymotrypsin-like enzymes, and other lysosomal enzymes. The major secondarily formed mediators are the leucotrienes (previously classified as SRS-A) and prostaglandins (Lewis & Austen 1981).

The role of preformed mediators
Monoamines

Although raised levels of plasma histamine were recorded by Stewart (1953) during the self-cure response against *H. contortus* in sheep, the only other comparable measurement was described by Jones and colleagues (1974) who recorded increased levels of histamine in the gut contents of guinea pigs at the time of expulsion of *T. colubriformis*. There have been numerous studies of monoamine levels in the gut wall of parasitized animals, but the results have been almost as confusing as the mast-cell data already described. Often little or no correlation was observed between histamine concentrations and worm expulsion (Keller 1971; Befus, Johnston & Bienenstock 1979), although the concentration of histamine in the mucosal tissue correlated well with mucosal mast-cell density (Befus, Johnston & Bienenstock 1979; Wingren et al. 1983). By contrast, in *T. colubriformis*-infected guinea pigs, there was a progressive increase in the concentration of histamine before the onset of worm expulsion and this coincided with the infiltration of the mucosa by basophils and mast cells (Jones et al. 1974; Rothwell & Dineen 1972; Rothwell 1975).

The concentration of 5HT in the rat also increased at about the time of expulsion of *N. brasiliensis* (Murray et al. 1971a; Wingren et al. 1983). Mucosal mast cells in normal rats apparently contain little or no 5HT

(Enerback 1981; Miller & Walshaw 1972; Wingren et al. 1983), but in rats immune to *N. brasiliensis* low levels of 5HT were detected in mucosal mast cells shortly after the completion of worm expulsion (Miller & Walshaw 1972; Wingren et al. 1983). This was confirmed by immunohistochemistry using an antiserum specific for 5HT (Wingren et al. 1983). There was a close correlation between the rise in histamine and 5HT levels, and it was concluded that up to one-third of mucosal 5HT was stored in mucosal mast cells (Wingren et al. 1983). The 5HT content in the intestines of *T. colubriformis*-infected guinea pigs fluctuated, with an initial increase followed by a fall at about the time of worm expulsion (Jones et al. 1974).

It must be stressed that none of these experiments was concerned with amine turnover at the site of infection. Ultrastructural and histochemical studies indicate that mucosal mast cells and globule leucocytes are depleted of amines and proteoglycans in parasitized rodents and ruminants (Miller, Murray & Jarrett 1967; Murray, Miller & Jarrett 1968; Miller 1971a; Miller & Walshaw 1972; Wingren et al. 1983), and there is evidence to suggest that rat mucosal mast cells release their granule contents by a nonexocytotic secretory process (Miller 1971a; Miller et al. 1983b). It is, therefore, more important to measure the rate of secretion of these granule constituents than the actual concentration of stored amines in the tissues. Recent studies have shown that histamine-synthesizing cells are recruited to sites of inflammation (Dy & Lebel 1983). As already mentioned, histamine-producing cells of bonemarrow origin can be recruited to the skin of mast cell-deficient W/W^v mice by the application of a tumour promoter (Taguchi et al. 1982). Since neither of these cell populations has mast-cell characteristics and presumably neither can store amines, the only method of assessing their activity would be to determine the rate of synthesis and turnover of amines in the sites to which they are recruited.

Efforts to dislodge nematode parasites by treatment of the host with amines or their precursors have produced equivocal results (Stewart 1953). The most detailed and successful studies have been those of Rothwell and colleagues, who showed that significant numbers of fourth-stage *T. colubriformis* larvae were expelled from nonimmune animals when 5HT and histamine were infused intraduodenally 5 to 9 days after challenge (Rothwell, Pritchard & Love 1974). However, 5HT, rather than histamine, depressed the fermentative metabolism of the parasites *in vitro*.

Numerous attempts have been made to demonstrate the functional role of monamines by treating parasitized animals with amine-depleting or amine-blocking drugs. The difficulties involved, both in experimental design and interpretation of results, have been discussed at length (Keller & Ogilvie 1972; Bell, McGregor & Adams 1982). In several cases, anti-histamine or anti-serotonin drugs have blocked worm expulsion, in agreement with Stewart's early observations (Urquhart et al. 1965; Sharp & Jarrett 1968; Kelly & Dineen 1972; Murray et al. 1971b). However, such treatments tended to be long term and may have had effects unrelated to their amine-blocking activities. For example, the anti-histaminic drug promethazine reduced the efficacy of sensitized lymphocytes in the adoptive transfer of resistance against intestinal nematodes (Kelly & Dineen 1972; Rothwell, Love & Dineen 1973). It has been suggested that this anti-histaminic may bind to regulatory T cells, thereby altering the functions of this cell subpopulation (Askenase 1977).

Keller and Ogilvie (1972) tested a wide range of amine antagonists in *N. brasiliensis*-infected rats and concluded that histamine was not involved in worm expulsion. Rothwell, Jones and Love (1974) obtained a contrasting result which supported their previous studies both of amine levels in the gut wall and of the direct effect of amines on *T. colubriformis* in guinea pigs. These authors showed that short-term reserpine treatment blocks the rapid expulsion of transplanted *T. colubriformis* and reduces the mucosal content of 5HT, but not histamine. Their results strongly support the hypothesis that a mast cell-basophil-derived mediator could be involved in the expulsion process, at least in one host-parasite system. To summarize, the precise role that monoamines play in the rejection of nematode parasites is poorly understood. There is sufficient indirect evidence to suggest that they are involved in worm expulsion, but a glance at the list compiled by Befus and Bienenstock (1982) which shows that, for example, histamine can also regulate immune responses, would indicate that the functional role even of this amine has yet to be defined.

Proteoglycans

The participation of either heparin or chondroitin sulphate in resistance to nematodes has received very little attention. An interesting observation by Justus and Morakote (1981) suggests that mast-cell heparin could bind nematode antigens and perhaps, in this manner, reduce their toxicity. Mucosal mast cells in the rat contain a nonheparin proteoglycan, possibly chondroitin sulphate (Jarrett & Haig 1984), as do murine mast cells grown in culture (Razin et al. 1983) and putatively identified as mucosal mast cells (Ginsburg et al. 1981; Crapper & Schrader 1983). Histochemical and ultrastructural studies indicate that proteoglycan is depleted from mucosal mast cells in both rats (Miller 1971a; Miller & Walshaw 1972) and ruminants (Murray, Miller & Jarrett 1968; Huntley et al. 1984) during enteric nematodiasis, but it is not at all clear what effector function is served by the release of this granule product.

Chymotrypsin-like enzymes

Although it has been known for a number of years that mast cells contain granule-specific chymotrypsin-like enzymes, their possible role in nematode parasite infections has only recently been considered. The presence of the rat mast-cell protease II has already been discussed, along with the fact that this enzyme is secreted into the systemic circulation during primary *N. brasiliensis* infections. We have now extended these observations, in collaboration with D. Wakelin, to show that primary expulsion of *T. spiralis* from the rat is also associated with the secretion into the blood vascular compartment of this mucosal mast cell-specific enzyme (Woodbury et al. in press). Furthermore, rat mast-cell protease II is quickly released into the bloodstream during the rapid expulsion of *N. brasiliensis* from primed rats (Miller et al. 1983b). These results show clearly that mucosal mast cells release granule products during the immune expulsion of nematodes.

Marked changes in intestinal mucosal permeability occur during primary *N. brasiliensis* infection in the rat (Murray, Jarrett & Jennings 1971; Nawa 1979), coinciding with the systemic release of rat mast-cell protease II (Nawa 1979; Woodbury et al. 1984). Intestinal permeability is greatly increased in *Nippostrongylus*-primed rats subjected to anaphylactic shock by intravenous injection of worm antigen (Miller et al. 1983b; King & Miller 1984). At the same time, there is substantial secretion of rat mast-cell protease II into the blood circulation and extremely rapid release of this enzyme into the gut lumen (King & Miller 1984). Since rat mast-cell protease II has proteolytic activity against type IV collagen of basement membrane (Sage, Woodbury & Bornstein 1979), and because enteric secretion of rat mast-cell protease II is associated with extensive epithelial shedding (Miller et al. 1983b; King & Miller 1984), we have proposed that this enzyme is a mediator of intestinal permeability (King & Miller 1984). It is thus possible that the changes in permeability which occur during *N. brasiliensis* infection in the rat (Murray, Jarrett & Jennings 1971; Murray 1972; Nawa 1979) and in *O. circumcincta*-infected sheep (Yakoob, Holmes & Armour 1983; Smith et al. 1984) share a common pathway involving serine protease.

Secondarily formed mast-cell mediators and nematode expulsion

Prostaglandins

Rat peritoneal mast cells release prostaglandin D2 upon interaction with IgE (Lewis & Austen 1981), but there is little evidence to suggest that mast cells are involved in the synthesis and secretion of other prostaglandins. For example, the mucosal concentration of prostaglandin E1 is increased early in infection with *N. brasiliensis* (Kelly & Dineen 1976). Furthermore, injection of prostaglandin E1 into the intestinal lumen of infected rats caused premature expulsion of adult worms (Kelly & Dineen 1976). Worms incubated *in vitro* in prostaglandin E were damaged, and it was proposed that prostaglandins were responsible for worm expulsion (Kelly & Dineen 1976). Others have failed to confirm the protective role of prostaglandin E, either in *N. brasiliensis*-infected rats (Kassai et al. 1980) or *T. colubriformis*-infected guinea pigs (Rothwell, Love & Goodrich 1977). The source of prostaglandin E described by Dineen and colleagues is not known, but one possibility is that other inflammatory cells secrete this mediator.

Leucotrienes

The class of mediator known as leucotrienes was formerly identified as SRS-A. Murine mast cells derived in culture and putatively identified as mucosal mast cells (Ginsburg et al. 1981; Crapper & Schrader 1983; Sredni et al. 1983) secrete leucotriene C_4 when stimulated (Razin et al. 1983). At present there is no data confirming that leucotriene C_4 is secreted by murine or rat mucosal mast cells *in vivo*. However, recent studies by Douch and colleagues (1983) have shown that extracts of ovine gastrointestinal mucous contain leucotriene-like material which apparently depressed motility of nematode larvae in an *in vitro* assay. This activity was detected in mucus extracted from resistant sheep, but not from normal uninfected sheep or parasitized nonresistant animals. Biochemical assays demonstrated the leucotriene nature of the mucous extract, although the subclass of leucotriene involved has yet to be characterized (Douch et al. 1983). The possibility that mucosal mast cells and globule leucocytes are the source of this leucotriene-like activity needs to be considered, especially in light of recent findings on the *in vitro* capacity of putative murine mucosal mast cells to secrete leucotriene C_4 (Razin et al. 1983).

Rapid expulsion of intestinal nematodes — a role for local anaphylaxis

The early studies by McCoy (1940), subsequently confirmed by Castro and colleagues (1976) and Love, Ogilvie and McLaren (1976), showed that infective *T. spiralis* larvae were rejected within 12 to 24 h of reinfecting *T. spiralis*-primed rats. Later studies demonstrated that the rate of expulsion of the larvae was extremely rapid, occurring within 15 to 30 min of challenge (Russell & Castro 1979). Rapid rejection of *T. spiralis* occurs within 6 h of challenging immune mice (Alizadeh & Wakelin 1982a), while *N. brasiliensis* is expelled within 4 h of intraduodenal infection in immune rats (Miller, Huntley & Wallace 1981), and both *O. circumcincta* and *H. contortus* are expelled within 48 h of infecting hyperimmune sheep (Miller et al. 1983a; Smith et al. 1984). Fourth-stage *T. colubriformis* larvae are expelled within 8 h of challenging immune guinea pigs (Rothwell, Jones & Love 1974).

There are a number of indications that localized immediate hypersensitivity reactions are involved in these rapid expulsion systems. For example, rejection of *T. spiralis* is associated with a net secretion of water into the gut lumen by the enteric epithelium (Castro, Hessel & Whalen 1979). Castro (1982) has hypothesized that vasoactive intestinal polypeptide released from mucosal mast cells could mediate this response. Clear-cut evidence of mucosal mast-cell involvement in the rapid rejection of *N. brasiliensis* comes from studies already mentioned on the release of rat mast-cell protease II into the blood circulation following challenge (Miller et al. 1983b).

More than 90% of a challenge infection with 50,000 *O. circumcincta* larvae were expelled from hyperimmune sheep within 48 h of challenge. The rejection process was temporally associated with an increase in gastric permeability, determined by monitoring the concentration of pepsinogen in gastric lymph (Smith et al. 1984). By contrast, the rejection response following challenge with 1,000 *O. circumcincta* was much less efficient and provoked no detectable permeability change. It was suggested that larval rejection was effective only when triggered by a sufficiently large antigenic stimulus and that the greatly increased population of mast cells and globule leucocytes in hyperimmune sheep was responsible for the changes in permeability (Smith et al. 1984).

Where rapid rejection has been studied in detail, the main event on the immune, as opposed to the nonimmune, mucosal surface is the exclusion of the larvae from their normal predilection sites attached to or within the mucosa (Russell & Castro 1979; Miller, Huntley & Wallace

1981; Miller et al. 1983a). Both *N. brasiliensis* and *T. spiralis* become enveloped and enmeshed within the superficial intestinal mucus in immune rats (Miller, Huntley & Wallace 1981; Lee & Ogilvie 1981). This phenomenon has been reviewed at length elsewhere (Miller 1984). Mucous trapping of *T. spiralis* may be an example of cooperative interaction between mucus and antibody (Lee & Ogilvie 1981) and it has been suggested that local permeability changes, mediated by anaphylactic antibodies, facilitate the translocation of serum-derived antibody (Miller, Huntley & Wallace 1981).

By contrast, repeated attempts to block rapid expulsion of *T. spiralis* with anti-allergic drugs were without effect, nor was it possible to induce rapid expulsion after heterologous intestinal priming with *N. dubius* and passive transfer of *T. spiralis*-immune serum rich in reagins. Corticosteroids and irradiation were, however, highly effective in blocking rapid expulsion of *T. spiralis*, and it was proposed that lymphoid, rather than anaphylactic, protective responses were involved in the expulsion process (Bell, McGregor & Adams 1982).

Our own experience with the corticosteroid suppression of rapid expulsion of *N. brasiliensis* is that immune exclusion and mucous trapping of the parasites are abrogated by treatment of primed rats, either with betamethasone (2 mg/kg) 24 h befoe challenge or with methyl prednisolone (25 mg/kg) 24 and 48 h before challenge (Miller & Huntley 1982). Both these short-term drug treatments decimated the mucosal mast-cell population in the intestinal mucosa and eliminated the secretory release of rat mast-cell protease II (S.J. King et al. unpublished). In fact, corticosteroids are well recognized for their anti-anaphylactic activity (Church & Miller 1978) and mucosal mast-cell sensitivity to steroids has been shown (Jarrett et al. 1967). The same drugs block the synthesis of leucotrienes (Wasserman 1983), inhibit the migration of eosinophils to parasitized gut (Sukhdeo & Croll 1981), alter the sensitivity of enteric smooth muscle to agonists (Farmer 1982) and block mucosal permeability (Murray, Jarrett & Jennings 1971). The wide range of biological activities of corticosteroids makes it difficult to pinpoint which of these effects is most important in the abrogation of resistance against nematodes. Nevertheless, nearly all of them can be considered in the context of local immediate hypersensitivity reactions and support the view that corticosteroids interfere with anaphylactic responses (Church & Miller 1978). Such results generally favour the concept that local anaphylactic reactions are involved in the rapid expulsion of gastrointestinal nematodes.

Overview of the relevance of immediate hypersensitivity reactions in the protective response

Despite the obvious gaps in our understanding of the role of immediate hypersensitivity responses, a diagrammatic representation of the part they might play in resistance to enteric nematodes is shown in Figure 1. This emphasizes the fact that immediate hypersensitivity is only part of a complex series of interactions involving thymus-dependent hyperplastic responses in the mucosa, immunoglobulin synthesis and recruitment of bloodborne effector cells.

Although there is a major association between IgE and mast cell-basophil function, it must be stressed that IgG isotypes also possess anaphylactic properties and that activation of the complement cascade promotes the generation of anaphylatoxins which themselves can mimic IgE-mediated responses. Similarly, a T cell-derived factor which causes histamine release from connective-tissue mast cells has recently been described in the mouse (Askenase & van Loveren 1983). The context of immediate hypersensitivity can, therefore, be extended beyond the IgE-mast cell and basophil axis.

What then is the function of mast cells and basophils? Both cell types are recruited to the site of infection during primary worm expulsion and, during rapid worm expulsion, an expanded population of sensitized mast cells is already *in situ*. In both instances, it is possible to demonstrate the release of mast cell-specific granule products at the time of worm expulsion. Products of release include preformed mediators such as

Figure 1. Diagrammatic representation of the events which are likely to be associated with the development of an effector response against nematode parasites in the mammalian gut.

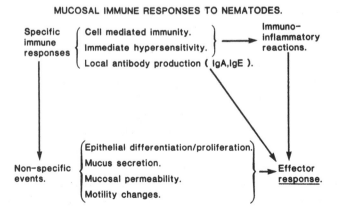

MUCOSAL IMMUNE RESPONSES TO NEMATODES.

histamine, 5HT, proteoglycan, serine proteases and chemotactic peptides, as well as secondarily formed or generated mediators such as prostaglandins and leucotrienes. These mediators have permeability-inducing and chemotactic activities, promoting both the translocation of plasma proteins across the mucosal epthelium and the recruitment of immuno-inflammatory cells. They may also promote secretion by epithelial absorptive cells and mucus-secreting cells.

A hypothetical view of events could be summarized as follows. Lumen-dwelling or migratory nematodes which had not yet attained their predilection sites in the mucosa would be directly susceptible to some of the products released from mast cells-basophils. For example, leucotrienes may have a paralysing or disorienting effect on the worms and, under certain circumstances, histamine and 5HT may possess similar activity. A second line of defence may be the mucous layer through which worms must penetrate in order to reach the mucosa itself. Inflammatory mediators may promote mucus secretion, epithelial cell secretion of water and sodium, and the translocation across the epithelium of plasma proteins. These proteins, intermingling with the secreted mucus, could both increase its viscosity and the levels of humoral antibodies and complement. Antibody and complement would interact cooperatively with the already viscous mucus and either entrap or block the passage of the worms or their secreted products, thus preventing their access to the mucosa and/or blocking their feeding processes. The increased fluid secreted by the epithelial cells may help to wash the worms down the 'slippery tube' created by the viscous mucus.

Parasites which entered the tissues would also be susceptible to the direct effects of mast cell-derived mediators, most notably histamine, 5HT and leucotrienes. Release of these products would promote local vascular permeability and flooding of the tissues with plasma proteins, including immunoglobulin and complement, and would recruit monocytes-macrophages and granulocytes. Whether such reactions are of significance for nematodes which dwell within the mucosal epithelium itself (reviewed in Miller 1984) is not clear. However, those parasites which migrate through or into the *lamina propria* or submucosa would be directly susceptible to IgE-, IgG- or complement-mediated cellular helminthotoxicity as macrophages or granulocytes adhered to the nematode surface via Fcε, Fcγ or complement receptors on their plasma membranes.

This hypothetical sequence of events is drawn from many published studies which have been extensively reviewed (Askenase 1980; Capron, Capron & Dessaint 1980; Castro 1980; Befus & Bienenstock 1982; Castro

1982; Jarrett & Miller 1982; Miller 1984) and is thus a synthesis of ideas and results derived from many authors. It may, however, provide a useful framework to be modified or built upon when further studies begin to reveal the overall significance of immediate hypersensitivity reactions in the rejection of gut nematodes.

References

Alizadeh, H. & Wakelin, D. (1982a). Comparison of rapid expulsion of *Trichinella spiralis* in mice and rats. *International Journal for Parasitology*, **12**, 65-73.

Alizadeh, H. & Wakelin, D. (1982b). Genetic factors controlling the intestinal mast-cell response in mice infected with *Trichinella spiralis*. *Clinical & Experimental Immunology*, **49**, 331-37.

Andrews, J.M. (1962). Parasitism and allergy. *Journal of Parasitology*, **48**, 3-12.

Askenase, P.W. (1977). Role of basophils, mast cells and vasoamines in hypersensitivity reactions with a delayed time course. *Progress in Allergy*, **23**, 199-320.

Askenase, P.W. (1980). Immunopathology of parasitic diseases: involvement of basophils and mast cells. *Springer Seminars in Immunopathology*, **2**, 417-42.

Askenase, P.W. & van Loveren, H. (1983). Delayed-type hypersensitivity: activation of mast cells by antigen-specific T-cell factors initiates the cascade of cellular interactions. *Immunology Today*, **9**, 259-64.

Barth, E.E.E., Jarrett, W.F.M. & Urquhart, G.M. (1966). Studies on the mechanism of the self-cure reaction in rats infected with *Nippostrongylus brasiliensis*. *Immunology*, **10**, 459-64.

Beavan, M.A., Aiken, D.L., Woldemussie, E. & Soll, A.M. (1983). Changes in histamine synthetic activity, histamine content and responsiveness to compound 48/80 with activation of rat peritoneal mast cells. *Journal of Pharmacology & Experimental Therapeutics*, **224**, 620-26.

Befus, A.D. & Bienenstock, J. (1979). Immunologically mediated intestinal mastocytosis in *Nippostrongylus brasiliensis*-infected rats. *Immunology*, **38**, 95-101.

Befus, A.D. & Bienenstock, J. (1982). Factors involved in symbiosis and host resistance at the mucosa-parasite interface. *Progress in Allergy*, **31**, 76-177.

Befus, A.D., Johnston, N. & Bienenstock, J. (1979). *Nippostrongylus brasiliensis*: mast cell and histamine levels in tissues of infected and normal rats. *Experimental Parasitology*, **48**, 1-8.

Befus, A.D., Pearce, F.L., Gauldie, J., Horsewood, P. & Bienenstock, J. (1982). Mucosal mast cells. 1. Isolation and functional characteristics of rat intestinal mast cells. *Journal of Immunology*, **128**, 2475-80.

Bell, R.G., McGregor, D.D. & Adams, L.S. (1982). Studies on the inhibition of rapid expulsion of *Trichinella spiralis* in rats. *International Archives of Allergy & Applied Immunology*, **69**, 73-80.

Capron, A., Capron, M. & Dessaint, J.-P. (1980). ADCC as primary mechanisms of defence against metazoan parasites. In *Progress in immunology IV* (ed. M. Fougereau & J. Dausset), pp. 782-93. London: Academic Press.

Castro, G.A. (1980). Regulation of pathogenesis in disease caused by gastrointestinal parasites. In *The host-invader interplay* (ed. H. van der Bossche), pp. 457-67. Amsterdam: Elsevier.

Castro, G.A. (1982). Immunological regulation of epithelial function. *American Journal of Physiology*, **243**, G321-29.

Castro, G.A., Hessel, J.J. & Whalen, G. (1979). Altered intestinal fluid movement in response to *Trichinella spiralis* in immunized rats. *Parasite Immunology*, **1**, 259-66.

Castro, G.A., Badial-Aceves, F., Adams, P.R., Copeland, E.M. & Dudrick, S.J. (1976). Response of immunized, parenterally nourished rats to challenge infection with the nematode, *Trichinella spiralis*. *Journal of Nutrition*, **106**, 1484-91.

Church, M.K. & Miller, P. (1978). Time courses of the anti-anaphylactic and anti-inflammatory effects of dexamethasone in the rat and mouse. *British Journal of Pharmacology*, **62**, 481-86.

Crapper, R.M. & Schrader, J.W. (1983). Frequency of mast-cell precursors in normal tissue determined by an *in vitro* assay: antigen induces parallel increases in the frequency of P-cell precursor and mast cells. *Journal of Immunology*, **131**, 923-28.

Crowle, P.K. & Reed, N.D. (1981). Rejection of the intestinal parasite *Nippostrongylus brasiliensis* by mast cell-deficient W/W^v anaemic mice. *Infection & Immunity*, **33**, 54-58.

Crowle, P.K. & Phillips, D.E. (1983). Characteristics of mast cells in Chedia-Higashi mice: light and electron microscopic studies of connective tissue and mucosal mast cells. *Experimental Cell Biology*, **51**, 130-39.

Curtain, C.C. (1969). A new immunoglobulin subclass in sheep. *Immunology*, **16**, 373-80.

Dessein, A.J., Parker, W.L., James, S.L. & David, J.R. (1981). IgE antibody and resistance to infection. 1. Selective suppression of the IgE response in rats diminishes the resistance and the eosinophil response to *Trichinella spiralis* infection. *Journal of Experimental Medicine*, **153**, 423-26.

Dobson, C. (1966). Immunofluorescent staining of globule leucocytes in the colon of the sheep. *Nature*, **211**, 875.

Douch, P.G.C., Harrison, G.B.L., Buchanan, L.L. & Greer, K.S. (1983). *In vitro* bioassay of sheep gastrointestinal mucus for nematode paralyzing activity mediated by a substance with some properties characteric of SRS-A. *International Journal for Parasitology*, **13**, 207-12.

Dy, M. & Lebel, B. (1983). Skin allografts generate an enhanced production of histamine and histamine-producing cell-stimulating factor (HCSF) by spleen cells in response to T-cell mitogens. *Journal of Immunology*, **130**, 3243-47.

Elek, P. & Durie, P.H. (1966). The histopathology of the reactions of calves to experimental infection with the nodular worm, *Oesophagostomum radiatum* (Rudolphi 1983). 1. Host reaction to reinfection with single large doses of larvae. *Australian Journal of Agricultural Research*, **17**, 807-19.

Enerback, L. (1981). The gut mucosal mast cell. *Monographs in Allergy*, **17**, 222-32.

Farmer, S.G. (1982). The effect of betamethasone on altered responsiveness of isolated intestine from rats infected with *Nippostrongylus brasiliensis*. *British Journal of Pharmacology*, **76**, 192P.

Ferguson, A. & Miller, H.R.P. (1979). The role of the mast cell in the defence against gut parasites. In *The mast cell: its role in health and disease* (eds. J. Pepys & A.M. Edwards), pp. 159-65. London: Pitman.

Gabriel, B.W. & Justus, D.E. (1979). Quantitation of immediate and delayed hypersensitivity responses in *Trichinella*-infected mice. *International Archives of Allergy & Applied Immunology*, **60**, 275-85.

Gershwin, L.J. & Dygert, B.S. (1983). Development of a semi-automated microassay for bovine immunoglobulin E: definition and standardization. *American Journal of Veterinary Research,* **44**, 891-95.

Ginsburg, H., Olson, E.C., Huff, T.F., Okudaira, H. & Ishizaka, T. (1981). Enhancement of mast-cell differentiation *in vitro* by T-cell factor(s). *International Archives of Allergy & Applied Immunology,* **66**, 447-58.

Gregory, M.W. (1979). The globule leucocyte and parasitic infection: a brief history. *Veterinary Bulletin,* **49**, 821-27.

Ha, T.-Y., Reed, N.D. & Crowle, P.K. (1983). Delayed expulsion of adult *Trichinella spiralis* by mast cell-deficient mice. *Infection & Immunity,* **41**, 445-47.

Haig, D.M., McKee, T.A., Jarrett, E.E.E., Woodbury, R.G. & Miller, H.R.P. (1982). Generation of mucosal mast cells is stimulated *in vitro* by factors derived from T cells of helminth-infected rats. *Nature,* **300**, 188-90.

Haig, D.M., McMenamin, C., Gunneberg, C., Woodbury, R.G. & Jarrett, E.E.E. (1983). Stimulation of mucosal mast-cell growth in normal and nude rat bonemarrow cultures. *Proceedings of the National Academy of Sciences of the USA,* **80**, 4499-4503.

Handlinger, J.H. & Rothwell, T.L.W. (1981). Studies of the response of basophil and eosinophil leucocytes and mast cells to the nematode *Trichostrongylus colubriformis*: comparison of cell populations in parasite resistant and susceptible guinea pigs. *International Journal for Parasitology,* **11**, 67-70.

Hogarth-Scott, R.S. (1969). Homocytotropic antibody in sheep. *Immunology,* **16**, 543-48.

Hogarth-Scott, R.S. (1972). Peripheral circulating antigens as a cause of loss of PCA reactivity in parasitized rats. *Immunology,* **24**, 503-9.

Huntley, J.F., Wallace, G.R. & Miller, H.R.P. (1982). Quantitative recovery of isolated mucosal mast cells and globule leucocytes from parasitized sheep. *Research in Veterinary Science,* **33**, 58-63.

Huntley, J.F., Newlands, G.F. & Miller, H.R.P. (1984). The isolation and characterization of globule leucocytes: their derivation from mucosal mast cells in parasitized sheep. *Parasite Immunology,* **6**, 371-90.

Huntley, J.F., McGorum, B., Newlands, G.F.J. & Miller, H.R.P. (1984). Granulated intraepithelial lymphocytes: their relationship to mucosal mast cells and globule leucocytes in the rat. *Immunology,* **53**, 525-35.

Huxtable, C.R. & Rothwell, T.L.W. (1975). Studies of the responses of basophil and eosinophil leucocytes and mast cells to the nematode *Trichostrongylus colubriformis*: ultrastructural changes in basophils and eosinophils at the site of infection. *Australian Journal of Experimental Biology & Medical Science,* **53**, 437-45.

Jacobson, R.H., Reed, N.D. & Manning, D.D. (1977). Expulsion of *Nippostrongylus brasiliensis* from mice lacking antibody production potential. *Immunology,* **32**, 867-74.

Jarrett, E.E.E. & Miller, H.R.P. (1982). Production and activities of IgE in helminth infection. *Progress in Allergy,* **31**, 178-233.

Jarrett, E.E.E. & Haig, D.M. (1984). Mucosal mast cells *in vivo* and *in vitro*. *Immunology Today,* **5**, 115-19.

Jarrett, E.E.E., Jarrett, W.F.H. & Urquhart, G.M. (1968). Quantitative studies on the kinetics of establishment and expulsion of intestinal nematode populations in susceptible and immune hosts: *Nippostrongylus brasiliensis* in the rat. *Parasitology,* **58**, 625-39.

Jarrett, E.E.E., Haig, D.M. & Bazin, H. (1976). Time-course studies on rat IgE production in *N. brasiliensis* infection. *Clinical & Experimental Immunology,* **24**, 346-51.

Jarrett, W.F.H., Jarrett, E.E.E., Miller, H.R.P. & Urquhart, G.M. (1967). Quantitative studies on the mechanism of self-cure in *Nippostrongylus brasiliensis* infections. In *The reaction of the host to parasitism* (ed. E.J.L. Soulsby), pp. 191-98. Marburg-Lahn: Elwert.

Jones, V.E., Edwards, A.J. & Ogilvie, B.M. (1970). The circulating immunoglobulins involved in protective immunity to the intestinal stage of *Nippostrongylus brasiliensis* in the rat. *Immunology*, **18**, 621-33.

Jones, W.O., Rothwell, T.L.W., Dineen, J.K. & Griffiths, D.A. (1974). Studies on the role of histamine and 5-hydroxytryptamine in immunity against the nematode *Trichostrongylus colubriformis*. 2. Amine levels in the intestine of infected guinea pigs. *International Archives of Allergy & Applied Immunology*, **46**, 14-27.

Justus, D.E. & Morakote, N. (1981). Mast-cell degranulation associated with sequestration and removal of *Trichinella spiralis* antigens. *International Archives of Allergy & Applied Immunology*, **64**, 371-84.

Kassai, T., Redl. P., Jecsai, G., Balla, E. & Harangozo, E. (1980). Studies on the involvement of prostaglandins and their precursors in the rejection of *Nippostrongylus brasiliensis* from the rat. *International Journal for Parasitology*, **10**, 115-20.

Keller, R. (1971). *Nippostrongylus brasiliensis* in the rat: failure to relate intestinal histamine and mast-cell levels with worm expulsion. *Parasitology*, **63**, 473-81.

Keller, R. & Ogilvie, B.M. (1972). The effects of drugs on worm expulsion in the *Nippostrongylus brasiliensis*-infected rat: a discussion of the interpretation of drug action. *Parasitology*, **64**, 217-27.

Kelly, J.D. & Dineen, J.K. (1972). The cellular transfer of immunity to *Nippostrongylus brasiliensis* in inbred rats (Lewis strain). *Immunology*, **22**, 199-210.

Kelly, J.D. & Dineen, J.K. (1976). Prostaglandins in the gastrointestinal tract: evidence for a role in worm expulsion. *Australian Veterinary Journal*, **52**, 391-97.

King, S.J. & Miller, H.R.P. (1984). Anaphylactic release of mucosal mast-cell protease and its relationship to gut permeability in *Nippostrongylus*-primed rats. *Immunology*, **51**, 653-60.

Lee, G.B. & Ogilvie, B.M. (1981). The mucous layer in intestinal nematode infections. In *The mucosal immune system in health and disease* (eds. P.L. Ogra & J. Bienenstock), pp. 175-83. Columbus, Ohio: Ross Laboratories.

Lewis, R.A. & Austen, K.F. (1981). Mediation of local homoeostasis and inflammation by leucotrienes and other mast cell-dependent compounds. *Nature*, **293**, 103-8.

Liu, S.-K. (1965). Pathology of *Nematospiroides dubius*. 2. Reinfections in Webster mice. *Experimental Parasitology*, **17**, 136-47.

Love, R.J., Ogilvie, B.M. & McLaren, D.J. (1976). The immune mechanism which expels the intestinal stage of *Trichinella spiralis* from rats. *Immunology*, **30**, 7-15.

Mayrhofer, G. (1979). The nature of the thymus dependency of mucosal mast cells. 2. The effect of thymectomy and of depleting recirculating lymphocytes on the response to *Nippostrongylus brasiliensis*. *Cellular Immunology*, **47**, 312-22.

McCoy, O.R. (1940). Rapid loss of *Trichinella* larvae fed to immune rats and its bearing on the mechanism of immunity. *American Journal of Hygiene*, **32**, 105-16.

Metcalfe, D.D., Kaliner, M. & Donlon, M.A. (1981). The mast cell. *CRC Critical Reviews in Immunology*, **3**, 23-74.

Miller, H.R.P. (1969). The intestinal mast cell in normal and parasitized rats. Ph.D. thesis. University of Glasgow.

Miller, H.R.P. (1971a). Immune reactions in mucous membranes. 3. The discharge of intestinal mast cell during helminth expulsion in the rat. *Laboratory Investigation,* **24**, 348-52.

Miller, H.R.P. (1971b). Immune reactions in mucous membranes. 2. The differentiation of intestinal mast cells during helminth expulsion in the rat. *Laboratory Investigation,* **24**, 339-47.

Miller, H.R.P. (1979). Passive transfer of the mucosal mast cell response: its relationship to goblet-cell differentiation. In *The mast cell: its role in health and disease* (eds. J. Pepys & A.M. Edwards), pp. 738-42. London: Pitman.

Miller, H.R.P. (1980). The origin, structure and function of mucosal mast cells: a brief review. *Biologie Cellulaire,* **39**, 229-32.

Miller, H.R.P. (1984). The protective mucosal response against gastrointestinal nematodes in ruminants and laboratory animals. *Veterinary Immunology & Immunopathology,* **6**, 167-259.

Miller, H.R.P. & Walshaw, R. (1972). Immune reactions in mucous membranes. 4. Histochemistry of intestinal mast cells during helminth expulsion in the rat. *American Journal of Pathology,* **69**, 195-206.

Miller, H.R.P. & Huntley, J.F. (1982). Intestinal mucus and protection against *Nippostrongylus brasiliensis*: the effect of corticosteroids in immune rats. *Molecular & Biochemical Parasitology,* **Suppl.**, 4.

Miller, H.R.P., Murray, M. & Jarrett, W.F.H. (1967). Globule leucocytes and mast cells. In *The reaction of the host to parasitism* (ed. E.J.L. Soulsby), pp. 198-210. Marburg-Lahn: Elwert.

Miller, H.R.P., Huntley, J.F. & Wallace, G.R. (1981). Immune exclusion and mucous trapping during the rapid expulsion of *Nippostrongylus brasiliensis* from primed rats. *Immunology,* **44**, 419-29.

Miller, H.R.P., Jackson, F., Newlands, G. & Apppleyard, W.T. (1983a). Immune exclusion: a mechanism of protection against the ovine nematode *Haemonchus contortus. Research in Veterinary Science,* **35**, 357-63.

Miller, H.R.P., Woodbury, R.G., Huntley, J.F. & Newlands, G. (1983b). Systemic release of mucosal mast-cell protease in primed rats challenged with *Nippostrongylus brasiliensis. Immunology,* **49**, 471-79.

Murray, M. (1972). Immediate hypersensitivity effector mechanisms. 2. *in vivo* reactions. In *Immunity to animal parasites* (ed. E.J.L. Soulsby), pp. 155-90. New York: Academic Press.

Murray, M., Miller, H.R.P. & Jarrett, W.F.H. (1968). The globule leucocyte and its derivation from the subepithelial mast cell. *Laboratory Investigation,* **19**, 222-34.

Murray, M., Jarrett, W.F.H. & Jennings, F.W. (1971). Mast cells and macromolecular leak in intestinal immunological reactions: the influence of sex of rats infected with *Nippostrongylus brasiliensis. Immunology,* **21**, 17-31.

Murray, M., Miller, H.R.P., Sanford, J. & Jarrett, W.F.H. (1971a). 5-Hydroxytryptamine in intestinal immunological reactions: its relationship to mast-cell activity and worm expulsion in rats infected with *Nippostrongylus brasiliensis. International Archives of Allergy & Applied Immunology,* **40**, 236-47.

Murray, M., Smith W.D., Waddell, A.H. & Jarrett, W.F.H. (1971b). *Nippostrongylus brasiliensis*: histamine and 5-hydroxytryptamine inhibition and worm expulsion. *Experimental Parasitology,* **30**, 58-63.

Musoke, A.J. & Williams, J.F. (1975). The immunological response of the rat to infection with *Taenia taeniaformis.* 5. Sequence of appearance of protective immunoglobulins and the mechanism of action of 7Sγ2a antibodies. *Immunology,* **29**, 855-66.

Musoke, A.J., Williams, J.F. & Leid, R.W. (1978). Immunological response of the rat to infection with *Taenia taeniaformis.* 6. The role of immediate hypersensitivity in resistance to reinfection. *Immunology,* **34**, 565-70.

Nawa, Y. (1979). Increased permeability of the gut mucosa in rats infected with *Nippostrongylus brasiliensis. International Journal for Parasitology,* **9**, 251-55.

Nawa, Y. & Miller, H.R.P. (1979). Adoptive transfer of the intestinal mast-cell response in rats infected with *Nippostrongylus brasiliensis. Cellular Immunology,* **42**, 225-39.

Ogilvie, B.M. (1967). Reagin-like antibodies in rats infected with the nematode parasite *Nippostrongylus brasiliensis. Immunology,* **12**, 113-31.

Ogilvie, B.M. & Love, R.J. (1974). Cooperation between antibodies and cells in immunity to a nematode parasite. *Transplantation Reviews,* **19**, 147-68.

Ogilvie, B.M., Askenase, P.W. & Rose, M.E. (1980). Basophils and eosinophils in three strains of rats and in athymic (nude) rats following infection with the nematodes *Nippostrongylus brasiliensis* or *Trichinella spiralis. Immunology,* **39**, 385-89.

Panter, H.C. (1969). The mechanism of immunity of mice to *Nematospiroides dubius. Journal of Parasitology,* **55**, 38-43.

Pearce, F.L., Befus, A.D., Gauldie, J. & Bienenstock, J. (1982). Mucosal mast cells. 2. Effects of anti-allergic compounds on histamine secretion of isolated intestinal mast cells. *Journal of Immunology,* **128**, 2481-86.

Pritchard, D.I., Williams, D.J.L., Behnke, J.M. & Lee, T.D.G. (1983). The role of IgG, hypergammaglobulinaemia in immunity to the gastrointestinal nematode *Nematospiroides dubius*: the immunochemical purification antigen specificity and *in vivo* anti-parasite effect of IgG from immune serum. *Immunology,* **49**, 353-65.

Razin, E., Mencia-Huerta, J.-M., Stevens, R.L., Lewis, R.A., Liu, F.-T., Corey, E.J. & Austen, K.F. (1983). IgE-mediated release of leucotriene C4, chondroitin sulphate E proteoglycan, B-hexosaminidase, and histamine from cultured bonemarrow-derived mouse mast cells. *Journal of Experimental Medicine,* **157**, 189-201.

Reed, N.D., Crowle, P.K. & Ha T.-Y. (1982). Use of mast cell-deficient W/Wᵛ mice to study host-parasite relationships. *Experimental Cell Biology,* **50**, 324-25.

Roth, R.L. & Levy, D.A. (1980). *Nippostrongylus brasiliensis*: peripheral leucocyte response and correlation of basophils with blood histamine concentrations during infection in rats. *Experimental Parasitology,* **50**, 331-41.

Rothwell, T.L.W. (1975). Studies of responses of basophil and eosinophil leucocytes and mast cells to the nematode *Trichostrongylus colubriformis.* 1. Observations during the expulsion of first and second infections by guinea pigs. *Journal of Pathology,* **116**, 51-60.

Rothwell, T.L.W. & Dineen, J.K. (1972). Cellular reactions in guinea pigs following primary and challenge infection with *Trichostrongylus colubriformis* with special reference to the roles played by eosinophils and basophils in rejection of the parasite. *Immunology,* **22**, 733-45.

Rothwell, T.L.W., Love, R.J. & Dineen, J.K. (1973). Studies on the inhibition of rejection of *Trichostrongylus colubriformis* in guinea pigs by promethazine. *Australian Journal of Experimental Biology & Medical Science,* **51**, 221-28.

Rothwell, T.L.W., Jones, W.O. & Love, R.J. (1974). Studies on the role of histamine and 5-hydroxytryptamine in immunity against the nematode *Trichostrongylus colubriformis*. 2. Inhibition of worm expulsion from guinea pigs by treatment with reserpine. *International Archives of Allergy & Applied Immunology*, **47**, 875-86.

Rothwell, T.L.W., Prichard, R.K. & Love, R.J. (1974). Studies on the role of histamine and 5-hydroxytryptamine in immunity against the nematode *Trichostrongylus colubriformis*. 1. *In vivo* and *in vitro* effects of the amines. *International Archives of Allergy & Applied Immunology*, **46**, 1-13.

Rothwell, T.L.W., Love, R.J. & Goodrich, B.S. (1977). Failure to demonstrate involvement of prostaglandins in the immune expulsion of *Trichostrongylus colubriformis* from the intestine of guinea pigs. *International Archives of Allergy & Applied Immunology*, **53**, 93-95.

Ruitenberg, E.J. & Elgersma, A. (1976). Absence of intestinal mast-cell response in congenitally athymic mice during *Trichinella spiralis* infection. *Nature*, **264**, 258-60.

Ruitenberg, E.J. & Elgersma, A. (1979). Response of intestinal globule leucocytes in the mouse during a *Trichinella spiralis* infection and its independence of intestinal mast cells. *British Journal of Experimental Pathology*, **60**, 246-51.

Russell, D.A. & Castro, G.A. (1979). Physiological characterization of a biphasic immune response to *Trichinella spiralis* in the rat. *Journal of Infectious Diseases*, **139**, 304-12.

Sage, H., Woodbury, R.G. & Bornstein, P. (1979). Structural studies on human type IV collagen. *Journal of Biological Chemistry*, **254**, 9893-900.

Sharp, N.C.C. & Jarrett, W.F.H. (1968). Inhibition of immunological expulsion of helminths by reserpine. *Nature*, **218**, 1161-62.

Smith, W.D., Jackson, F., Jackson, E., Williams, J. & Miller, H.R.P. (1984). Manifestations of resistance to ovine ostertagiasis associated with immunological response in the gastric lymph. *Journal of Comparative Pathology*, **94**, 591-600.

Sredni, B., Friedman, M.M., Bland, C.E. & Metcalfe, D.D. (1983). Ultrastructural, biochemical, and functional characterics of histamine-containing cells cloned from mouse bonemarrow: tentative identification as mucosal mast cells. *Journal of Immunology*, **131**, 915-22.

Stewart, D.F. (1950). Studies on resistance of sheep to infections with *Haemonchus contortus* and *Trichostrongylus* spp. and on the immunological reactions of sheep exposed to infection. 4. The antibody response to natural infestation in grazing sheep and the 'self-cure' phenomenon. *Australian Journal of Agricultural Research*, **1**, 427-39.

Stewart, D.F. (1953). Studies on resistance of sheep to infestation with *Haemonchus contortus* and *Trichostrongylus* spp. and on the immunological reactions of sheep exposed to infestation. *Australian Journal of Agricultural Research*, **4**, 100-17.

Stewart, D.F. (1955). 'Self-cure' in nematode infestations of sheep. *Nature*, **176**, 1273-74.

Stoll, N.R. (1929). Studies with the strongyloid nematode *Haemonchus contortus*. 1. Acquired resistance of hosts under natural reinfection conditions out of doors. *American Journal of Hygiene*, **10**, 384-418.

Strobel, S., Miller, H.R.P. & Ferguson, A. (1981). Human intestinal mucosal mast cells: evaluation of location and staining techniques. *Journal of Clinical Pathology*, **34**, 851-58.

Sukhdeo, M.V.K. & Croll, N.A. (1981). The location of parasites within their hosts: factors affecting longitudinal distribution of *Trichinella spiralis* in the small intestine of mice. *International Journal for Parasitology*, **11**, 163-68.

Taguchi, Y., Tsuyama, K., Watanabe, T., Wada, H. & Kitamura, Y. (1982). Increased histidine decarboxylase activity in skin of genetically deficient W/ Wᵛ mice after application of pharbol myristate 13-acetate: evidence for histamine-producing cells without basophilic granules. *Proceedings of the National Academy of Sciences of the USA*, **79**, 6837-41.

Taliaferro, W.H. & Sarles, M.P. (1939). The cellular reactions in the skin, lungs and intestine of normal and immune rats after infection with *Nippostrongylus muris*. *Journal of Infectious Diseases*, **64**, 157-92.

Uber, C.L., Roth, R.L. & Levy, D.A. (1980). Expulsion of *Nippostrongylus brasiliensis* by mice deficient in mast cells. *Nature*, **287**, 226-28.

Urbina, C., Ortiz, C. & Hurtado, I. (1981). A new look at basophils in mice. *International Archives of Allergy & Applied Immunology*, **66**, 158-60.

Urquhart, G.M., Mulligan, W., Eadie, R.M. & Jennings, F.W. (1965). Immunological studies on *Nippostrongylus brasiliensis* infection in the rat: the role of local anaphylaxis. *Experimental Parasitology*, **17**, 210-17.

Wakelin, D. (1978). Immunity to intestinal parasites. *Nature*, **273**, 617-20.

Wasserman, S.I. (1983). Mediators of immediate hypersensitivity. *Journal of Allergy & Clinical Immunology*, **72**, 101-15.

Wingren, U., Enerbck, L., Ahlman, H., Allenmark, S. & Dahlstrm, A. (1983). Amines of the mucosal mast cell of the gut in normal and nematode-infected rats. *Histochemistry*, **77**, 145-58.

Woodbury, R.G. & Miller, H.R.P. (1982). Quantitative analysis of mucosal mast-cell protease in the intestines of *Nippostrongylus*-infected rats. *Immunology*, **46**, 487-95.

Woodbury, R.G., Gruzenski, G.M. & Lagunoff, D. (1978). Immunofluorescent localization of a serine protease in rat small intestine. *Proceedings of the National Academy of Sciences of the USA*, **75**, 2785-89.

Woodbury, R.G., Katanuma, N., Kobayashi, K., Titani, K. & Neurath, M. (1978). Structure of a group-specific protease from rat small intestine. *Biochemistry*, **17**, 811-19.

Woodbury, R.G., Miller, H.R.P., Hntley, J.F., Newlands, G.F.J., Palliser, A.C. & Wakelin, D. (1984). Mucosal mast cells are functionally active during spontaneous explusion of intestinal nematode infections in rat. *Nature*, **312**, 450-2.

Yakoob, A., Holmes, P.H. & Armour, J. (1983). Pathophysiology of gastrointestinal trichostrongyles in sheep: plasma losses and changes in plasma pepsinogen levels associated with parasite challenge of immune animals. *Research in Veterinary Science*, **34**, 305-9.

27

Host responses which control *Trypanosoma brucei* parasites

S.J. BLACK

Two host responses control *Trypanosoma brucei* parasitaemia in mice: an immune system-independent process which regulates the rate of parasite multiplication, and an antibody response specific for external epitopes ...on parasite-attached variable surface glycoprotein (VSG), which determines the rate of parasite destruction. Parasite multiplication appears to be stimulated by host-derived growth factors which have a molecular mass of 10^5 to greater than 10^6 daltons. These molecules are not lipoproteins or immunoglobulins and serve to maintain *T. brucei* parasites as dividing slender forms. When the molecules become limiting, the parasites differentiate to nondividing stumpy forms. Nondividing stumpy-form *T. brucei* parasites degenerate in their mammalian host, giving rise to fragments which stimulate antibody responses. In the absence of these immunostimulatory fragments, no antibody responses are induced, i.e. antibody responses are not induced by multiplying organisms. The efficiency of antibody responses in animals infected with the parasites is modified by an immunodepressive reaction. Immunodepression is associated with rapid disruption of splenic organization into red and white pulp areas, and the generation of plasma cells which have highly distended sacs of endoplasmic reticulum and do not secrete antibody. The information on which these conclusions are based is reviewed and the relevance of the studies to the development of strategies for the control of African trypanosomes in their mammalian hosts is discussed.

Introduction

African trypanosomes are flagellated, spindle-shaped protozoa which live extracellularly in the mammalian host and cause sleeping sickness in humans (*Trypanosoma (Trypanozoon) brucei*) and trypanosomiases of domestic animals (*T. vivax, T. congolense, T. simiae, T. brucei, T. suis*). The organisms are generally transmitted from mammal to mammal by tsetse flies in which they undergo a series of differentiation events (Hoare 1970) and perhaps mate (Tait 1980). Occasionally, trypanosome transmission from the blood of one mammal to another can be mediated by other biting flies which act as natural syringes (Hoare 1970). Trypanosomiasis can be chronic or acute, depending on the susceptibility of the host and the parasite species, serodeme or clone used to initiate the infection (Murray, Morrison & Whitelaw 1982). Wild animals show a range of susceptibility to the disease but are generally more competent to control infections than domestic animals. In susceptible hosts, the main features of the disease are recurring waves of parasitaemia, anaemia, tissue lesions and immunodepression which renders the animal susceptible to secondary infections (Urquhart 1980).

African trypanosomes have coevolved with wild animals and in the process have fixed genetic traits which increased the efficiency of cyclical transmission from animal to animal by tsetse flies. Several strategies have been adopted by the parasites to ensure a high chance of transmission. Of these, two are of primary importance: the parasites have a broad host range and can give rise to chronic infections of long duration. The parasite adaptations which influence host range have not been characterized; however, it has been speculated that induction, perception or utilization of host-derived parasite growth-stimulating molecules might be the primary factors which determine this parameter of parasite behaviour (Black et al. 1985). Parasite mechanisms which promote chronic infections have been extensively investigated. They fall into two main categories: mechanisms related to the evasion of efficient immune responses and mechanisms related to inhibition of immune responses.

Evasion of immune responses

Metacyclic trypanosomes of tsetse flies and all mammalian bloodstream-form trypanosomes are covered with a glycoprotein coat (Vickerman 1971; Cross 1975). Each bloodstream parasite can give rise to parasites with a different surface glycoprotein (Cross 1975). The variable surface glycoprotein (VSG) repertoire of bloodstream-form parasites is greater than 100 (Capbern et al. 1977) and is considerably more diverse than that of the metacyclic organisms which are infective for mammals (Le Ray, Barry & Vickerman 1978). Metacyclic parasites rapidly switch to bloodstream-form parasites in the mammalian host, and thereafter the sequence of bloodstream-form VSG expression is not random (Gray & Luckins 1976). The restricted metacyclic VSG repertoire and somewhat ordered appearance of bloodstream-form VSG types ensures that the host sequentially encounters different VSG types, and promotes chronic infections, since new VSG types evade preceding immune responses.

Inhibition of immune responses

Trypanosome-induced immunodepression limits the efficiency of the immune response against the parasites (Black et al. in press; Sacks & Askonas 1980) and reduces the capacity of an infected host to mount responses against unrelated antigens (Urquhart 1980). Several mechanisms contribute to the immunodepression, including an alteration in macrophage activity (Grosskinsky & Askonas 1981), the generation of suppressor T cells (Eardley & Jayawardena 1977), an impairment in the capacity of activated B cells to mature to antibody-secreting cells (Black et al. 1985) and a reduction in the half-life of serum immunoglobulins (Nielsen et al. 1978). The relationships between these phenomena have not been defined.

The evolutionary commitment of African trypanosomes to evading or subverting the host immune system suggests that VSG-specific immune responses are likely to play a major role in controlling infections. Parasite VSG-specific antibody responses appear to be responsible for the elimination of trypanosomes (Murray & Urquhart 1977), and several studies have addressed the question of how trypanosomes stimulate antibody responses (Black, Hewett & Sendashonga 1982; Sendashonga & Black 1982; Black et al. 1983).

Parasite VSG-specific antibody responses are stimulated by fragments of trypanosomes but not by multiplying organisms

Mice infected with strains of *T. brucei* parasites which give rise to chronic infections develop high titres of serum antibodies which react with surface-accessible (but not other) epitopes on parasite-attached VSG, with plasma membrane antigens and with a variety of antigens not detectable in or on the parasites (Sendashonga & Black 1982). These latter bystander antibody responses are thought to arise as a result of trypanosome-induced polyclonal B-cell activation. When mice are immunized with purified trypanosome VSG, the majority of VSG-specific antibodies elicited do not bind to surface-accessible epitopes of parasite-attached VSG, but rather react with epitopes accesssible on soluble VSG or VSG on acetone-fixed parasites (Black, Hewett & Sendashonga 1982; Sendashonga & Black 1982). The specificity of the VSG-binding antibodies produced during infections and the concomitant responses against trypanosome plasma membrane antigens (Sendashonga & Black 1982) suggest that B cells in infected animals respond to fragments of trypanosomes on which VSG organization is intact and on which plasma membrane antigens are exposed.

In contrast, mice infected with clones of *T. brucei* parasites which grow exponentially until death of the host do not mount detectable serum antibody responses against parasite-specific or bystander antigens (Sendashonga & Black 1982). Antibodies cannot be detected on the surface of parasites collected from these mice, washed and reacted with fluorescein-conjugated anti-mouse Ig reagents (Sendashonga 1983). Furthermore, splenic lymphocytes collected from the infected animals and fused with myeloma cells do not give rise to VSG-specific, antibody-producing hybrids or an increase in bystander antigen-specific hybrids, compared with spleen cells from uninfected mice (Sendashonga & Black 1982).

The clones of *T. brucei* parasites which give rise to such very acute infections do not appear to depress B-cell responses in infected mice. Infected mice can mount normal antibody responses against sheep erythrocytes and, when coinfected with *T. brucei* parasites which give rise to chronic infections, mount normal immune responses against the VSG of those parasites (Sendashonga & Black 1982; MacAskill et al. 1981). Trypanocidal drug treatment of mice infected with acute parasites elicits a rapid external VSG epitope-specific antibody response (Sendashonga & Black 1982; Sendashonga 1983) as does inoculation of intact mice with lethally irradiated *T. brucei* parasites of acute clones (Sendashonga & Black 1982). These studies suggest that the failure of very acute *T. brucei* parasites to trigger specific and bystander antibody responses in infected mice is not because the parasites are either intrinsically nonimmunogenic or potently immunodepressive.

T. brucei parasites which give rise to very acute or chronic infections in mice, differ in one major respect: bloodstream parasites of chronic clones differentiate from rapidly multiplying, slender-form organisms to nondividing, stumpy-form organisms during the course of infection (Sendashonga & Black 1982), whereas parasites belonging to very acute clones do not (Sendashonga & Black 1982). The former parasites are termed pleomorphic, whilst the latter are termed monomorphic (Vickerman 1965). The morphological transition from slender-form to stumpy-form *T. brucei* parasites is accompanied by biochemical and physiological shifts in the synthesis of respiratory enzymes and in carbohydrate metabolism towards aerobic pathways of energy utilization (Flynn & Bowman 1973). The biochemical changes are thought to prepare the parasite for cyclical transmission through its tsetse fly vector (Wijers & Willett 1960). However, if stumpy-form organisms are not taken up by tsetse flies, they die in the mammalian host (Black, Hewett & Sendashonga 1982).

Neither multiplying slender-form nor living stumpy-form *T. brucei* parasites release VSG into culture medium or host plasma under physiological conditions (Black, Hewett & Sendashonga 1982). On the basis of the evidence presented here, it seems likely that immunostimulatory parasite fragments are derived from senescent stumpy-form organisms. The conclusion raises the possibility that the kinetics of *T. brucei* parasite differentiation from dividing to senescent forms controls the rate of development of host-protective antibody responses.

Host responses which influence parasite differentiation and the rate of generation of immunostimulatory parasite fragments

The rate of differentiation of *T. brucei* parasites from rapidly dividing slender forms to senescent stumpy forms is not an immutable property of the parasites themselves. The kinetics of parasite differentiation differs in inbred strains of mice simultaneously infected with the same number of a cloned parasite population which is homogeneous for VSG expression (Black et al. 1985; Black et al. 1983). Parasite differentiation is accelerated if mice are pre-exposed to dead (*Corynebacterium parvum*) organisms a few days before infection (M. Murray & S.J. Black unpublished). Similarly, parasite differentiation is accelerated if infected mice are given daily injections of acetylsalicylic acid or indomethacin (Jack et al. 1984). The latter studies suggest that prostaglandins might influence parasite differentiation; however, it was found that caprofen and flurbiprofen, drugs which are potent prostaglandin synthetase inhibitors, had no effect on parasite differentiation rates *in vivo* (S.J. Black unpublished). Hence, the mechanism by which *C. parvum*, indomethacin and acetylsalicylic acid accelerate parasite differentiation is unknown.

A number of immunodepressive treatments appear to retard the rate of parasite differentiation and/or result in higher parasitaemia. These include γ-irradiation of the host before infection (Black et al. 1985; Black et al. 1983) or administration of cyclophosphamide (Hudson & Terry 1979) or hydrocortisone acetate (Ashcroft 1957). These observations suggested to some investigators (Ashcroft 1957; Balber 1972) that components of developing immune responses might play a role in accelerating the net accummulation of nondividing parasites in the blood, either by selectively destroying dividing slender-form organisms (Balber 1972) or by providing them with morphogenetic stimuli (Ashcroft 1957). However, several studies indicate that developing immune responses are unlikely to influence the rate of parasite differentiation: the rate of parasite differentiation is similar in intact and athymic mice with the same genetic background (S.J. Black unpublished). Parasite differentiation rates are almost identical in γ-irradiated and γ-irradiated, syngeneic spleen cell-reconstituted mice (Black et al. 1985; Black et al. 1983). The rates of parasite differentiation and parasitaemia are similar in intact and splenectomized mice, although splenectomized mice mount delayed antibody responses and remit their parasites some days later than the intact mice (Black et al. 1985; Black et al. 1983). Furthermore, during the period in which stumpy-form parasites accumulate in the blood of resistant mice there appears to be no selective destruction of slender-form parasites, as measured by the clearance rates of ^{75}Se-methionine labeled homologous slender and stumpy-form parasites (Black et al. 1983). Parasitaemia reaches similar levels in intact and anti-IgM suppressed mice (Campbell, Esser & Weinbaum 1977).

Taken together, these studies suggest that, although several immunodepressive treatments do retard parasite differentiation rates *in vivo*, they do so by interfering with the efficiency of immune responses or by stimulating mechanisms unrelated to immune responses. Elucidation of these mechanisms awaits the identification of molecules which control *T. brucei* parasite differentiation rates.

Parasites which give rise to very acute infection in mice do not differentiate to stumpy forms and do not generate immunostimulatory fragments. However, these parasites — when inoculated into cattle, eland or buffalo — do give rise to stumpy forms and do stimulate antibody responses (Black, Jack & Morrison 1983). The same parasites, when cloned back into mice from any of the early parasitaemic waves in the cattle, give rise to acute infections and do not generate stumpy forms. Furthermore, plasma collected from cattle during the first few parasitaemic waves and injected into mice with very acute clones of *T. brucei* parasites does not stimulate the generation of stumpy-form organisms (Black, Jack & Morrison 1983). It was concluded that parasite differentiation was negatively regulated by the host, i.e. that the host provides molecules which stimulate parasite multiplication and hence inhibit parasite differentiation. A series of *in vitro* analyses suggests that multiplication of *T. brucei* organisms is stimulated by serum in the presence of fibroblasts (Hirumi, Doyle & Hirumi 1977; Hill et al. 1978; Brun et al. 1981), epithelial cells (Hirumi, Doyle & Hirumi 1977) or adipocytes (Balber 1983), but not in the presence of macrophages, T cells or B cells (Black et al. in press). The serum components which promote parasite multiplication are neither lipoproteins nor immunoglobulins and have a molecular mass range of 10^5 to greater than 10^6 daltons (Black et al. in press). At present it is unclear how these macromolecules interact with fibroblasts, epithelial cells or adipocytes to stimulate *T. brucei* multiplication.

The relationship between the rate of generation of immunostimulatory fragments, the rate of parasite population expansion and the efficiency of the host immune response

Mice of different inbred strains have differing susceptibilities to infections initiated with the same chronic *T. brucei* parasites (Black et al. 1983). Susceptible mice (e.g. C3H/He) have a higher first-wave parasitaemia, a higher parasitaemia during the trough period between parasitaemic waves, an inability to control second-wave parasitaemia and a shorter time to death, when compared with similarly infected resistant mice (e.g. C57Bl/6) (Black et al. 1983). Susceptibility correlates with a slow rate of parasite differentiation to stumpy forms and delayed and inefficient or absent parasite-specific or bystander antibody responses (Black et al. 1985; Black et al. 1983). The bystander responses arise as a result of nonspecific B-cell activation. The retarded immune responses might relate to slow generation of immunostimulatory parasite fragments. However, the concomitant high parasitaemia and eventual high level of stumpy-form parasites and immunostimulatory fragments that arise in susceptible mice infected with *T. brucei* would be expected to lead to very high-titred antibody responses. This does not occur.

The poor antibody responses mounted by susceptible mice infected with chronic *T. brucei* parasites might result from the immunodepressive influence of the high load of trypanosome membrane on the host's immune system (Clayton et al. 1979). However, treatment with the trypanocidal drug diminazene aceturate stimulates rapid and very high-titred antibody responses (parasite-specific and bystander) in the infected susceptible mice (S.J. Black, P. Webster & S.J. Shapiro unpublished). Diminazene aceturate eliminates parasites by 10 h after treatment; the antibody response is detectable by 16 h after treatment and peaks by 30 h after treatment. Thus the antibody response is at its most efficient when the load of parasite membrane presented to the immune system is very high. This suggests that the inefficient antibody response mounted by untreated susceptible mice infected with chronic *T. brucei* parasites is unlikely to result from immunodepression induced by parasite membrane.

Current experiments using a clone of *T. brucei* parasites which induces a serologically detectable antibody response in resistant (C57Bl/6), but not susceptible (C3H/He), strains of mice have shown that T cells and B cells of both strains are stimulated to mount similar DNA-synthetic responses. Furthermore, immunofluorescence analyses of spleen cell

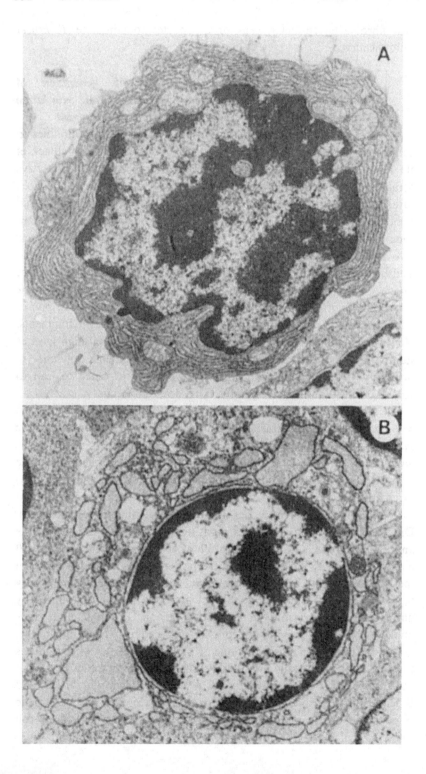

preparations from both strains of mice have shown that after infection with the parasites, the absolute number of cytoplasmic Ig-containing cells in the spleen increases considerably, reaching a peak of 15 to 20% of total spleen cells by 7 to 8 days after infection. However, these cytoplasmic Ig-containing cells differ considerably in intracellular morphology in the two mouse strains. Those present in infected resistant mice have extensive rough-surface endoplasmic reticulum (RSER) which in cross-section appears as a series of parallel tubes with narrow cisternae (Figure 1A). In contrast, cytoplasmic Ig-containing cells from infected susceptible mice have markedly distended RSER which in cross section appears as irregular blocks (Figure 1B). These studies suggest that the inefficient antibody response mounted by susceptible mice infected with *T. brucei* parasites results, not from a failure of B cells to mature to antibody-forming cells, but rather from a parasite-induced Ig-transport disorder. Plasma cells with distended RSER also arise in resistant strains of mice after the first parasitaemic wave. By 3 to 6 weeks after infection of C57Bl/6 mice, between 30 and 50% of total splenic plasma cells have distended RSER. The events which lead to the generation of cells with this phenotype might then occur in isolated splenic niches or might be clonotype specific.

The development of abnormal plasma cells in the spleens of susceptible and resistant mice correlates with a disruption in splenic organization. In susceptible C3H/He mice, the definition between red and white pulp regions becomes progressively reduced during the first parasitaemic wave. This reduced definition occurs at a much later time in resistant C57Bl/6 mice (M. Murray & N. McHugh personal communication). The development of abnormal plasma cells is also not restricted to infections initiated with *T. brucei* parasites. It has been shown that infection of

Figure 1. C57Bl/6 and C3H/He mice were infected intraperitoneally with 10^3 *T. brucei* GUTat 3.1a parasites. After 8 days, spleen cells were collected, pelleted and fixed in 2.5% gluteraldehyde in 0.1 M sodium cacodylate buffer (pH 7.4) then postfixed in 1% osmium tetroxide in the same buffer. *En bloc* staining was carried out in 1% uranyl acetate in 0.05 M sodium maleate buffer (pH 5.4) before dehydration in ethanol, embedding in Epon-Araldite and sectioning. Sections, mounted on formvar carbon-coated grids, were stained with uranyl acetate and lead citrate then photographed with a Zeiss EM 10A electron microscope. A is of a representative plasma cell in C57Bl/6 mice, while B is a representative plasma cell in C3H/He mice. In both cases the final magnification is x 12,000.

C3H/He mice with *T. vivax* organisms leads to extensive B-cell activation to undergo DNA synthetic responses, the development of large numbers of splenic B cells with cytoplasmic Ig and with distended RSER, and the absence of serologically detectable parasite-specific or bystander antigen-specific antibody responses (S. Mahan & S.J. Black unpublished). The development of abnormal B-cell responses in C3H/He mice infected with *T. vivax* parasites is also accompanied by rapid disorganization of the red and white pulp regions of the spleen (S. Mahan & S.J. Black unpublished). The block in Ig secretion by plasma cells of trypanosome-infected mice is therefore one component of a broad pathogenic syndrome. It is, however, a component which is amenable to further investigation.

Some conclusions

Research in mice has led to the hypothesis that parasitaemia in animals infected with *T. brucei* is likely to be regulated by two host responses: an immune system-independent process which controls the rate of parasite division, and a trypano-destructive antibody response specific for external epitopes on parasite-attached VSG. *T. brucei* parasite division rates appear to be positively regulated by the host, i.e. the parasites are induced to divide and remain as slender forms following exposure to host-derived macromolecular growth factors, and they differentiate to nondividing stumpy forms when the growth factors become limiting (Black et al. 1985). A similar mechanism influences multiplication of progenitor cells for erythrocytes and leucocytes in mammals (Dexter, Whetton & Brazill 1984). Identification of the growth factors involved in stimulation of *T. brucei* multiplication and the parasite receptors to which they bind might lead to the development of efficient immune or chemoprophylactic reagents which would limit parasite growth and induce a quick, effective immune response to eliminate the parasites.

The efficiency of host-protective antibody responses, the second process which limits *T. brucei* parasitaemia, appears to be under the control of an immunodepressive reaction which is maximally induced in the most susceptible host (Selkirk & Sacks 1980) and which inhibits antibody responses both against the parasite (Sacks & Askonas 1980; Black et al. 1983; Levine & Mansfield 1984) and against unrelated antigens (Mansfield 1978). Immunodepression in murine trypanosomiasis has been most extensively investigated in chronic infections of resistant mice and has been shown to be associated with macrophage (Grosskinsky & Askonas 1981) and suppressor T-cell (Eardley & Jayawardena 1977) activities.

Recent studies, described above, show that immunodepression, which appears very rapidly in susceptible mice and hence may be mediated by mechanisms not detected in resistant mice, is associated with a rapidly induced syndrome which results both in a disruption in splenic red and white pulp organization and in a failure of Ig secretion by plasma cells. The syndrome is reversed on elimination of trypanosomes, as indeed is immunodepression (Murray et al. 1974) and splenic disorganization (Roelants et al. 1979) detected in chronic infections in more resistant animals. It is tempting to speculate that living or dying trypanosomes produce or induce production of short-lived molecules which change the surfaces of lymphocytes, leading to abnormal migration patterns and altered secretory activity. As-yet-unidentified host polymorphisms influence the degree to which the host is susceptible to these events. Identification of the relevant parasite components and host polymorphisms may lead to methods for improving host responses against trypanosomes, e.g. vaccination against the trypanosome components involved in the disruption of lymphocyte function or selective breeding of efficient responders.

It is perhaps optimistic to consider that experiments conducted in inbred laboratory rodents will provide insights pertinent to the control of trypanosomiasis in humans and domestic animals. However, the phenomenon of trypanosome interference (Morrison et al. 1982) —which has been reported in cattle and in which infection with one trypanosome type can inhibit the growth of a second trypanosome type — suggests that regulation of parasite multiplication rates is pertinent to control of parasitaemia in large animals. Furthermore, it is significant that Russell body-containing plasma cells appear early after infection and are a striking feature of trypanosomiasis in man (Mott 1907) and in cattle (Murray et al. 1980). Russell bodies are distended sacs of endoplasmic reticulum which contain Ig (Avrameas & Leduc 1970) and which one could envisage to arise through a block in Ig secretion. Analyses of the mechanisms which control parasitaemia in mice, namely regulation of parasite multiplication and induction of efficient antibody responses, are therefore likely to lead to an understanding of responses which control parasitaemia in other species of host. It is hoped that understanding will, in turn, lead to elucidation of strategies for the biological control of African trypanosomes in mammals.

References

Ashcroft, M.T. (1957). The polymorphism of *Trypanosoma brucei* and *Trypanosoma rhodesiense*, its relationship to relapses and remissions of infections in white rats, and the effect of cortisone. *Annals of Tropical Medicine & Parasitology*, **51**, 301-12.

Avrameas, S. & Leduc, E.H. (1970). Detection of simultaneous antibody synthesis in plasma cells and specialized lymphocytes in rabbit lymph nodes. *Journal of Experimental Medicine*, **131**, 1137-68.

Balber, A.E. (1972). *Trypanosoma brucei*: fluxes of the morphological variants in intact and X-irradiated mice. *Experimental Parasitology*, **31**, 307-19.

Balber, A.E. (1983). Primary murine bonemarrow cultures support continuous growth of infectious human trypanosomes. *Science*, **220**, 421-23.

Black, S.J., Hewett, R.S. & Sendashonga, C.N. (1982). *Trypanosoma brucei* variable surface coat is released by degenerating parasites but not by actively dividing parasites. *Parasite Immunology*, **4**, 233-44.

Black, S.J., Jack, R.M. & Morrison, W.I. (1983). Host:parasite interactions which influence the virulence of *Trypanosoma (Trypanozoon) brucei brucei* organisms. *Acta Tropica*, **40**, 11-18.

Black, S.J., Sendashonga, C.N., Borowy, N.K., Webster, P. & Murray, M. (1985). Regulation of parasitaemia in mice infected with *Trypanosoma brucei*. *Current Topics in Microbiology & Immunology*, **117**, 93-118.

Black, S.J., Sendashonga, C.N., Lalor, P.A., Whitelaw, D.D., Jack, R.M., Morrison, W.I. & Murray, M. (1983). Regulation of the growth and differentiation of *Trypanosoma (Trypanozoon) brucei brucei* in resistant (C57Bl/6) and susceptible (C3H/He) mice. *Parasite Immunology*, **5**, 465-78.

Brun, R., Jenni, L., Schonenberger, M. & Schell, K.F. (1981). *In vitro* cultivation of bloodstream forms of *Trypanosoma brucei, T. rhodesiense* and *T. gambiense*. *Journal of Protozoology*, **28**, 470-79.

Campbell, G.H., Esser, K.M. & Weinbaum, F.I. (1977). *Trypanosoma rhodesiense* infection in B cell-deficient mice. *Infection & Immunity*, **18**, 434-38.

Capbern, A., Giroud, C., Baltz, T. & Mattern, P. (1977). *Trypanosoma equiperdum*: tude des variations antigniques au cours de la trypanosomose experimentale du lapin. *Experimental Parasitology*, **42**, 6-13.

Clayton, C.E., Sacks, D.L., Ogilvie, B.M. & Askonas, B.A. (1979). Membrane fractions of trypanosomes mimic the immunosuppressive and mitogenic effects of living parasites on the host. *Parasite Immunology*, **1**, 241-49.

Cross, G.A.M. (1975). Identification, purification and properties of clone-specific glycoprotein antigens constituting the surface coat of *Trypanosoma brucei*. *Parasitology*, **71**, 393-417.

Dexter, T.M., Whetton, A.D. & Brazill G.W. (1984). Haemopoietic cell growth factor and glucose transport: its role in cell survival and the relevance of this in normal haemopoiesis and leukaemia. *Differentiation*, **27**, 163-67.

Eardley, D.D. & Jayawardena, A.N. (1977). Suppressor cells in mice infected with *Trypanosoma brucei*. *Journal of Immunology*, **119**, 1029-33.

Flynn, I.W. & Bowman, I.B.R. (1973). The metabolism of carbohydrate by pleomorphic African trypanosomes. *Comparative Biochemistry & Physiology*, **45B**, 25-42.

Gray, A.R. & Luckins, A.G. (1976). Antigenic variation in Salivarian trypanosomes. In *Biology of the kinetoplastida* (eds. W.H.R. Lumsden & D.A. Evans), vol. 1, pp. 493-542. London: Academic Press.

Grosskinsky, C.M. & Askonas, B.A. (1981). Macrophages as primary target cells and mediators of immune dysfunction in African trypanosomiasis. *Infection & Immunity*, **33**, 149-55.

Hill, G.C., Shimer, S., Caughey, B. & Sauer, S. (1978). Growth of infective forms of *Trypanosoma (T) brucei* on buffalo lung and Chinese hamster lung tissue culture cells. *Acta Tropica*, **35**, 201-7.

Hirumi, H., Doyle, J.J. & Hirumi, K. (1977). African trypanosomes: cultivation of animal infective *Trypanosoma brucei in vitro*. *Science*, **196**, 992-94.

Hoare, C.A. (1970). Systematic description of the mammalian trypanosomes of Africa. In *The African trypanosomiases* (ed. H.W. Mulligan), pp. 24-29. London: George Allen & Unwin.

Hudson, K.M. & Terry, R.J. (1979). Immunodepression and the course of infection of a chronic *Trypanosoma brucei* infection in mice. *Parasite Immunology*, **1**, 317-26.

Jack, R.M., Black, S.J., Reed, S.L. & Davis, C.E. (1984). Indomethacin promotes differentiation of *Trypanosoma brucei*. *Infection & Immunity*, **43**, 445-48.

Le Ray, D., Barry, J.D. & Vickerman, K. (1978). Antigenic heterogeneity of metacyclic forms of *Trypanosoma brucei*. *Nature*, **273**, 300-2.

Levine, R.F. & Mansfield, J.M. (1984). Genetics of resistance to the African trypanosomes. 3. Variant-specific antibody responses of H-2-compatible resistant and susceptible mice. *Journal of Immunology*, **133**, 1564-69.

MacAskill, J.A., Holmes, P.H., Jennings, F.W. & Urquhart, G.M. (1981). Immunological clearance of ⁷⁵Se-labeled *Trypanosoma brucei* in mice. 3. Studies in animals with acute infections. *Immunology*, **43**, 691-98.

Mansfield, J.M. (1978). Immunobiology of African trypanosomiasis. *Cellular Immunology*, **39**, 204-10.

Morrison, W.I., Wells, P.W., Moloo, S.K., Paris, J. & Murray, M. (1982). Interference in the establishment of superinfections with *Trypanosoma congolense* in cattle. *Journal of Parasitology*, **68**, 755-64.

Mott, F.W. (1907). Histological observations on the changes in the nervous system in trypanosome infections, especially sleeping sickness, and their relation to syphilic lesions of the nervous system. *Archives of Neurology*, **3**, 581-646.

Murray, M. & Urquhart, G.M. (1977). Immunoprophylaxis against African trypanosomiasis. In *Immunity to blood parasites of animals and man* (eds. L.H. Miller, J.A. Pino & J.J. McKelvey Jr.), pp. 209-41. New York: Plenum.

Murray, M., Morrison, W.I. & Whitelaw, D.D. (1982). Host susceptibility to African trypanosomiasis: trypanotolerance. *Advances in Parasitology*, **21**, 1-68.

Murray, M., Morrison, W.I., Emery, D.L., Akol, G.W.O., Masake, R.A. & Moloo, S.K. (1980). Pathogenesis of trypanosome infections in cattle. In *Isotope and radiation research on animal diseases and their vectors*, pp. 15-32. Vienna: International Atomic Energy Agency.

Murray, P.K., Jennings, F.W., Murray, M. & Urquhart, G.M. (1974). The nature of immunodepression in *Trypanosoma brucei* infections in mice. 2. The role of T and B lymphocytes. *Immunology*, **27**, 825-40.

Nielsen, K., Sheppard, J., Holmes, W. & Tizard, I. (1978). Experimental bovine trypanosomiasis: changes in the catabolism of serum immunoglobulins and complement components in infected cattle. *Immunology*, **35**, 811-16.

Roelants, G.E., Pearson, T.W., Morrison, W.I., Mayor-Withey, K.S. & Lundin L.B. (1979). Immune depression in trypanosome-infected mice. 4. Kinetics of suppression and alleviation by the trypanocidal drug Berenil. *Clinical & Experimental Immunology*, **37**, 457-69.

Sacks, D.L. & Askonas, B.A. (1980). Trypanosome-induced suppression of anti-parasite responses during experimental African trypanosomiasis. *European Journal of Immunology*, **10**, 971-74.

Selkirk, M.E. & Sacks, D.L. (1980). Trypanotolerance in inbred mice: an immunological basis for variation in susceptibility to infection with *Trypanosoma brucei*. *Tropenmedizin und Parasitologie*, **31**, 435-38.

Sendashonga, C.N. (1983). Induction, expression and regulation of antibody responses against *Trypanosoma brucei brucei* in mice. Ph.D. thesis. Universit Libre de Bruxelles.

Sendashonga, C.N. & Black, S.J. (1982). Humoral immune responses against *Trypanosoma brucei* variable surface antigens are induced by degenerating parasites. *Parasite Immunology*, **4**, 245-57.

Tait, A. (1980). Evidence for diploidy and mating in trypanosomes. *Nature*, **287**, 536-38.

Urquhart, G.M. (1980). The pathogenesis and immunology of African trypanosomiasis in domestic animals. *Transactions of the Royal Society for Tropical Medicine & Hygiene*, **74**, 726-29.

Vickerman, K. (1965). Polymorphism and mitochondrial activity in sleeping sickness trypanosomes. *Nature*, **108**, 762-66.

Vickerman, K. (1971). Morphological and physiological considerations of extracellular blood protozoa. In *Ecology and physiology of parasites* (ed. A.M. Fallis), pp. 51-58. Toronto: University Press.

Wijers, D.J.B. & Willett, K.C. (1960). Factors that may influence the infection rate of *Glossina palpalis* with *Trypanosoma gambiense*. 2. The number and morphology of the trypanosomes present in the blood at the time of the infected feed. *Annals of Tropical Medicine & Parasitology*, **54**, 31-36.

28

Studies on the protection of cattle against *Babesia bovis* infection

D.F. MAHONEY

The presence of protective antibodies to *Babesia bovis* was demonstrated by the passive transfer of immunity with serum from carriers of the organism to highly susceptible splenectomized calves. The effector mechanism was mediated by antibodies which reacted with the parasitized erythrocytes and emerging parasites. It was concluded that opsonization was probably the basis of protection by the system. Variation of antigen(s) occurred within strains of the parasite but this had little effect on the efficiency of the host's immune response. However, there was no cross-protection between the antibodies against different strains. These interstrain differences in antibody specificity were reconciled with other observations that cross-immunity commonly occurs between different strains in infected animals. It was concluded that the mechanism of cross-immunity relied on priming of the host's immune system by the protective antigen(s) of one strain so that a secondary response against the heterologous strain occurred soon after challenge. These conclusions led to experiments on the induction of immunity in cattle with antigen extract from *B. bovis*-infected erythrocytes. Protection was as strong using crude antigen preparations as when induced by natural infection. Protective antigens have been produced in varying degrees of purity with the ultimate goal of synthesizing protective antigen by recombinant DNA technology.

Introduction

Babesia bovis is a protozoan parasite that infects the red blood cells of cattle and is transmitted by the one-host tick, *Boophilus microplus*. The parasite is inoculated with saliva as the larval tick feeds, and immediately invades the red cells of the host. It grows from a single form, divides by budding into pear-shaped organisms which eventually separate, leave the cell and invade new ones. The vertebrate cycle is asexual and repetitive. Infected red cells ingested by the engorging adult female tick initiate the invertebrate life cycle. This is characterized by the production of a large motile form that multiplies by schizogony in the cells of the tick's gut. It enters the egg and then infects the gut cells of the developing larva. After the larva hatches and starts feeding on a new host, the parasite invades the cells of the salivary gland and undergoes further schizogony to produce the forms infective for cattle.

Antibodies against *Babesia bovis* were first demonstrated serologically in cattle by Mahoney (1962; 1964). This was foreshadowed by Hall (1960; 1963), who found that the calves of infected mothers were more resistant to *B. bovis* infection than the calves of noninfected mothers for at least 2 months after birth. Presumably, resistance was transferred to the calves by antibodies in colostrum. Protection by the serum of infected animals was later demonstrated (Mahoney 1967a). However, there was no evidence that serologically detectable antigen-antibody reactions were involved in protection, and it was concluded that the 'protective' antigen(s) of *B. bovis* were either weakly antigenic for the host or, in quantitative terms, minor components of the parasite's antigenic profile. These results generated new interest in the mechanisms of immunity to *B. bovis* and caused speculation on the feasibility of developing a dead vaccine against the disease.

The role of antibodies in protection against *B. bovis*
Experimental methods
A passive transfer system has been developed to analyse the effect of antibodies on the parasite *in vivo*. Splenectomized calves were used as test animals because parasitaemia developed quantitatively after infection by intravenous inoculation and provided the basis for a reproducible test system. The experiments consisted first of collecting and storing bovine serum from infected and uninfected donors. Then uninfected calves weighing approximately 100 to 150 kg were splenectomized and infected with virulent *B. bovis* parasites. After 4 days, parasites were detectable in thick films of jugular blood, and the calves were then given serum by intravenous infusion. The test calves received serum from *B. bovis*-infected donors and the controls received serum from uninfected donors. The doses of sera varied from 2 to 45 ml/kg body weight. The number of parasites per µl of jugular blood was recorded daily for 10 to 12 days. Differences in parasitaemia were assessed between calves that received *B. bovis* antiserum and controls. Some experiments were varied by (1) counting parasitaemia every half hour after administering serum for kinetic analysis of parasite removal, and (2) using purified γ-globulin extracted from the antiserum to show that the protective activity resided in the antibody fraction.

An *in vitro* test system was also designed. It was first established that a minimum dose of 1 to 10 infected erythrocytes is required to establish infection by intravenous inoculation in splenectomized calves. Doses of 10, 100 and 1000 infected erythrocytes were incubated at 37°C for 1 h in 10 ml of antiserum and inoculated into calves for tests of infectivity, together with appropriate controls. In some experiments fresh bovine complement was added to the antiserum, and in others the bovine erythrocytes in the inoculum were lysed with guinea pig complement and sheep antiserum raised against normal bovine erythrocytes, before exposure to the *B. bovis* antiserum.

The effect of antiserum on parasitaemia

When the infected calves were treated with antiserum, the rate and degree to which parasitaemia was affected depended on the dose and type of antiserum administered. Serum from donors that had been infected several times at intervals of 3 to 6 months (i.e. hyperimmune serum) was highly protective in low doses (e.g. 2 ml/kg live weight), as

Figure 1. (a) Parasitaemia and survival of *B. bovis*-infected splenectomized calves after treatment with 2.2, 4.4, 8.8 and 17.6 ml of hyperimmune antiserum per kg bodyweight. Each plot is the mean of two animals. Figures in brackets represent proportion died. Taken from Mahoney et al. (1979). (b) Parasitaemia and survival of *B. bovis*-infected splenectomized calves after treatment with hyperimmune serum and serum taken from donors 2 weeks after recovery from initial infection (convalescent). Each plot is the mean of two animals. Figures in brackets represent proportion survived.

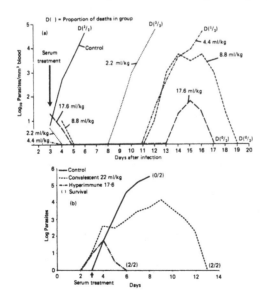

shown in Figure 1a. The parasites began to disappear from the blood within 1 h after treatment with hyperimmune serum and fell below the detectable level within 24 to 48 h. Antiserum taken soon after recovery from initial infection (convalescent serum) was less effective, and higher doses (20 ml/kg) were required to demonstrate protection (Figure 1b).

Mixtures of IgG_1 and IgG_2 prepared from hyperimmune serum were also effective. A mathematic model, based on the assumption that antibodies attack infected erythrocytes and emerging parasites, gave the best fit of kinetic data on the removal of parasites from the blood after serum administration (Mahoney et al. 1979). From this it was deduced that the target antigens were located on the surface of infected erythrocytes and on the parasites. However, the antibody clearly required the environment of the living host to function: exposure of parasitized erythrocytes to antibody *in vitro* had no effect on infectivity for splenectomized calves (Mahoney 1972). The antibody probably acts as an opsonin, the organisms being killed by phagocytes. Although no definitive work has been performed on the phagocytosis of *B. bovis*, Rogers (1974) correlated opsonic activity of antiserum to *B. rodhaini* with its ability to protect rats in passive transfer tests. The suitability of splenectomized calves for passive transfer tests was not regarded as contrary to this hypothesis: because of its size, the liver is a more important phagocytic organ than the spleen (Taliaferro 1956), and splenectomized animals were thus expected to retain a significant part of their phagocytic capability. In addition, both splenectomized and intact animals were used during preliminary work on the development of the passive transfer system (Mahoney 1967a). No qualitative differences were observed in the reactions in the two types of animal, while the splenectomized groups provided a test of higher sensitivity.

Relapse of parasitaemia was a feature of the passive transfer experiments when hyperimmune serum was given in low dosage (2 to 4 ml/kg). Higher doses (10 to 20 ml/kg) caused a progressive diminution in the height and severity of relapses, while relapses did not occur in most calves that received over 15 ml/kg (Figure 1a). The relapses were simply the result of the quantitative relationship between factors, such as the number of parasites in the blood at the time of serum administration, the initial concentration of antibody and its rate of wastage on antigen-antibody reactions and normal catabolism. Relapses were not caused by antigenic variation of the parasite population induced by the action of protective antibody because the relapse parasites were susceptible to the same serum on re-exposure (Mahoney et al. 1979).

*The relationships between antigenic and strain variation and the
activity of protective antibodies*

A population of parasites isolated at one time from the field was
designated as a strain and named after the locality from which it came.
Each isolate (strain) was first passaged in a noninfected, splenectomized
calf and then maintained as a separate population in the laboratory by
cryopreservation in liquid nitrogen. Three strains — Lismore (L),
Samford (S) and Helidon (H) — were used to infect splenectomized calves,
and hyperimmune serum against two, Gayndah (G) and L, was used for
treatment of the infected animals. All sera were tested for activity against
the homologous parasites before use in the experiments.

In two experiments, 17.6 ml/kg of antiserum was administered. In the
first, four animals were infected with the L strain and treated with G
strain antiserum. In the second, two groups of two animals were infected
with H and S strains respectively, and treated with L strain antiserum.
In one additional experiment, two animals infected with L strain were
given G strain antiserum at a dose of 44 ml/kg. In all cross-protection
experiments, the antiserum treatment had no effect on parasites of
heterologous strains.

Another type of variation in the antigenic character of *B. bovis* was
described by Curnow (1968; 1973). This occurred within strains during
the period of chronic infection in individual animals. Mahoney (1962)
showed that chronic infection with *B. bovis* lasts for several years and is
characterized by subclinical relapses of detectable parasitaemia in
peripheral blood at irregular intervals. Curnow showed that each relapse
is associated with a different agglutinogen on the surface of the infected
red blood cells. He hypothesized that the variable agglutinogens are
targets for protective antibodies and that the variation of specificity is
the parasite's strategy to avoid the host's immune response and maintain
chronic infection.

The relationship between the protective strain-specific immunity
observed in passive transfer experiments and the variant-specific
serological activity during chronic infection had to be determined because
of implications for the development of a killed vaccine. If the two
phenomena were associated, the likelihood of developing a killed vaccine
of relatively simple antigenic composition seemed remote. However,
experiments suggested that the two immunological events are
manifestations of different antigen-antibody reactions (Mahoney et al.
1979). In passive transfer tests, there is no difference in the specificity of
protective antibodies taken from a chronically infected animal at different
times in terms of protection against relapse substrains from the same

animal. Thus protective antibodies generated by one variant population from within a strain are effective against all other variant populations produced by that strain.

The use of antigen to induce protection

Experimental methods

The first requirement for studies on *B. bovis* antigens is to obtain a concentrated source of the parasite. This requirement was met by the separation of infected from uninfected cells in blood (Mahoney 1967b). The technique is based on the observation that infected erythrocytes are less susceptible to hypotonic lysis than uninfected cells. Thus it is possible to select a concentration of salt solution that lyses all uninfected cells and leaves the infected ones intact to be recovered by differential centrifugation. The method is rapid, reproducible and applicable on a preparative scale. It was used to produce suspensions of 95 to 100% infected erythrocytes from blood with parasitaemia in the range of 5 to 15%. Extracts of crude antigen were prepared from these infected cell suspensions by sonic disintegration for 2 to 4 min, followed by centrifugation at 145,000 g for 60 min at 4°C. The supernatant fluid, containing crude soluble antigen, was fractionated by precipitation with protamine sulphate, by gel filtration and by affinity chromatography using antibodies from immune cattle and monoclonal antibodies produced by the techniques established by Kohler and Milstein (1975). A flow diagram for fractionation procedures is shown in Figure 2.

The immunogenic activity of antigens was tested by subcutaneous inoculation of the antigen as a water-in-oil emulsion with Freund's complete adjuvant into 2-year-old steers. In early work, three inoculations were given at 2-week intervals, but in later studies the number of inoculations was reduced to two and to one. Two to 4 weeks after immunization, the cattle were challenged by intravenous inoculation with a strain of *B. bovis* different from the one used to prepare the antigen. The immune response was then assessed by comparing daily rectal temperatures, levels of parasitaemia and decreases in packed cell volume in vaccinated and control groups. Controls were sham-immunized with the adjuvant alone.

Protection by crude antigen

Mahoney and Wright (1976) showed that immunization with crude antigen, prepared from infected erythrocytes, protects cattle against heterologous *B. bovis* infection as effectively as immunity associated with active infection (Figure 3a). The crude antigen was a mixture of all particulate and soluble components contained in infected erythrocytes. It was fractionated into soluble and insoluble components by ultracentrifugation and both fractions induced similar protection (Figure 3b), demonstrating that protective antigen is released into solution.

Fractionation of the crude soluble antigen

The fractionation procedure was based on studies by Goodger (1971; 1973; 1976) who showed that the antigens extracted from *B. bovis*-infected erythrocytes contain several different specificities which he classified into three groups. The first group is composed of two antigens associated with the stroma of the erythrocyte; these are responsible for staining the erythrocyte membrane by fluorescein-labeled antibodies

Figure 2. Flow diagram of the fractionation procedure for bovine erythrocytes infected with *B. bovis*.

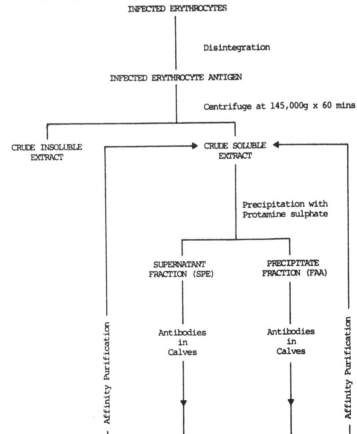

546   D. F. Mahoney

(Ludford 1967). One of these antigenic components is located as a dense band in or under the erythrocyte membrane, and the other has a granular distribution throughout the stroma. Characterization of the stromal antigens showed that they are altered fibrinogen molecules conjugated with a number of peptides, some of which are of babesial origin and others of host origin (Goodger et al. 1980). Separation of the fibrinogen-associated antigen complexes was achieved by methods that specifically precipitated fibrinogen (Goodger 1976). The second group of antigens is located on or in the parasite. They differ in specificity from those in the erythrocyte stroma and membrane, but little is known of their physical

Figure 3. (a) Comparison of the changes in packed cell volume (PCV) in nonsplenectomized, adult cattle vaccinated with crude *B. bovis* antigen and in similar cattle immunized by *B. bovis* infection, after challenge with virulent heterologous *B. bovis* organisms. Taken from Mahoney and Wright (1976). (b) The changes in PCV in nonsplenectomized, adult cattle vaccinated with soluble and insoluble fractions of crude *B. bovis* antigen and challenged with a heterologous strain of the organism.

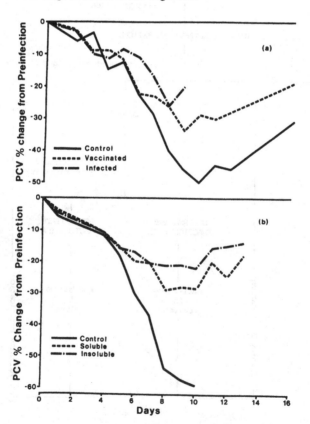

and chemical properties. The third antigenic component is found in the cytoplasm of the infected erythrocyte and was extracted from the haemoglobin solution obtained after lysis of the erythrocytes in distilled water.

The fibrinogen-associated antigens were precipitated from the crude soluble extract with protamine sulphate; this step effectively separated those antigens that were associated with the erythrocyte stroma from those located on the parasite and in the erythrocyte cytoplasm. The response in calves inoculated with each fraction seemed to confirm this broad separation: antibodies from calves immunized with fibrinogen-associated antigen stained the stroma of infected erythrocytes in the indirect fluorescent antibody (IFA) tests, while antibodies from calves immunized with the antigen(s) left in solution after the removal of the precipitate stained only the parasites (Mahoney, Wright & Goodger 1981).

Both precipitate and supernatant fractions protected cattle against challenge. One interpretation of this result is that an antigenic component, common to both fractions but not distinguishable by the IFA test, is responsible for protection. However, it seems equally plausible that there could be a number of antigens involved in protection against *B. bovis* which might be located on the membranes of infected erythrocytes and in the parasites.

The immunoglobulins from the serum of calves inoculated with the precipitate or the supernatant were purified and coupled to cyanogen bromide-activated sepharose 4B columns. These columns were then used to extract antigen from the crude soluble material (Figure 2). The concentration of protein in the extracts obtained from the affinity columns was approximately 200 µg/ml and contained, in addition to *B. bovis*-specific material, components from normal bovine erythrocytes which had nonspecifically absorbed to the sepharose columns. The fractions obtained using antisera to precipitate or supernatant antigens each contained three antigenic components from *B. bovis*. No reactions of identity or even partial identity were detected between the two sets of antigens in immunodiffusion tests. Cattle were immunized with each fraction by a single injection. Dose levels of 100 and 500 µg of protein were used. With the antigens from the supernatant fraction, the dose of 100 µg gave protection (Figure 4a), but the higher dose was less protective. The reason for this unexpected result is not clear, but at the higher dose level, nonprotective antigen in the fraction may have induced immunosuppression or sensitized the animals, causing shock after challenge. With the antigens isolated from crude soluble material by

548   *D. F. Mahoney*

antibodies to the precipitate, the higher dose gave more effective protection than the lower dose (Figure 4b).

The use of monoclonal antibodies

Antigens obtained from the crude soluble extract by affinity chromatography, utilizing antibodies from calves previously inoculated with the supernatant fraction, were used to immunize BALB/c mice. The spleen cells of the mice were then fused with P3-NSI-Ag4-1 mouse myeloma cells. After screening the supernatants of the hybrid-cell lines for antibody to babesial antigens by radioimmunoassay and IFA tests, three clones were obtained which produced antibodies that stained *B. bovis* differently in IFA tests. Each clone was used to produce ascitic fluid in BALB/c mice and γ-globulin was extracted from the fluid and used to isolate antigen from crude soluble extract by affinity chromatography. The specificity of each antibody is shown in Table 1 as determined by IFA analysis.

Electrophoresis of this material on polyacrylamide gel showed that each eluate contained a number of proteins. However, one antigen was detectable in each by Western transfer analysis using bovine antiserum

Figure 4. Parasitaemia and survival in nonsplenectomized, adult cattle vaccinated with different doses (100 or 500 mg) of protein isolated from crude soluble material by affinity chromatography using bovine antibodies to (a) the supernatant fraction of *B. bovis* (SPE) and (b) to the fibrinogen-like precipitate (FAA) and challenged with a virulent heterologous strain of the organism.

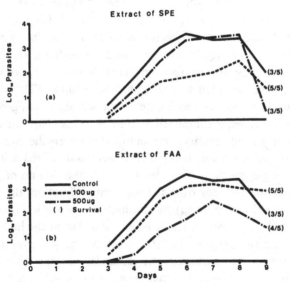

and a peroxidase-labeled anti-bovine γ-globulin. Each eluate was then used to immunize groups of four splenectomized calves. Each calf was given two doses of 100 μg protein in Freund's complete adjuvant at 4-week intervals and then challenged with homologous parasite strains 2 weeks after the last inoculation. Only the calves immunized with the proteins isolated by the 15B1 antibody showed evidence of protection. The control calves died within 10 days of inoculation, whereas three of the four calves which received antigen isolated by the 15B1 antibody survived for 14 days. Levels of parasitaemia in the 15B1 group were also significantly lower than those in the controls from day 5 onwards.

Discussion

There is no doubt that antibodies provide an important effector mechanism in protection against secondary infection with *B. bovis*. Protective activity, albeit of low efficiency, was demonstrated in convalescent antiserum, but the effectiveness of antiserum increased with the duration of chronic infection and/or after further antigenic challenge with the same strain of parasite. This is readily explained on the basis of increasing protective antibody concentration and/or avidity resulting from continued exposure of the host to parasite antigens. However, the mechanism that protects the infected host against challenge with a different strain of the organism is more difficult to explain. It cannot be the circulating antibody produced against the primary infection because of the highly strain-specific nature of the protection conferred by passive transfer of immune serum. Nevertheless, field evidence suggests that, regardless of the multiplicity of different immunological strains in the environment, few breakdowns of immunity occur. This was supported by experimental evidence in which inoculation of cattle with antigens extracted from one strain protected against infection with another.

The strain-specific nature of protection by antibody suggests that the determinants of the protective antigens are different for each parasite strain. Experiments on the immune response after heterologous challenge showed that animals immunized either by antigen or by active infection react in a manner similar to the susceptible controls for 4 to 6 days after infection, and then the protective reaction occurs. Recovery is probably triggered by the production of protective antibody specific for the antigen of the challenge strain. To accomplish this in time to prevent disease, the host must have been primed for a secondary reaction against the apparently unrelated antigenic determinant(s) of the new strain. This suggests that there is a relationship between the protective antigens of different *B. bovis* strains, probably involving the part of the molecules not directly concerned in the reaction with protective antibody, i.e. the carrier moiety. The role of the carrier in inducing T-helper cells and thus priming the host for a secondary antibody response to hapten-carrier conjugates is well recognized (Mitchison 1971a; 1971b). Furthermore, Terres, Habicht and Stoner (1974), in studies on the response of mice to antigens with related carriers but different determinant groups, found that priming with antigen-antibody complexes of one antigen enhanced the antibody response to the unrelated determinant of another. The strain-specific *B. bovis* antigens might behave similarly in giving cross-protection. This question of the nature of the protective antigens can only be answered by their isolation from cloned strains of *B. bovis* and comparison of amino-acid sequences.

The feasibility of vaccination against *B. bovis* with crude, killed antigen derived from infected erythrocytes has been established. Antigenic variation of the parasite does not circumvent the host's protective response, and the nature of the antigenic differences between strains, although poorly understood at the molecular level, obviously does not prevent cross-protection in the field. The question remains: can molecules be isolated that will give the same degree of protection as the crude material?

In the last 5 years, *in vitro* culture systems for *B. bovis* have been developed, using either bovine erythrocytes (Erp et al. 1978; Levy & Ristic 1980) or cultured embryonic cells of the tick vector *B. microplus* (Bhat, Mahoney & Wright 1979). Antigen prepared from the asexual blood stages grown in tick cells does not confer immunity on susceptible cattle (D.F. Mahoney & U.K.M. Bhat unpublished), but a degree of protection is obtained with antigen contained in the supernatant fluids of erythrocyte-based cultures (Smith, et al. 1979; 1981; Kuttler et al. 1982; Timms et al. 1983). However, in further experiments, protection provided by establishment of infection in the animal was markedly superior to the protection provided by culture supernatant fluid. This was clearly demonstrated by Timms and colleagues (1983) who used heterologous challenge to assess the degree of immunity. In contrast, Mahoney and Wright (1976) found that antigen prepared from concentrated, naturally infected erythrocytes gave protection against heterologous challenge comparable to that conferred by infection. However, it is difficult to compare the results of artificial immunization with those obtained through exposure to active infection because of quantitive differences in antigen dosage, although these results suggest that more attention could be given to the biochemical analysis of the antigens contained in cultured merozoites.

A successful vaccine derived from *B. bovis*-infected erythrocytes should have the following characteristics: (1) prevention of clinical disease; (2) efficacy against all immunological strains of the parasite; (3) protection induced by one or two inoculations; (4) protection lasting a minimum of 6 months; (5) no concurrent immunization against host blood-group antigens; (6) availability in large quantities; (7) stability on storage. The first three criteria have been fulfilled. Protection after immunization with culture-supernatant antigen has been shown to last at least 6 months (Kuttler, Levy & Ristic 1983). Callow and colleagues (1974) found that immunity to reinfection remained for a similar period after the elimination of infection by chemotherapy.

The molecular nature of the protective antigens, their mass production and their efficacy after purification are the important unknown factors in relation to vaccine development. The growth of parasites in calves would not provide enough antigenic material for even a small vaccination program. Although culture systems for *B. bovis* in bovine blood have been developed, the difficulties of producing large quantities of antigen and of contamination of antigen preparations with erythrocyte antigens remain (Timms et al. 1983). The alternative is to isolate and characterize the protective antigen(s) from the parasite using monoclonal antibodies, and from this step proceed to the production of the antigen(s) by recombinant DNA methods. This approach should solve both erythrocytic contamination and production problems and should provide enough material for the exhaustive testing that will be required before a killed vaccine for *B. bovis* can be made available for widespread use.

The purification of a protective antigen seems imminent and the knowledge accumulated so far needs to be reassessed. For example, it is not yet known whether a single purified antigen will confer protection under all circumstances or whether several antigens are needed to give artificially induced immunity, meeting the essential criteria outlined above. The optimum conditions for the administration of a purified antigen may be different from those established for the crude material. Further complications are the possible interactions if purified antigens must be combined in a polyvalent vaccine.

In the work described here, antigen material was partially purified by absorption to antibodies from cows immunized with crude soluble and precipitate fractions. Two preparations were produced, each containing three babesial antigens that were all different on analysis by immunodiffusion tests. Both preparations contained impurities of host origin, but nevertheless small doses protected cattle against disease.

These results suggest that *B. bovis* contains more than one antigen

Table 1. Specificity of monoclonal antibodies raised against *B. bovis*.

Hybridoma clone No.	Specificity of monoclonal antibodies assessed by immunofluorescence	Strain specificity of of antigen
2C3	parasites + erythrocyte stroma	multiple
15B1	parasites only	multiple
18A5	parasites only with preference for blunt end	multiple

capable of conferring protection. The three antigens detected by immunodiffusion in crude soluble fractions might be contained in the fractions identified by monoclonal antibodies 2C3, 15B1 and 18A5 (Table 1), but only the fraction identified by 15B1 gave a degree of protection. This fraction may, therefore, contain a protective antigen that has a role in immunization. The protective antigen(s) in the precipitate fraction should also be identified, and it needs to be determined whether this antigen(s) can, on its own, provide an effective protection system. If the protective activity of the antigenic fraction defined by 15B1 is confirmed by further tests, the fraction should be rigorously purified and the protective antigen cloned. It will then be available in sufficient quantity for more intensive study, including work on the molecular basis of its antigenicity.

References
Bhat, U.K.M., Mahoney, D.F. & Wright, I.G. (1979). The invasion and growth of *Babesia bovis* in tick tissue culture. *Experientia*, **35**, 752-53.
Callow, L.L., McGregor, W., Parker, R.J. & Dalgliesh, R.J. (1974). The immunity of cattle to *Babesia argentina* after drug sterilization of infections of varying duration. *Australian Veterinary Journal*, **50**, 6-11.
Curnow, J.A. (1968). *In vitro* agglutination of bovine erythrocytes infected with *Babesia argentina*. *Nature*, **217**, 267-68.
Curnow, J.A.(1973). Studies on antigenic changes and strain differences in *Babesia argentina* infections. *Australian Veterinary Journal*, **49**, 279-83.
Erp, E.E., Gravely, S.M., Smith, R.D., Ristic, M., Osorno, B.M. & Carson, C.A. (1978). Growth of *Babesia bovis* in bovine erythrocyte cultures. *American Journal of Tropical Medicine & Hygiene*, **27**, 1061-64.
Goodger, B.V. (1971). Preparation and preliminary assessment of purified antigens in the passive haemagglutination test for bovine babesiosis. *Australian Veterinary Journal*, **47**, 251-56.
Goodger, B.V. (1973). *Babesia argentina*: intraerythrocytic location of babesial antigen extracted from parasite suspensions. *International Journal for Parasitology*, **3**, 387-91.
Goodger, B.V. (1976). *Babesia argentina*: studies on the nature of an antigen associated with infection. *International Journal for Parasitology*, **6**, 213-16.
Goodger, B.V., Wright, I.G. & Mahoney, D.F. (1981). The use of

pathophysiological reactions to assess the efficacy of the immune response
to *Babesia bovis* in cattle. *Zeitschrift für Parasitenkunde,* **66,** 41-48.

Goodger, B.V., Wright, I.G., Mahoney, D.F. & McKenna, R.V. (1980).
Babesia bovis (argentina): studies on the composition and location of
antigen associated with infected erythrocytes. *International Journal for
Parasitology,* **10,** 33-36.

Hall, W.T.K. (1960). The immunity of calves to *Babesia argentina* infection.
Australian Veterinary Journal, **36,** 361-66.

Hall, W.T.K. (1963). The immunity of calves to tick-transmitted *Babesia
argentina* infection. *Australian Veterinary Journal,* **39,** 386-89.

Kohler, G. & Milstein, C. (1975). Continuous cultures of fused cells secreting
antibody of predefined specificity. *Nature,* **256,** 495-97.

Kuttler, K.L., Levy, M.G. & Ristic, M. (1983). Cell-culture derived *Babesia
bovis* vaccine: sequential challenge exposure of protective immunity during a
6-month postvaccination period. *American Journal of Veterinary Research,*
44, 1456-59.

Kuttler, K.L., Levy, M.G., James, M.A. & Ristic, M. (1982). Efficacy of a
nonviable culture-derived *Babesia bovis* vaccine. *American Journal of
Veterinary Research,* **43,** 281-84.

Levy, M.G. & Ristic, M.(1980). *Babesia bovis:* continuous cultivation in a
microaerophilous stationary phase culture. *Science,* **207,** 1218-20.

Ludford, C.G. (1967). Studies on *Babesia rodhaini:* its morphology, course of
infection and immunity in rats, with observations on *Babesia* affecting
cattle. Ph.D. thesis. University of Queensland.

Mahoney, D.F. (1962). Bovine babesiosis: diagnosis of infection by a
complement-fixation test. *Australian Veterinary Journal,* **38,** 48-52.

Mahoney, D.F. (1964). Bovine babesiosis: an assessment of the significance of
complement-fixing antibody based upon experimental infection. *Australian
Veterinary Journal,* **40,** 369-75.

Mahoney, D.F. (1967a). Bovine babesiosis: the passive immunization of calves
against *Babesia argentina* with special reference to the role of complement-
fixing antibodies. *Experimental Parasitology,* **20,** 119-24.

Mahoney, D.F (1967b). Bovine babesiosis: preparation and assessment of
complement-fixing antigens. *Experimental Parasitology,* **20,** 232-41.

Mahoney, D.F. (1972). Immune response of haemoprotozoa. 2. *Babesia* spp.
In *Immunity to animal parasites* (ed. E.J.L. Soulsby), pp. 301-41. New
York: Academic Press.

Mahoney, D.F. & Wright, I.G. (1976). *Babesia argentina:* immunization of
cattle with a killed antigen against infection with a heterologous strain.
Veterinary Parasitology, **2,** 273-82.

Mahoney, D.F., Wright, I.G. & Goodger, B.V. (1981). Bovine babesiosis: the
immunization of cattle with fractions of erythrocytes infected with *Babesia
bovis* (syn. *B. argentina*). *Veterinary Immunology & Immunopathology,* **2,**
145-56.

Mahoney, D.F., Kerr, J.D., Goodger, B.V. & Wright, I.G. (1979). The
immune response of cattle to *Babesia bovis* (syn. *B. argentina*): studies on
the nature and specificity of protection. *International Journal for
Parasitology,* **9,** 297-306.

Mitchison, N.A. (1971a). The carrier effect in the secondary response to
hapten-protein conjugates. 1. Measurement of the effect with transferred
cells and objections to the local environment hypothesis. *European Journal
of Immunology,* **1,** 10-17.

Mitchison, N.A. (1971b). The carrier effect in the secondary response to
hapten-protein conjugates. 2. Cellular cooperation. *European Journal of
Immunology,* **1,** 18-27.

Rogers, R.J. (1974). Serum opsonins and the passive transfer of protection in *Babesia rodhaini* infections of rats. *International Journal for Parasitology*, **4**, 197-201.

Smith, R.D., James, M.A., Ristic, M., Aikawa, M. & Vega y Murguia, C.A. (1981). Bovine babesiosis: protection of cattle with culture-derived soluble *Babesia bovis* antigen. *Science*, **212**, 335-38.

Smith, R.D., Carpenter, J., Cabrera, A., Gravely, S.M., Erp, E.E., Osorno, M. & Ristic, M. (1979). Bovine babesiosis: vaccination against tick-borne challenge exposure with culture-derived *Babesia bovis* immunogens. *American Journal of Veterinary Research*, **40**, 1678-82.

Taliaferro, W.H. (1956). Functions of the spleen in immunity. *American Journal of Tropical Medicine & Hygiene*, **5**, 391-410.

Terres, G., Habicht, G.S. & Stoner, R.D. (1974). Carrier-specific enhancement of the immune response using antigen-antibody complexes. *Journal of Immunology*, **112**, 804-11.

Timms, P., Dalgliesh, R.J., Barry, D.N., Dimmock, C.K. & Rodwell, B.J. (1983). *Babesia bovis*: comparison of culture-derived parasites, nonliving antigen and conventional vaccine in the protection of cattle against heterologous challenge. *Australian Veterinary Journal*, **60**, 75-77.

29

Protective immune responses in bovine theileriosis

W.I. MORRISON, D.L. EMERY, A.J. TEALE and
B.M. GODDEERIS

The protozoan parasite *Theileria parva* is the causal agent of an acute, usually fatal, disease of cattle characterized by widespread parasitism and destruction of cells of the lymphoid system. Animals which recover from infection are immune to challenge with the homologous stock of the parasite. Furthermore, immunity can be induced by various infection and treatment regimes. There are two potential levels at which protective immunity may operate, namely the sporozoite and the macroschizont-infected cell. It has been shown that serum from immune cattle and anti-sporozoite monoclonal antibodies can neutralize sporozoite infectivity. However, it is still not clear whether such antibodies are sufficiently efficient *in vivo* to prevent infection. Methods of immunization currently in use appear to depend on the establishment of active infection and development of the parasite to the macroschizont stage in recipient cattle. However, three out of four cattle have been successfully immunized with a cell membrane fraction prepared from autologous *Theileria*-infected cell lines. During immunization by infection and treatment or following challenge of immune cattle, cell-mediated cytotoxic responses are generated against macroschizont-infected cells. The activity of these cells is restricted by class I major histocompatibility (MHC) antigens, and their appearance shows a close correlation with the development of immunity. These findings suggest that the parasite induces antigenic changes on the surface of infected host cells and that cell-mediated immune responses against cell-surface antigens are important in the acquisition of protective immunity.

Introduction

East Coast fever (ECF) is a disease of cattle caused by the protozoan parasite *Theileria parva*. The disease, which is transmitted transstadially by the three-host tick *Rhipicephalus appendiculatus*, is prevalent throughout large areas of Eastern Africa where it causes high morbidity and mortality in susceptible cattle (reviewed in Irvin &

Morrison in press; Morrison et al. 1986). Control of ECF relies largely on regular dipping or spraying of cattle with acaricides, sometimes as frequently as two or three times a week. Thus, ECF has a major impact on cattle production in East Africa, not only by the mortalities it causes but also by the heavy costs incurred in implementing effective tick control.

There are three subtypes of *T. parva*, *T. p. parva*, *T. p. lawrencei* and *T. p. bovis*, which are morphologically and serologically indistinguishable. Their classification is based principally on clinical and epidemiological parameters (Irvin & Morrison in press). In this chapter the term *T. parva* will often be used to encompass all three subtypes, although most of the data cited is derived from studies of *T. p. parva*. An important factor in the epidemiology of ECF is that the African buffalo (*Syncercus caffer*) is invariably a carrier of the infection (particularly *T. p. lawrencei*), without showing overt clinical signs of disease.

T. parva undergoes a complex series of developmental stages within the tick vector culminating in the production of sporozoites in a specific cell type within the salivary glands (Schein, Warnecke & Kirmse 1977; Fawcett, Buscher & Doxsey 1982). Sporozoites are released into the saliva several days after the tick has commenced feeding on the mammalian host. In susceptible cattle, the sporozoites rapidly gain entry into lymphocytes in which they develop to macroschizonts. The infected cells are induced to proliferate and, as the parasite has the ability to divide synchronously with the host cell (Hulliger et al. 1964), there is rapid clonal expansion of the parasitized cell population. In the majority of cattle, this results in overwhelming infection of the lymphoid system and eventually lymphocytolysis and death of the host within 2 to 4 weeks of infection. During the later stages of the disease the number of schizont nuclei increases and a proportion of the macroschizonts develop to microschizonts. The latter give rise to merozoites which, upon release from the lymphocytes, enter erythrocytes and develop to piroplasms which are infective to the tick.

Following inoculation with low doses of sporozoites, a proportion of cattle recover from the infection and are subsequently immune to challenge with the same parasite stock (Cunningham et al. 1974). Furthermore, it is possible to immunize cattle experimentally using a number of infection and treatment regimes (Radley 1981; Dolan et al. 1984a). In this chapter, we will review current knowledge of the mechanisms of immunity to theileriosis. Particular emphasis will be given to studies conducted over the past 5 years which have provided new insight into how immunity may operate and indicate potential alternative approaches to immunization.

Kinetics of parasite development in the lymphoid system

As the emphasis in this chapter will be on immune responses to schizont-infected lymphocytes, it is worthwhile to consider, briefly, the kinetics of multiplication and dissemination of this stage of the parasite within the infected animal. Evidence from *in vitro* studies indicates that sporozoites can bind to and enter target lymphocytes in less than 10 min (Fawcett et al. 1982). The parasites then undergo a period of 2 to 3 days maturation before detection of typical multinucleate schizonts and induction of proliferation of the host cell. Depending on the number of sporozoites inoculated into an animal, schizont-infected cells may be detected initially in the regional lymph node from 4 to 14 days after inoculation. Parasitized cells are detected in other lymph nodes 2 to 3 days later. Amputation of the tick feeding sites on the ears shortly after infected ticks start to inoculate sporozoites (Wilde, Hulliger & Brown 1966), or extirpation of the regional lymph node as early as 48 h after inoculation of sporozoites (Emery 1981a), have been found to have no significant effect on the subsequent course of infection. Thus, there is widespread dissemination of the parasitized cells in the very early stages of infection. The earlier detection of infected cells in the regional lymph node, therefore, probably merely reflects a relative enrichment of infected cells in this location due to its proximity to the site of inoculation.

Within 2 to 3 days of initial detection of parasitized cells in the regional lymph node, the node increases 3- to 6-fold in size. There is a marked increase in cell output in efferent lymph and an increase in blast-cell content of the lymph from less than 10% to greater than 50% (Morrison et al. 1981b; Emery 1981a). These changes are occurring at a time when parasitosis is still less than 1%, so that they cannot be accounted for merely by increases in the parasitized cell population. There is obviously a dramatic proliferative reaction involving nonparasitized cells at this stage of the infection. Similar changes are observed in other lymph nodes and in the spleen, although they occur a few days later and are quantitatively less pronounced. In tissue sections stained by immunofluorescence with anti-schizont antibody, there does not appear to be preferential localization of the parasitized cells within any particular compartment of the lymph nodes or spleen, although they tend to be more numerous in the T cell-dependent areas. It is not known to what extent the parasitized cells recirculate through the lymphoid tissues.

As the levels of parasitosis in the lymphoid organs increase to 5 to 10%, the tissues start to take on a markedly disorganized appearance which, in many respects, resembles that produced by multicentric lymphoid tumours. Foci of lymphocytosis appear and, as the disease progresses, more extensive areas of lymphocytolysis are observed (Morrison et al. 1981a; Irvin & Morrison in press). This results in a marked decrease in cellularity of the lymphoid tissues and profound decreases in white blood cells and cellular content of lymph, both of which may fall to less than 10% of normal values (Morrison et al. 1981b; Emery 1981a).

The pathological changes in the secondary lymphoid tissues are compounded by changes in the primary lymphoid organs and various nonlymphoid tissues. Numerous parasitized cells are found in the thymus, bonemarrow and Peyer's patches and are associated with marked atrophy of the normal lymphoid elements in these tissues. Furthermore, large numbers of infected cells accummulate in tissues such as the lungs and gastrointestinal tract (Irvin & Morrison in press), through which there is normally a high rate of lymphocyte traffic. Indeed, these infected cells are associated with severe pulmonary pathology in the terminal stages of the disease and sometimes also ulceration of the gastrointestinal tract.

The observations on the course of infection with *T. parva* in cattle would indicate that any immune response which the animal mounts is outpaced by the rapid growth of the parasitized cells. Undoubtedly, at one stage of the infection, there is a pronounced proliferative response of uninfected lymphocytes, but it is difficult to say whether this impedes or potentiates growth of the parasitized cells. As the infection progresses, there is destruction and disorganization in the lymphoid tissues which almost certainly result in severely impaired lymphopoiesis, so that during the later stages of infection the immune system is profoundly compromised.

General features of immunity

Several different methods have been used to induce protective immunity in cattle against *T. parva*. Broadly, these methods fall into two categories. First, cattle may be inoculated with a dose of sporozoites which is normally lethal, accompanied by treatment either with long-acting tetracycline at the time of infection (Radley et al. 1975b) or with theileriotoxic drugs 8 to 12 days later (Dolan et al. 1984a). Treatment with tetracycline results in attenuation of the infection; the mechanism by which this is achieved is unknown. Second, cattle may be inoculated with large numbers of schizont-infected cells derived either from the tissues of infected cattle (Spreull 1914; Pirie, Jarrett & Crighton 1970) or from cell lines established *in vitro* (Brown 1981). Approximately 10^8 allogeneic macroschizont-infected cells are required to immunize the majority of cattle.

These methods of immunization involve establishment of infection in the cells of recipient cattle. Attempts to immunize with irradiated noninfective sporozoites, sporozoites in adjuvants, noninfective macroschizont-infected cells or semipurified schizont or piroplasm antigens in adjuvants have all been unsuccessful (Wilde et al. 1968; Cunningham et al. 1973; Purnell et al. 1974; Wagner, Duffus & Burridge 1974; Emery et al. 1981b). Even in the case of immunization with macroschizont-infected cells, there is evidence that the infection must transfer from the donor cells into those of the recipient for successful immunization (Wilde, Hulliger & Brown 1966; Brown 1981; Morrison et al. 1981a). The fact that such large numbers of cells are required to immunize indicates that the frequency of transfer of infection between cells is very low. Furthermore, it has been found that induction of antibody to the macroschizont before inoculation of allogeneic macroschizont-infected cells can block the transfer of infection and induction of immunity (Emery et al. 1981b). Apart from the large number of cells required, the other main disadvantage of immunization with macroschizont-infected cells is that occasionally cattle develop severe or lethal infections. This is probably because such animals are of a very closely related major histocompatibility (MHC) type to the donor cells and, therefore, do not reject them. This is supported by the finding that relatively small numbers of macroschizont-infected cells from *in vitro* established cell lines are required to produce infection in autologous cattle (Buscher, Morrison & Nelson 1984). As few as 10^2 autologous cells were found to initiate infection and at doses of up to 10^5 cells the majority of animals developed mild transient infection and were immune to subsequent challenge. However, doses of 10^6 or more autologous cells usually resulted in fatal infections. It has also been shown that when the inoculated parasitized cells are matched, with respect to A-locus MHC antigens, with the recipient animals, smaller numbers of cells are required to infect and immunize than when using A-locus mismatched cells (Dolan et al. 1984b).

As a consequence of using live parasites for immunization, a low- level carrier status of infection is often established which may persist for many months (Young, Leitch & Newson 1981). This carrier state is usually only detectable by transmission of infection with ticks fed on immunized cattle. In cattle which recover from ECF, immunity to the homologous parasite has been shown to persist for up to 3.5 years in the absence of challenge (Burridge et al. 1972). It is likely that the carrier state contributes significantly to this long-lasting immunity.

Apart from the practical problems associated with the use of infective organisms for immunization, the major limitation to exploitation of presently available methods of immunization is that immunity generated using one stock (i.e. uncloned isolate) of parasite does not always extend to all other stocks (Radley et al. 1975a; Irvin et al. 1983). The number of strains of the parasite which exist is not known but is probably limited. A striking feature of the results of cross-protection experiments is that breakthrough infections do not occur reciprocally between stocks. Furthermore, when breakthroughs do occur, they usually involve only a proportion of animals. These observations might relate either to there being multiple strains within the stocks of parasites being studied or, perhaps more likely, to a bias of the protective immune response either to common or strain-specific parasite antigenic determinants with different parasite strains or in different individual animals.

The lack of correlation between the presence of antibodies to macroschizonts or piroplasms and immune status means that immunity is probably mediated by cellular mechanisms. This belief has been strengthened by the fact that attempts to transfer immunity passively with serum from immune cattle have been unsuccessful (Muhammed, Lauerman & Johnson 1975). Moreover, immunity has been transferred adoptively between two sets of chimaeric twin calves with thoracic duct lymphocytes (Emery 1981b).

At what level does immunity operate?

Infection with *T. parva* presents a number of problems to the immune system. First, the rapidity with which the sporozoites enter their target cells (Fawcett et al. 1982) means that this stage of the parasite is directly accessible to the immune system for only a short period of time. Second, the ability of the macroschizont to multiply synchronously with the host cells and thus to remain intracellular (Hulliger et al. 1964) also renders this stage of the parasite inaccessible to immune responses, at least during the early stages of infection. However, there is evidence that infection with macroschizonts results in antigenic changes on the surface of the host cell, which stimulate cell-mediated immune responses (Pearson et al. 1979; Emery et al. 1981a). Since piroplasm development occurs relatively late in the course of infection, this stage of the parasite is unlikely to play an important role in induction of protective immunity. Thus, immunity against infections with *T. parva* might operate at two levels: against the sporozoite and against the macroschizont-infected cell.

Evidence that immunity might operate against the sporozoite stems from the observation that cattle repeatedly challenged with large numbers of sporozoites produce antibody which neutralizes the infectivity of sporozoites (Musoke et al. 1982). Neutralization was demonstrated by incubation of serum with suspensions of sporozoites before inoculation into susceptible cattle. Cattle immunized on a single occasion by infection and treatment with tetracycline exhibited only low levels of neutralizing antibody, mainly of the IgM class, whereas immune animals repeatedly challenged with large numbers of infected ticks produced high levels of neutralizing antibody, predominantly IgG_2 (Musoke et al. 1982).

Monoclonal antibodies have also been produced which can neutralize the infectivity of sporozoites (Musoke et al. 1984; Dobbelaere et al. 1984). One of these antibodies has been shown to be directed against a surface-coat antigen on the sporozoite (Dobbelaere, Shapiro & Webster 1985) and, by binding to the surface of the organisms, to prevent their entry into target lymphocytes. An important feature of these monoclonal antibodies and of antibodies produced in cattle is that they can neutralize sporozoites from a range of different stocks of *T. parva*, some of which do not cross-protect.

Thus, it would appear that antibody directed against the sporozoite could potentially prevent establishment of infection and be effective against all parasite strains. However, the efficacy of such antibodies *in vivo* has yet to be determined. Significant protection could only be achieved by neutralization of most, if not all, sporozoites at the site of inoculation. In view of the numbers of organisms which may be inoculated (an estimated 10^4 sporozoites per infected tick salivary gland acinar cell) and the rapidity with which the organisms can invade the target cells, the antibody would have to be present in the tissues at the site of inoculation and to act extremely efficiently in order to be effective. Furthermore, immunity of significant duration would require sustained high levels of circulating antibody. Nevertheless, it is likely that the anti-sporozoite antibodies will at least reduce the initial level of establishment of infection.

A number of features of immunity to *T. parva* suggest that protection is mediated by immune responses against the macroschizont-infected cell. All of the successful methods of immunization using live parasites involve establishment and development of infection to the macroschizont stage. Moreover, immune cattle challenged with sporozoites often exhibit transient low levels of schizont parasitosis before the parasite is eliminated. It has also been shown that cattle immunized by infection and treatment are solidly immune to challenge with large numbers (up to 5×10^8) of autologous schizont-infected cells (Eugui & Emery 1981; Emery et al. 1981b). Finally, the fact that cattle can be immunized by inoculation with macroschizont-infected cells indicates that protective immune responses are directed against this stage of the parasite, particularly since there is no evidence of expression of sporozoite surface antigens on other parasite stages.

Immune responses against the parasite

Cattle which recover from infection with *T. parva* or have undergone immunization mount immune responses to all three stages of the parasite. The antibody responses to the sporozoite and their possible relevance to immunity have already been discussed. Thus, as there is no indication that antibody responses to the piroplasm are important in immunity, in this section we will concentrate on the responses to the macroschizont.

Anti-macroschizont antibodies, both IgM and IgG, can be detected in cattle undergoing immunization as soon as the immunizing infection is brought under control and reach peak titres 5 to 10 days later (Duffus & Wagner 1974; Wagner et al. 1975). However, there is no compelling evidence that these antibodies play an important role in protective immunity. Indeed, it has been shown that cattle inoculated with heat-killed allogeneic macroschizont-infected cells (Emery et al. 1981b) or semipurified preparations of schizont antigens (Wagner, Duffus & Burridge 1974) produce antibody of comparable titre to that accompanying immunization with live parasites, but are fully susceptible to challenge with homologous sporozoites.

Cell-mediated immune responses to the macroschizont-infected cell have been examined both during lethal infections and following immunization by various methods. Such studies were made possible by the development of techniques to infect normal lymphocytes with sporozoites *in vitro* (Brown et al. 1973), thus enabling the derivation, from each animal under study, of an autologous parasitized cell line against which immune responses of the animal could subsequently be monitored.

In cattle suffering from fatal infections, it was found that in the later stages of the infection cytotoxic cells were generated which were capable of killing a range of different target cells (Emery et al. 1981a). These included allogeneic parasitized and nonparasitized lymphoblasts as well as xenogeneic lymphoblasts, but not the autologous parasitized cells. By contrast, in cattle immunized by infection and treatment with tetracycline or animals which fortuitously recovered from infection, cytotoxic cells specific for the autologous parasitized cells were generated (Eugui & Emery 1981; Emery et al. 1981a). These effectors were detectable in peripheral blood mononuclear cells for only 2 to 4 days immediately following elimination of the immunizing infection. This observation provided the first direct evidence that immune responses are directed against antigenic changes induced on the surface of the parasitized cell. Cytotoxic activity resided in a population of peripheral blood mononuclear cells which were negative for surface immunoglobulin and receptors for immunoglobulin Fc and complement and positive for peanut agglutinin and soyabean agglutinin receptors (Emery, Tenywa & Jack 1981). The findings were consistent with the effectors being genetically restricted cytotoxic T cells analogous to those generated against virus-infected cells, as described in other species (Zinkernagel & Doherty 1979). As the genetic restriction of the latter is determined principally by polymorphic determinants on class I MHC molecules, we have carried out a series of experiments to investigate the role of class I MHC determinants in restriction of bovine anti-*Theileria* cytotoxic cells. Polymorphic determinants on MHC antigens in cattle have been defined for class I antigens coded for by one locus (the A locus) on the basis of

Figure 1. MHC restriction of the cell-mediated cytotoxic response in a *T. parva* (Muguga)-immune animal (C-25) following challenge with homologous sporozoites. Results were obtained at the peak of the response on day 9 after challenge. Cytotoxicity was measured in a 4-h ^{51}Cr-release assay using labeled macroschizont-infected cells obtained from cell lines. The results were calculated from the percentage of ^{51}Cr released at an effector-to-target ratio of 40:1. Levels of cytotoxicity of greater than 10% were obtained only on cell lines which shared the w6 and/or w8 A-locus determinants with the animal undergoing immunization.

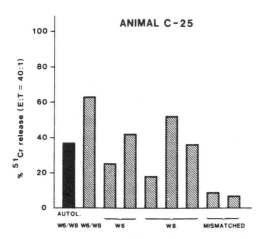

typing with absorbed alloantisera (Spooner this volume). Our own studies also indicate that, as in other species, alloreactive cytotoxic bovine cells generated *in vitro* are directed principally against the class I antigens (Teale et al. this volume). Cytotoxic cells generated *in vivo* in four *T. parva*-immune cattle of known A-locus phenotype following challenge with sporozoites have been assayed on panels of parasitized cells which were either matched, semimatched or mismatched for A locus-encoded antigens with the animal undergoing immunization (Figure 1). In all four animals, it was found that the cytotoxic cells killed not only the autologous target but also targets which were matched or semimatched for A-locus antigens. Interestingly, in three of the animals the response was found to be strongly biased towards the product of one of the alleles. These findings provide convincing evidence that bovine A-locus determinants are involved in the genetic restriction of anti-*Theileria* cytotoxic cells.

Similar genetically restricted cytotoxic T cells have been detected in cattle following immunization with autologous or allogeneic parasitized cell lines (Emery et al. 1982). However, these findings are possibly complicated by responses to culture antigens acquired by the parasitized cells *in vitro* (Morrison et al. 1986). Nevertheless, the temporal relationship of the cytotoxic cell response with acquisition of immunity, along with the fact that immunity can be adoptively transferred with cells, suggests that the cytotoxic response is important in protection. This would explain why attempts to immunize with inactivated allogeneic macroschizont-infected cells have not been successful, since immunity would only be expected if the parasite is presented in the animals' own cells or in cells of an MHC-related animal. In a recent experiment, D.L. Emery, W.I. Morrison and R.M. Jack (1986) were able to immunize cattle with a cell membrane fraction prepared from autologous parasitized cells. In cattle which received three or four inocula of cell membranes derived from a total of 6 to 12 x 10^{10} autologous infected cells, three out of four were immune to homologous sporozoite challenge, whereas two allogeneic cattle similarly immunized with the same membrane antigen were fully susceptible. Although this method of immunization would have little practical use, the results provide strong supportive evidence that genetically restricted immune responses are important in immunity and have shown that it is feasible to immunize cattle with subcellular fractions of macroschizont-infected cells.

If genetically restricted immune responses are important in immunity, it would be expected that heterogeneity in parasite strains may be related to differences in the cell-surface determinants recognized by cytotoxic cells. Preliminary evidence that this may be so has been reported by Eugui and others (1981), who showed that cytotoxic cells generated *in vivo* in three animals during immunization with one stock of *T. parva* differed in their capacity to kill autologous cell lines infected with several different stocks of the parasite. Further studies are currently in progress to confirm this observation using cloned parasites.

Attempts to identify the target antigen(s) against which the cytotoxic cell response is directed have so far proven unsuccessful. Antibody to the surface of the parasitized cell has not been detected in the serum of immune cattle (Creemers 1982) and, of over 100 monoclonal antibodies raised against *Theileria*-infected cells, none have been found to be specific for the parasitized cell (J. Newson & J. Naessens personal communication). Furthermore, one-dimensional and two-dimensional gel analyses of extracts from infected cell membranes have not revealed additional proteins on the cell surface in comparison with uninfected blast cells. These results are perhaps not surprising in view of recent findings that virus-specific cytotoxic cells in mice may be directed against viral products which are not easily detected on the cell surface (Townsend et al. 1984).

Parasite strain-specific antigenic determinants have been identified on the intracellular macroschizont by derivation of a series of monoclonal antibodies (Pinder & Hewett 1980; Minami et al. 1983), and on the basis of reactivity with these monoclonal antibodies it has been possible to classify stocks of *T. p. parva* into three groups. Cross-challenge experiments with cattle immunized with these stocks have demonstrated that stocks from the same group always give cross-protection, whereas challenge with a stock belonging to a different group from that used for immunization sometimes, but not always, results in breakthrough infections (Irvin et al. 1983). Whether or not these parasite-specific determinants are in any way related to the antigenic changes on the cell surface detected by cytotoxic cells has yet to be determined. To resolve this question, it will be necessary to isolate the relevant genes and transfect them into bovine cells of the appropriate MHC type.

One of the major problems in looking at the fine specificity of the anti-*Theileria* cytotoxic cells is the difficulty of generating specific effectors *in vitro* reproducibly. *Theileria*-infected cells elicit strong proliferative responses in autologous PBM *in vitro*, but the cytotoxic cells generated often include a population of effectors which is not *Theileria* specific (Pearson et al. 1982; Emery & Kar 1983) and not MHC restricted (Emery & Kar 1983; Goddeeris, Lalor & Morrison this volume). Further attempts are being made to propagate *Theileria*-specific cytotoxic cells *in vitro*, since, if such cells could be cloned, they would provide cellular reagents which may be of value in identification of relevant antigenic changes on the surface of parasitized cells and in distinguishing between noncross-protective strains of the parasite.

Future approaches to immunization

An effective and easily applied vaccine against ECF would have a major impact on the cattle industry in East Africa. The main obstacles at present are the requirement for use of live parasites and the problem of parasite strain heterogeneity. Nevertheless, recent field trials have shown that infection and treatment methods of immunization can be applied successfully, using judiciously chosen local strains of the parasite (Morzaria et al. 1985).

With regard to the possibility of an inactivated vaccine, the use of sporozoite antigens would theoretically appear to be ideal, since they would induce responses which operate at the initial level of interaction of the parasite with the host and may be effective against most, if not all, parasite strains. Furthermore, studies with one of the monoclonal antibodies which neutralizes sporozoites have shown that it precipitates a protein antigen of approximately 68,000 daltons (Dobbelaere, Shapiro & Webster 1985). Thus, it should be possible to apply molecular biological techniques to isolate the relevant parasite DNA and produce the antigen or an antigen component for immunization. However, as already discussed, the question of whether antibody against the sporozoite can be completely effective *in vivo* has yet to be resolved. Nevertheless, it is likely that the antibodies will at least reduce the level of establishment of the infection. In such circumstances, immunization with sporozoite antigens might be of use in complementing other methods of immunization orientated towards induction of responses to the macroschizont-infected cell.

The observations outlined here indicate that the critical factor in immunity against the macroschizont-infected cell is that the relevant immune response is directed against antigenic changes on the surface of the parasitized cell and is genetically restricted by MHC antigens. Thus, the limiting factor in immunization with allogeneic macroschizont-infected cells is the low frequency with which the parasite can transfer into the cells of the recipient animal. Attempts to identify cloned cell lines with an increased capacity to transfer the parasite have so far been unsuccessful (P.A. Lalor & W.I. Morrison unpublished).

Approaches to the development of an inactivated vaccine effective against the macroschizont-infected cell depend on defining the nature of the antigenic changes induced on the surface of the parasitized cell. Such changes may be related to the expression of a parasite-derived antigen on the cell surface or to an intracellular parasite product which causes expression of abnormal self-antigens. It has not been possible, to date, to identify such parasite products using conventional biochemical and immunochemical techniques. However, it is possible that the application of molecular biological techniques to identify DNA sequences which differ between noncross-protective strains of the parasite might be more rewarding. If relevant parasite products can be identified, the problem remains of how to present them to the host so that they are expressed in a similar manner as in the infected cell. One possible approach is to insert the gene coding for the antigen of interest into a virus vector which is capable of infecting bovine cells and giving appropriate expression of the inserted gene product. This technique has been applied successfully to obtain appropriate expression of isolated genes from human viruses following insertion into vaccinia virus (Smith, Mackett & Moss 1983; Bennink et al. 1984).

References

Bennink, J.R., Yewdell, J.W., Smith, G.L., Moller, C. & Moss B. (1984). Recombinant vaccinia virus primes and stimulates influenza haemagglutinin-specific cytotoxic T cells. *Nature,* **311**, 578-79.

Brown, C.G.D. (1981). Application of *in vitro* techniques to vaccination against theileriosis. In *Advances in the control of theileriosis* (eds. A.D. Irvin, M.P. Cunningham & A.S. Young), Current Topics in Veterinary Medicine & Animal Science 14, pp. 104-19. The Hague: Nijhoff.

Brown, C.G.D., Stagg, D.A., Purnell, R.E., Kanhai, G.K. & Payne, R.C. (1973). Infection and transformation of bovine lymphoid cells *in vitro* by infective particles of *Theileria parva. Nature,* **245**, 101-3.

Burridge, M.J., Morzaria, S.P., Cunningham, M.P. & Brown, C.G.D. (1972). Duration of immunity to East Coast fever (*Theileria parva*) infection in cattle. *Parasitology,* **64**, 511-15.

Buscher, G., Morrison, W.I. & Nelson, R.T. (1984). *Theileria parva*: titration of infectivity and immunogenicity of autologous infected cells lines in cattle. *Veterinary Parasitology,* **15**, 29-38.

Creemers, P. (1982) Lack of reactivity of sera from *Theileria parva*-immune and recovered cattle against cell membrane antigens of *Theileria parva*-transpassed cell lines. *Veterinary Immunology & Immunopathology,* **3**, 427-38.

Cunningham, M.P., Brown, C.G.D., Burridge, M.J., Musoke, A.J., Purnell, R.E. & Dargie, J.D. (1973). East Coast fever of cattle: ⁶⁰Co-irradiation of infective particles of *Theileria parva. Journal of Protozoology,* **20**, 298-300.

Cunningham, M.P., Brown, C.G.D., Burridge, M.J., Musoke, A.J., Purnell, R.E., Radley, D.E. & Sempebwa C. (1974). East Coast fever: titration in cattle of suspensions of *Theileria parva* derived from ticks. *British Veterinary Journal,* **130**, 336-45.

Dobbelaere, D.A.E., Shapiro, S.Z. & Webster, P. (1985). Identification of a surface antigen on *Theileria parva* sporozoites by monoclonal antibody. *Proceeedings of the National Academy of Sciences of the USA,* **82**, 1771-5.

Dobbelaere, D.A.E., Spooner, P.R., Barry, W.C. & Irvin, A.D. (1984). Monoclonal antibody neutralizes the sporozoite stage of different *Theileria parva* stocks. *Parasite Immunology*, **6**, 361-70.

Dolan, T.T., Linyangi, A., Mboga, S.K. & Young, A.S. (1984a). Comparison of long-acting oxytetracycline and parvaquone in immunization against East Coast fever by infection and treatment. *Research in Veterinary Science*, **37**, 175-78.

Dolan, T.T., Teale, A.J., Stagg, D.A., Kemp, S.J., Cowan, K.M., Young, A.S., Groocock, C.M., Leitch, B.L., Brown, C.G.D. & Spooner, R.L. (1984b). A histocompatibility barrier to immunization against East Coast fever using *Theileria parva*-infected lymphoblastoid cell lines. *Parasite Immunology*, **6**, 243-50.

Duffus, W.P.H. & Wagner, G.G. (1974). The specific immunoglobulin response in cattle immunized with *Theileria parva* (Muguga) stabilate. *Parasitology*, **69**, 31-41.

Emery, D.L. (1981a). Kinetics of infection with *Theileria parva* (East Coast fever) in the central lymph of cattle. *Veterinary Parasitology*, **9**, 1-16.

Emery, D.L. (1981b). Adoptive transfer of immunity to infection with *Theileria parva* (East Coast fever) between cattle twins. *Research in Veterinary Science*, **30**, 364-67.

Emery, D.L. & Kar, S.K. (1983) Immune responses of cattle to *Theileria parva* (East Coast fever): specificity of cytotoxic cells generated *in vivo* and *in vitro*. *Immunology*, **48**, 723-31.

Emery, D.L., Tenywa, T. & Jack, R.M. (1981). Characterization of the effector cell that mediates cytotoxicity against *Theileria parva* (East Coast fever) in immune cattle. *Infection & Immunity*, **32**, 1301-4.

Emery, D.L., Eugui, E.M., Nelson, R.T. & Tenywa, T. (1981a). Cell-mediated immune responses to *Theileria parva* (East Coast fever) during immunization and lethal infections in cattle. *Immunology*, **43**, 323-36.

Emery, D.L., Morrison, W.I. & Jack, R.M. (1986). Induction of immunity against infection with *Theileria parva* (East Coast Fever) in cattle using plasma membranes from parasitised lymphoblasts. *Veterinary Parasitology*, **19**, 321-7.

Emery, D.L., Morrison, W.I., Nelson, R.T. & Murray, M. (1981b). The induction of cell-mediated immunity in cattle inoculated with cell lines parasitized with *Theileria parva*. In *Advances in the control of theileriosis* (eds. A.D. Irvin, M.P. Cunningham & A.S. Young), Current Topics in Veterinary Medicine & Animal Science 14, pp. 295-310. The Hague: Nijhoff.

Emery, D.L., Morrison, W.I., Buscher, G. & Nelson, R.T. (1982). Generation of cell-mediated cytotoxicity to *Theileria parva* (East Coast fever) after inoculation of cattle with parasitized lymphoblasts. *Journal of Immunology*, **128**, 195-200.

Eugui, E.M. & Emery, D.L. (1981) Genetically restricted cell-mediated cytotoxicity in cattle immune to *Theileria parva*. *Nature*, **290**, 251-54.

Eugui, E.M., Emery, D.L., Buscher, G. & Khaukha, G. (1981). Specific cellular immune response to *Theileria parva* in cattle. In *Advances in the control of theileriosis* (eds. A.D. Irvin, M.P. Cunningham & A.S. Young), Current Topics in Veterinary Medicine & Animal Science 14, pp. 289-94. The Hague: Nijhoff.

Fawcett, D.W., Buscher, G. & Doxsey, S. (1982). Salivary gland of the tick vector of East Coast fever. 3. The ultrastructure of sporogony in *Theileria parva*. *Tissue & Cell*, **14**, 183-206.

Fawcett, D.W., Doxsey, S., Stagg, D.A. & Young, A.S. (1982). The entry of sporozoites of *Theileria parva* into bovine lymphocyte *in vitro*: electron microscopic observations. *European Journal of Cell Biology*, **27**, 10-21.

Hulliger, L., Wilde, J.K.H., Brown, C.G.D. & Turner, L. (1964). Mode of multiplication of *Theileria* in cultures of bovine lymphocytic cells. *Nature*, **203**, 728.

Irvin, A.D. & Morrison, W.I. (in press). Immunology, immunopathology and immunoprophylaxis of *Theileria* infections. In *Immunology, immunopathology & immunoprophylaxis of parasitic infections* (ed. E.J.L. Soulsby). West Palm Beach, Florida: CRC Press.

Irvin, A.D., Dobbelaere, D.A.E., Mwamachi, D.M., Minami, T., Spooner, P.R. & Ocama, J.G.R. (1983). Immunization against East Coast fever: correlation between monoclonal antibody profiles of *Theileria parva* stocks and cross-immunity *in vivo*. *Research in Veterinary Science*, **35**, 341-46.

Minami, T., Spooner, P.R., Irvin, A.D., Ocama, J.G.R., Dobbelaere, D.A.E. & Fuginaga, T. (1983). Characterization of stocks of *Theileria parva* by monoclonal antibody profiles. *Research in Veterinary Science*, **35**, 334-40.

Morrison, W.I., Lalor, P.A., Goddeeris, B.M & Teale, A.J., (1986). Theileriosis: antigens and host-parasite interactions. In *Parasite antigens: Toward new strategies for vaccines*, (ed. T.W. Pearson), pp. 167-212. New York: Marcel Dekker.

Morrison, W.I., Buscher, G., Emery, D.L., Nelson, R.T. & Murray, M. (1981a). The kinetics of infection with *Theileria parva* in cattle and the relevance to the development of immunity. In *Advances in the control of theileriosis* (eds. A.D. Irvin, M.P. Cunningham & A.S. Young), Current Topics in Veterinary Medicine & Animal Science 14, pp. 311-26. The Hague: Nijhoff.

Morrison, W.I., Buscher, G., Murray, M., Emery, D.L., Masake, R.A., Cook, R.H. & Wells, P.W. (1981b). *Theileria parva*: kinetics of infection in the lymphoid system of cattle. *Experimental Parasitology*, **52**, 248-60.

Morzaria, S.P., Irvin, A.D., Taracha, E. & Spooner, P.R. (1985). East Coast fever immunization trials in the Coast Province of Kenya. In *Immunization against theileriosis in Africa* (ed. A.D. Irvin), pp. 76-8. Nairobi: ILRAD.

Muhammed, S.I., Lauerman, L.H. & Johnson, L.W. (1975). Effect of humoral antibodies on the course of *Theileria parva* infection (East Coast fever) of cattle. *American Journal of Veterinary Research*, **36**, 399-402.

Musoke, A.J., Nantulya, V.M., Rurangirwa, F.R. & Buscher, G. (1984). Evidence for a common protective antigenic determinant on sporozoites of several *Theileria parva* strains. *Immunology*, **52**, 231-38.

Musoke, A.J., Nantulya, V.M., Buscher, G., Masake, R.A. & Otim, B. (1982). Bovine immune response to *Theileria parva*: neutralizing antibodies to sporozoites. *Immunology*, **45**, 663-68.

Pearson, T.W., Dolan, T.T., Stagg, D.A. & Lundin, L.B. (1979). Cell-mediated immunity to *Theileria*-transformed cell lines. *Nature*, **281**, 678-80.

Pearson, T.W., Hewett, R.S., Roelants, G.E., Stagg, D.A. & Dolan, T.T. (1982). Studies on the induction and specificity of cytotoxicity to *Theileria*-transformed cell lines. *Journal of Immunology*, **128**, 2509-13.

Pinder, M. & Hewett, R.S. (1980). Monoclonal antibodies detect antigenic diversity in *Theileria parva* parasites. *Journal of Immunology*, **124**, 1000-1.

Pirie, H.W., Jarrett, W.F.H. & Crighton, G.W. (1970). Studies on vaccination against East Coast fever using macroschizonts. *Experimental Parasitology*, **27**, 243-49.

Purnell, R.E., Brown, C.G.D., Burridge, M.J., Cunningham, M.P., Emu, H., Irvin, A.D., Ledger, M.A., Njuguna, C.M., Payne R.C. & Radley, D.E. (1974). East Coast fever: ^{60}Co-irradiation of *Theileria parva* in its tick vector *Rhipicephalus appendiculatus*. *International Journal of Parasitology*, **4**, 507-11.

Radley, D.E. (1981). Infection and treatment method of immunization against theileriosis. In *Advances in the control of theileriosis* (eds. A.D. Irvin, M.P. Cunningham & A.S. Young), Current Topics in Veterinary Medicine & Animal Science 14, pp. 227-37. The Hague: Nijhoff.

Radley, D.E., Brown, C.G.D., Burridge, M.J., Cunningham, M.P., Kirimi, I.M., Purnell, R.E. & Young A.S. (1975a). East Coast fever. 1. Chemoprophylactic immunization of cattle against *Theileria parva* (Muguga) and five theilerial strains. *Veterinary Parasitology*, **1**, 35-42.

Radley, D.E., Brown, C.G.D., Cunningham, M.P., Kimber, C.D., Musisi, F.L., Purnell, R.E., Stagg, S.M. & Punyua, D.K. (1975b). East Coast fever: homologous challenge of immunized cattle in an infected paddock. *Veterinary Record*, **96**, 525-27.

Schein, E., Warnecke, M. & Kirmse, P. (1977). Development of *Theileria parva* (Theiler 1904) in the gut of *Rhipicephalus appendiculatus* (Newmann 1901). *Parasitology*, **75**, 309-16.

Smith, G.L., Mackett, M. & Moss, B. (1983). Infectious vaccinia virus recombinants that express hepatitis B virus surface antigen. *Nature*, **302**, 490-95.

Spreull, J. (1914). East Coast fever inoculation in the Transleian territories, South Africa. *Journal of Comparative Pathology & Therapeutics*, **27**, 229-304.

Townsend, A.R.M., Skehel, J.J., Taylor, P.M. & Palese, P. (1984). Recognition of influenza A virus nucleoprotein by an H-2 restricted cytotoxic T-cell clone. *Virology*, **133**, 456-59.

Wagner, G.G., Duffus, W.P.H. & Burridge, M.J. (1974). The specific immunoglobulin response in cattle immunized with isolated *Theileria parva* antigens. *Parasitology*, **69**, 43-53.

Wagner, G.G., Duffus, W.P.H., Akwabi, C., Burridge, M.J. & Lule, M. (1975). The specific immunoglobulin response in cattle to *Theileria parva* (Muguga) infection. *Parasitology*, **70**, 95-102.

Wilde, J.K.H., Hulliger, L. & Brown, C.G.D. (1966). Some recent East Coast fever research. *Bulletin of Epizootic Diseases for Africa*, **14**, 29-35.

Wilde, J.K.H., Brown, C.G.D., Hulliger, L., Gall, D. & MacLead, G. (1968). East Coast fever: experiments with the tissues of infected ticks. *British Veterinary Journal*, **124**, 196-208.

Young, A.S., Leitch, B.L. & Newson, R. (1981). The occurrence of a *Theileria parva* carrier state in cattle from an East Coast fever endemic area of Kenya. In *Advances in the control of theileriosis* (eds. A.D. Irvin, M.P. Cunningham & A.S. Young), Current Topics in Veterinary Medicine & Animal Science 14, pp. 60-62. The Hague: Nijhoff.

Zinkernagel, R.M. & Doherty, P.C. (1979). MHC-restricted T cells: studies on the biological role of polymorphic major transplantation antigens determining T cell-restriction-specificity function and responsiveness. *Advances in Immunology*, **27**, 51-177.